INTRODUCTION TO APPLIED GEOPHYSICS

EXPLORING THE SHALLOW SUBSURFACE

H. Robert Burger
SMITH COLLEGE

Anne F. Sheehan
UNIVERSITY OF COLORADO

Craig H. Jones
UNIVERSITY OF COLORADO

CAMBRIDGE
UNIVERSITY PRESS

Shaftesbury Road, Cambridge CB2 8EA, United Kingdom

One Liberty Plaza, 20th Floor, New York, NY 10006, USA

477 Williamstown Road, Port Melbourne, VIC 3207, Australia

314–321, 3rd Floor, Plot 3, Splendor Forum, Jasola District Centre, New Delhi – 110025, India

103 Penang Road, #05-06/07, Visioncrest Commercial, Singapore 238467

Cambridge University Press is part of Cambridge University Press & Assessment, a department of the University of Cambridge.

We share the University's mission to contribute to society through the pursuit of education, learning and research at the highest international levels of excellence.

www.cambridge.org
Information on this title: www.cambridge.org/highereducation/isbn/9781009433129

DOI: 10.1017/9781009433112

A catalogue record for this publication is available from the British Library

A Cataloging-in-Publication data record for this book is available from the Library of Congress.

ISBN 978-1-009-43312-9 Paperback

Additional resources for this publication at www.cambridge.org/burger

Cambridge University Press & Assessment has no responsibility for the persistence or accuracy of URLs for external or third-party internet websites referred to in this publication and does not guarantee that any content on such websites is, or will remain, accurate or appropriate

Acknowledgments

I would like to thank my fellow geophysicists for granting permission to reproduce figures from their publications. I also want to thank my undergraduate students for feedback on many of the chapters, illustrations, and tables, and for their enthusiasm for this project.

HRB

Thanks to Carolyn Ruppel, Jeff Lucius, Gaspar Monsalve, Koni Steffen, Don Steeples, Joya Tetreault, Derald Smith, Sander Huisman, Larry Conyers, Bruce Luyendyk, Steven Sloan, Steve Park, Doug Wiens, Payson Sheets, Sarah Kruse, H. P. Marshall, and Penny Axelrad for helpful input, figures, and other materials. Thanks also to our students at the University of Colorado at Boulder for providing feedback on the material.

AFS AND **CHJ**

About the Authors

H. ROBERT BURGER (Ph.D. 1966, Indiana University) is Achilles Professor of Geology at Smith College in Northampton, Massachusetts. His research focuses on the evolution of ancient mountain belts in southwestern Montana, applying geophysics to further elucidate the structural evolution of the Connecticut Valley in Massachusetts, and using Geographic Information Systems (GIS) to mitigate natural hazards. In addition to research articles and papers, Burger also co-authored *An Introduction to Structural Methods*.

ANNE F. SHEEHAN (Ph.D. 1991, MIT) is Associate Professor of Geological Sciences and Fellow of the Cooperative Institute for Research in Environmental Sciences (CIRES) at the University of Colorado at Boulder. Her research focuses on the study of the crust and upper mantle of the Earth and its relation to tectonic deformation. She has conducted field studies in many regions including the Rocky Mountains, the Himalaya, and New Zealand. Dr. Sheehan teaches undergraduate and graduate courses in geological sciences and geophysics, and has served as Director of the University of Colorado Geophysics Ph.D. program.

CRAIG H. JONES (Ph.D. 1987, MIT) is Associate Professor of Geological Sciences and Fellow of the Cooperative Institute for Research in Environmental Sciences (CIRES) at the University of Colorado at Boulder. His research has focused on understanding the tectonics of continental areas, especially the western U.S., and on obtaining geophysical observations of those features. He teaches undergraduate and graduate courses in geology and geophysics and has developed and maintained geophysical software for more than 15 years.

Brief Contents

Preface   xix

Preface to Computer Materials   xxii

1   Approaching the Subsurface   1

2   Seismic Exploration: Fundamental Considerations   7

3   Seismic Exploration: The Refraction Method   65

4   Seismic Exploration: The Reflection Method   149

5   Electrical Resistivity   265

6   Exploration Using Gravity   349

7   Exploration Using the Magnetic Method   429

8   Electromagnetic Surveying   499

Appendixes   A1

Brief Contents

Preface x

Preface to Computer Materials xxii

1 Approaching the Subsurface 1

2 Seismic Exploration: Fundamental Consideration 7

3 Seismic Exploration: The Reflection Method 56

4 Seismic Exploration: The Reflection Method 146

5 Electrical Resistivity 265

6 Exploration Using Gravity 349

7 Exploration Using the Magnetic Method 426

8 Electromagnetic Surveying 498

Appendices A1

Contents

Preface   xix

Preface to Computer Materials   xxii

1	Approaching the Subsurface	1
1.1	What Are the Options?	1
1.2	Some Fundamental Considerations	2
1.3	Defining Objectives	3
1.4	Limitations	4
1.5	The Advantage of Multiple Methods	5

2	Seismic Exploration: Fundamental Considerations	7
2.1	Seismic Waves and Wave Propagation	7
2.1.1	Wave Terminology	9
2.1.2	Elastic Coefficients	11
2.1.3	Seismic Waves	14
2.1.4	Seismic Wave Velocities	18
2.2	Ray Paths in Layered Materials	21
2.2.1	Huygens' Principle	21
2.2.2	Fermat's Principle	22
2.2.3	Reflection	23
2.2.4	Refraction	25
2.2.5	Snell's Law	27
2.2.6	Critical Refraction	28
2.2.7	Diffraction	29
2.2.8	Wave Arrivals at the Surface	35

2.3	**Wave Attenuation and Amplitude**	**39**
2.3.1	Spherical Spreading	40
2.3.2	Absorption	40
2.3.3	Energy Partitioning	42
2.3.4	Additional Factors	46
2.4	**Energy Sources**	**47**
2.4.1	Source Types	47
2.4.2	Source Considerations	49
2.5	**Seismic Equipment**	**50**
2.5.1	Signal Detection	50
2.5.2	Signal Conditioning	53
2.5.3	Signal Recording	56
2.6	**Summary**	**60**
	Problems	**61**
	References Cited	**63**

3	**Seismic Exploration: The Refraction Method**	**65**
3.1	**A Homogeneous Subsurface**	**65**
3.2	**A Single Subsurface Interface**	**67**
3.2.1	Derivation of a Travel-Time Equation	67
3.2.2	Analysis of Arrival Times	70
3.2.3	Determining Thickness	72
3.2.4	Crossover Distance	73
3.2.5	Critical Distance	74
3.2.6	Constructing a Travel-Time Curve from a Field Seismogram	76
3.2.7	Using REFRACT	78
3.2.8	The Mohorovicic Discontinuity	79
3.3	**Two Horizontal Interfaces**	**81**
3.3.1	Derivation of a Travel-Time Equation	81
3.3.2	Determining Thickness	83
3.3.3	Critical Distance	85
3.3.4	Analyzing a Second Field Seismogram	87
3.4	**Multiple Interfaces**	**90**
3.5	**Dipping Interfaces**	**91**
3.5.1	Analyzing the Problem	91
3.5.2	Derivation of a Travel-Time Equation	97
3.5.3	Determining Thickness	99
3.6	**Multiple Dipping Interfaces**	**100**
3.6.1	Travel-Time Equation	100
3.6.2	Analyzing Field Seismograms	102

3.7	**The Nonideal Subsurface**	**106**
3.7.1	Hidden Zones: The Low-Velocity Layer	107
3.7.2	Hidden Zones: The Thin Layer	111
3.7.3	Laterally Varying Velocity	114
3.7.4	Interface Discontinuities	116
3.8	**The Delay-Time Method**	**120**
3.9	**Other Methods**	**127**
3.9.1	Wave Front Method	127
3.9.2	Ray Tracing and the Generalized Reciprocal Method	129
3.10	**Field Procedures**	**129**
3.10.1	Site Selection and Planning Considerations	129
3.10.2	Equipment Considerations	130
3.10.3	Geophone Spread Geometries and Placements	131
3.10.4	Corrections to Data	134
3.11	**Applications Using Seismic Refraction**	**136**
3.11.1	Whately, Massachusetts	136
3.11.2	Southeastern New Hampshire	137
3.11.3	Waste Disposal Site	138
3.11.4	Maricopa Area, Arizona	139
	Problems	**141**
	References Cited	**147**
	Suggested Reading	**148**

4   Seismic Exploration: The Reflection Method   149

4.1	**A Single Subsurface Interface**	**150**
4.1.1	Using REFLECT	150
4.1.2	Derivation of a Travel-Time Equation	151
4.1.3	Analysis of Arrival Times	156
4.1.4	Normal Move-Out	160
4.1.5	Determining Velocity and Thickness	163
4.1.6	Applying the $x^2 - t^2$ Method to a Field Seismogram	165
4.2	**Multiple Horizontal Interfaces**	**167**
4.2.1	The Dix Equation	170
4.2.2	Determining Velocities	170
4.2.3	Determining Thicknesses	174
4.2.4	Further Discussion of the Dix Method	175
4.2.5	Analyzing a Field Seismogram Containing Multiple Reflections	181
4.3	**Dipping Interface**	**182**
4.3.1	Derivation of a Travel-Time Equation	183
4.3.2	Determining Dip, Thickness, and Velocity	186

4.3.3	Determining Dip, Thickness, and Velocity—Another Approach	190
4.3.4	A Return to Normal Move-Out	192
4.3.5	Determining Dip, Thickness, and Velocity—Yet Another Approach	197
4.4	**Acquiring and Recognizing Reflections from Shallow Interfaces**	**200**
4.4.1	The Optimum Window	200
4.4.2	Multiple Reflections	206
4.4.3	Diffractions	209
4.5	**Common Field Procedures**	**217**
4.5.1	Equipment Considerations	217
4.5.2	Geophone Spreads	219
4.5.2.1	Split Spread	219
4.5.2.2	Common Offset	221
4.5.2.3	Common Depth Point	225
4.6	**Computer Processing of Reflection Data**	**229**
4.6.1	The Static Correction	229
4.6.2	Correcting for Normal Move-Out	232
4.6.2.1	Velocity Analysis A	233
4.6.2.2	Velocity Analysis B	235
4.6.2.3	Velocity Analysis C	238
4.6.3	Stacking CDP Gathers	238
4.6.4	Migration	241
4.6.5	Waveform Adjustments	244
4.6.6	Seismic Sections: Time Sections and Depth Sections	245
4.7	**Applying the Seismic Reflection Method**	**248**
4.7.1	Whately, Massachusetts	248
4.7.2	Meers Fault, Oklahoma	249
4.7.3	Cavity Detection	249
4.7.4	Other Applications	252
	Problems	**254**
	References Cited	**262**
	References Cited	**263**

5	**Electrical Resistivity**	**265**
5.1	**Introduction**	**265**
5.1.1	Applied Currents	265
5.1.2	Natural Currents	266
5.1.3	A Brief History	266
5.1.4	Chapter Goals	267
5.2	**Basic Electricity**	**267**
5.3	**Current Flow in a Homogeneous, Isotropic Earth**	**270**

5.3.1	Point Current Source	270
5.3.2	Two Current Electrodes	271
5.3.3	Two Potential Electrodes	276
5.4	**A Single Horizontal Interface**	**280**
5.4.1	Current Distribution	280
5.4.2	Current Flow Lines and Current Density	284
5.4.3	Apparent Resistivity	287
5.4.4	Qualitative Development of the Resistivity Pattern over a Horizontal Interface	288
5.4.5	Quantitative Development of the Resistivity Pattern over a Horizontal Interface	290
5.4.6	Using RESIST	297
5.5	**Multiple Horizontal Interfaces**	**302**
5.6	**Vertical Contact**	**305**
5.6.1	Constant-Spread Traverse	306
5.6.2	Expanding-Spread Traverse	309
5.7	**Two Vertical Contacts, Hemispherical Structures, and Dipping Interfaces**	**312**
5.7.1	Two Vertical Contacts	312
5.7.2	Hemispherical Structures	313
5.7.3	Dipping Interfaces	314
5.8	**Field Procedures**	**314**
5.8.1	Equipment	315
5.8.2	Electrode Configurations	317
5.8.3	Surveying Strategies	321
5.8.4	Other Considerations	323
5.9	**Quantitative Interpretation of Apparent Resistivity Curves**	**323**
5.9.1	Electrical Resistivities of Geologic Materials	324
5.9.2	Empirical Methods	325
5.9.3	Analytical Methods—Curve Matching	326
5.9.4	Analytical Methods—Automated Curve Matching	329
5.10	**Applications of Electrical Resistivity Surveying**	**331**
5.10.1	Applications Related to Aquifers	332
5.10.2	Applications Related to Contamination	335
5.10.3	Applications in Mapping Karst and Geologic Structures	335
5.10.4	Other Applications	338
5.11	**Other Electrical Methods**	**338**
5.11.1	Induced Potential (IP)	338
5.11.2	Spontaneous Potential (SP)	339
5.11.3	Telluric and Magnetotelluric Methods	340
	Problems	**341**
	References Cited	**346**
	Suggested Reading	**347**

6	Exploration Using Gravity	349
6.1	**Fundamental Relationships**	**350**
6.1.1	Gravitational Acceleration	350
6.2	**Measuring Gravity**	**351**
6.2.1	Relative Measurements Using a Pendulum	351
6.2.2	Relative Measurements Using a Gravimeter	352
6.2.3	Absolute Measurements	355
6.2.4	International Gravity Standardization Net 1971 (IGSN71)	355
6.3	**Adjusting Observed Gravity**	**356**
6.3.1	Variation in g as a Function of Latitude	356
6.3.2	Correcting for the Latitude Effect	357
6.3.3	Elevation Correction 1: The Free-Air Correction	358
6.3.4	Elevation Correction 2: The Bouguer Correction	360
6.3.5	Elevation Correction 3: The Terrain Correction	364
6.3.6	The Isostatic Anomaly	369
6.4	**Basic Field Procedures**	**370**
6.4.1	Drift and Tidal Effects	370
6.4.2	Establishing Base Stations	373
6.4.3	Determining Elevations	373
6.4.4	Determining Horizontal Position	375
6.4.5	Selecting a Reduction Density	376
6.4.6	Survey Procedure	376
6.5	**Gravity Effects of Simple Geometric Shapes**	**378**
6.5.1	Rock Densities	378
6.5.2	Gravity Effect of a Sphere	379
6.5.3	Gravity Effect of a Horizontal Cylinder	383
6.5.4	Gravity Effect of a Vertical Cylinder	386
6.5.5	Gravity Effect of an Inclined Rod	388
6.5.6	Gravity Effect of a Horizontal Sheet	390
6.5.7	GRAVMAG	397
6.6	**Analyzing Anomalies**	**399**
6.6.1	Regionals and Residuals	399
6.6.2	Trend Surfaces	402
6.6.3	Upward and Downward Continuation	403
6.6.4	Second Derivatives	407
6.6.5	Filtering	409
6.7	**Gravity Interpretation**	**409**
6.7.1	Half-Maximum Technique	410
6.7.2	Second Derivative Techniques	411
6.7.3	Revisiting Some Bouguer Anomaly Values	413
6.8	**Applications of the Gravity Method**	**415**

6.8.1	Bedrock Depths	415
6.8.2	Subsurface Voids	416
6.8.3	Landfill Geometry	417
	Problems	**420**
	References Cited	**426**
	Suggested Reading	**427**

7	**Exploration Using the Magnetic Method**	**429**
7.1	**Fundamental Relationships**	**430**
7.1.1	Magnetic Force	431
7.1.2	Magnetic Field Strength	431
7.1.3	Magnetic Moment	432
7.1.4	Intensity of Magnetization	433
7.1.5	Magnetic Susceptibility	434
7.1.6	Magnetic Potential	437
7.2	**The Earth's Magnetic Field**	**438**
7.2.1	Field Elements	438
7.2.2	Dipolar Nature of the Earth's Field	439
7.2.3	Variations of the Earth's Field	441
7.2.4	Dipole Equations	444
7.3	**Measuring the Magnetic Field**	**446**
7.3.1	Flux-Gate Magnetometer	446
7.3.2	Proton-Precession Magnetometer	447
7.3.3	Total-Field Anomalies	447
7.4	**Basic Field Procedures**	**450**
7.4.1	Magnetic Cleanliness	450
7.4.2	Diurnal Corrections	450
7.4.3	Elevation Corrections	451
7.4.4	Correcting for Horizontal Position	452
7.5	**Magnetic Effects of Simple Geometric Shapes**	**454**
7.5.1	Rock Susceptibilities	454
7.5.2	Magnetic Effect of an Isolated Pole (Monopole)	456
7.5.3	Magnetic Effect of a Dipole	457
7.5.4	Magnetic Effect of a Sphere	465
7.5.5	Magnetic Effect of a Thin, Horizontal Sheet	470
7.5.6	Magnetic Effects of Polygons with Infinite Strike Length (Using GRAVMAG)	478
7.6	**Interpretation of Magnetic Data**	**482**
7.6.1	Disadvantages and Advantages	482
7.6.2	Quantitative Interpretation Techniques	484
7.6.2.1	Half-Maximum Techniques	484

7.6.2.2	Slope Methods	485
7.6.2.3	Computer Modeling	488
7.7	**Applications of the Magnetic Method**	**488**
7.7.1	Archaeological Surveys	488
7.7.2	Detection of Voids and Well Casings	490
7.7.3	Defining Landfill Geometry	493
	Problems	**494**
	References Cited	**496**
	Suggested Reading	**497**

8   Electromagnetic Surveying   499

8.1	**Electromagnetic Waves**	**500**
8.1.1	Wavelengths	502
8.1.2	AC/DC	504
8.1.3	Electrical Properties of Geologic Materials	504
8.1.3.1	Electrical Resistivity and Conductivity	505
8.1.3.2	Dielectric Properties	505
8.1.4	Absorption and Attenuation	507
8.2	**EM Sounding**	**509**
8.2.1	Near-Field Continuous-Wave Methods of Frequency Domain Electromagnetics (FDEM)	512
8.2.1.1	Moving Transmitter-Plus-Receiver System (Slingram)	512
8.2.1.2	Noncontacting Ground Conductivity Measurements	514
8.2.2	Other FDEM Systems	519
8.2.3	Time Domain Electromagnetics (TDEM)	519
8.3	**EM Field Techniques**	**521**
8.3.1	Profiling versus Sounding	521
8.3.1.1	Sounding	521
8.3.1.2	Profiling	523
8.3.2	Interpretation	523
8.4	**Ground-Penetrating Radar**	**524**
8.4.1	Radar Velocity	525
8.4.2	Data Acquisition	527
8.4.3	GPR Velocity Analysis	533
8.4.3.1	Burial of Known Object	533
8.4.3.2	Walkaway Test	533
8.4.3.3	Diffraction Hyperbola	537
8.5	**Applications of Electromagnetic Surveying**	**539**
8.5.1	Archaeological Surveys	539
8.5.1.1	Aztec Ruins, New Mexico	539

8.5.1.2	Ceren, El Salvador	541
8.5.2	Geologic Applications	543
8.5.3	Snow and Ice Mapping	546
8.5.4	Environmental and Engineering Applications	546
	Problems	**548**
	References Cited	**553**
	Suggested Reading	**554**

Appendixes

A	**Instructions for Using REFRACT**	**A1**
B	**Instructions for Using REFLECT**	**A6**
C	**Instructions for Using RESIST**	**A13**
D	**Instructions for Using GRAVMAG**	**A18**
E	**Instructions for Using DIFFRACT**	**A25**
	Index	A33

Preface

This book is based on exploration geophysics courses that we have taught for a combined total of 35 years. The level is directed at advanced undergraduates who typically are majoring in geology. However, this text also is appropriate for geology graduate students and students majoring in other disciplines (hydrology, engineering, archaeology) who require some knowledge of geophysical methods.

Because of the increasing interest in the shallow subsurface and the ability of undergraduate classes to conduct their own field experiments, we focus on methods and approaches that are appropriate for the shallow target. Examples, illustrations, applications, and problem sets all emphasize shallow exploration—as contrasted with the more typical geophysics text, which contains material directed more toward targets at depth.

The organization of this book follows, in a general way, the relative utility and frequency of use of various geophysical methods for exploring the shallow subsurface. Seismic methods are discussed first. After a brief chapter devoted to some basics (propagation of elastic waves, instrumentation, and so forth), we present refraction seismology because this still is the most common method employed for shallow work. The longest chapter in the book concentrates on reflection seismology. The use of this technique in shallow work is relatively recent, and its shallow applications are not covered in most existing texts. Electrical resistivity methods follow seismic methods because these often are used in conjunction with seismic work for investigations at shallow depths. The next chapter is concerned with gravity. Although not as commonly used for shallow surveys, gravity, especially microgravity investigation, has quite a lot to offer in obtaining useful information; so applications to shallow as well as deep surveys are elucidated, compared, and contrasted. Magnetic surveying is discussed next to last because this is the least common method currently used for shallow investigations other than for metals exploration and locating buried iron and steel objects. The last chapter of the book covers electromagnetic methods. These methods build on the physical principles introduced in the seismic, electrical resistivity, and magnetics chapters. Electromagnetic techniques include ground conductivity measurements and ground-penetrating radar. Both of these techniques are used extensively in engineering and environmental applications.

We introduce each geophysical method by presenting the basic theory and considerations necessary for understanding its application in exploring a given target. Although the fundamental equations for each method always are developed in detail, mathematical sophistication on the part of the student is not required. Calculus is used sparingly but is included where necessary. Each chapter includes a brief discussion of the instruments used in field surveys. This discussion is intended to convey the fundamental principles of design to support discussions of field procedures and data and interpretation limitations. Data collection and reduction procedures are discussed in sufficient detail to enable the student to plan well-conceived field surveys. Approaches and techniques for analyzing field data follow the section on field operations. This discussion typically covers the classic methods but also attempts to pay sufficient attention to the inherent uncertainties in a given method and some common pitfalls. We consider the fact that uncertainties in interpretation are common by discussing the goals of the field survey, the accuracy of results likely with a given method, and the time needed in the field to gather data. The final material presented for each exploration method focuses on applications and illustrates data and interpretations. A problem set accompanies each chapter.

Because geophysics exploration depends heavily on computer processing and analysis, students must be introduced to hands-on experience as soon as possible. Accompanying this text is a series of computer programs, written for both Macintosh and Windows operating systems, that perform forward and inverse modeling and support all the major exploration methods covered in the book. These programs are as similar to one another in operation as possible, so they are easy to learn to use, and data input is quick and flexible. Students can access these programs to explore concepts discussed in the text (exercises occasionally are suggested in boxed inserts) as well as to work on many of the problems in the problem sets. A total of five programs are available: REFLECT for reflection seismology, REFRACT for refraction seismology, RESIST for electrical resistivity, GRAVMAG for gravity and magnetics, and DIFFRACT for ground-penetrating radar.

The programs are standard routines that easily can be incorporated into geophysics work. They are not designed as research tools for a geophysicist. However, their capabilities are more than adequate for students using this text and will support the field surveys that commonly accompany a course of this sort. The strength of the programs is their ease of use and uniformity of design.

Because of the utility and graphics capability of these computer programs, fundamental concepts are illustrated and supported with examples created by them. Therefore, students can reproduce the diagrams used in the text to elucidate concepts and can test variations in starting assumptions as their curiosity dictates. For example, discussions of the hidden layer in refraction surveying and of the optimum offset in reflection surveying are accompanied by suggestions for "discovering" these relationships through experimentation with the software. Problem sets also emphasize program use and include samples of real field data for reduction and analysis. It is important to realize, however, that the text is written to stand alone and does not depend on using these programs. Rather the programs add an important dimension that is missing from most existing texts.

Almost all the tables in the book are produced from spreadsheet templates and are referred to as *dynamic tables*. These Microsoft® Excel templates are also on the resources website accompanying this book. As a student works through text material, she or he can access the template, change values in certain cells, and study the effect. Most tables contain already constructed graphs, so the result of changing variables is conveniently displayed.

Our main objective in designing our own courses and this book was to cover the fundamental principles of common methods of exploration geophysics, while at the same time making it possible for a student to interact with and explore these ideas. Such hands-on experience has proved successful for both learning and retention of basic concepts; but even more important, it fosters an atmosphere in which students not only ask questions but often seek their own answers by using the computer templates and programs. Creating an environment for such independent investigation is our goal, and we hope this text will facilitate independent inquiry.

H. ROBERT BURGER
NORTHAMPTON, MASSACHUSETTS

ANNE F. SHEEHAN
BOULDER, COLORADO

2005

Preface to Computer Materials

The computer applications contained on the digital resources website were developed to allow a student to perform forward and inverse modeling for the geophysical methods covered in the text. Most are familiar because they are standard programs used in many geophysics courses. These programs are designed for student use to aid comprehension of the topics presented in the text and to work with data collected during field exercises typically accompanying a course in exploration geophysics. To keep the programs easy to use (and to keep development time within reasonable bounds) these programs are not intended to substitute for the full-featured programs necessary for research applications. Instructions for using these applications are in the appendixes.

These programs have been thoroughly tested. Their calculations agree with those from other programs using the same data and, therefore, are believed to be accurate. However, the programs are included with this text as supplementary material and should be treated as such. Bugs and errors may still exist. We hope these will be reported directly to us so we can correct them in future versions.

The Excel templates that also are included on the resources website were created with Microsoft® Excel X. Templates are included for almost all tables in the text. Cell values shown in boldface in the tables can be changed by the user. Results of calculations are automatically displayed by the graphs accompanying most tables.

We hope both the applications and the templates will prove valuable in your exploration geophysics course. They constitute a rich teaching and learning resource that should be utilized if possible.

CRAIG H. JONES
BOULDER, COLORADO

H. ROBERT BURGER
NORTHAMPTON, MASSACHUSETTS

DOUGLAS C. BURGER
AUSTIN, TEXAS

2005

CHAPTER 1

Approaching the Subsurface

1.1 WHAT ARE THE OPTIONS?

Presumably you are reading this book because you want to learn about exploring the subsurface. Your interest may be an active one in that you intend to actually go into the field and collect information; or you may be interested only in learning how subsurface information is gathered so you will be better equipped to judge recommendations for action based on field surveys. In either case you probably intend to lend your talents to improving humanity's lot by identifying sources for much-needed resources, protecting water supplies, providing safe building sites, locating areas for safe disposal of certain wastes, or extending knowledge of our forebears by finding where they lived and worked. We hope that this book will provide you with tools to accomplish your objectives and that you will be stimulated to continue your quest.

This first chapter is not presented as a survey of what is to come. Rather we mention a few important points about the practice of exploration geophysics that we hope you will keep in mind as you work through the pages ahead. If you continue to refer back to these points frequently as you cover new topics, you likely will emerge as a more reasoned observer or practitioner of the discipline.

Various approaches are available to gather information about the subsurface. The best is direct observation of the sediments and rocks themselves. Of course, this is rarely possible to the extent that we would like. It's not often that numerous, deep valleys cut through an area of interest to the necessary depths so that all we need to do is correlate geologic sections from one position to another. It also is relatively unusual for an investigation site to contain numerous wells that were logged by a professional geologist and are uniformly distributed. More commonly, when subsurface information is necessary, we acquire physical measurements on the surface and use these to deduce subsurface geology. The value of geophysics is its ability to acquire information about the subsurface over a substantial area in a reasonable time frame and in a cost-effective manner.

Geophysics essentially is the measurement of contrasts in the physical properties of materials beneath the surface of the Earth and the attempt to deduce the nature and distribution of the materials responsible for these observations. Variations in elastic moduli and density cause seismic waves to travel at different speeds through different materials. By timing the arrivals of these waves at surface observation points, we can deduce a great deal about the nature and distribution of subsurface bodies. Variations in the electrical conductivity of rocks and sediments produce varying values of apparent resistivities as the distance between measuring probes is increased or as the positions of the probes are changed on the surface. Density variations in the subsurface lead to variations in gravitational acceleration at surface instrument stations, and variations in magnetic susceptibilities produce measurable differences in the magnetic field at field observation sites. These variations in physical properties must be sufficiently large so our instruments can determine their effects. A frequent problem is that insufficient contrasts exist to detect the subsurface target of interest. Other times the presence of nearby bodies of great contrasts creates effects that mask those created by our target.

To the uninformed it seems almost magical that these contrasts in physical properties permit the explorationist to map in detail bedrock topography that is present at depths of perhaps 150 m. Equally impressive is our ability to extract the form and distribution of alluvial sand and gravel deposits or to trace a buried river channel. Perhaps you will find the stone walls of an ancient city or determine the distribution of contaminated waters flowing from a waste disposal site. Maybe you will find a placer deposit or develop a new technique to interpret electrical resistivity data. Or your contribution may be in defining guidelines for groundwater use based on the results of geophysical surveys. Whatever future awaits you, if you continue to deal with matters depending on or affected by subsurface geology, we hope this text helps you make informed decisions.

1.2 SOME FUNDAMENTAL CONSIDERATIONS

Throughout the chapters that follow we tend to treat the Earth and the subsurface as an ideal subject for study. Subsurface geology often is visualized as constituted by bodies of constant thickness with perfectly planar contacts. Dipping beds are only infrequently pictured, and even then dips rarely depart from those of constant inclination. Lateral variations in physical properties are mentioned as something to be aware of, but they rarely are treated in as much detail as homogeneous bodies. Lateral variations also always seem to be pictured as abrupt and with vertical boundaries; gradational, inclined contacts are not the rule in geophysics texts. Every geologic unit pictured always has a sufficient physical property contrast to enable it to be differentiated from all adjacent units. The Earth's surface is almost always horizontal, and differences in materials at the surface are nonexistent. Finally, vibrations from wind and traffic or currents induced by electrical transmission lines are never illustrated.

Of course, the real world departs significantly from the ideal of the next several chapters. We must make certain assumptions to deduce the effects produced by the variations in physical properties that we have mentioned. We also take pains to point out how the nonideal subsurface affects our assumptions and results. On occasion definite techniques are developed to deal with these departures from uniformity. However, there is no question that the parade of diagrams depicting planar contacts and homogeneous strata creates a subtle mental trap. If you are not careful, you will tend to focus on the world of the text and will try to force your observations into its orderly geology. Try to avoid this trap. Always try to determine what information about nonuniformity you can extract from your observations rather than trying to decide how far you can stretch the assumption of homogeneity to enable interpretation of "less than ideal" data.

1.3   DEFINING OBJECTIVES

As you begin your work in geophysics, your excitement and curiosity will quite naturally make you impatient to go into the field and begin to accumulate observations for analysis. Without sufficient exposure to proper exploration techniques, it is easy to go somewhere close by, lay out some seismic cable, and begin taking measurements. If this approach is followed too often, it will develop an undesirable operational mode. Always have a specific survey objective in mind, and define it as explicitly as possible. This should be true even as you are learning a specific method such as electrical resistivity profiling. Even the most standard elements of survey procedure will vary somewhat depending on what subsurface information is of primary importance. For example, suppose you elected to use the seismic reflection method to map buried bedrock topography. To do this as well as possible, you may have to sacrifice some details from within the sediments above the bedrock surface. You therefore need to be sure that this degree of detail is truly warranted in view of what you will lose.

In the real world there always will exist constraints, which necessitate that survey objectives be defined well and in detail. Funding always will be tight, and time is money. You will not enjoy the luxury of laying out a few more seismic lines. Transportation costs, crew salaries, prices for expendables (such as explosives), and deadlines will impose severe restrictions. Even if you are working alone on a summer research project, you may have only two months and $1000 to define the geometry of a local aquifer.

The first step, of course, is to understand the fundamentals of the various exploration methods so you know what each can accomplish. Ask what is most important and what constitutes an added dividend but is not essential. Before planning the data acquisition stage, be sure to determine what information already exists. It's not particularly exciting to search through town archives for drilling logs, but in some localities that's where you will find them; and even one or two good well logs could be valuable for the interpretative phase of your project. After survey objectives are carefully defined and existing information researched, then plan your field design, carefully and in detail. Think about a lot of "What if?" questions and prepare contingency plans for unforeseen but likely difficulties during data acquisition (such as a farmer plowing a field near your most important seismic line).

If you adopt this approach now, essentially at the beginning, it will become second nature, and you will become one of those individuals others always envy because of your success.

1.4 LIMITATIONS

Although geophysical methods are extremely useful and in the majority of cases provide valuable information about the subsurface, it is important to recognize their limitations. A common limitation is the lack of sufficient contrast in physical properties. Even though seismic wave velocities vary for saturated sediments of different compositions, they typically are similar enough in value that layers of different compositions cannot be differentiated by the seismic refraction method.

A second common limitation is the nonuniqueness of many interpretations. In geophysics we tend to engage in two types of modeling: forward and inverse. In *forward modeling* we develop an equation to describe the effect of the specific physical property variation we are studying. This equation then tells us what effects this variation produces. In many cases such an equation can be derived and leads to unambiguous results. However, we most often use the reverse process of observing effects and then modeling the cause. This is referred to as *inverse modeling* and typically is more difficult and more ambiguous. The computer programs included with this book support both modeling types.

Ambiguity tends to be most obvious in gravity work, when it often is difficult to differentiate the effect of a small body near the surface from that of a larger body at depth. However, many apparent resistivity curves can be due to a variety of subsurface layering configurations, and many magnetic curves do not lead to only one possible explanation for the observed magnetic anomaly. A good geological background and knowledge of the area being studied are essential if you are to arrive at the correct interpretation. You will perform better as an exploration geophysicist if you are a good geologist.

Another limitation is resolution. All methods are saddled with this restriction. Because seismic wavelengths typically are many meters, there is a lower limit to stratigraphic thicknesses that can be resolved by reflection surveys. Thin beds also may not contribute their signature to an apparent resistivity curve unless they create an especially high resistivity contrast with overlying or underlying strata. Resolution limits often can be reduced but normally at increased cost.

Finally, we must mention "noise." Almost all geophysical data contain some undesired signal (or noise) to a greater or lesser extent. Today most noise is not due to equipment design. It is relatively straightforward to reduce noise due to operator error to a very minor amount. Noise due to human activities and natural phenomena (such as water pumps and wind) normally can be eliminated by careful survey planning. What "noise" remains carries important information about ways in which the subsurface varies from our idealized picture. The continuing challenge is to develop ways to use such information. Perhaps someday you will make a contribution of this sort.

1.5 THE ADVANTAGE OF MULTIPLE METHODS

Often specific survey objectives cannot be met by applying only one geophysical method. In seismic reflection surveys, seismic refraction lines provide important information about velocities of near-surface materials that make it possible to apply needed corrections to reflection travel times. Because electrical resistivity surveys depend on different physical properties than seismic surveys, the two taken together may provide important cross-checks. For example, consider a survey in which an accurate bedrock depth is essential. Drilling is out of the question. Seismic refraction results are perfect, but the nagging question of the "hidden" layer and its effect on computed depths remains. However, resistivity results are not affected by the hidden layer and could be used to confirm the seismic thickness value.

Assume that we are mapping the confined aquifer diagrammed in Figure 1.1. A thick sequence of saturated clay overlies saturated sand that overlies a heavily fractured bedrock surface. Our goal is to map the thickness of the sand. The clay–sand contact does not have a sufficient velocity contrast to be detected by seismic refraction, but the resistivity contrast is sufficient for mapping by the electrical resistivity method. Conversely, the fractured bedrock surface is filled with water and possesses a resistivity similar to the saturated sand. However, a sufficient velocity contrast is present here for bedrock depth to be determined by the refraction method.

As we learn about the basic methods of exploration geophysics, keep in mind the suggestions in this brief chapter. Learn the strengths and weaknesses of each

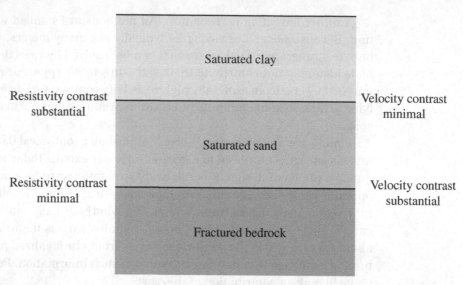

FIGURE 1.1 Determining the thickness of a confined aquifer by using both seismic refraction and electrical resistivity.

method, and determine how and when you will benefit by employing more than one method. Remember to define field survey objectives carefully beginning with your first excursion. Recognize the limitations of geophysics, but don't be discouraged by the ambiguities inherent in the various methods. Of course, after you analyze your field observations, your conclusions must address this lack of certainty. But learn to seek clues and to make maximum use of preexisting data to keep the degree of uncertainty to a minimum. Don't be misled by illustrations in this and other texts, and never forget that the real world is messy and that your data reflect this. Rather strive to solve the sticky problems. The easy ones aren't as much fun. Geophysics plays a more and more important role in our lives with each passing year as populations increase and resources dwindle. Your journey is therefore an important one. We hope we have provided the means for you to enjoy it.

CHAPTER 2

Seismic Exploration: Fundamental Considerations

2.1 SEISMIC WAVES AND WAVE PROPAGATION

The physical basis on which the science of exploration seismology rests is remarkably straightforward. A sudden disturbance, or displacement, occurs at a small localized region in a mass of rock. This displacement propagates outward from its origin as a spherical wave front and eventually arrives at distant points, where it can be recognized by the motion it induces in the materials present. You surely are familiar with two very different phenomena, each of which illustrates just such an event: the impact of a pebble in a body of water and an earthquake.

When a pebble is thrown into a lake, the impact momentarily displaces a volume of water. This disturbance then radiates outward in the form of circular wave fronts or ripples. If the toss occurs on a day when the lake surface is quiet, the regularity of the ripple pattern is quite striking. The arrival of these ripples at distant points on the lake is recognized by the motion the water particles induce as the wave fronts pass by. Many inexperienced young boys and girls, fishing for the first time, almost assuredly have linked the up-and-down motion of their bobs to causes other than the motion of the water particles induced by the passage of ripples. Although it doesn't occur to most people admiring the almost hypnotic appeal of orderly trains of ripples spreading across a lake or pond, these could be used to determine some useful information. It would be fairly simple to calculate the speed with which the ripples are moving. Then, with the help of a companion, you could use this new bit of information to calculate various distances of interest.

Similarly, the sudden, violent displacement of two masses of rock across a fault plane is a disturbance that sends seismic waves propagating through the Earth. Anyone who has taken an introductory geology course knows that these waves are used to elucidate the details of Earth structure and, in conjunction with other measurements of Earth properties, provide insights into the nature of materials at depth. In a general sense, seismologists are attacking a much more sophisticated version of the lake distance problem. Determining the time of the disturbance (earthquake) and measuring the arrival of the waves is not difficult if needed equipment is available to measure the small ground motions present at significant distances from the earthquake. Because many of the seismic waves that are received at recording stations far from the earthquake have passed through the Earth, it seems evident that to make deductions about internal structures and compositions, we must understand the types of waves with which we are dealing, what factors govern their transmission, and the speeds with which they travel.

In this book our immediate focus is on the materials just beneath the surface. Usually we will be concerned with those at the relatively shallow depths of 0 to 500 m even though most of the principles we establish apply to exploration efforts at considerably greater depths. For the moment, let us consider a simplified model for our discussion of the generation and transmission of seismic waves (Figure 2.1). In this case we assume an isotropic and homogeneous body of granite that is separated from the atmosphere by a perfectly planar surface. A disturbance is produced at a point on the surface by a blow with a heavy hammer. This disturbance propagates outward as a series of wave fronts quite analogous to the ripples on the lake. The passage of the wave fronts by each point on the surface is marked by the motion of that surface, which can be measured and recorded by sufficiently sensitive detecting and recording instruments. As in our lake discussion, we can calculate the distance of the recording instrument from the hammer blow if we know the time interval between the impact and the arrival of the first wave front at the instrument and if we can calculate the speed of the wave. Such an exercise obviously is trivial because at the scale of our investigation we easily could measure this distance directly. Determining the speed of the wave is not quite so trivial: We cannot directly observe its passage as we could in the case of ripples on a lake surface. However, given the distance and assuming the presence of an instrument that can measure the time interval between impact and wave arrival, the value for the velocity of the wave is easy to obtain.

At this point, as a curious scientist, you probably have many questions about this procedure and analogy. What kinds of wave speeds are we likely to encounter?

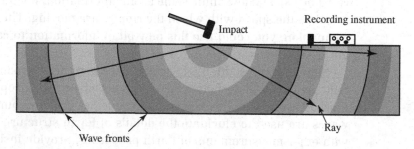

FIGURE 2.1 A hammer blow creates a disturbance that propagates through homogeneous granite as seismic waves. Wave fronts and rays are indicated.

Would these speeds be different if the surface material was sand and gravel instead of granite? How does the ground move when a wave front arrives at our sensing instrument? How does the design of the instrument take advantage of this motion? Is only one type of wave generated by the hammer blow? What if the material beneath the surface is not homogeneous? Can we detect this? Can we identify the composition of material beneath the surface if the seismic waves arriving at the surface have passed through this material? Let's begin to seek some answers to these questions.

2.1.1 Wave Terminology

Before pursuing these questions in greater detail, we need to be sure that we are familiar with the terminology used to describe various aspects of waveform and wave motion. For the following discussion, we will assume that we are dealing with periodic waves of sinusoidal shape (Figure 2.2).

The *wavelength* λ is the distance between two adjacent points on the wave that have the same phase or similar displacements. The distance between successive ripple crests is one wavelength. The *amplitude A* of the wave is the maximum displacement associated with the particle motions that occur as the wave passes through the material. For our ripple crests this would be the height of a ripple crest above the undisturbed lake surface. In granite the amplitude would be the maximum particle displacement from the undisturbed position.

Wavelength and amplitude provide information concerning the shape of the waveform, but we also require information concerning wave movement through a material. Figure 2.3 illustrates how a water particle in a lake behaves as ripples pass through its location. This diagram could equally well represent the behavior of a granite particle as seismic waves pass through a region. Imagine that the wave train is moving and that the arrow marked "observation point" is your point of reference. The *period T* is the time it takes for two successive wave crests to

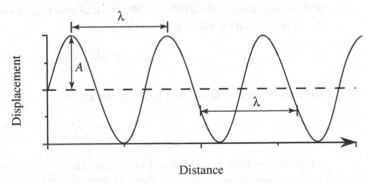

FIGURE 2.2 Pattern of particle displacements during passage of wave created by hammer blow illustrated in Figure 2.1. λ = wavelength; A is amplitude of maximum particle displacement from normal position.

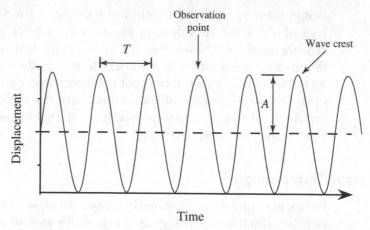

FIGURE 2.3 Motion of a single water particle in a lake or a particle in granite as waves pass by. T (wave period) is the time for two successive wave crests to pass the observation point.

pass this reference point and, therefore, for the motion to complete one cycle. When working with seismic waves, we are usually interested in how often such repetitions or cycles occur in a given unit of time. In other words, what is the repetition rate? The numbers of repetitions per unit of time is defined as *frequency f.* Consider a seismic wave that has a period $T = 0.1$ s. Wave crests must pass a reference point every tenth of a second. The frequency, or repetitions per second, is then 10. It should be clear that frequency and period are related in the form

$$f = \frac{1}{T} . \tag{2.1}$$

The unit of frequency is the *Hertz* (Hz), which is the number of repetitions per second.

If we refer back to lake ripples, it is evident that we can determine the speed with which the ripple crest or wave front is moving by noting the time it takes the wave front to traverse a known distance:

$$V = \frac{distance}{time} . \tag{2.2}$$

Similarly, we can determine speed if we know the wavelength and the frequency:

$$V = f\lambda . \tag{2.3}$$

If this relationship is not immediately clear, consider two separate waveforms, each with the same cycle of repetition or frequency but with vastly different wavelengths. The wave with the greater wavelength must also have a greater velocity to have the same frequency.

Even though wavelength, amplitude, and frequency are straightforward terms, you should constantly be solving simple mental problems using these relationships as you read through the material in this and subsequent chapters. Calculate the frequency or the wavelength of the waves with which you are working whenever possible. What kinds of amplitudes can you expect? What wavelength do a given speed and frequency imply? What frequencies are characteristic for the waves with which you are working? Once these questions and their answers become second nature, they will be valuable as you plan, evaluate, and interpret field studies using the seismic methods to be outlined in subsequent chapters.

As the ripples spread across the lake, we could draw a surface through a given ripple crest. Along this surface all water particles are in the same phase of motion. This surface is referred to as a *wave front*. It moves along as the disturbance that initiated it spreads outward. If the medium through which the wave is moving is isotropic and homogeneous, like the granite in Figure 2.1, the direction of propagation is always perpendicular to the wave front. Thus we could draw lines perpendicular to the wave front, and these lines would indicate the direction in which the wave is moving through the material. Such lines are called *rays* and are used extensively in deriving the relationships for the seismic refraction and reflection methods. Rays will appear in most diagrams in this book rather than wave fronts. Nevertheless they are a simplification, and you will be well served to constantly imagine the wave front to which these rays are perpendicular.

2.1.2   Elastic Coefficients

If we refer to Figure 2.1 and consider the waves spreading out from the impact point of the hammer blow, we might wonder what determines how fast the waves travel. Considering the physical state of the granite before the hammer blow and after the waves have passed through the rock, it seems logical to propose that the waves are propagated by displacements of rock particles in a fashion analogous to water waves being propagated by displacements of water particles. Following the passage of the disturbance the particles return to their original state (except perhaps at the exact point of impact of the hammer blow). In other words, the rock has been subjected to a stress (force/area), there has been a shape change (strain), and the rock returns to its original state (shape). Such behavior is termed *elastic*. Perhaps the manner and speed with which seismic waves travel through materials are controlled by the materials' elastic properties.

To determine if this proposition makes intuitive sense, let's examine several basic relationships between the application of *stress* (a given force divided by the area over which it is applied) to an object and its resultant change in shape (*strain*). When determining these basic elastic constants, we assume Hookean behavior. Materials exhibiting *Hookean behavior* are elastic and exhibit an instantaneous linear relation between stress and strain (Means 1976, 240). In other

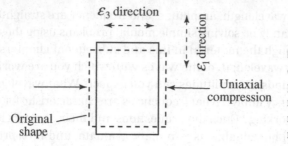

FIGURE 2.4 A uniaxial compression produces a positive elongation $+\varepsilon_1$ and a negative elongation $-\varepsilon_3$. The ratio of these two strains is referred to as *Poisson's ratio μ*.

words, when a stress is applied to such materials, the resulting strain is immediate. When the stress is removed, the material returns to its original shape.

If a Hookean material is subjected to a uniaxial compression or tension, the linear relationship between the applied stress σ and resulting strain ε is given by

$$\sigma = E\varepsilon \tag{2.4}$$

where the constant of proportionality E is termed *Young's modulus*. Because this constant directly relates resultant strain to a given stress, it seems probable that rocks with different values for E may have different velocities.

Elongation ε is the change in length of a line in its final or deformed state (l_f) divided by its original length (l_o):

$$\varepsilon = \frac{l_f - l_o}{l_o} = \frac{\Delta l}{l_o}. \tag{2.5}$$

If, as is illustrated in Figure 2.4, a Hookean solid is subjected to a uniaxial compression, it will shorten in the direction of the applied stress but at the same time will lengthen in a direction at right angles to the compression. Elongations can be measured for each direction, and their ratio is referred to as *Poisson's ratio μ*:

$$\mu = -\frac{\varepsilon_1}{\varepsilon_3} \quad \text{where } \mu \le 0.5. \tag{2.6}$$

Two additional elastic coefficients also are important. Given an isotropic material, subject it to a general pressure change (Figure 2.5). A volume change will occur, and the change in volume ($V_f - V_o$) divided by the original volume V_o, compared to the pressure change (ΔP) that induced it, is termed the *bulk modulus K*:

$$K = \frac{\Delta P}{\Delta} \quad \text{where} \quad \Delta = \frac{V_f - V_o}{V_o}. \tag{2.7}$$

The bulk modulus is a measure of the incompressibility of the material.

Finally, it is possible to deform a solid by simple shear (Figure 2.6). In this case a *shear strain γ* is induced by applying a *shear stress σ_s*. The ratio of these quantities is the *rigidity modulus G*:

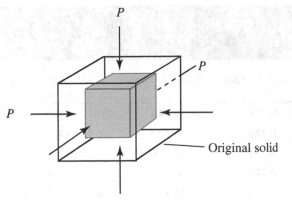

FIGURE 2.5 A change in volume (reduction) produced by a change in pressure (increase). The ratio of the pressure change to the volume change is a measure of the incompressibility of the material and is known as the *bulk modulus K*.

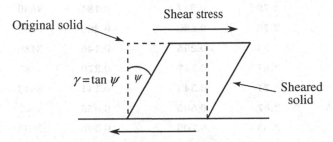

FIGURE 2.6 A measure of the shearing resistance of a material is the ratio of the shear stress to the shear strain, which is termed the *rigidity modulus G*. The *angular shear* is ψ.

$$G = \frac{\sigma_s}{\gamma}. \tag{2.8}$$

If we pause for a moment to consider Hookean materials, it seems reasonable to propose that the quantities E, μ, K, and G define elastic behavior with reasonable thoroughness and therefore must be important factors in governing the speeds with which disturbances travel through Earth materials. Fortunately, values for all of these quantities can be determined in the laboratory. Some representative values for common rock types are given in Table 2.1. As you might guess, these four quantities are not mutually independent. Specifically, G and K can be defined in terms of E and μ:

$$G = \frac{E}{2(1+\mu)} \tag{2.9}$$

$$K = \frac{E}{3(1-2\mu)}. \tag{2.10}$$

TABLE 2.1 | ELASTIC COEFFICIENTS AND SEISMIC VELOCITIES FOR SELECTED COMMON ROCKS

Rock Type	Density ρ	Young's Modulus E	Poisson's Ratio μ	V_p (m/s)	V_s (m/s)	V_p/V_s	V_s as %V_p
Shale (AZ)	2.67	0.120	0.040	2124	1470	1.44	69.22%
Siltstone (CO)	2.50	0.130	0.120	2319	1524	1.52	65.71%
Limestone (PA)	2.71	0.337	0.156	3633	2319	1.57	63.84%
Limestone (AZ)	2.44	0.170	0.180	2750	1718	1.60	62.47%
Quartzite (MT)	2.66	0.636	0.115	4965	3274	1.52	65.96%
Sandstone (WY)	2.28	0.140	0.060	2488	1702	1.46	68.42%
Slate (MA)	2.67	0.487	0.115	4336	2860	1.52	65.96%
Schist (MA)	2.70	0.544	0.181	4680	2921	1.60	62.41%
Schist (CO)	2.70	0.680	0.200	5290	3239	1.63	61.24%
Gneiss (MA)	2.64	0.255	0.146	3189	2053	1.55	64.38%
Marble (MD)	2.87	0.717	0.270	5587	3136	1.78	56.13%
Marble (VT)	2.71	0.343	0.141	3643	2355	1.55	64.65%
Granite (MA)	2.67	0.605	0.055	3967	2722	1.46	68.62%
Granite (MA)	2.63	0.705	0.096	3693	2469	1.50	66.85%
Gabbro (PA)	3.05	0.727	0.162	5043	3203	1.57	63.51%
Diabase (ME)	2.96	1.020	0.271	6569	3682	1.78	56.05%
Basalt (OR)	2.74	0.630	0.220	5124	3070	1.67	59.91%
Andesite (ID)	2.57	0.540	0.180	4776	2984	1.60	62.47%
Tuff (OR)	1.45	0.014	0.110	996	659	1.51	66.20%

Units for Young's modulus are $(N/m^2) \times 10^{11}$. Velocities computed from ρ, E, and μ. Values selected from Press (1966, 97–173).

2.1.3 Seismic Waves

Using advanced analysis, we can derive equations of motion demonstrating that, for an isotropic homogeneous body of infinite extent deforming elastically, it is possible for two types of waves to be transmitted. One type of wave is transmitted by particle movements back and forth along the direction of propagation of the wave and is referred to as a *longitudinal wave*. The other wave type is termed a *transverse wave* because particle motions are transverse to the direction of movement of the wave front, or perpendicular to the ray. It is, of course, somewhat satisfying that these two wave types are observed in laboratory experiments and in the field.

The particle motions that transmit longitudinal waves consist of a series of compressions and dilations that can be envisioned as the centers of rock particles being moved closer together than normal and then moved farther apart than normal (Figure 2.7). The familiar analogy is a coiled spring under tension in which an oscillation is created and then propagated along the spring by the coils moving together and pulling apart. Longitudinal waves usually are termed *P-waves* (*primary waves*) because, as we shortly will demonstrate, they have the greatest speed and therefore appear first on traces illustrating wave arrivals at a specific location. P-waves also may be referred to as *compressional waves* due to the particle compressions during their transport.

As a transverse wave moves through material, particles are subjected to shearing stresses as adjacent points move in a plane at right angles to the direction of propagation of the wave (Figure 2.8). These *S-waves* have speeds less than P-waves, appear on seismograms after P-waves, and may be referred to as *secondary waves* or *shear waves*. The familiar analogy for this type of wave is a rope anchored at one end and then set into oscillation by a flick of the hand holding the rope. The wave travels along the rope, but the rope particles move in a direction at right angles to the propagation direction. The rope analogy can be taken a step further in that particle motion can be generated in any direction in a plane perpendicular to the direction of propagation of the wave, depending on the direction in which the hand holding the rope displaces it. Similarly, for transverse waves traveling through the Earth, the particle motion is not constrained to a specific direction as in the case of longitudinal waves and may be in any

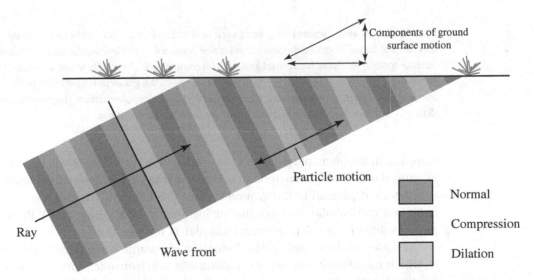

FIGURE 2.7 P-wave incident on surface. Particle motion is parallel to direction of travel. Because the wave front plane is canted to the horizontal plane, ground surface motion will have horizontal and vertical components.

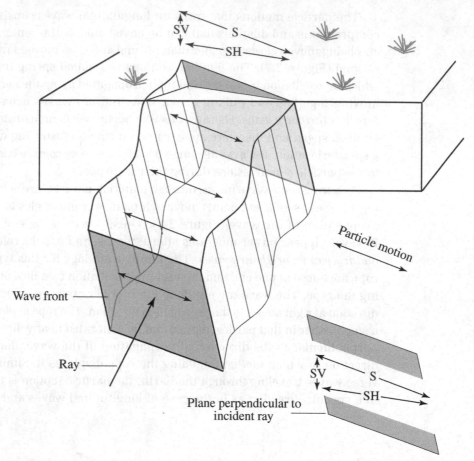

FIGURE 2.8 S-wave incident on ground surface. Particle motion is at right angles to the direction of travel. Because the wave front plane is canted to the horizontal plane, ground surface motion will have horizontal and vertical components. Normally shear wave motion S is resolved into two components, which are contained in the plane perpendicular to the incident ray: (1) an SH component, which is parallel to the ground surface (horizontal), and (2) an SV component in a vertical plane that contains the incident ray.

direction in the plane perpendicular to propagation direction. Often, especially in advanced analysis, S-wave motion is resolved into two components (Figure 2.8): a component parallel to the ground surface (SH component) and a component lying in a vertical plane containing the incident ray (SV component). Both components must be in the plane perpendicular to the ray.

As stated at the outset of this discussion of seismic wave types, the material in which the disturbance occurs is assumed to be isotropic, homogeneous, and infinite in extent. Because most of our efforts utilizing the seismic method involve generating energy at or close to the surface, this assumption clearly is not valid. The presence of this surface permits other types of waves to be generated. These are referred to as *surface waves* because they are propagated within and confined to the region of the bounding surface. In contrast, P- and S-waves often are referred to as *body waves*. Two types of surface waves are observed: *Rayleigh* waves and *Love* waves. They are named after the scientists (Lord J.W.S. Rayleigh, 1842–1919, and A.E.H. Love, 1863–1940) who proved theoretically that these waves could exist.

Rayleigh waves propagate by particle motion that is confined to a vertical plane, is retrograde elliptical (Figure 2.9(a)), and is in the direction of wave travel. Particle displacements are greatest at the surface and decrease exponentially

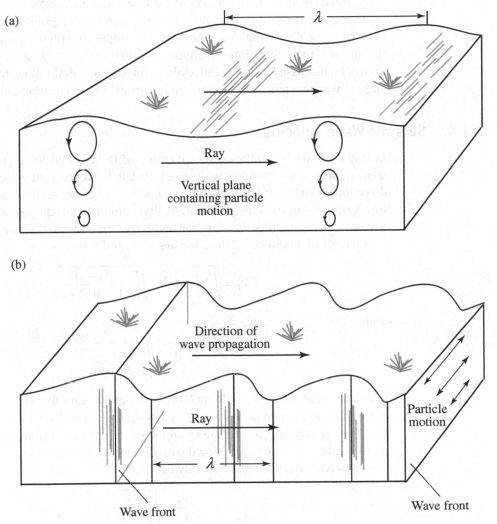

FIGURE 2.9 (a) Rayleigh waves. All particle motion is contained in a vertical plane, and disturbance decreases downward. (b) Ground surface motion associated with the passage of Love waves. All particle motion is in the horizontal plane and decreases downward.

downward (Richter 1958, 241). Because of this substantial vertical displacement at the surface, Rayleigh waves produce a pronounced signature on exploration seismic records. For this reason, surface waves often are referred to as *ground roll* by those involved in field studies. Rayleigh waves illustrate *dispersion*, which means that velocity is not constant but varies with wavelength. Love waves travel by a transverse motion of particles that is parallel to the ground surface (Figure 2.9(b)) and in this sense are similar to S-waves.

You might ask, and rightly so, why we are so concerned with particle motions connected with the various wave types. Remember that our observations usually are restricted to the surface, and our sources of information concerning the subsurface consist predominantly of the times at which wave energies arrive at recording instruments and the direction and magnitude of ground surface motions. Clearly, those displacements depend on the angles at which the waves arrive at the surface, their vibration directions, and their relative energies. These factors all will be discussed or reviewed again in the sections that follow, but our knowledge of wave types and behavior will be critical in interpreting field data.

2.1.4   Seismic Wave Velocities

In this book we will follow the convention of Dobrin and Savit (1988, 32) and treat velocity synonymously with speed. Recall, however, that velocity possesses direction as well as magnitude (speed) and is, therefore, actually a vector quantity. A result of the mechanical analysis that demonstrates the existence of P- and S-waves are equations that give velocities in terms of the density ρ and elastic coefficients of a material. The velocities of P- and S-waves are

$$V_P = \sqrt{\frac{K + 4/3G}{\rho}} = \sqrt{\frac{E}{\rho} \frac{(1-\mu)}{(1-2\mu)(1+\mu)}} \qquad (2.11)$$

and

$$V_S = \sqrt{\frac{G}{\rho}} = \sqrt{\frac{E}{\rho} \frac{1}{2(1+\mu)}} \quad . \qquad (2.12)$$

If we note that the bulk modulus K and the rigidity modulus G are always positive and recall that Poisson's ratio μ is less than or equal to 0.5, then it is evident that the velocity of P-waves always must be greater than that of S-waves by a considerable factor. Establishing the actual ratio consists simply of applying the previous two equations and simplifying:

$$\frac{V_P}{V_S} = \sqrt{\frac{1-\mu}{1/2-\mu}} \quad . \qquad (2.13)$$

Because G is equal to zero for liquids, the velocity of S-waves in liquids goes to zero. In other words, shear waves cannot be propagated by liquids. It is this fact and the observation that direct S-waves are not recorded in the Earth's "shadow zone" that constitute a primary line of evidence for a liquid outer core.

Refer once again to Table 2.1. P- and S-wave velocities calculated from the appropriate values of elastic coefficients and densities are listed, as are the velocity ratios.

> Table 2.1 is a dynamic table, which means that it is a spreadsheet template on the resources website accompanying this book. You can insert values in any cell that is in boldface, and values will be recalculated. To gain an appreciation for how density, Young's modulus, and Poisson's ratio affect velocities, insert a range of values similar to those in Table 2.1 and note the effect on P- and S-wave velocities.

Because compressional waves are used routinely in seismic exploration, especially at shallow depths, we focus mainly on P-wave velocities in the rocks and sediments normally encountered in our work. Velocities are determined by laboratory measurements, velocity logging in wells, or seismic field methods. All have advantages and disadvantages. However, for much exploration work only approximate values are needed for planning the actual survey because more refined values are determined during the survey itself.

Table 2.2 lists velocity ranges for a range of common materials. Many ranges overlap, so there are no unique values for rocks or sediments. However, a few general rules are suggested by these values:

- Unsaturated sediments have lower values than saturated sediments.
- Unconsolidated sediments have lower values than consolidated sediments.
- Velocities are similar in saturated, unconsolidated sediments.
- Weathered rocks have lower values than similar rocks that are unweathered.
- Fractured rocks have lower values than similar rocks that are unfractured.

In general, on the basis of velocity alone, you should have no trouble in distinguishing dry sediments from wet sediments, sediments from rocks, and sedimentary rocks from igneous and metamorphic rocks. Of course, it likely would be impossible to distinguish a wet, compacted glacial till from a weathered sandstone or poorly consolidated sandstone. Although we hesitate to suggest specific values, when in the first stages of field survey design, it is not unreasonable to use 500 m/s for dry unconsolidated materials, 1500 m/s for wet unconsolidated materials, 4000 m/s for sedimentary rocks, and 6000 m/s for unweathered igneous and metamorphic rocks.

The velocity of the air wave also is listed in Table 2.2. When hammer blows or explosives located near the surface are used as energy sources, a compressional

TABLE 2.2	REPRESENTATIVE P-WAVE VELOCITIES				
Unconsolidated Materials		Consolidated Materials		Other	
Weathered layer	300–900	Granite	5000–6000	Water	1400–1600
Soil	250–600	Basalt	5400–6400	Air	331.5
Alluvium	500–2000	Metamorphic rocks	3500–7000		
Clay	1100–2500	Sandstone and shale	2000–4500		
Sand		Limestone	2000–6000		
Unsaturated	200–1000				
Saturated	800–2200				
Sand and gravel					
Unsaturated	400–500				
Saturated	500–1500				
Glacial till					
Unsaturated	400–1000				
Saturated	1700				
Compacted	1200–2100				

Velocities in m/s.

Velocity ranges compiled from Press (1966, 195–218).

wave is formed that travels through the air to the seismometers or geophones. This is referred to as the *air wave* and has a velocity that can be slightly greater than some common surface materials. We can expect it to appear in much of our work and always should look for it on our records.

In some field studies we also will want to estimate velocities for S-waves and Rayleigh waves. Love waves typically are not generated during exploration work, so we have little interest in estimating their velocities. Here are some general rules of thumb for such estimates:

$$V_S = 0.6V_P \qquad \text{for crystalline rocks}$$
$$V_S = 0.5V_P \qquad \text{for sedimentary rocks}$$
$$V_S = 0.4V_P \qquad \text{for soils and unconsolidated materials} \qquad (2.14)$$
$$V_R = 0.9V_S.$$

Earlier in this discussion we noted the relation between wavelength, velocity, and frequency. Energy sources typically used to generate seismic waves for exploration produce waves in the 10–100 Hz range. Now that we have some representative P-wave velocities, we can calculate what types of wavelengths might be expected. As we will discover in our discussions of reflection seismology, wavelength will affect both our horizontal and vertical resolution. If we take our "typical" value for saturated unconsolidated materials to be 1500 m/s, then a wave with a frequency of 10 Hz will have a wavelength of 150 m, and a 100 Hz wave will have a wavelength of 15 m. These certainly are substantial wavelengths, and often in our field surveys, other factors being equal, we will try to work with values in the lower range.

2.2   RAY PATHS IN LAYERED MATERIALS

The preceding discussion was based on simple assumptions for the generation and propagation of seismic waves. For the most part we assumed an infinite, homogeneous, and isotropic medium. Our greatest complication was to allow for a horizontal, planar interface between rock (or water) and the atmosphere. Of course, these assumptions are unbelievably simplistic when we consider the actual subsurface, but they are necessary to manage the physics of the phenomena. Once again, based on your own past experiences, you should be curious about what happens to these waves when they encounter "obstructions." You may have observed ripples striking a planar surface and reflecting and probably have heard or read about the bending, or refraction, of light waves when they pass from air into water. Even if you have studied relatively little physics or geology, it probably is clear that we can expect seismic waves to reflect and refract as they travel through sediment or rock and encounter surfaces bounding materials with different elastic coefficients and densities.

Two insights are important in developing the relationships for refracting and reflecting waves that we will use throughout our explanation of seismic exploration methods. These are Huygens' principle and Fermat's principle.

2.2.1   Huygens' Principle

Christian Huygens (1629–1695) was a Dutch mathematician, physicist, and astronomer who formulated a conceptually simple but remarkably powerful principle as part of his development of the wave theory of light. His principle states that all points on a wave front can be considered as point sources for the generation of spherical secondary *wavelets*; after a time t the new position of the wave front is the surface of tangency to these wavelets. If we apply this principle to the wave front at time t_1 in Figure 2.10, we can construct the wave front at time t_2. For simplicity in the example we assume constant velocity throughout the medium. We next select a few point sources on the initial wave front, calculate the

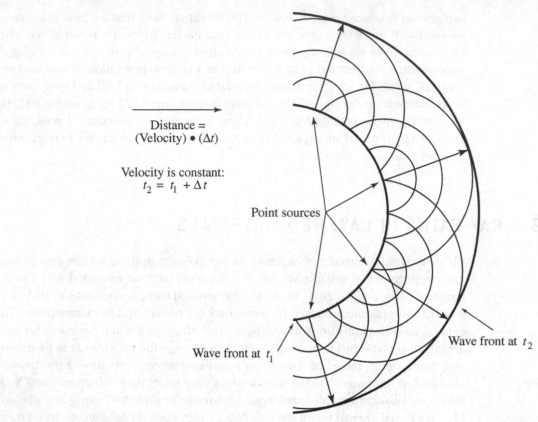

Distance =
(Velocity) • (Δt)

Velocity is constant:
$t_2 = t_1 + \Delta t$

Point sources

Wave front at t_2

Wave front at t_1

FIGURE 2.10 Applying Huygens' principle to determine the position of the wave front at t_2 after a time interval Δt. Given the position of a wave front at time t_1 and applying Huygens' principle, the position of the wave front at time t_2 is determined.

radius of the secondary wavelets based on the velocity of the medium and the elapsed time, Δt, construct these wavelets, and draw a surface tangent to them. This produces the wave front at time t_2.

2.2.2 Fermat's Principle

The Frenchman Pierre de Fermat (1601–1665) was an outstanding mathematician who developed what has come to be known as the *principle of least time*. This principle states that in the propagation of waves, the wave path between any two fixed points is the one along which the travel time is the least of all possible paths.

In our discussion of reflection and refraction, we assume that velocity in a given medium remains constant. By applying Fermat's principle we see that the wave path or ray must be a straight line in a medium of constant velocity. Because a straight line is the shortest distance between two points and because velocity is constant, a straight line between two points also is the one for which the time of transit is a minimum.

In previous diagrams wave fronts often were illustrated as spherical surfaces, but in many of the derivations in following sections the wave front is drawn as a plane. As a wave increases its distance from its center of origin, the radius of curvature increases to such an extent that we can treat the wave front as a plane near any particular point of interest.

2.2.3   Reflection

As anyone who has purchased a pair of sunglasses should know, a wave front is reflected when it encounters a surface bounding two materials with differing velocities of wave transmission. For our discussion we assume a planar wave front that strikes a planar, horizontal surface separating materials with seismic velocities V_1 and V_2. The angle between a ray and a normal to the horizontal surface is termed the *angle of incidence* and is labeled θ_1 in all diagrams. If we refer to Figure 2.11 or 2.12, we can ask, "What is the relationship of angle θ_2 to θ_1?" In other words, "What path will the reflected ray take?"

It is possible to answer this question by applying either Huygens' principle or Fermat's principle. Because it is instructive to use both of these principles, we will take a look at both in developing the basic geometries of reflection and refraction of waves.

First, let us examine reflection through an application of Huygens' principle. Consider a wave front with velocity V_1 approaching a horizontal interface (Figure 2.11). When ray 1 strikes the interface, it creates a disturbance that spreads outward into both the upper and lower materials (for the moment we are concerned

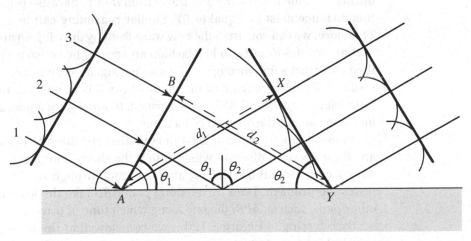

FIGURE 2.11 Using Huygens' principle to demonstrate that the angle of incidence equals the angle of reflection ($\theta_1 = \theta_2$). Analysis is detailed in the text.

(a)

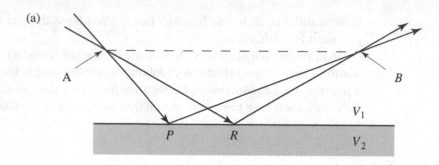

(b)

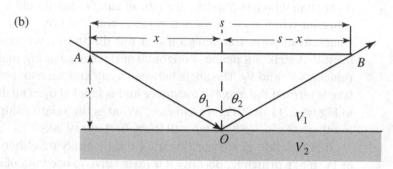

FIGURE 2.12 Using Fermat's principle to demonstrate that the angle of incidence equals the angle of reflection ($\theta_1 = \theta_2$). (a) Which path takes the least time? (b) Symbols used for analysis in text.

only with the propagation of the reflected wave into the upper material). At some later time ray 2 strikes the interface, and still later, ray 3 arrives at point Y. The secondary wave front generated at A must travel outward a certain distance during the time it takes ray 3 to travel from B to Y. Because velocity is constant, this distance must be equal to BY. Similar reasoning can be applied to ray 2. Therefore, we can construct the new wave front by drawing an arc with an origin at A and radius $d_1 = d_2$. In like fashion an arc can be constructed for ray 2. We then construct a line through Y that also is tangent to these arcs and locates the position of the wave front at the moment ray 3 is at point B. Because the two right triangles ABY and AXY are congruent, θ_1 and θ_2 are equal, and the angle of incidence equals the angle of reflection.

In using the principle of least time to discover the relationship between θ_1 and θ_2, we want to discover which path is the shortest for a ray passing through A, reflecting from the interface, and emerging through B. Figure 2.12(a) illustrates the problem. Does travel along path ARB take the least time or is some other path, such as APB, the one along which time of travel is at a minimum?

By referring to Figure 2.12(b), we can state that the time (path distance/velocity) for a ray to travel from A through some point O to B is

$$t = \frac{(x^2 + y^2)^{\frac{1}{2}}}{V_1} + \frac{((s-x)^2 + y^2)^{\frac{1}{2}}}{V_1}. \tag{2.15}$$

To determine the minimum value of time t, we take the first derivative of the function and set it equal to zero:

$$\frac{dt}{dx} = \frac{x}{V_1(x^2 + y^2)^{\frac{1}{2}}} - \frac{(s-x)}{V_1((s-x)^2 + y^2)^{\frac{1}{2}}} = 0. \tag{2.16}$$

Using the relationships

$$\sin\theta_1 = \frac{x}{(x^2 + y^2)^{\frac{1}{2}}} \quad \text{and} \quad \sin\theta_2 = \frac{(s-x)}{((s-x)^2 + y^2)^{\frac{1}{2}}}, \tag{2.17}$$

we see that

$$\frac{\sin\theta_1}{V_1} - \frac{\sin\theta_2}{V_1} = 0 \text{ and, therefore, } \theta_1 = \theta_2. \tag{2.18}$$

Thus the path for which the time of travel is least is the one for which the angle of incidence equals the angle of reflection.

2.2.4 Refraction

The approach in developing the relationship between incident and refracted rays is almost identical to that utilized for reflected rays. Let us apply Huygens' principle first and then employ Fermat's principle.

The construction in Figure 2.13 is essentially the same as in Figure 2.11 except here we deal with refracted rays. When ray 1 arrives at point A, it creates a disturbance in the material with velocity V_2. The disturbance spreads outward in this layer and will travel a distance d_2 during the time t_1 it takes ray 3 to travel from point X to point Y (a distance d_1). We construct the position of the new wave front at the moment ray 3 arrives at Y by drawing a line connecting Y and B (which is the tangent to the wavelet with radius d_2 that was generated at A). Because

$$\sin\theta_2 = \frac{d_2}{AY} \quad \text{and} \quad \sin\theta_1 = \frac{d_1}{AY} \tag{2.19}$$

$$d_1 = t_1 V_1 \quad \text{and} \quad d_2 = t_1 V_2, \tag{2.20}$$

it holds that

$$AY = \frac{t_1 V_1}{\sin\theta_1} = \frac{t_1 V_2}{\sin\theta_2} \quad \text{and} \quad \frac{\sin\theta_1}{\sin\theta_2} = \frac{V_1}{V_2}. \tag{2.21}$$

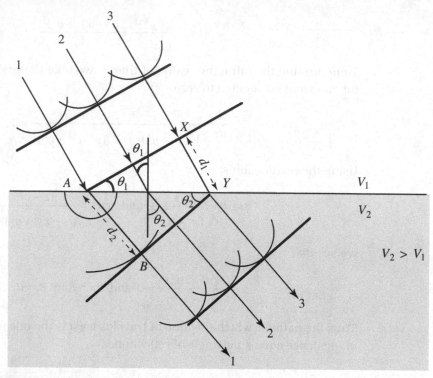

FIGURE 2.13 Using Huygens' principle to demonstrate the relationship between the angles of incidence and refraction. Analysis is detailed in the text.

Thus, approaching the geometry of refraction using Huygens' principle, we learn that the ratio of the sines of the angle of incidence and angle of refraction is equal to the ratio of the velocities of the two materials.

When applying Fermat's principle, we can ask the same sort of questions that we did for the reflection problem. When examining Figure 2.14(a), we would like to know what distinguishes the path of least time of a ray traveling from A to C as it passes through an interface separating materials of differing velocities. Once again we write an equation expressing the time it takes for a ray to travel from A through B to C:

$$t = \frac{(x^2 + y^2)^{\frac{1}{2}}}{V_1} + \frac{((s-x)^2 + z^2)^{\frac{1}{2}}}{V_2}.$$ (2.22)

Continuing, we take the derivative and arrive at this expression:

$$\frac{dt}{dx} = \frac{x}{V_1 (x^2 + y^2)^{\frac{1}{2}}} - \frac{(s-x)}{V_2 ((s-x)^2 + z^2)^{\frac{1}{2}}} = 0 .$$ (2.23)

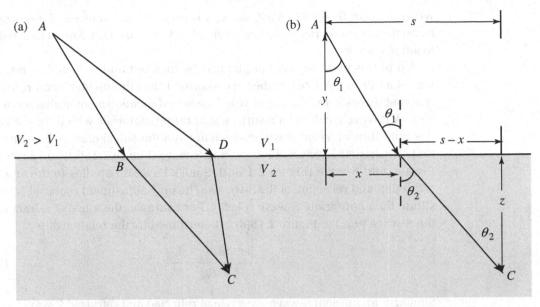

FIGURE 2.14 Using Fermat's principle to demonstrate the relationship between the angles of incidence and refraction. (a) What are the distinguishing features of the path of least time (*ABC*) as opposed to others (*ADC*)? (b) Symbols for analysis used in text.

In taking the derivative, note that z and y are constants because they retain the same value regardless of where B is placed along the interface. As before, we can use identities to arrive at the final form relating angles and velocities:

$$\sin\theta_1 = \frac{x}{(x^2+y^2)^{\frac{1}{2}}} \quad \text{and} \quad \sin\theta_2 = \frac{(s-x)}{((s-x)^2+z^2)^{\frac{1}{2}}}, \tag{2.24}$$

$$\frac{\sin\theta_1}{V_1} - \frac{\sin\theta_2}{V_2} = 0 \quad \text{and, therefore,} \quad \frac{\sin\theta_1}{\sin\theta_2} = \frac{V_1}{V_2}. \tag{2.25}$$

As you probably expected, this is the same relationship derived using Huygens' principle and, if V_1 is substituted for V_2, is the same as that for reflected rays. Those interested in a more detailed treatment of the material discussed in this and the preceding sections should consult Slotnick (1959).

2.2.5 Snell's Law

The relationship $\sin\theta_1/\sin\theta_2 = V_1/V_2$ is commonly referred to as *Snell's Law* and is fundamental in deriving other expressions to utilize the reflection and refraction of seismic waves in deducing subsurface relationships. Snell's Law often is written as $\sin \theta_i / \sin \theta_r = V_i / V_r$ where i represents incidence and r represents refraction. Because ray paths are confined to V_i in reflection, Snell's Law reduces to $\sin \theta_i = \sin \theta_{reflection}$.

Up to this point we did not identify the incident and reflected or refracted waves as P- or S-waves; rather, we assumed that the incident and refracted/reflected waves were of the same type. The actual situation is somewhat more complex. A P-wave incident on a surface separating materials with different velocities (and thus different elastic coefficients and densities) creates a disturbance that gives rise to a reflected P-wave and S-wave (Figure 2.15(b)) and a refracted P-wave and S-wave (Figure 2.15(d)). Snell's Law still applies to the angles of refraction and reflection of the S-waves. The only adjustment required is to substitute the appropriate S-wave velocity. For example, the angle of refraction for the S-wave (θ_{rfr_S}) in Figure 2.15(d) is determined by the relationship

$$\frac{\sin \theta_{i_P}}{\sin \theta_{rfr_S}} = \frac{V_{1_P}}{V_{2_S}}. \tag{2.26}$$

Similarly, an incident S-wave gives rise to reflected and refracted P-waves as well as reflected and refracted S-waves (Figure 2.15(c) and (e)). Specific examples with values are illustrated in Figure 2.15(b), (c), (d) and (e). You might want to apply Snell's Law to these situations as practice in calculating the appropriate angles. Table 2.3 also lists angles for incident P- and S-waves, their velocities, and the calculated angles of reflection and refraction.

The only case for which an incident S-wave does not give rise to reflected and refracted P-waves is when the S-wave is entirely of the SH type. In this case particle motions are confined to the horizontal plane and are, therefore, parallel to the velocity discontinuity.

Table 2.3 also is a dynamic table. Experiment by inserting various velocities in place of those present or by altering the value of the *first* cell in the column labeled "Angle of Incidence."

2.2.6 Critical Refraction

You may have noticed when examining Table 2.3 that in three of the columns for refracted rays angles increase to a maximum of 87° or 88° and then are replaced by asterisks. If you experimented with the dynamic table, you noticed these cells are labeled *#NUM!*. This message appears in a cell when an unacceptable argument is used for a function. If you changed values for the angles of incidence or velocities, the *#NUM!* message may have disappeared or perhaps appeared in even more cells. What is the significance of these relationships?

Because Snell's Law was used to calculate the angles of refraction, we should examine the quantities in the equation with more care. First, for simplicity, let us restrict our discussion to compressional waves. As is most common in shallow exploration, the velocity of the layer below the interface is greater than the velocity above the interface ($V_2 > V_1$). As the angle of incidence increases, the value of the sine of the angle of refraction also must increase to maintain the equality of the ratios in Snell's Law. At some point the angle of incidence will be such that $\sin \theta_i = V_1/V_2$, which requires that the sine of the angle of refraction = 1.0 (and the angle = 90°). If θ_i is increased beyond this value, an error ensues because the sine of an angle cannot take on values greater than 1, and therefore the equality cannot be maintained. What this means in physical terms is that the angle of refraction increases as the angle of incidence increases until rays are refracted parallel to the interface between the two materials. If the angle of incidence increases beyond this special value, then no refraction occurs, and the ray is *totally reflected*.

The refraction seismic method is based on this special case, and so we speak of the *critical angle* and *critical refraction*. If θ_{i_C} denotes the critical angle, its value can be determined by

$$\theta_{i_C} = \sin^{-1}\left(\frac{V_1}{V_2}\right). \tag{2.27}$$

As before, it is trivial to prove that this general form also holds for all cases of refraction such as refracted shear waves produced by incident compressional waves.

We can visualize this critically refracted ray as traveling parallel to the interface, but in the V_2 medium, and producing a disturbance in V_2 at the V_1/V_2 interface (Figure 2.16). Because the V_1 material must move in phase with the V_2 material, there is a disturbance in the V_1 material moving at V_2. This generates secondary wavelets as the refracted wave passes, which return to the surface through the V_1 medium. We once again can utilize Huygens' principle to construct the wave front traveling upward at velocity V_1. For example, at time t, point P is the center for a wavelet. After Δt, the disturbance is at Q, and we have the wave front RQ. This wave generally is referred to as the *head wave*, and any of its rays also are at the critical angle (Figure 2.16) because $\sin \theta = V_1 \Delta t / V_2 \Delta t$ (see Equation 2.27).

The phenomena of refraction and reflection, critical refraction, and total reflection are summarized in Figure 2.17.

2.2.7   Diffraction

For the most part we will consider only continuous planar surfaces and other simple geometries in our investigations of the seismic reflection and refraction methods, even though the subsurface is rarely so ideal. However, one important exception to this concerns the phenomenon of *diffraction*. If a disturbance is generated at the surface and encounters a sudden change of curvature at a velocity

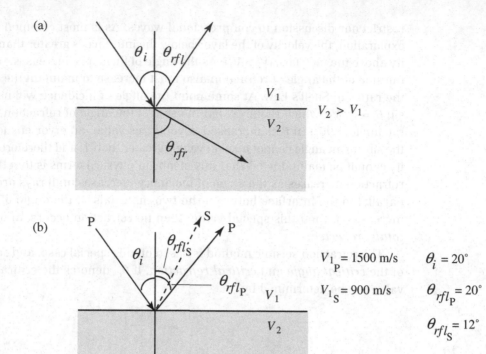

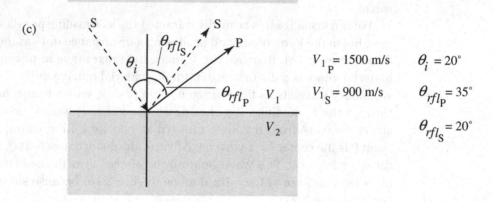

FIGURE 2.15 (a) General terminology for subsequent diagrams: θ_i = angle of incidence, θ_{rfl} = angle of reflection, and θ_{rfr} = angle of refraction. V_1 is the velocity of a given wave in the material above the interface; V_2 is the velocity below the interface. (b and d) An incident compressional wave produces a reflected P-wave, a reflected S-wave, a refracted P-wave, and a refracted S-wave. The angles of reflection and refraction of these waves depend on the angle of incidence and the velocities of the waves in the upper and lower materials. (c and e) An incident shear wave produces both reflected and refracted P- and S-waves. (f) This summary diagram is intended to reinforce the idea that an incident P-wave or an incident S-wave

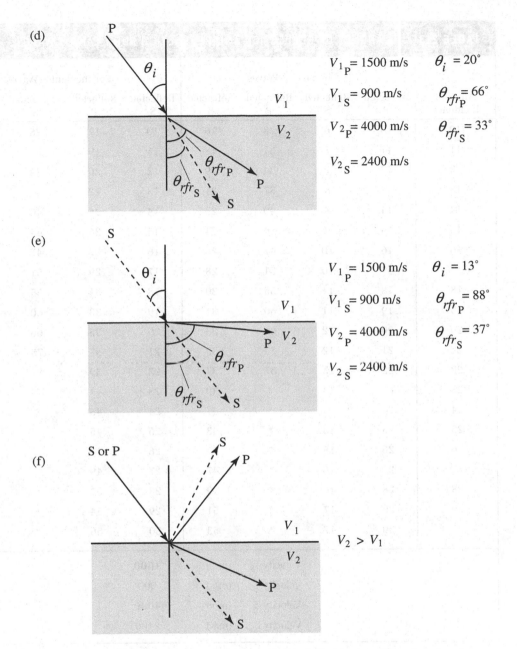

(d)

$V_{1_P} = 1500$ m/s $\theta_i = 20°$

$V_{1_S} = 900$ m/s $\theta_{rfr_P} = 66°$

$V_{2_P} = 4000$ m/s $\theta_{rfr_S} = 33°$

$V_{2_S} = 2400$ m/s

(e)

$V_{1_P} = 1500$ m/s $\theta_i = 13°$

$V_{1_S} = 900$ m/s $\theta_{rfr_P} = 88°$

$V_{2_P} = 4000$ m/s $\theta_{rfr_S} = 37°$

$V_{2_S} = 2400$ m/s

(f)

$V_2 > V_1$

produces both reflected and refracted compressional and shear waves. Note, however, by studying (b) through (e), that the ray paths of reflected and refracted P-waves due to an incident P-wave are not the same as the ray paths of reflected and refracted P-waves due to an incident S-wave. This also is true for reflected and refracted S-waves produced by incident P- and S-waves.

TABLE 2.3 ANGLES OF REFLECTION AND REFRACTION FOR P- AND S-WAVES

Angle of Incidence	For Incident P-Waves				For Incident S-Waves			
	Reflected P	Reflected S	Refracted P	Refracted S	Reflected S	Reflected P	Refracted S	Refracted P
10	10	6	28	16	10	17	28	51
11	11	7	31	18	11	19	31	58
12	12	7	34	19	12	20	34	68
13	13	8	37	21	13	22	37	88
14	14	8	40	23	14	24	40	*
15	15	9	44	24	15	26	44	*
16	16	10	47	26	16	27	47	*
17	17	10	51	28	17	29	51	*
18	18	11	55	30	18	31	55	*
19	19	11	60	31	19	33	60	*
20	20	12	66	33	20	35	66	*
21	21	12	73	35	21	37	73	*
22	22	13	87	37	22	39	87	*
23	23	14	*	39	23	41	*	*
24	24	14	*	41	24	43	*	*
25	25	15	*	43	25	45	*	*
26	26	15	*	45	26	47	*	*
27	27	16	*	47	27	49	*	*
28	28	16	*	49	28	51	*	*
29	29	17	*	51	29	54	*	*
30	30	17	*	53	30	56	*	*

Velocity 1_P	(m/s)	**1500**
Velocity 1_S	(m/s)	900
Velocity 2_P	(m/s)	**4000**
Velocity 2_S	(m/s)	2400

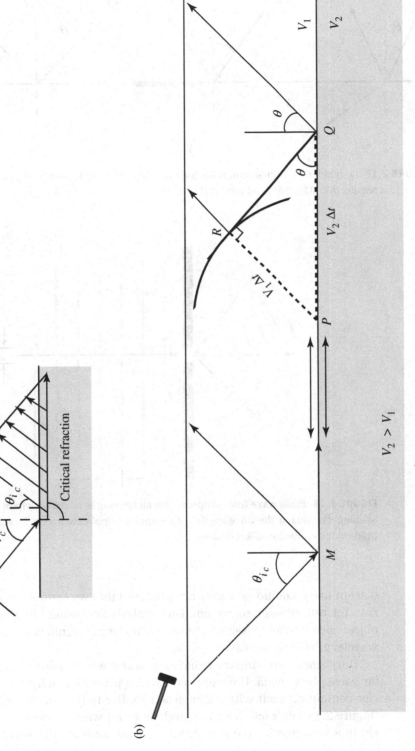

FIGURE 2.16 A wave that strikes an interface at its critical angle (θ_{ic}) is refracted parallel to the interface, producing what is commonly referred to as a *head wave*. (b) Explanation of head wave generation. Analysis is detailed in the text.

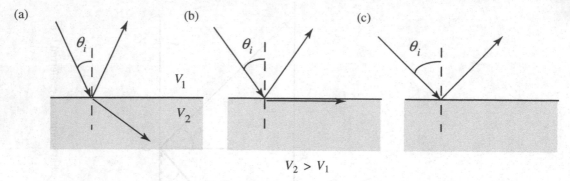

$$V_2 > V_1$$

FIGURE 2.17 (a) Incident, reflected, and refracted rays. (b) Increasing θ_i results in a critically refracted ray. (c) Increasing θ_i still further produces total reflection.

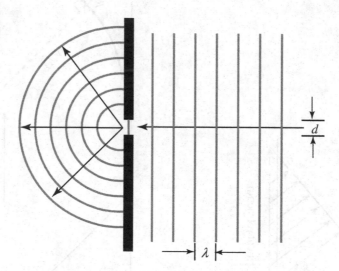

FIGURE 2.18 Plane wave fronts approaching an opening in a barrier where the width of the opening d is equal to the wavelength λ of the approaching wave fronts. Huygens' principle predicts the diffraction effect observed.

discontinuity, additional waves are produced that we cannot predict by drawing rays for reflection or refraction. Our analysis fails when the dimensions of an object encountered by seismic waves are no longer significantly greater than the wavelengths of these waves.

Once again, studying the behavior of water waves helps us visualize particular wave phenomena. If water waves with a planar wave front encounter a barrier containing a slit with a dimension similar to the wavelength of the waves (Figure 2.18), diffraction occurs, and the water waves spread outward from the slit in a semicircular pattern. Actually, if we consider the small portion of the

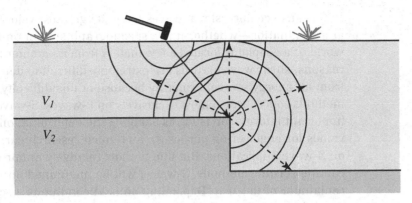

FIGURE 2.19 Diffracted waves are produced when wave fronts encounter a sudden change in curvature of an interface.

wave that passes through the slit and apply Huygens' principle (every point on the wave front can be considered as the source of a secondary wavelet), the reason for the diffraction pattern is clear.

Although the process that occurs at a sharp discontinuity, such as that diagrammed in Figure 2.19, is somewhat more complex, we can explain the general pattern quite well by considering the sharp change in curvature as a source of secondary wavelets and thereby reconstructing the wave front by Huygens' method. Under what conditions are diffractions likely to occur in the subsurface? Refer again to Figure 2.18. As the size of the slit in the barrier is increased, the diffraction effect is decreased and is barely perceptible when the slit dimension is several times the wavelength. This suggests that diffraction will be most important when sharp changes in curvature have radii that are similar in dimension to seismic wavelengths. Because seismic wavelengths typically are several tens of meters in length, diffracted waves certainly will not be uncommon. Because of the very different approaches in refraction and reflection seismology, considerations of diffraction phenomena play a much greater role in the reflection technique. This does not mean, however, that diffraction effects are not present on refraction records.

2.2.8 Wave Arrivals at the Surface

In the next section we will discuss how wave energy is diminished as the wave propagates through materials and is reflected and refracted by velocity discontinuities. Such a discussion is undertaken primarily to establish the magnitude of amplitudes and ground motion expected at the surface for different types of events. Before we begin that discussion, however, it seems appropriate to consider the important aspects of the wave events already discussed, and to question, in a general way, what a surface observer might expect.

Because compressional waves have the greatest velocities, they arrive first at a destination—whether it is a seismograph station on the other side of the world or a geophone located a few meters from a hammer blow. For a variety of reasons, shear wave arrivals are extremely difficult to detect at short distances from an energy source. Primarily because of this difficulty, seismic exploration methods concentrate almost exclusively on P-waves. S-waves provide additional important information (such as enabling the determination of elastic coefficient values for engineering purposes), and much research currently is taking place on S-wave applications. But due to their relatively minor role in common and routine seismic methods, S-waves will be mentioned only occasionally in the remainder of this text. Depending on field conditions, geometries, and equipment, however, S-waves will contribute their signature on seismic records, so we cannot ignore them entirely.

What general major arrivals might we expect to see in a configuration such as that illustrated in Figure 2.20? If energy is imparted at or near the surface, a P-wave (the ground-coupled *air wave*) will activate sensors (geophones) placed at intervals along the surface. Other compressional waves will follow a direct path through the uppermost layer from the energy source to the sensors (the *direct wave*). P-wave reflections and the critical refraction also will arrive and produce ground motion. Finally, due to the significant ground motion they induce, Rayleigh waves (*ground roll*) certainly will impart their distinctive signature.

An easy way to gain some initial insight into the distribution in time and space of the arrivals of these waves at the surface is to construct a spreadsheet form. One such possibility is provided in Table 2.4. Assume that geophones are placed at even intervals from a shot point (or hammer blow). Decide on this interval and also select compressional wave velocities for the layers above and below a planar interface as well as the depth of the interface. If we calculate the path distances for each wave and the velocities over that path, we can determine the time of arrival at each surface geophone. Although these equations

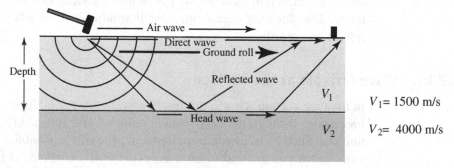

FIGURE 2.20 Generalized diagram illustrating ray paths for waves included in Table 2.4. Included are the direct wave, reflection, critical refraction (head wave), air wave, and ground roll. All waves are compressional with the exception of ground roll.

| TABLE 2.4 | TIME—DISTANCE VALUES FOR SELECTED WAVES |

	Arrival Times (ms)				
Distance from Shot (m)	Direct Wave	Reflected Wave	Head Wave	Air Wave	Ground Roll
0	0.00	20.00	18.54	0.00	0.00
3	2.00	20.10	19.29	8.96	4.44
6	4.00	20.40	20.04	17.91	8.89
9	6.00	20.88	20.79	26.87	13.33
12	8.00	21.54	21.54	35.82	17.78
15	10.00	22.36	22.29	44.78	22.22
18	12.00	23.32	23.04	53.73	26.67
21	14.00	24.41	23.79	62.69	31.11
24	16.00	25.61	24.54	71.64	35.56
27	18.00	26.91	25.29	80.60	40.00
30	20.00	28.28	26.04	89.55	44.44
33	22.00	29.73	26.79	98.51	48.89
36	24.00	31.24	27.54	107.46	53.33
39	26.00	32.80	28.29	116.42	57.78
42	28.00	34.41	29.04	125.37	62.22
45	30.00	36.06	29.79	134.33	66.67
48	32.00	37.74	30.54	143.28	71.11
51	34.00	39.45	31.29	152.24	75.56
54	36.00	41.18	32.04	161.19	80.00
57	38.00	42.94	32.79	170.15	84.44
60	40.00	44.72	33.54	179.10	88.89

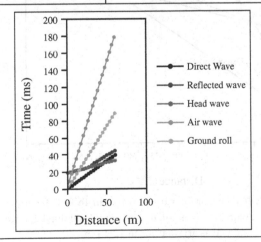

Velocity 1_P (m/s)	**1500**
Velocity 2_P (m/s)	**4000**
Depth (m)	**15**
Shot offset (m)	**0**
Geophone interval (m)	**3**
No head wave before:	12.14

will be developed in subsequent chapters, for present purposes they simply are included in the spreadsheet. As in previous instances, Table 2.4 is included as a template on disk so you can experiment with various geophone intervals, velocities, and depths if you wish.

The pattern of the wave arrivals at points on the surface is much easier to analyze if we construct a graph of arrival time versus distance from the shot point for each wave type. Figure 2.21 was produced by placing the values in Table 2.4 into a computer graphing program. Such a graph, usually referred to as a *travel-time curve*, provides a clear sense of arrival relationships at various positions along the surface. Although it is somewhat premature, you still should study this graph and begin to analyze the arrival patterns, especially for the air wave, direct wave, and ground roll. Figure 2.22 represents a further step in our introduction to wave arrival patterns. This computer-produced diagram simulates the wave arrivals in Table 2.4 and Figure 2.20 at each geophone position.

Even in this simulated seismogram, many of the relationships so evident in Figure 2.21 are much more difficult to discern, although myriad factors that would further obscure a real field seismogram are not included. Note especially that it is virtually impossible to deduce individual events at geophones 1–10 in the time range of 0–50 ms. An important task will be to formulate methods that make it possible to extract such information for interpretative purposes.

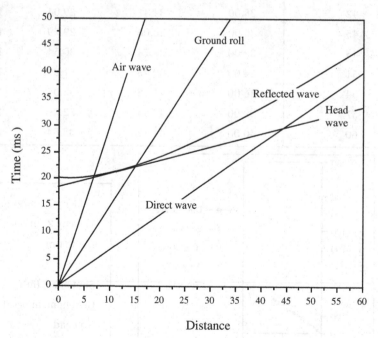

FIGURE 2.21 A plot of time against distance for the various waves in Table 2.4. Only compressional waves are considered with the exception of the ground roll (Rayleigh waves). Although projected back to 0 m, no head waves arrive prior to 12.14 m.

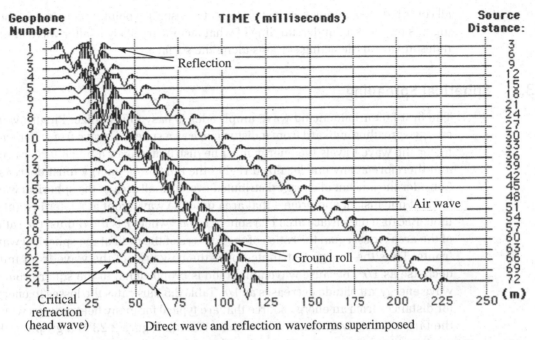

FIGURE 2.22 A computer-generated diagram that simulates how the waves illustrated in Figure 2.20 would appear on a field record.

It is quite helpful to develop a feel for the arrival patterns and relationships calculated in Table 2.4 and graphed in Figure 2.21. Because Table 2.4 is a dynamic table, you can change the depth of the interface and the velocities of the layers above and below the interface. You can vary the distance from the shot point or hammer blow to the first geophone and can change the geophone interval. Also, because this spreadsheet program has a graph already available, you can instantly observe the results of any changes you make for any waves you want to investigate. Thus this table is a powerful tool to help develop insights into the arrivals of these waves, and you should spend some time experimenting with the values in boldface if the appropriate computer software is available to you.

2.3 WAVE ATTENUATION AND AMPLITUDE

The major goal in seismic exploration is the interpretation of wave arrival patterns at the surface. It already is clear that this will not be a trivial task, even when Earth materials are homogeneous and only one interface is present. The variety of seismic wave types, variations in densities and elastic coefficients, reflection and refraction at interfaces, and the conversion of waves at interfaces all contribute to a complex arrival pattern at any surface point. An aid in deciphering this puzzle is an understanding of what factors are likely to alter wave amplitudes at various detecting stations along the surface.

2.3.1　Spherical Spreading

The first factor influencing wave amplitudes is easy to visualize. Picture wave energy spreading outward from a disturbance as a spherical wave front. For each meter the wave travels the energy must be distributed over a much larger area, and therefore at any one localized region the amount of energy must decrease. Consider the amount of energy distributed over a small area of the spherical wave front at one instant and project that area onto the wave front at a later instant in time. Because the surface area of a sphere is $4\pi r^2$ where r is the radius, the ratios of the areas must equal the ratios of the squares of the radii of the spherical wave fronts. Thus the energy distributed over the new area of the wave front must decrease as $1/r^2$. Because wave amplitude is proportional to the square root of wave energy, amplitude decreases as $1/r$. Table 2.5 illustrates the losses in energy for distances from an energy source that are typical for many field surveys where the target is relatively shallow (less than 100 m). Figure 2.23 is a graph of the relationships between rows 1 and 2 of the table.

2.3.2　Absorption

As seismic waves propagate, they do so by distorting (straining) the materials through which they pass. Heat energy thus generated reduces the total energy of a wave. This reduction in the elastic energy of the wave is termed *absorption*. The equation for this process is

$$I = I_0 e^{-qr} \tag{2.28}$$

where I is defined as the energy intensity (the amount of energy passing through a unit area in a unit time), q is the absorption coefficient, and r is distance. Given the energy intensity at a point I_0, I represents the energy intensity at a point a distance r from I_0. The absorption coefficient q is expressed in decibels/wavelength (db/λ). Table 2.6 illustrates energy losses for a situation where q is given as 0.55 db/λ. This value for q is in the midrange of published values for various Earth materials, but the important aspect of the table is not the absolute values but the comparisons at various frequencies and distances.

These figures vividly illustrate that energy losses due to absorption are much greater at higher frequencies than at lower frequencies at a given distance from the energy source. Note the marked difference between the 10 Hz wave and the 100 Hz wave at 240 m. Because many natural seismic pulses contain a broad array of frequency components, the higher-frequency components will be progressively

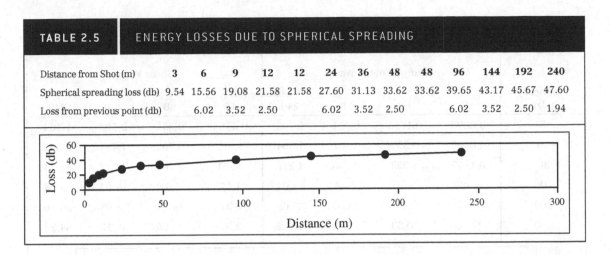

Distance from Shot (m)	3	6	9	12	12	24	36	48	48	96	144	192	240
Spherical spreading loss (db)	9.54	15.56	19.08	21.58	21.58	27.60	31.13	33.62	33.62	39.65	43.17	45.67	47.60
Loss from previous point (db)		6.02	3.52	2.50		6.02	3.52	2.50		6.02	3.52	2.50	1.94

TABLE 2.5   ENERGY LOSSES DUE TO SPHERICAL SPREADING

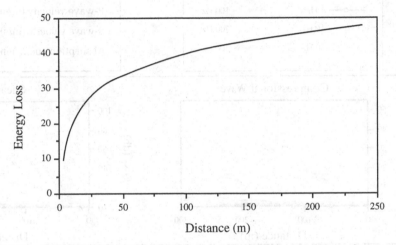

FIGURE 2.23 Energy losses in decibels due to spherical spreading at various distances from an energy source.

reduced with distance so that the form of the seismic pulse changes to a smoother and broader shape. This is why the Earth is often referred to as a *low-pass filter*. Lower frequencies are transmitted with less energy loss, and higher frequencies are progressively attenuated. If a survey is interested in high-frequency energy (as we will be when attempting shallow reflection work), we need to find ways to increase that component when generating energy and to be particularly sensitive to detecting those higher frequencies rather than the stronger lower ones.

All other factors being equal, waves traveling at lower velocities will lose energy more quickly from absorption than waves traveling at higher velocities. On this basis alone, S-waves will lose energy more quickly than P-waves. Remembering that wave amplitude is proportional to the square root of wave energy, we can see that amplitudes of S-waves will diminish much more quickly than amplitudes of P-waves.

TABLE 2.6	ENERGY LOSSES (IN DECIBELS) DUE TO ABSORPTION							
	Compressional Wave				Shear Wave			
Frequency (Hz)	Distance (m)							
	10	60	120	240	10	60	120	240
5	0.0293	0.1755	0.351	0.702	0.0585	0.351	0.702	1.404
10	0.0585	0.351	0.702	1.404	0.117	0.702	1.404	2.808
30	0.1755	1.053	2.106	4.212	0.351	2.106	4.212	8.424
100	0.585	3.51	7.02	14.04	1.17	7.02	14.04	28.08
200	1.17	7.02	14.04	28.08	2.34	14.04	28.08	56.16
300	1.755	10.53	21.06	42.12	3.51	21.06	42.12	84.24

—○— 5 Hz	—●— 100 Hz	P-wave velocity (m/s)	**4000**
—●— 10 Hz	—●— 200 Hz	S-wave velocity (m/s)	**2000**
—●— 30 Hz	—●— 300 Hz	Absorption coefficient = 0.55 db/λ	

Compressional Wave — Shear Wave

Loss (db) vs. Distance (m)

You can experiment with the values in boldface in both Table 2.5 and Table 2.6. It is especially interesting to vary the velocities in Table 2.6 and to note the amount of energy loss due to absorption as you increase or decrease velocity.

2.3.3 Energy Partitioning

You already realize that P- and S-waves striking a velocity discontinuity may not only be reflected and refracted but can generate other P- and S-waves as well. Because the total amount of energy must remain constant between the incident and reflected and refracted waves, amplitudes of refracted and reflected waves will be diminished relative to the incident wave.

Given the amplitude of an incident compressional wave, we can calculate the amplitudes of reflected and refracted P- and S-waves using equations developed by Zoeppritz (1919). If the velocities and densities of materials above and below an interface are known, these equations provide solutions for all angles of incidence. Essentially, we are dealing with four equations with four unknowns (the amplitudes), and although the equations are not that complicated, we will not list them here because they are available in several advanced texts such as Telford, Geldart, and Sheriff (1990, 155–56).

However, if P-wave incidence is normal to the interface, the equations reduce to a simple form. No S-waves are generated under normal incidence. The ratios of the amplitudes A are

$$\frac{A_{rfl}}{A_i} = \frac{\rho_2 V_2 - \rho_1 V_1}{\rho_2 V_2 + \rho_1 V_1} \qquad \frac{A_{rfr}}{A_i} = \frac{2\rho_1 V_1}{\rho_2 V_2 + \rho_1 V_1} \tag{2.29}$$

where the quantities are as diagrammed in Figure 2.24. These equations often are shortened to the form

$$\frac{A_{rfl}}{A_i} = \frac{Z_2 - Z_1}{Z_2 + Z_1} \qquad \frac{A_{rfr}}{A_i} = \frac{2Z_1}{Z_2 + Z_1} \tag{2.30}$$

where $Z_1 = \rho_1 V_1$ and $Z_2 = \rho_2 V_2$.

Values for these amplitude ratios are given in Table 2.7 for one set of density–velocity relationships in the cells labeled "Reflection coefficient" and "Refraction coefficient." These terms are most commonly used to denote the amplitude ratios. Some geophysicists also use the same terminology to refer to the energy ratios of the refracted and reflected waves. Recall the general relationship between energy and amplitude. As you might expect, the equations for the energy ratios are quite similar in appearance to those for amplitudes:

$$\frac{I_{rfl}}{I_i} = \left(\frac{\rho_2 V_2 - \rho_1 V_1}{\rho_2 V_2 + \rho_1 V_1}\right)^2 \qquad \frac{I_{rfr}}{I_i} = \frac{4\rho_1 V_1 \rho_2 V_2}{(\rho_2 V_2 + \rho_1 V_1)^2} \tag{2.31}$$

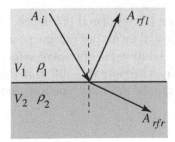

FIGURE 2.24 Illustration of quantities used in the simple form of Zoeppritz's equations.

TABLE 2.7 REFLECTION COEFFICIENTS FOR P-WAVE AT NORMAL INCIDENCE

Density (g/cm^3)—layer 1	**2.00**	Z_1	3000
Density (g/cm^3)—layer 2	**2.60**	Z_2	11,700
P-wave velocity (m/s)—layer 1	**1500**		
P-wave velocity (m/s)—layer 2	**4500**		

Reflection coefficient	0.59	Energy fraction reflected	0.35
Refraction coefficient	0.41	Energy fraction refracted	0.65

or

$$\frac{I_{rfl}}{I_i} = \left(\frac{Z_2 - Z_1}{Z_2 + Z_1}\right)^2 \qquad \frac{I_{rfr}}{I_i} = \frac{4Z_1 Z_2}{(Z_2 + Z_1)^2} \,. \tag{2.32}$$

Solutions to these equations also are presented in Table 2.7.

The densities and velocities in Table 2.7 are representative for a layer of saturated sediment overlying bedrock. In this case, a large fraction of incident energy is reflected, which diminishes the energy transmitted into layer 2. It is instructive to insert various velocity and density values to determine if this relationship is maintained. You should discover that as the velocities become similar (as in a sequence of sedimentary rocks), the energy fraction reflected becomes quite small. If a negative sign appears, this designates that the waveform has undergone a 180° phase reversal. Also, insert values such that $\rho_1 V_1 = \rho_2 V_2$. As long as these two products are equal, all the energy is transmitted. The greater the difference between the products, the greater the percentage of the energy that is reflected.

Of course, not all cases of interest have normal incidence, and Zoeppritz's equations must be solved in their original form. For our purposes, however, it is sufficient to illustrate the results of a solution for one set of velocities and densities, although the form of the curves can become very different depending on the values used. Figure 2.25 is based on graphs in Tooley, Spencer, and Sagoci (1965).

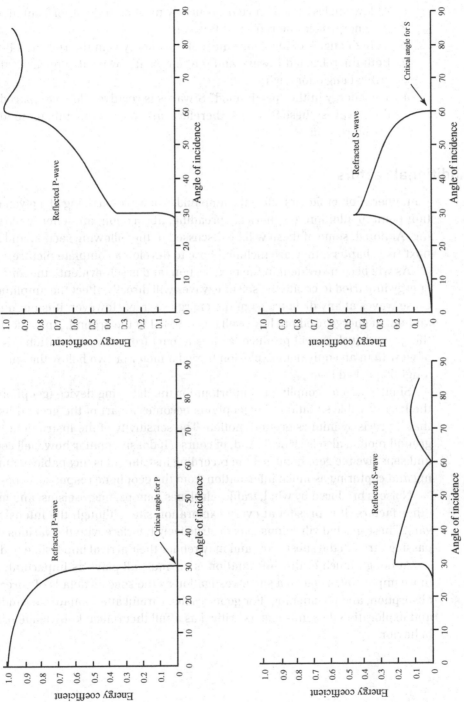

FIGURE 2.25 Partitioning of energy at an interface from an incident P-wave. Because critical angles exist for both P- and S-waves, the velocity V_1 must be less than V_2. For this particular case $V_2/V_1 = 2$ and $\rho_2/\rho_1 = 0.5$.

The most useful observations that can be drawn from these curves are these:

1. At low angles of incidence (rays nearly normal to the interface) most of the energy is in the refracted P-wave.
2. At high angles of incidence most of the energy is in the reflected P-wave.
3. Both the refracted P-wave and the refracted S-wave disappear at the critical angle for each.
4. The energy in the "partitioned" S-waves is relatively low compared to the P-waves; these S-waves, therefore, are likely to be difficult to identify on records.

2.3.4 Additional Factors

A number of other factors affect the amplitudes of waves arriving at a given detection point in addition to spherical spreading, absorption, and energy partitioning. Although some of these will be discussed in the following section and in the next two chapters, they are included here to develop a complete picture.

As will become evident in the next section (and is self-evident), the amount of energy imparted to create the seismic waves will directly affect the amplitudes of these waves at any distance from the energy source. However, it is not only the amount of energy used but how well it is coupled to the ground. An explosion on the ground surface will produce far less impact from the explorationist's point of view than an equivalent explosion buried a meter or two below the surface in a tightly packed hole.

Similarly, good coupling is important for the detecting device (geophone). In the best possible situation, the geophone becomes a part of the ground itself so that it precisely mimics ground motion. The sensitivity of the instrument to this ground motion also is critical. And, of course, it doesn't matter how well coupled and sensitive the geophone is if the recording instrument is incapable of amplifying and capturing as much information from the geophone response as possible.

Noise is produced by wind, traffic, airplanes, pumps, microseisms, and myriad other factors. It is present at every exploration site, although the intensity will vary. These ground vibrations may destructively interfere with the various waves passing through the subsurface and may render their arrival impossible to detect.

Although much of this information is obvious, all plays as important, if not more important, a role in a site investigation as the specific details of spreading, absorption, and partitioning. For good results, careful attention to site conditions and exploration design is just as critical as a full theoretical knowledge of wave behavior.

2.4   ENERGY SOURCES

A wide variety of energy sources have been and still are used in seismic exploration. Because our emphasis is on the shallow subsurface, we will concentrate on the most appropriate sources for elucidating targets at shallow depths. What are the characteristics of the ideal energy source? First, it must impart sufficient energy to provide reflections and refractions from the contacts between different materials whether they are rock units or sediments. The amplitudes must be sufficient to be detected and displayed reasonably well for interpretative purposes. Thus not only the total amount of energy imparted is important but also its frequency content and the shape of the waveform at the source. For most purposes we do not want to generate a broad waveform of complex shape.

2.4.1   Source Types

In land-based work most energy sources fit into one of three categories: weight drops, explosive, or vibratory. Of these, weight drops and explosives are most commonly used in shallow work. The simplest example of the weight drop is the hammer source. A sledgehammer, usually 5.4 or 7.3 kg, is swung against a metal plate to more efficiently couple the energy transfer. Obviously the energy imparted is not great, and in the past a hammer source was used only for very shallow investigations. The hammer blow became a more useful source with the advent of relatively inexpensive instruments that could sum ground motion for a number of impacts and thus enhance the waveforms on records. Under good conditions a hammer source routinely can detect the overburden–bedrock interface at depths of 50 meters or slightly more. Most other weight drop systems depend on a truck or trailer to transport the weight, lift it to a height of 2 meters or more, and release it. The weights vary from steel plates to leather bags filled with birdshot and can weigh over 250 kg.

Explosive sources run the gamut from dynamite to 50-caliber rifles. What we tend to consider "true" explosives most frequently are items similar to Primacord® (an explosive in the form of a rope from which lengths can be cut and detonated) and containers of ammonium nitrate. These materials can be rigged in the field for a wide range of energy inputs, but all require placement in drilled holes typically several meters in depth. In many cases the target depths are sufficiently large or ground conditions are sufficiently poor that only large explosive charges will produce interpretable results. However, with the advent of the signal enhancement seismograph, the shotgun source has become widely used in shallow work (Pullan and MacAulay 1987). It clearly is superior to the hammer source and rivals or exceeds the smaller weight drop sources (Figure 2.26). Note that the vertical scale in Figure 2.26 is logarithmic, so the energy contrast between the 7.3 kg hammer and the 8-gauge shotgun is substantial.

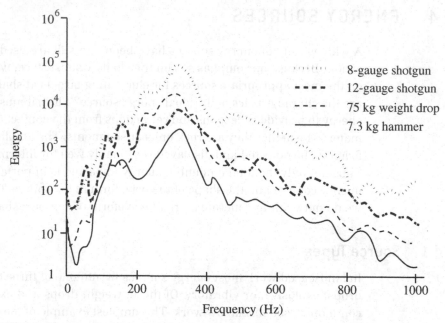

FIGURE 2.26 A comparison of relative energies and frequency content for shotgun and weight drop seismic sources.

Vibratory sources are used mainly for deep exploration targets in the range of several hundreds of meters to tens of kilometers. They typically require a large investment in field equipment and computer processing capabilities. Although this type of source is not of direct interest for shallow work, its approach is so different from weight drops or explosives that a few words of description seem warranted. Weight drops and explosions apply a momentary force at the source. The typical vibratory system, vibroseis, is oscillatory in character and lasts for a few seconds. Essentially, a large mass is placed in contact with the ground surface and vibrated according to a specific pattern. During the several seconds' duration of signal generation, the signal frequency is continuously varied. This long variable signal results in a final record of reflections and other events that cannot be interpreted directly but requires computer processing to separate the events of interest. Because the exact form of the source signal is known, such extraction is possible although complicated.

In marine conditions hammer blows are ineffective, and other sources must be used. The most common of these are air guns and sparkers. Both methods create a shock wave surrounding an air bubble in the water that is similar to one created by an explosive source but with somewhat more control and much less hazard. This shock wave, produced by rapidly compressing the water molecules in the vicinity of the source, spreads out and travels through the sediment and rock underlying the water column. Sparkers create an air bubble by an electric discharge

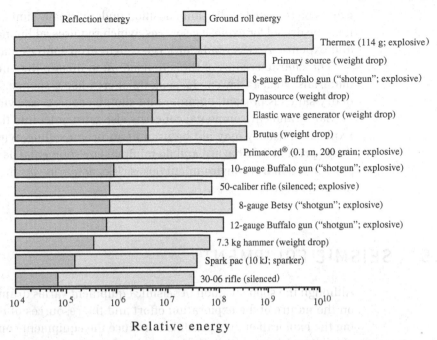

□ Reflection energy   □ Ground roll energy

Thermex (114 g; explosive)
Primary source (weight drop)
8-gauge Buffalo gun ("shotgun"; explosive)
Dynasource (weight drop)
Elastic wave generator (weight drop)
Brutus (weight drop)
Primacord® (0.1 m, 200 grain; explosive)
10-gauge Buffalo gun ("shotgun"; explosive)
50-caliber rifle (silenced; explosive)
8-gauge Betsy ("shotgun"; explosive)
12-gauge Buffalo gun ("shotgun"; explosive)
7.3 kg hammer (weight drop)
Spark pac (10 kJ; sparker)
30-06 rifle (silenced)

10^4 10^5 10^6 10^7 10^8 10^9 10^{10}

Relative energy

FIGURE 2.27 A plot of relative energies from various sources determined under uniform conditions at an experimental site. Based on information in Miller et al. (1986, 2067–92).

that vaporizes the water in the vicinity of the discharge. Air guns are designed to suddenly release a volume of air under high pressure into the water, thereby creating the bubble. Variations of these methods have been applied to land-based surveys, but for a variety of reasons they are not in common use.

Figure 2.27 compares the energy output for a variety of sources measured under identical conditions. Be sure to note that the scale is logarithmic. In addition to the energy output of the source, the frequency content of the signal (see Figure 2.26) and the amount of ground roll generated also are important. However, a number of other factors must be taken into account before a source is selected for a particular study.

2.4.2   Source Considerations

Because almost all seismic exploration programs work within specific budget constraints, cost becomes an important factor. In typical surveys many impacts or shots take place. We must therefore consider repeat costs as well as initial cost. A source with high initial cost but with low cost per event may be preferable to one with a low initial cost but high repeat cost.

Convenience and efficiency also enter into an energy source decision. What is the portability of a given source? How easy is it to get the source to the exploration site? Does it take a long time to set up, break down, and move? Can energy events be repeated at the same position under virtually similar conditions? Holes must be drilled for explosive sources, which requires additional equipment, time, and expense. Access roads for large weight drop sources may not be present.

The characteristics of the site itself often dictate the energy source. Some materials, such as dry sands, soak up energy like a sponge. If a thick sequence of dry sands is present, hammer and shotgun sources are virtually useless.

Safety and environmental concerns also must enter into the energy equation. Explosives usually may not be used in an urban setting, whereas large weight drops present little hazard and do minimal environmental damage. Using explosives in the marine environment poses risks to marine life and should be undertaken with care.

2.5   SEISMIC EQUIPMENT

Although the sophistication of seismic equipment varies dramatically depending on the nature of the exploration effort and the resources of the party undertaking the exploration, it is possible to reduce the equipment configuration to three elements: detection of ground motion, conditioning of the weak signal from the detectors, and recording of the conditioned signal.

2.5.1   Signal Detection

As you already are aware, the devices that detect the ground motion from arriving wave energy commonly are called *geophones* or *seismometers*. These are simple but elegant in design (Figure 2.28), small (Figure 2.29), and quite rugged. Geophone design essentially consists of a cylindrical coil of fine wire suspended in a cylindrical cavity in a magnet. As illustrated in Figure 2.28, the coil is held in place by leaf springs. The coil and magnet are mounted within a protective case, usually plastic, which also has a long metal spike to anchor the geophone to the ground. When ground motion occurs, the geophone moves, inducing relative motion between the magnet and the coil due to the inertia of the coil. This motion generates a voltage that is proportional to the amount of ground displacement. This voltage is returned to the equipment and constitutes the "raw" signal before electronic processing.

Knowing most of the details of geophone design is not required to understand and apply the basics of seismic exploration techniques, but there are a few additional important considerations. First, it is critical to be sure the geophone is coupled as well as possible to the ground. Thus the spike should be inserted as far as possible into the ground but should not be rocked from side to side during insertion because this will loosen the contact between the spike and ground. Also, in the standard geophone the relative motion between coil and magnet is constrained

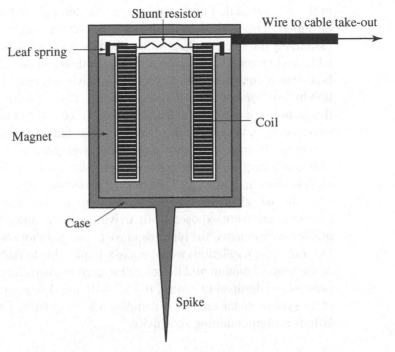

FIGURE 2.28 Schematic cross-section of typical geophone elements.

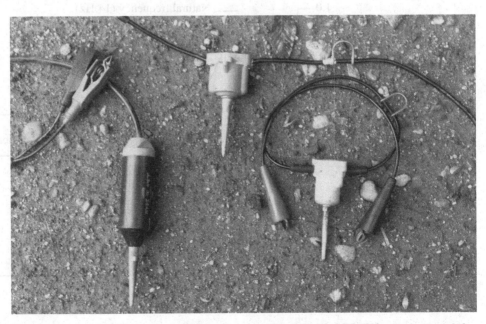

FIGURE 2.29 Photograph of typical geophones used for shallow exploration work. From left to right: 14 Hz phone, 30 Hz phone, and 50 Hz phone with marsh case. Each geophone has a spike for insertion into the ground and two clips for connection to seismic cable.

to the vertical so that horizontal ground motion cannot be measured. Geophones designed to measure shear waves are readily available, however, and function essentially the same as already described except that motion is constrained parallel to the horizontal. In marine surveying, unless geophones are placed on the lake bed or seafloor, conventional geophones obviously cannot be used. In this situation hydrophones are the detector of choice. These contain piezoelectric materials that produce a potential difference when subjected to pressure. Pressure differences occur in the water due to arriving reflected and refracted waves.

One of the most important aspects of a given geophone is its natural frequency. This is the frequency for which the output of the geophone has the greatest value, as illustrated in Figure 2.30. Once such an oscillatory system is activated, it will continue to oscillate even when ground motion has ceased. If such a freely oscillating system were exposed to an arriving seismic pulse composed of many frequency components, the high response at the geophone's natural frequency and its continuing oscillations would make it impossible to faithfully record the details of the ground motion and therefore the arriving waveform. For this reason geophones are damped to control the oscillations. Damping reduces the sensitivity of the system, so the amount of damping is a compromise between controlling oscillations and maintaining sensitivity.

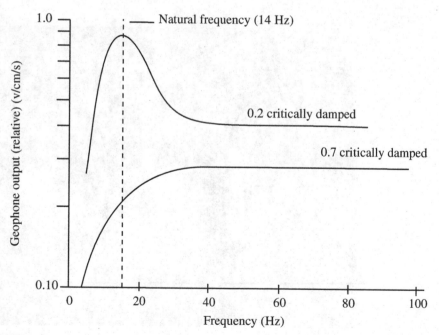

FIGURE 2.30 Response curves for 14 Hz geophone. 0.2 critical damping still produces a much greater response at the geophone's natural frequency than at other frequencies. 0.7 critical damping produces a more or less equal response to all frequencies above the natural frequency.

A natural damping already occurs due to the interaction between the magnetic field of the magnet and the magnetic field produced by current flowing in the coil. Additional damping is achieved by connecting a shunt resistor across the coil of the geophone (see Figure 2.28). Resistors with higher values produce less damping than resistors with lower values. If a geophone is damped to a point where a gentle tap fails to produce an oscillation, the geophone is said to be critically damped. The damping factor usually chosen during manufacturing is 0.7 of the critical damping value. This value produces what is termed a "flat response," which means the geophone responds with the same sensitivity to all frequencies above the natural frequency (Figure 2.30).

Depending on the seismic method and objectives, geophone selection is important. Recalling that we have characterized the Earth as a low-pass filter, you should remember that if we plot relative transmitted amplitude of seismic waves against frequency (Figure 2.31(a)), maximum amplitudes will be at low frequencies. For reasons we soon will discover, refraction surveying design usually seeks the largest possible signal for waves that arrive first at the geophone. Thus for refraction work we should select geophones with a natural frequency in this same low-frequency range. A typical refraction geophone sold for shallow work has a natural frequency of 14 Hz and can sample the frequency spectrum illustrated in Figure 2.31(b).

In our earlier discussion of seismic wavelengths we learned that the shorter wavelengths have higher frequencies and therefore higher resolution. In most shallow reflection work we are interested in as much resolution as possible, and therefore we want to isolate the higher-frequency component of the seismic pulse. For detailed shallow work we might choose geophones with natural frequencies of 50 or 100 Hz. These would reduce the low-frequency signals with high amplitudes (Figure 2.31(b)), emphasize the high-frequency pulse, and increase resolution.

In a typical low-budget, shallow-target seismic survey, geophones are used in groups of 12, 24, or, less frequently, 48. The output of each geophone is transferred to amplifying and recording instruments by means of a multiconductor cable. Each geophone connects to the cable with two clips at "take-outs" that are spaced at uniform intervals along the cable to facilitate placing the geophones at equal spacings along the ground (Figure 2.32).

2.5.2 Signal Conditioning

The term *signal conditioning* is used in several contexts in exploration geophysics. Here we use the term to include signal amplification and filtering. The amplifier is an electronic device that receives and amplifies the very small signals transmitted by the geophones. Most amplification systems in modern seismographs treat the incoming signal so as to maximize the information in the signal. The main objective is to suppress somewhat the early strong signals, especially at geophones close to the energy source, while increasing the signal in the latter part of the record or at geophones at substantial distances from the energy source. Many

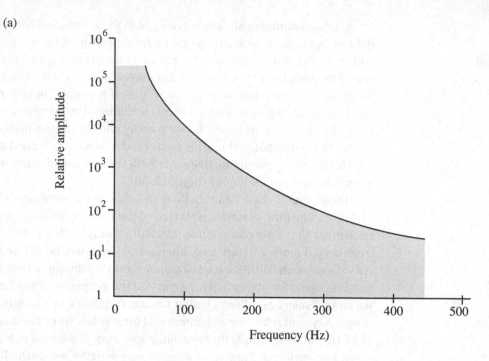

(a)

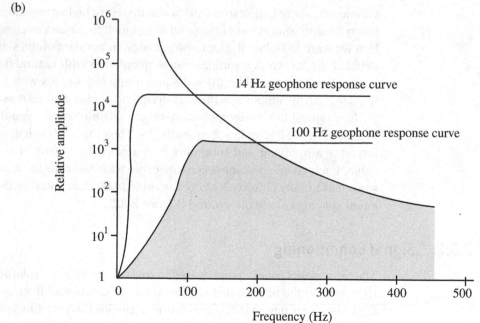

(b)

FIGURE 2.31 (a) Relative amplitude of seismic waves transmitted by Earth materials plotted against frequency. (b) Contrast in parts of frequency spectrum sampled by 100 Hz geophone (shaded) and 14 Hz geophone (shaded and unshaded areas beneath curve).

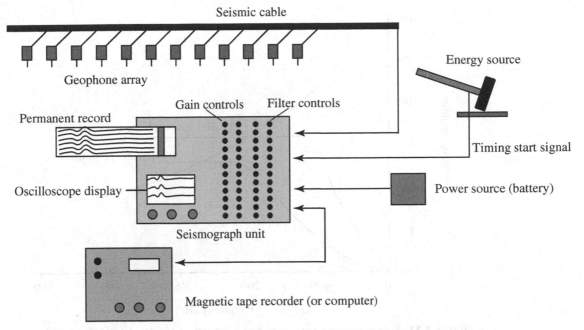

FIGURE 2.32 Diagrammatic representation of basic components of seismic exploration equipment.

engineering seismographs have gain controls that permit amplifier gain to be set at low levels for geophones near the energy source and to be systematically increased for geophones farther along the cable (Figure 2.32). Some systems offer AGC (automatic gain control), which attempts to keep amplified output at a more or less constant level. If you are interested in a more detailed discussion of seismic amplifiers, Telford, Geldart, and Sheriff (1990, 197–202) present a concise and readable discussion of both analog and digital amplification strategies.

Another important procedure in electronic treatment of the geophone signal is filtering. Simply put, electronic filters cut or filter out certain frequencies and pass along others for recording. This capability is useful for two major reasons: (1) the ability to remove unwanted noise and (2) the ability to restrict the seismic pulse frequencies recorded and eventually displayed for interpretation. At some field sites unwanted ground vibrations (noise) generated by one or more of many possible sources (such as pumps or wind) are sufficiently great to mask the desired seismic signal. Often this noise has greater amplitudes in a limited frequency range. By cutting or filtering these frequencies we can sometimes reduce noise to acceptable levels. A typical engineering seismograph (Figure 2.32) can display ground motion on an oscillograph screen prior to energy input. Various filter values can be selected and their effect viewed on the screen to determine if filtering will be effective in reducing the noise level.

In shallow reflection work, as noted previously, we are interested in high resolution and therefore in the higher-frequency portion of the seismic signal. We

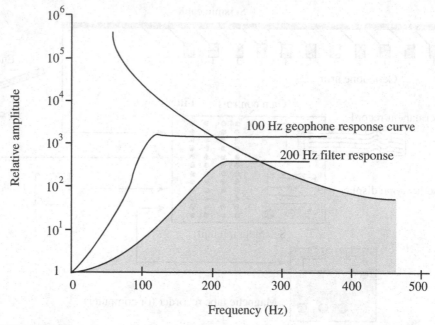

FIGURE 2.33 Frequency spectrum sampled by combination of a 100 Hz geophone and a 200 Hz filter.

already demonstrated the advantages of geophones with natural frequencies of 50 or 100 Hz. An additional strategy is to employ high-pass or low-cut filters. The capabilities of the typical equipment used for shallow seismic exploration, while impressive (especially compared to what was available just a few years ago), are still limited. Because the amplitudes of the lower-frequency signals are so large relative to the higher frequencies, the limits of the amplifiers are reached (saturated) by these lower frequencies, and the higher frequencies are not recorded. However, by filtering the lower-frequency component, we can preserve the higher-frequency component. Figure 2.33 is patterned after Figure 2.31 except that here we show diagrammatically the effect of a 200 Hz, high-pass filter. Figure 2.34 also illustrates the effect of filtering. This diagram plots relative energy against frequency for a filtered (300 Hz high-pass) and nonfiltered (all-pass) field trial using an 8-gauge shotgun source.

2.5.3 Signal Recording

Virtually all seismic equipment used in serious exploration endeavors today has some form of digital recording capability. Older equipment had only analog capability. The difference between analog and digital is straightforward. In analog recording the amplified and filtered geophone signal is essentially routed directly to a recording device that is activated by continually varying voltages. Although

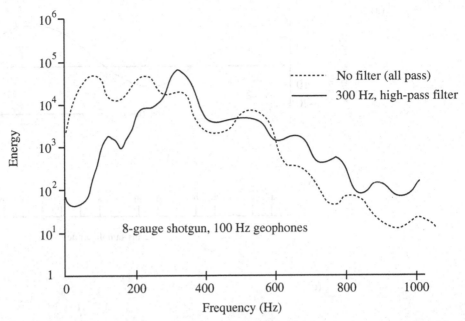

FIGURE 2.34 Effect of filtering on relative energies for the frequency spectrum produced by an 8-gauge shotgun source.

it is not normally utilized in field seismographs, a strip-chart recorder is an excellent example of this process. The constantly moving ink pen draws a curve on a paper roll due to constant rotation of the paper roll and continual deflection of the pen due to incoming voltage levels. In the early days of seismic exploration most analog records were recorded on photographic paper, which was then replaced by magnetic tape. In all cases, however, in analog recording, the detail of the signal is limited by the capabilities of the recording device.

These limitations are not present in digital recording. The main difference in digital recording occurs after signal conditioning. The conditioned signal is routed through an analog to digital converter (A/D converter), where the analog signal is converted to digital form. This is accomplished by sampling the signal at discrete intervals (perhaps 2 ms) and representing the amplitude of the signal at that moment by a signed number (Figure 2.35). Such a process theoretically places no limit on the ratio of the greatest to least signal amplitude that can be recorded (the dynamic range) because the signal at any moment in time is represented by a number, no matter how large. The sampling rate is important, however. If the sampling intervals are too far apart, the true form of the signal is not preserved and the higher frequencies are lost (Figure 2.35). Thus, in practice, the constraint placed on a digital system is primarily available memory in the digital recorder because more memory is required to store larger numbers and to sample at higher rates (more frequently).

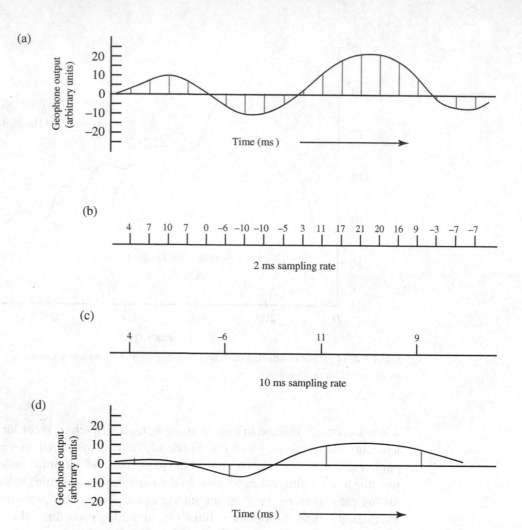

FIGURE 2.35 (a) Analog output of geophone (arbitrary units). (b) Digital representation of geophone response in (a) taken at 2 ms intervals. (c) Digital representation of same waveform at 10 ms sampling interval. (d) Reconstruction of waveform from data in (c) illustrates effect of inadequate sampling rate.

Several benefits accrue from digital recording in addition to greater signal fidelity. The data are stored in digital form on magnetic media (tape or disk) and can be read directly by computer for further display and processing. The data can be displayed at any time by digital to analog conversion. Also, many engineering seismographs possess signal enhancement capabilities. Because the data are in digital form, the waveforms from several hammer blows or shots can be added together. This process tends to enhance or increase the wave amplitude of the true reflection and refraction arrivals because their arrival pattern is always the same for the same shot point. Thus amplitudes that are too small to be iden-

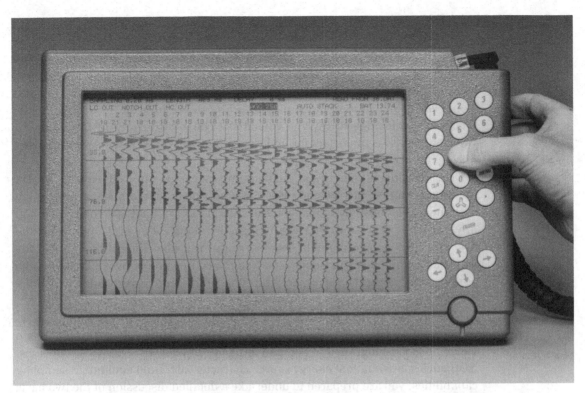

FIGURE 2.36 Photograph of oscilloscope display on seismic unit. Horizontal lines are timing lines at 10 ms intervals. Each vertical trace represents one geophone's response after amplification and filtering.

tified because of small energy sources may eventually be distinguished because their waveforms add constructively. Background noise, however, may have a substantial randomness component such that noise interference is reduced somewhat by destructive interference.

Most modern engineering seismographs have, in addition to digital recording capability, an oscilloscope display so that the seismic signal from each geophone can be viewed immediately after each shot (Figure 2.36). This is valuable in monitoring the quality of the data being recorded and also forms a critical component of a number of field procedures to be developed in subsequent chapters. Most instruments also can produce a permanent record on electrostatic or light-sensitive paper. These records (*seismograms*) include a trace from each geophone and timing lines spaced at equal intervals (usually 5 or 10 ms). Usually an option is present to plot the traces simply as a wiggly-line trace (Figure 3.21) that resembles the standard analog record or as a wiggly-line trace with variable-area shading superimposed (Figure 4.9). The standard wiggly-line record tends to be used more for refraction surveys, whereas the wiggly-line/variable-area trace is used more frequently in reflection surveys.

2.6 SUMMARY

After reading the preceding material, you now should have some idea of how seismic waves are generated, how they are propagated, and what controls their velocities. In this context it is important to remember the velocities for compressional waves of common Earth materials and the ratio of shear wave and surface wave velocities to P-wave velocities. The relationship of frequency to wavelength also is important for understanding why certain techniques are used during data collection.

The relationships between incident and refracted and reflected rays form the foundation for the seismic refraction and reflection methods, and working with these relationships should become second nature. Having a good command of what types of seismic energy are expected along a line of geophones is the first step in interpreting field seismograms. You also should have a feel for expected amplitudes and frequencies and the connection between the latter, the energy source characteristics, and the effect of the materials through which the energy passes.

Finally, you should be comfortable in discussing the role of geophone selection and filter selection in sampling the arriving seismic waveform. If you can view a field seismogram and begin to think in terms of wave arrivals, velocities, frequencies, attenuation, energy source characteristics, and equipment sampling capabilities, you are prepared to undertake a detailed discussion of the two basic seismic exploration techniques: refraction and reflection seismology.

Before proceeding to the next chapter, however, work through the problems that follow. This will help to reinforce your knowledge and will increase your expertise when it is time to collect and interpret field data.

PROBLEMS

2-1 A common compressional wave velocity for saturated, unconsolidated materials is 1400 m/s. If a seismic wave with a frequency of 10 Hz is traveling through this material, what is the wavelength of the wave? What is the wavelength of a wave with a frequency of 100 Hz?

2.2 What is the wavelength of a 10 Hz wave propagating through bedrock with a velocity of 4500 m/s? What is the wavelength of a wave with a frequency of 100 Hz?

2.3 Calculate the velocity of a compressional wave in a homogeneous rock layer with a density of 2.60 g/cm^3, a Young's modulus of 0.39×10^{11} N/m^2, and a Poisson's ratio of 0.11.

2.4 Calculate the velocity of a shear wave in a homogeneous rock layer with a density of 2.04 g/cm^3, a Young's modulus of 0.07×10^{11} N/m^2, and a Poisson's ratio of 0.14.

2.5 Estimate the velocity of Rayleigh waves in a homogeneous rock layer with a density of 2.52 g/cm^3, a Young's modulus of 0.14×10^{11} N/m^2, and a Poisson's ratio of 0.16.

2.6 In most common rocks density varies from 2.0 to 3.0 g/cm^3, Young's modulus varies from 0.12 to 1.1×10^{11} N/m^2, and Poisson's ratio varies from 0.04 to 0.3. Using the Table 2.1 spreadsheet template, determine whether density variations or variations in elastic coefficients have the greatest effect on P-wave velocities.

2.7 A sandstone has a density of 2.28 g/cm^3, a Young's modulus of 0.14×10^{11} N/m^2, and a Poisson's ratio of 0.06. A geophone is located 50 m from a hammer blow. How long will it take the air wave, the direct compressional wave, and the surface (Rayleigh) wave to travel from the hammer blow to the geophone?

2.8 An arkose has a density of 2.62 g/cm^3, a Young's modulus of 0.16×10^{11} N/m^2, and a Poisson's ratio of 0.29. Twelve geophones are arranged along a line at 10 m intervals. The shot point is located 5 m from the first geophone in the line (15 m from the second geophone, and so on). Construct a graph that illustrates time of arrival (vertical axis) against geophone position (horizontal axis) for the air wave, direct P-wave, direct S-wave, and surface wave.

2.9 Use Huygens' principle to demonstrate that plane waves encountering an opening in a barrier with a dimension similar to that of the wavelength of the waves produce the diffraction pattern illustrated in Figure 2.18. You need only demonstrate the *general* form of the diffraction pattern.

2.10 Demonstrate that the relationship specified in Equation 2.26 between an incident P-wave and a refracted S-wave is valid (use Huygens' principle).

2.11 Demonstrate that the relationship specified in Equation 2.26 between an incident S-wave and a refracted P-wave is valid (use Fermat's principle).

2.12 A ray from a sonic boom strikes the Earth's surface at an incident angle of 20°. The material beneath the surface has a P-wave velocity of 500 m/s and an S-wave velocity of 200 m/s. Determine the angles of refraction of the refracted P- and S-waves.

2.13 An incident P-wave strikes an interface between two different rock types. The upper layer has a compressional wave velocity of 1200 m/s. The lower layer has a compressional wave velocity of 3800 m/s and a shear wave velocity of 1900 m/s. The incident angle is 18°. Calculate the angles of refraction for the refracted P- and S-waves.

2.14 A P-wave with an angle of incidence of 18° encounters a contact between two different rock types. The upper layer has a compressional wave velocity of 2500 m/s. The lower layer has a compressional wave velocity of 3800 m/s and a shear wave velocity of 1400 m/s. Calculate the angles of refraction for the refracted P- and S-waves.

2.15 A high-velocity rock unit (2500 m/s) overlies a low-velocity rock unit (1400 m/s). Calculate the angle of refraction for a P-wave incident at 30°.

2.16 Examine Snell's Law (Equation 2.25), and the relationships you calculated in Problems 2.14 and 2.15. Can you formulate a general statement concerning the angles of incidence and refraction for compressional waves when (a) the material above an interface has a lower velocity than the material below the interface and (b) the material above an interface has a higher velocity than the material below the interface?

2.17 A P-wave is traveling through a rock unit (velocity = 1200 m/s). This rock unit is in contact with another unit that has greater seismic wave velocities (P-wave = 3800 m/s; S-wave = 1900 m/s). At what angle must a ray strike the contact between these two units to produce a compressional head wave? A shear head wave?

2.18 Derive a general relationship to compute the minimum distance from shot to detector at which a head wave can be received. Refer to the accompanying diagram as a guide, and use the same symbols in your equation.

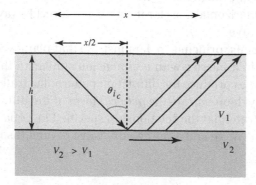

PROBLEM 2.18

REFERENCES CITED

Birch, Francis. 1966. Compressibility; elastic constants. In *Handbook of physical constants*, rev. ed., ed. S.P. Clark Jr., 97–173. Geological Society of America Memoir 97.

Dobrin, Milton B., and Carl H. Savit. 1988. *Introduction to geophysical prospecting*, 4th ed. New York: McGraw-Hill.

Means, W.D. 1976. *Stress and strain*. New York: Springer-Verlag.

Miller, R.D., S.E. Pullan, J.S. Waldner, and F.P. Haeni. 1986. Field comparison of shallow seismic sources. *Geophysics*, v. 51, 2067–92.

Press, Frank. 1966. Seismic velocities. In *Handbook of physical constants,* rev. ed., ed. S.P. Clark Jr., 195–218. Geological Society of America Memoir 97.

Pullan, S.E., and H.A. MacAulay. 1987. An in-hole shotgun source for engineering seismic surveys. *Geophysics*, v. 52, 985–96.

Richter, Charles F. 1958. *Elementary seismology*. San Francisco: W.H. Freeman and Company.

Slotnick, Morris Miller. 1959. *Lessons in seismic computing*, ed. Richard A. Geyer. *The Society of Exploration Geophysicists*, 70–76.

Telford, W.M., L.P. Geldart, and R.E. Sheriff. 1990. *Applied geophysics*, 2nd ed. Boston: Cambridge University Press.

Tooley, R.D., T.W. Spencer, and H.F. Sagoci. 1965. Reflection and transmission of plane compressional waves. *Geophysics*, v. 30, 552–70.

Zoeppritz, K. 1919. Über reflexion und durchgang seismischer wellen durch Ünstetıgkerlsfläschen: Über Erdbebenwellen VII B. Nachrichten der Königlichen Gesellschaft der Wissenschaften zu Göttingen, math.–phys. K1, Berlin, 57–84.

CHAPTER 3

Seismic Exploration:
The Refraction Method

Refraction seismology has a long, distinguished history. Although seismographs were developed in the latter part of the nineteenth century, it was not until the early twentieth century that major discoveries resulted from refraction techniques.

In 1909 A. Mohorovicic determined that a velocity discontinuity exists at the base of the crust based on his interpretation of wave arrivals from a nearby earthquake. Then, in 1913, B. Gutenberg determined the depth to the Earth's core. Thus refraction seismology was responsible for our early knowledge of the basic structure of the Earth.

By the 1920s refraction techniques were being vigorously and successfully applied to oil exploration. As seismic techniques and equipment improved, reflection eventually replaced refraction as the dominant method applied in oil exploration. However, refraction continues to be the most frequently applied seismic technique for shallow subsurface investigations. It is mainly for this reason that we consider it first in this text.

3.1   A HOMOGENEOUS SUBSURFACE

Before beginning to examine the refraction method, let's review waves spreading throughout a homogeneous subsurface. As the hemispherical wave front passes by a string of equally spaced geophones, each records the ground displacement due to this wave. The time of transit of this wave front from the energy source (shot point) to each geophone can be determined from a field seismogram. Because we know the geophone spacing (geophone interval) and the distance from the shot point to the first geophone (shot offset), we can construct a graph (travel-time curve) that plots time against distance (Figure 3.1).

Because the wave is traveling at a constant velocity and the geophones are equally spaced, this time–distance plot will be a straight line. Although somewhat trivial in this case, it is instructive to begin an analysis that we will follow

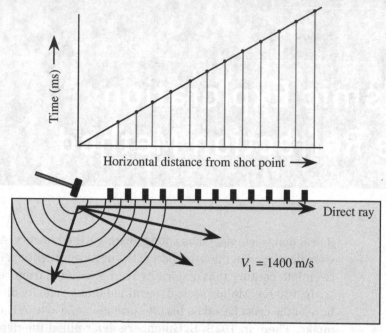

FIGURE 3.1 Generalized diagram illustrating ray paths in a uniform material with no discontinuities. Time–distance relationships for the portion of the wave traveling directly from energy source to receiver are shown in the travel-time curve.

throughout most of the rest of our seismic studies. Let's see if we can write an equation for this line. The form of the equation will be

$$\text{Time} = \frac{\text{Path}}{\text{Velocity}} \tag{3.1}$$

or

$$t = \frac{x}{V_1}. \tag{3.2}$$

For an equation this simple, it is fairly obvious that we can determine the velocity of the homogeneous material because we know the time and distance values. It's useful to review (or present) a bit of calculus at this point. Taking the first derivative of this equation with respect to x gives us the slope of the line. Thus

$$\frac{dt}{dx} = \frac{1}{V_1} \tag{3.3}$$

and

$$\text{Slope} = \frac{1}{V_1} \tag{3.4}$$

or

$$V_1 = \frac{1}{\text{Slope}}.$$

(3.5)

This means that if we have a travel-time curve similar to that in Figure 3.1, we can determine the slope (rise/run), take the inverse, and multiply by 1000 if the units for our graph are in meters and milliseconds. Following this procedure, we determine that the velocity of the direct wave in the travel-time curve in Figure 3.2 is 286 m/s. Although this insight into the subsurface is minimal, it gives us some information about the material near the surface because of its low velocity (see Table 2.2). However, as we soon shall see, the amount of useful information we can extract from typical travel-time curves is much greater.

3.2   A SINGLE SUBSURFACE INTERFACE

You realize, of course, that the subsurface usually is not homogeneous. We normally expect to have several interfaces present. Based on our discussion in Chapter 2, it is clear that such interfaces will produce reflections, refractions, and wave conversions. At the moment we will limit our analysis to refractions—in particular, to refractions that return energy to the surface where it may be sensed and recorded. You should recall the relationship that leads to critical refraction and produces a head wave. This might be a propitious time to review Snell's Law, the critical angle, and critical refraction.

Figure 3.3 illustrates the path from energy source E to geophone G that one such critically refracted ray follows. A compressional wave generated at E and traveling at V_1 strikes the interface between materials with different velocities, V_1 and V_2. The ray that strikes this interface at the critical angle, θ_{i_c}, is refracted parallel to the interface and travels at V_2. A head wave is generated as energy is returned to the surface along rays such as that traveling from N to G. Several possible questions arise from an analysis of Figure 3.3. Can we detect the arrivals of this refracted energy? How long does it take for the energy generated at E to arrive at G? How would the head wave arrivals plot on a travel-time curve? Can we extract any information about the subsurface from the head wave assuming we can identify it on seismograms?

3.2.1   Derivation of a Travel-Time Equation

A logical first step is to follow a procedure such as we undertook for the direct wave analysis in the preceding section. After we derive a travel-time equation for the refracted wave, we can analyze the equation and see what is possible. In deriving an equation similar to Equation 3.1, we will use the relationships and quantities in Figure 3.3.

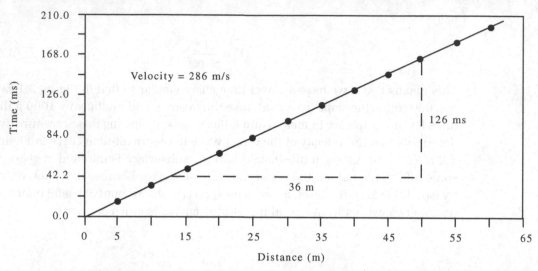

FIGURE 3.2 Travel-time curve (time–distance graph) illustrating only direct wave arrivals. The velocity is 286 m/s as determined from the inverse of the slope.

The first step is a simple one. The total time of travel must be

$$\text{Time} = \frac{EM}{V_1} + \frac{MN}{V_2} + \frac{NG}{V_1}. \tag{3.6}$$

Because

$$\cos\theta_{i_c} = \frac{h_1}{EM} \quad \text{and} \quad EM = NG, \quad \text{then}$$

$$EM = NG = \frac{h_1}{\cos\theta_{i_c}}. \tag{3.7}$$

Also,

$$EA = BG = h_1 \tan\theta_{i_c} \quad \text{and} \quad MN = x - 2h_1 \tan\theta_{i_c}, \quad \text{so}$$

$$\text{Time} = \frac{h_1}{V_1 \cos\theta_{i_c}} + \frac{x - 2h_1 \tan\theta_{i_c}}{V_2} + \frac{h_1}{V_1 \cos\theta_{i_c}}. \tag{3.8}$$

Equation 3.8 is the basic formula to determine time of travel for a critically refracted ray at a single interface. If we know h_1, V_1, and V_2, we then can compute the travel time for any distance x from energy source to receiver. We could deduce several interesting relationships at this point, but Equation 3.8 is more useful if we recast it in a slightly different form. Let's take care of this task before proceeding to an analysis. First we rearrange terms:

$$\text{Time} = \frac{2h_1}{V_1 \cos\theta_{i_c}} - \frac{2h_1 \tan\theta_{i_c}}{V_2} + \frac{x}{V_2}. \tag{3.9}$$

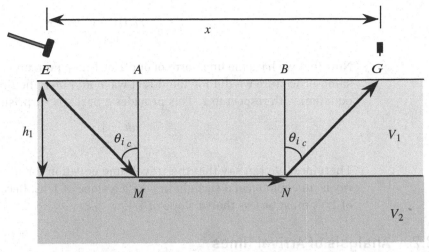

FIGURE 3.3 Diagram illustrating symbols used in derivation of time of travel for critically refracted ray.

Next we use the relationships

$$\tan\theta_{i_c} = \frac{\sin\theta_{i_c}}{\cos\theta_{i_c}} \quad \text{and} \quad \sin\theta_{i_c} = \frac{V_1}{V_2}$$

to change Equation 3.9 to

$$\text{Time} = \frac{2h_1}{V_1\cos\theta_{i_c}} - \frac{2h_1\sin^2\theta_{i_c}}{V_1\cos\theta_{i_c}} + \frac{x}{V_2}. \tag{3.10}$$

Consolidating, we obtain

$$\text{Time} = \frac{2h_1(1-\sin^2\theta_{i_c})}{V_1\cos\theta_{i_c}} + \frac{x}{V_2}. \tag{3.11}$$

Using the familiar relationship

$$\sin^2\theta_{i_c} + \cos^2\theta_{i_c} = 1,$$

and the identities noted previously, we can substitute to obtain

$$\text{Time} = \frac{2h_1\cos\theta_{i_c}}{V_1} + \frac{x}{V_2} \tag{3.12}$$

$$\text{Time} = \frac{2h_1\left(1-\left(\frac{V_1^2}{V_2^2}\right)\right)^{\frac{1}{2}}}{V_1} + \frac{x}{V_2} \tag{3.13}$$

$$\text{Time} = \frac{2h_1(V_2^2 - V_1^2)^{\frac{1}{2}}}{V_2 V_1} + \frac{x}{V_2}. \tag{3.14}$$

Now that we have the final form of our travel-time Equation 3.14, let's follow the same procedure we did for the direct wave and take the first derivative of the equation with respect to x. This provides a perhaps surprisingly simple result:

$$\frac{dt}{dx} = \frac{1}{V_2}. \tag{3.15}$$

Therefore, we can say that the travel-time equation for the critically refracted ray is an equation of a straight line with a slope of $1/V_2$. Because $V_2 > V_1$, a slope of $1/V_2$ must be less than a slope of $1/V_1$.

3.2.2 Analysis of Arrival Times

This is sufficient information to decide if we can use the critically refracted ray in resolving subsurface geology. In our analysis we will use the geometry in Figure 3.4 and the values in Table 3.1. Because we have derived the travel-time equa-

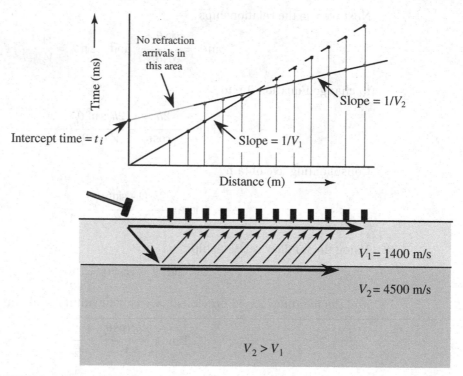

FIGURE 3.4 Generalized diagram illustrating ray paths in a material with one horizontal discontinuity. Time–distance relationships for both the direct and refracted rays are shown in the travel-time curve.

tions for the direct and critically refracted ray, we can place these in a spreadsheet and calculate travel times for various geophone placements. Table 3.1 shows the results for one combination of geophone intervals and geology.

If we scan the arrival times in Table 3.1, we see that the direct wave arrives first for distances up to 27 m from the shot, but that the refracted wave arrives before the direct wave at all geophones at distances greater than 27 m. Thus, if we had a geophone at each distance position in Table 3.1, the first event on a field seismogram for geophones 1–10 would be the direct wave arrival, and the first event for geophones 11–24 would be the refracted wave arrival. What is a simple physical explanation of this relationship? If we examine the subsurface model in Figure 3.4, we see that the direct ray travels from left to right at 1400 m/s. The critically refracted ray travels to the interface at 1400 m/s, is refracted and travels parallel to the interface at 4500 m/s, and then returns to the surface at 1400 m/s. At significant distances from the energy source, the greatest portion of the refracted wave's path is at a velocity more than three times as great as the direct wave. At some surface point, therefore, the refracted wave must arrive before the direct wave, which is traveling at a much slower velocity.

Note that Table 3.1 also is a dynamic table and that values shown in boldface may be changed. Experiment with the velocity and thickness values to convince yourself that the refracted wave ultimately will arrive before the direct wave. If you select velocities that are similar (remember the condition that $V_2 > V_1$), or if you select a very large depth to the interface, the times at the 24 geophone positions in the table may not illustrate this relationship. The reason is that the most distant geophone is not far enough from the shot point for the refracted wave to arrive first. If this happens, increase the geophone interval until you are satisfied that the relationship holds. (Actually, if you have this difficulty, you have discovered one of the limitations of the refraction method, which will be discussed shortly.)

Is this information sufficient for us to reach a decision on the utility of the critically refracted wave? If we plot the arrival times in Table 3.1 against distance from the energy source (geophone positions), we produce a graph similar to the travel-time curve in Figure 3.4. The direct wave times plot along a straight line with slope = $1/V_1$ that must pass through the origin (assuming the energy source to be at the surface), and the refracted wave times plot along a straight line with slope $1/V_2$. If we had only these times and distances (as would be the case when analyzing field data), we could deduce that we were dealing with a two-layer case (single interface) because we have two travel-time segments, and we could compute V_1 and V_2. The velocity values would give us some insight into the nature of the subsurface materials, but we still would not know the thickness of the upper layer. However, the travel-time equation we derived contains a layer thickness term h_1, so there must be some way to solve for h_1 if we know V_1 and V_2.

TABLE 3.1 | ARRIVAL TIMES FOR DIRECT AND REFRACTED WAVES

	Arrival Times	
Distance from Shot (m)	Direct Wave	Refracted Wave
0	0.00	13.58
3	2.14	14.24
6	4.29	14.91
9	6.43	15.58
12	8.57	16.24
15	10.71	16.91
18	12.86	17.58
21	15.00	18.24
24	17.14	18.91
27	19.29	19.58
30	21.43	20.24
33	23.57	20.91
36	25.71	21.58
39	27.86	22.24
42	30.00	22.91
45	32.14	23.58
48	34.29	24.24
51	36.43	24.91
54	38.57	25.58
57	40.71	26.24
60	42.86	26.91
63	45.00	27.58
66	47.14	28.24
69	49.29	28.91

Velocity 1$_P$ (m/s)	**1400**
Velocity 2$_P$ (m/s)	**4500**
Depth (m)	**10**

Shot offset (m)	0
Geophone interval (m)	3

No refracted wave arrivals before (ms):	6.55

3.2.3 Determining Thickness

Examine Figure 3.4 once again. The straight line passing through the arrival times for the critically refracted ray can be extended until it intersects the time axis. This time is termed the *intercept time*, t_i. Recall that it has no real physical significance because no refractions arrive at the energy source ($x = 0$). However, at $x = 0$ our travel-time Equation 3.14 reduces to

$$\text{Time} = t_i = 2h_1 \frac{(V_2^2 - V_1^2)^{\frac{1}{2}}}{V_2 V_1} \qquad (3.16)$$

and, therefore,

$$h_1 = \frac{t_i}{2} \frac{V_2 V_1}{(V_2^2 - V_1^2)^{\frac{1}{2}}}. \qquad (3.17)$$

Thus for a single horizontal interface, if we can determine times for direct and refracted arrivals from a field seismogram, we can calculate the thickness of the material above the interface and the velocities of the materials above and below the interface.

3.2.4 Crossover Distance

Before proceeding further, it is perhaps a good time to examine two other matters of interest. First, although we now have all the information necessary to solve problems related to a single horizontal interface, there is an additional approach that occasionally is used, and we now develop it for completeness. If you examine the travel-time curve in Figure 3.4, you see that the straight lines for the direct wave times and refracted wave times intersect at a point. The horizontal coordinate of this point, x_{co}, is referred to as the *crossover distance*. Crossover distance can be used instead of t_i to develop a solution for h_1, the depth to the interface. At x_{co} the times of travel for the direct wave and the refracted wave are equal, and, therefore,

$$\text{Time}_{\text{direct wave}} = \frac{x_{co}}{V_1}, \qquad \text{(see 3.2)}$$

$$\text{Time}_{\text{refracted wave}} = \frac{2h_1(V_2^2 - V_1^2)^{\frac{1}{2}}}{V_2 V_1} + \frac{x_{co}}{V_2} \qquad \text{(see 3.14)}$$

and

$$\frac{x_{co}}{V_1} = \frac{2h_1(V_2^2 - V_1^2)^{\frac{1}{2}}}{V_2 V_1} + \frac{x_{co}}{V_2}. \qquad (3.18)$$

If we rearrange terms, we arrive at

$$\frac{h_1(V_2^2 - V_1^2)^{\frac{1}{2}}}{V_2 V_1} = \frac{x_{co}}{2}\left(\frac{V_2 - V_1}{V_2 V_1}\right) \qquad (3.19)$$

and

$$h_1 = \frac{x_{co}}{2} = \left(\frac{V_2 - V_1}{V_2 V_1}\right)\left(\frac{V_2 V_1}{(V_2^2 - V_1^2)^{\frac{1}{2}}}\right). \qquad (3.20)$$

The V_2V_1 terms cancel, and remembering that $(V_2{}^2 - V_1{}^2) = (V_2 - V_1)(V_2 + V_1)$, we can simplify Equation 3.20 to

$$h_1 = \frac{x_{co}}{2}\left(\frac{V_2 - V_1}{V_2 + V_1}\right)^{\frac{1}{2}}.$$

(3.21)

It is mainly a matter of convenience whether we use Equation 3.21 or Equation 3.17 to determine layer thickness. However, because determination of x_{co} involves fitting *two* lines to field data whereas t_i requires only one line, t_i usually can be determined with more accuracy. In the computer program REFRACT that accompanies this book we use t_i (Equation 3.17).

3.2.5 Critical Distance

The second item of interest is one you have seen several times before (see Problem 2.18). You should recall that, based on the geometry of the critical refraction, there is a finite distance from the energy source to the first point at which this refracted energy can be received. This distance is referred to as the *critical distance* and is equal to x_{crit} in Figure 3.5. Please note that in many refraction modeling programs and in the tables and dynamic tables used in this text, refraction times are given for distances less than the critical distance. These times obviously cannot be valid but are given to avoid the complexity of programming a spreadsheet to exclude them. You must remember, however, the geometry of critical refraction, especially when planning and interpreting field surveys.

Derivation of an equation to calculate critical distance is straightforward (which you already know if you solved Problem 2.18):

$$\tan\theta_{i_c} = \frac{x_{crit}/2}{h_1}$$

(3.22)

and, because

$$\sin\theta_{i_c} = \frac{V_1}{V_2},$$

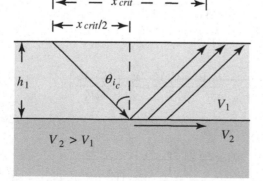

FIGURE 3.5 This diagram illustrates the relationships for calculating *critical distance:* the minimum distance from the energy source at which the first critical refraction can be received.

$$\tan\left(\sin^{-1}\left(\frac{V_1}{V_2}\right)\right) = \frac{x_{crit}/2}{h_1} \tag{3.23}$$

and

$$x_{crit} = 2h_1 \tan\left(\sin^{-1}\left(\frac{V_1}{V_2}\right)\right). \tag{3.24}$$

Equation 3.24 is quite sufficient for our purposes in this day of inexpensive hand calculators that handle trigonometric functions with ease. However, it is possible to use the trigonometric identities we employed when deriving the travel-time equation for head waves to arrive at a different form for Equation 3.24:

$$x_{crit} = \frac{2h_1}{\left(\left(\frac{V_2}{V_1}\right)^2 - 1\right)^{\frac{1}{2}}}. \tag{3.25}$$

Table 3.2 presents some critical distance values for a range of layer thicknesses. The velocities used to calculate these values are similar to those for unsaturated sands overlying saturated sands (see Table 2.2). Note that the percentage of critical distance relative to thickness is substantial. In a case such as this the positioning of geophones is important if we are to gather the maximum amount of information in the most economical fashion.

Try a variety of velocity values to investigate the role velocity contrast has in affecting the critical distance. Examine Table 2.2 in order to use realistic velocities.

TABLE 3.2	CRITICAL DISTANCE VALUES FOR VARIOUS THICKNESSES

Thickness	Critical Distance		
2	1.53		
4	3.06	Velocity 1_P	500
6	4.59	Velocity 2_P	1400
8	6.12		
10	7.65		
12	9.18	Thickness increment	2
14	10.71		
16	12.24		
18	13.76	$xcrit$ as % of h_1	76%
20	15.29		

You may have noticed that the path of the first critically refracted ray to arrive at the surface is the same as a reflection that strikes the interface at the critical angle. Thus the reflected and refracted waves arrive at the same time at x_{crit}. However, as we will later demonstrate, all other reflections arrive later than the critical refraction and so will not interfere with our analysis, which depends on the first wave arriving at each receiver.

3.2.6 Constructing a Travel-Time Curve from a Field Seismogram

Now that you have a good grasp of the relationships that enable us to determine velocities and thickness for a single horizontal interface separating two differing materials, let's examine a field seismogram from a site with this subsurface geology. Figure 3.6 illustrates a straightforward example. Pertinent information is contained in the caption for this figure. Recall that we are interested in the time at which the first energy arrives at each geophone. In Figure 3.6 this is indicated by the first sharp downturn in each geophone's output and is marked by arrows to help you recognize this first break.

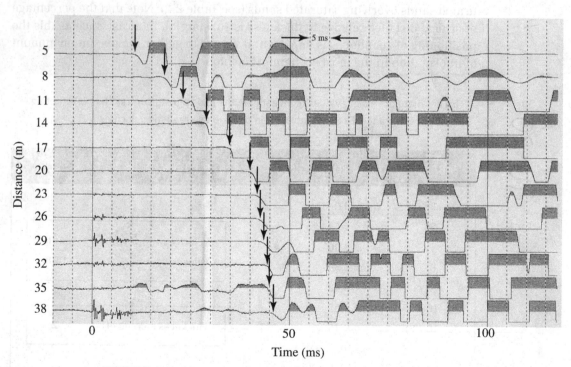

FIGURE 3.6 Field seismogram from the Connecticut Valley, Massachusetts. First breaks for each geophone trace are indicated by a downward-directed arrow. This seismogram exhibits a classic two-layer pattern.

TABLE 3.3	TIME–DISTANCE VALUES FOR SEISMOGRAM IN FIGURE 3.6											
Geophone	1	2	3	4	5	6	7	8	9	10	11	12
Distance (m)	5	8	11	14	17	20	23	26	29	32	35	38
Time (ms)	11.0	18.0	23.0	29.0	34.5	40.0	42.0	42.5	43.5	44.5	45.0	46.0

Construct a table similar to Table 3.3 in which you record the time of each first break at each geophone. Then graph these data to create a time–distance plot similar to that in Figure 3.7(a). Use a sufficiently large sheet of cross-ruled paper (at least 8½ × 11 inches) so that slopes and the intercept time can be determined accurately. When measuring first break times, it's a good idea to construct a template on a transparent sheet on which you have drawn lines at 0.5 or 1 ms intervals. Place this sheet on the seismogram to read values accurately and consistently.

Your next task is to interpret the data points. The first five data points define a straight line reasonably well (Figure 3.7(b)). The line for the direct wave must pass through the origin if the energy source was located on the surface, and this line satisfies that criterion. The last seven data points also lie along a straight line. This plot looks like a familiar case in which we receive the direct wave at the closest geophones and then the head wave at the most distant geophones. Next we measure the slopes of these lines and from the slopes compute the velocities. The intercept time can be read from the vertical axis (34.5 ms), and the subsurface structure can be computed from Equation 3.17. Our computations suggest an 8.3 m thick layer with a velocity of 480 m/s overlying material with a velocity of 3380 m/s.

(a)

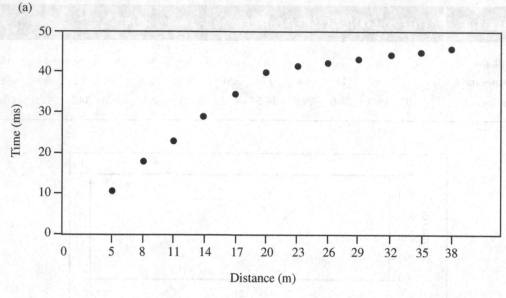

(b)

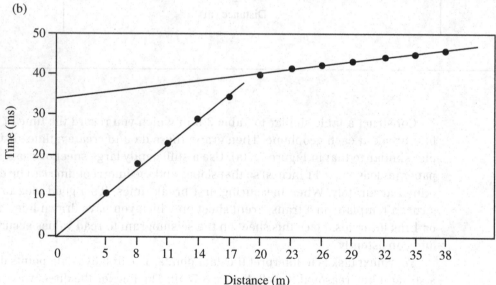

FIGURE 3.7 (a) Time–distance data from the seismogram in Figure 3.6. (b) Lines drawn through data points in (a).

3.2.7 Using REFRACT

Now that you have learned some of the basics of the refraction method, it is an appropriate time to introduce one of the computer programs included on the disk that accompanies this textbook. Instructions for using this program are in Appendix A.

REFRACT produces time–distance data for subsurface models that you designate. As such, it is extremely useful for investigating various relationships as we discuss them in the pages that follow. The goal of this chapter is to enable you to interpret travel-time curves based on the arrival of refracted waves, and the more experience you have with curves representing different subsurface conditions, the better your interpretations will be. So use REFRACT freely and often. Note that REFRACT plots time–distance data as if energy sources were placed at both ends of a string of geophones. Such practice is common, but we have not yet discussed the reasons for this procedure.

REFRACT also plots time–distance data that you enter directly. You then fit straight lines to the data by drawing with a mouse. The power of this program is in its ease of use. Because you can draw lines quickly and subsurface structures are computed quickly, you easily can attempt several interpretations when straight lines do not pass perfectly through all field data points. Such multiple interpretations are encouraged to gain some notion of the range of models that fit your data.

From time to time in the following pages, suggestions are offered for exploration using these computer programs. Spend some time investigating these suggestions because by doing so you will substantially improve your interpretative abilities.

> Use REFRACT to investigate a number of subsurface variations for the single interface case. First hold layer thickness constant and try a variety of velocity contrasts to gain an appreciation of velocities and related slopes. Remember that we assume $V_2 > V_1$. Next hold velocities constant and vary the thickness from very small to very large. Note how the travel-time curves change. Can you formulate a number of situations in the field for which you might not recognize that two layers (one interface) are present?

3.2.8   The Mohorovicic Discontinuity

Virtually everyone who has taken a physical geology course is familiar with the velocity discontinuity that separates the Earth's crust from the mantle. Most remember this interface as the "Moho" or M-discontinuity; a few recall that this terminology incorporates (to some extent) the name of its discoverer Andrija Mohorovicic; and fewer still are aware of the evidence advanced by Mohorovicic to demonstrate the existence of the *Mohorovicic discontinuity*.

Following a strong earthquake in October 1909, which affected his place of work at the Zagreb Meteorological Observatory, Mohorovicic plotted the main deflections on seismograms recorded by 29 seismological observatories located at distances from very near the epicenter out to 2400 km from the epicenter (Bonini and Bonini 1979). Considering only the P-wave arrivals, Mohorovicic

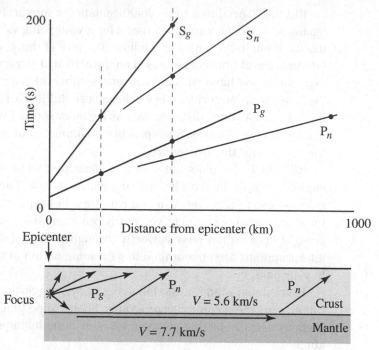

FIGURE 3.8 Travel-time curves simplified from Mohorovicic's original plots. P_g is the direct compressional wave, and P_n is the refracted compressional wave. Terminology for shear waves (S) is similar. The section below the travel-time curve illustrates Mohorovicic's interpretation.

noted that on seismograms recorded at stations close to the epicenter only one pulse appeared, but on seismograms from more distant stations there were two clearly distinguishable pulses. This same pattern held for S-wave arrivals (see Figure 3.8). Mohorovicic reasoned that the early pulses recorded at the more distant stations must be due to waves that traveled at a greater velocity over a considerable portion of their path distance. He computed velocities from the travel-time curves and calculated that the velocity discontinuity was located at a depth of 50 km. This was quite a remarkable achievement so early in the history of seismology.

If you examine Figure 3.8, you should recognize our classic two-layer case (one horizontal interface). The direct wave curves for P- and S-waves (labeled P_g and S_g) do not pass through the origin because the earthquake occurred not on the surface but at depth. Travel from the focus to the epicenter took a finite amount of time. Also, curves for both P- and S-waves are shown (S-ray paths are not shown). Because the distances traversed by P- and S-waves are measured in kilometers in this case (rather than in meters as in most exploration work), sufficient time is present for the disturbance due to the arrival of the P-wave to die down before the arrival of the S-wave. Therefore, it usually is possible to recognize S-wave arrivals. Although this presentation is simplified somewhat, you should be able to understand the essential elements of the evidence for the Mohorovicic discontinuity.

Depending on the objectives of your field study and the length of your geophone spread, you often will encounter a two-layer case. As you are well aware, however, most subsurface geology is more complex than a single horizontal interface. Let's take a look at the next step up in complexity.

3.3 TWO HORIZONTAL INTERFACES

It is not difficult to envisage a subsurface configuration in which there are two horizontal interfaces (often referred to as a three-layer case). For instance, consider an alluvial sequence of sands and gravels overlying bedrock. The sands and gravels near the surface normally will be above the water table, will be dry, and will possess a relatively low velocity. The water table forms the first interface because the saturated sands and gravels will have a substantially higher velocity than their dry counterparts. Finally, the top of the bedrock surface forms the second interface because bedrock velocities are higher than saturated sediments.

In attempting to derive a travel-time equation for this situation, we proceed in precisely the same manner as for a single interface.

3.3.1 Derivation of a Travel-Time Equation

Referring to Figure 3.9, we see that for this geology a direct wave will travel from E to G, and that wave energy also will traverse the path $EPQG$. This is identical to the situation of the single horizontal interface. Wave energy will strike the V_2–V_3 interface. Assuming $V_3 > V_2$, critical refraction will occur, so we are interested in a wave path that follows $EPRSQG$. We already know that $\sin \theta_{i_c} = V_2/V_3$. The major question we must answer before proceeding is how to determine the θ_i required for ray PR to be critically refracted. Figure 3.9 is drawn so as to call your attention to the answer to this question. Study the drawing carefully before continuing.

We know from Equation 2.25 (Snell's Law) that

$$\frac{\sin \theta_i}{\sin \theta_r} = \frac{V_1}{V_2}$$

and, because $\theta_r = \theta_{i_c}$,

$$\sin \theta_r = \sin \theta_{i_c} = \frac{V_2}{V_3} \quad \text{and, therefore,} \quad \sin \theta_i = \frac{V_1}{V_3} \, . \tag{3.26}$$

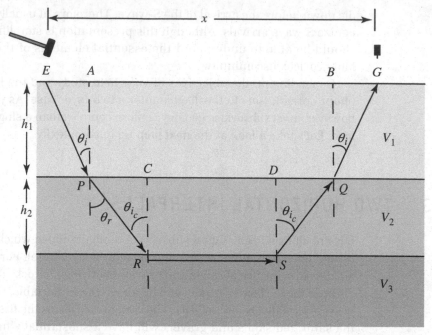

$$V_3 > V_2 > V_1$$

FIGURE 3.9 Diagram illustrating symbols used in derivation of time of travel for ray critically refracted along the second interface in a three-layer case.

Everything we now derive follows the same procedure as for the single interface. The total travel-time equation is

$$\text{Time} = \frac{EP}{V_1} + \frac{PR}{V_2} + \frac{RS}{V_3} + \frac{SQ}{V_2} + \frac{QG}{V_1}. \tag{3.27}$$

As before

$$EP = QG = \frac{h_1}{\cos\theta_i} \quad \text{and} \quad PR = SQ = \frac{h_2}{\cos\theta_{i_c}}.$$

Also,

$$EA = BG = h_1 \tan\theta_1 \quad \text{and} \quad PC = DQ = h_2 \tan\theta_{i_c} \quad \text{so that}$$

$$RS = x - 2h_1 \tan\theta_i - 2h_2 \tan\theta_{i_c}. \tag{3.28}$$

These relationships can be substituted into Equation 3.27 so that our travel-time equation becomes

$$\text{Time} = \frac{2h_1}{V_1 \cos\theta_i} + \frac{2h_2}{V_2 \cos\theta_{i_c}} + \frac{x - 2h_1 \tan\theta_i - 2h_2 \tan\theta_{i_c}}{V_3}. \tag{3.29}$$

It's worthwhile to present the major steps in simplifying this equation; but because the steps are similar to those used to arrive at Equation 3.14, they are presented without comment. The same identities that we used previously are adequate.

$$\text{Time}_{RS} = \frac{x}{V_3} - \frac{2h_1 \tan\theta_i}{V_3} - \frac{2h_2 \tan\theta_{i_c}}{V_3} \tag{3.30}$$

$$\text{Time} = \frac{x}{V_3} + \frac{2h_1}{V_1 \cos\theta_i} + \frac{2h_2}{V_2 \cos\theta_{i_c}} - \frac{2h_1 \sin^2\theta_i}{V_1 \cos\theta_i} - \frac{2h_2 \sin^2\theta_{i_c}}{V_2 \cos\theta_{i_c}} \tag{3.31}$$

$$\text{Time} = \frac{x}{V_3} + \frac{2h_1 - 2h_1 \sin^2\theta}{V_1 \cos\theta_i} + \frac{2h_2 - 2h_2 \sin^2\theta_{i_c}}{V_2 \cos\theta_{i_c}} \tag{3.32}$$

$$\text{Time} = \frac{x}{V_3} + \frac{2h_1 \cos\theta_i}{V_1} + \frac{2h_2 \cos\theta_{i_c}}{V_2} \tag{3.32}$$

and, finally

$$\text{Time} = \frac{x}{V_3} + \frac{2h_1(V_3^2 - V_1^2)^{\frac{1}{2}}}{V_3 V_1} + \frac{2h_2(V_3^2 - V_2^2)^{\frac{1}{2}}}{V_3 V_2}. \tag{3.33}$$

Once again we finish with an equation for a straight line. As you probably have observed by now, if we again take a derivative—voilà!

$$\frac{dt}{dx} = \frac{1}{V_3}. \tag{3.34}$$

Although it is now beginning to sound a little repetitious, we can continue with the same type of analysis as we did for the single interface. Examination of Figure 3.10 should convince you that at some point the ray that has a part of its path along the second interface eventually will be the first to arrive at one of the distant geophones. We see that the velocity of the third layer can be determined by taking the inverse of the slope of the straight line passing through the arrival times from the ray that was critically refracted along the second interface.

3.3.2 Determining Thickness

The line corresponding to refractions from the second interface can be extended back to the vertical axis (Figure 3.10) and is referred to as the *second intercept time* or t_{i_2}. If we use Equation 3.33 and solve for h_2 at $x = 0$, we arrive at

$$h_2 = \left(t_{i_2} - \frac{2h_1(V_3^2 - V_1^2)^{\frac{1}{2}}}{V_3 V_1} \right) \frac{V_3 V_2}{2(V_3^2 - V_2^2)^{\frac{1}{2}}}. \tag{3.35}$$

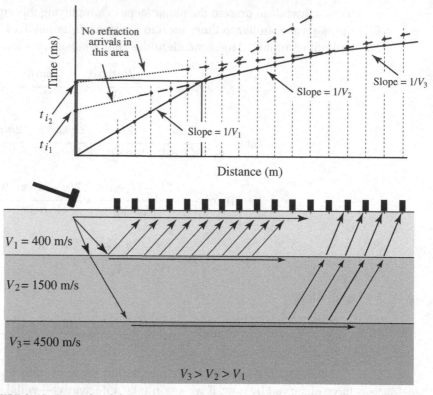

FIGURE 3.10 Generalized diagram illustrating ray paths in a material with two horizontal interfaces. Time–distance relationships for the direct and two critically refracted rays are shown in the travel-time curve.

We compute V_1, V_2, and V_3 from the inverse of the slopes of the three line segments on the travel-time curve in Figure 3.10. The thickness of the first layer h_1 is computed using t_{i_1} and Equation 3.17. The second layer thickness is then calculated using t_{i_2} and Equation 3.35.

Based on our discussions to this point, we interpret a travel-time curve such as that in Figure 3.10 as representing a three-layer case because it has three straight-line segments, each with a lower slope than the one closer to the energy source. Velocities for each layer are available from the curve, as are the thicknesses of the first two layers (and, of course, the depths to the two interfaces). From this information alone and some general knowledge of the local geology, we can arrive at a reasonably good model for the subsurface geology at the exploration site.

In the preceding analysis, we used only intercept times to develop an equation for the thickness of the second layer. We could develop an equation using crossover distances as in the case of the single interface, but for reasons cited previously and because most computerized solutions use intercept times, it does not seem necessary to produce yet more equations. This derivation is left for those interested in such an exercise. Because of implications for fieldwork, however, it seems wise to once again turn to the topic of critical distance.

3.3.3 Critical Distance

The critical distance x_{crit} for the head wave produced by critical refraction along the second interface is EG in Figure 3.11. The ray path for this geometry is $EPRQG$. Thus x_{crit} is equal to $EA + PC + CQ + BG$. Because $EA = BG = h_1 \tan \theta_i$ and $PC = CQ = h_2 \tan \theta_{i_c}$, and $\sin \theta_i = V_1/V_3$ and $\sin \theta_{i_c} = V_2/V_3$,

$$x_{crit} = 2 \left(h_1 \frac{\sin \theta_i}{(1 - \sin \theta_i{}^2)^{\frac{1}{2}}} + h_2 \frac{\sin \theta_{i_c}}{(1 - \sin \theta_{i_c}{}^2)^{\frac{1}{2}}} \right) \tag{3.36}$$

$$x_{crit} = 2 \left(h_1 \frac{V_1/V_3}{\left(1 - \left(V_1/V_3\right)^2\right)^{\frac{1}{2}}} + h_2 \frac{V_2/V_3}{\left(1 - \left(V_2/V_3\right)^2\right)^{\frac{1}{2}}} \right) \tag{3.37}$$

and

$$x_{crit} = 2 \left(h_1 \frac{V_1}{(V_3^2 - V_1^2)^{\frac{1}{2}}} + h_2 \frac{V_2}{(V_3^2 - V_2^2)^{\frac{1}{2}}} \right). \tag{3.38}$$

To investigate the implications of this formula for critical distance, let's create a table similar to Table 3.2.

The velocities in Table 3.4 are fairly representative of what might be expected in an area with unsaturated and saturated sediments lying above bedrock. Although the magnitude of the second critical distance (the distance from the energy source to the point where refractions from the second interface can be received) is more than half the depth to the second interface (sum of h_1 and h_2), this should not be unexpected to anyone who has examined Figure 3.11 closely.

> The magnitude of the second critical distance relative to the depth to the second interface varies according to the relative magnitudes of V_1, V_2, and V_3. Use Table 3.4 as a dynamic table and insert realistic velocity values for a case in which the values are reasonably similar in magnitude and for a case in which the velocities are very different. Let the values in Table 3.4 represent the standard or general case for comparison. Do velocity ratios affect the critical distances to any significant extent?

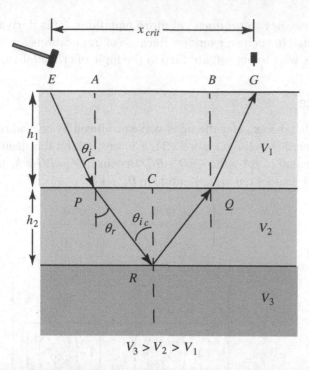

FIGURE 3.11 Diagram illustrating symbols used in derivation of critical distance for ray critically refracted along the second interface in a three-layer case.

$V_3 > V_2 > V_1$

TABLE 3.4	CRITICAL DISTANCES FOR TWO HORIZONTAL INTERFACES		
h_1 (m)	h_2 (m)	First X_{crit} (m)	Second X_{crit} (m)
1	4	0.76	2.84
2	8	1.53	5.68
3	12	2.29	8.53
4	16	3.06	11.37
5	20	3.82	14.21
6	24	4.59	17.05
7	28	5.35	19.90
8	32	6.12	22.74
9	36	6.88	25.58
10	40	7.65	28.42

Velocity 1_P (m/s)	**500**	h_1 increment (m)	**1**
Velocity 2_P (m/s)	**1400**	h_2 increment (m)	**4**
Velocity 3_P (m/s)	**4500**	Second X_{crit} as %of $h_1 + h_2$	**57%**

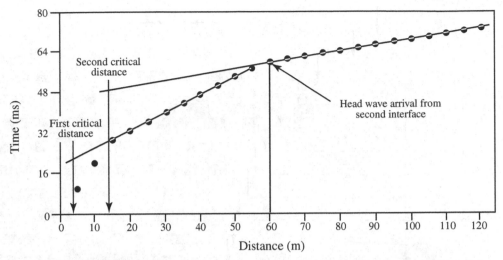

FIGURE 3.12 Travel-time curve based on a REFRACT plot using velocity values in Table 3.4 with $h_1 = 5$ m and $h_2 = 20$ m. Note that the first refraction from the second interface to be a *first arrival* is located at a distance considerably greater than the second critical distance.

Although critical distance is important, remember that the refraction method uses the first arrival at each geophone. Refer to Figure 3.11. It should be obvious that a wave traveling path *EPRQG* will take longer than a wave traveling from *E* to *G* that is refracted along the first interface. A wave refracted parallel to the second interface must travel a considerable distance at V_3 before overcoming its longer path distance and arriving before the head wave generated by the first interface. We can verify this impression by utilizing REFRACT. Figure 3.12 illustrates a plot using the same velocities as in Table 3.4 and layer thicknesses $h_1 = 5$ m and $h_2 = 20$ m. Table 3.4 tells us that the second critical distance for such values is 14.21 m, whereas Figure 3.12 demonstrates that it is approximately 60 m before the head wave from the second interface arrives prior to the head wave from the first interface.

Critical distance is of primary practical importance when we are trying to recognize later arrivals on a field seismogram. If we have a reasonable model for the subsurface, we can calculate critical distance and compare waveforms from geophones at distances less than and greater than the critical distance. If conditions are good, we might be able to recognize events in addition to the first arrivals. These additional data points then would enable us to refine our interpretation.

3.3.4 Analyzing a Second Field Seismogram

Now that we have increased our knowledge of the refraction method somewhat, let's interpret another field seismogram. The seismogram in Figure 3.13 was recorded approximately 0.5 km north of the field location for the seismogram in

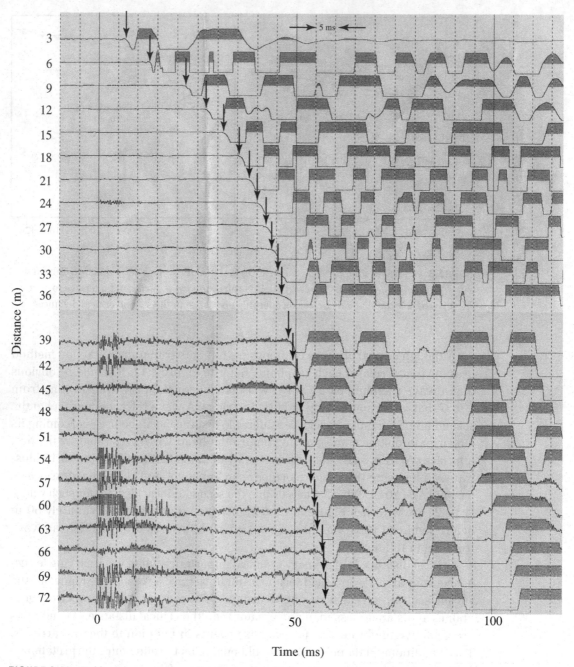

FIGURE 3.13 Field seismogram from the Connecticut Valley, Massachusetts. First breaks for each trace are indicated by a downward-directed arrow.

Figure 3.6. Note that the energy source–geophone offset and the geophone intervals are different. Figure 3.13 is reasonably straightforward to interpret. The first breaks for channels 1–12 are quite distinct. Amplifier gains on channels 12–24 were quite high so that background "noise" is evident and makes it a little more difficult to locate first breaks. Nevertheless, the first breaks are reasonably clear and consistent. Proceed in the same manner as before to gather your time–distance data.

When you are ready, either use REFRACT or work by hand to interpret your data. Figure 3.14 illustrates our preferred interpretation. Energy in this case was created by a hammer blow on the surface, so the direct wave should pass through the origin of the graph. As is usually the case, the points do not exactly fit straight lines, so we visually fit a line that passes through or close to the maximum number of points. In fitting these lines we also could disregard or weight less heavily points for which first breaks are not especially clear (such as trace 12).

Our analysis calculates a thickness of 5.1 m and a velocity of 424 m/s for the first layer, a thickness of 9.7 m and a velocity of 1441 m/s for the second layer, and a velocity of 3115 m/s for the third layer. The field site is in an area where glacial sands and gravels overlie Triassic arkoses. We, therefore, interpret the first interface to represent the water table and the second interface to represent the contact between sands and gravels and bedrock.

Another reminder about the utility of computer programs such as REFRACT is in order here. Fitting lines to the field data from Figure 3.13 is not completely

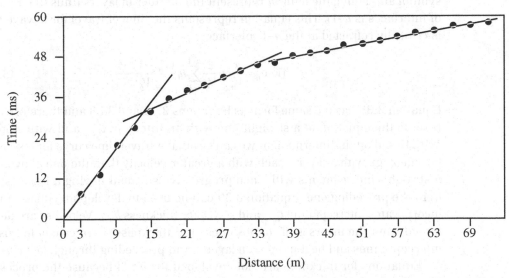

FIGURE 3.14 Preferred interpretation of time–distance data taken from Figure 3.13.

unambiguous. Therefore, you should attempt several different interpretations to determine how much variation is present in the subsurface configuration computed for each interpretation. This takes only a few more minutes with REFRACT (or a similar program) but gives you a good sense of how much uncertainty is present in the final geologic solution you adopt.

3.4 MULTIPLE INTERFACES

Although the three-layer case represents a configuration with multiple interfaces (two),we might ask what a travel-time curve looks like where there are more than two interfaces and velocity increases for each layer. Must we proceed through analyses like those just discussed in detail for a four-layer case, a five-layer case, and an n-layer case? If we reproduce Equations 3.14 and 3.33 here with the terms slightly rearranged, you can't help but notice a progressive similarity:

$$\text{Time} = \frac{x}{V_2} + \frac{2h_1(V_2^2 - V_1^2)^{\frac{1}{2}}}{V_2 V_1} \tag{3.14}$$

$$\text{Time} = \frac{x}{V_3} + \frac{2h_1(V_3^2 - V_1^2)^{\frac{1}{2}}}{V_3 V_1} + \frac{2h_2(V_3^2 - V_2^2)^{\frac{1}{2}}}{V_3 V_2} \tag{3.33}$$

We can rewrite these equations in a slightly different form using a summation symbol and letting the letter n represent the number of layers (thus the number of interfaces is $n-1$). This equation represents the time of travel for a wave that is critically refracted at the $n-1$ interface:

$$\text{Time}_n = \frac{x}{V_n} + \frac{2}{V_n} \sum_{i=1}^{n-1} h_i \frac{(V_n^2 - V_i^2)^{\frac{1}{2}}}{V_i} \tag{3.39}$$

Equation 3.39 has the same form as Equations 3.14 and 3.33 and therefore represents the equation of a straight line with an intercept $t_{i_{n-1}}$ and with slope = $1/V_n$. Based on this information, we can say that a travel-time curve for a subsurface geology with n layers, each with a greater velocity than the one above, has n straight-line segments with each progressive segment having a lesser slope than the preceding one. Equation 3.39 can be used to develop an equation that incorporates intercept time $t_{i_{n-1}}$ and solves for thickness h_{n-1}. Velocities are determined from the inverses of the n slopes, and thicknesses are calculated using intercept times and beginning with layer 1 and proceeding through layer $n-1$.

Equations for thickness are not developed for $n > 3$ because the process is identical to what already has been presented. Also, for most investigations in the shallow subsurface it is not especially common for more than three line segments to be present. There are a number of reasons for this, some of which we will mention later. One common reason is the similar velocities of saturated sediments and the velocity contrast between them and bedrock. This results in two obvious interfaces that we saw in the preceding exercise: the water table and the sediment–bedrock contact.

Fortunately, the iterative nature of the process lends itself nicely to computer solution. The process just described is the one used by REFRACT.

3.5 DIPPING INTERFACES

Although it would be simpler if the subsurface consisted of only horizontal interfaces between rocks and sediments, it would not be very interesting or challenging. Previously in our discussion of the refraction method, we assumed horizontal interfaces. We all can visualize numerous natural situations in which interfaces are inclined or are irregular. Let's examine the case of a single dipping interface and see how we might deal with this first departure from our idealized world.

3.5.1 Analyzing the Problem

To begin, study Figure 3.15. The geology in (a) and (b) is identical except that the interface in (b) dips at 4° down to the left. A refraction survey conducted at (a) produces a travel-time curve with two straight-line segments, which is exactly the same type of plot that a survey conducted at (b) would produce. There is no way to distinguish an inclined interface from a horizontal interface based on this type of field survey and the resulting travel-time curves. How then should we proceed? Although Figure 3.15 does not solve our problem, part (c) contains a clue. The slope of the line labeled "horizontal interface" is equal to $1/V_2$ as you already know. However, the slope of the lined labeled "dipping interface" is less than the prior value, and therefore gives a velocity that is greater than the true velocity. We refer to this as an *apparent velocity*.

Our clue lies in the reason for the different arrival times for the head wave in the two subsurface configurations. Notice in Figure 3.15(c) that the arrival of the refracted wave at each geophone is early relative to similar geophone positions in the case of the horizontal interface. The reason for this is evident when we compare (a) to (b). All rays in (a) returning to the surface from the interface have identical path distances to traverse, and all are longer than those in (b). Because the surface in (b) is inclined upward to the right, each ray returning to the surface through V_1 traverses a shorter distance than the ray immediately to its left. Thus the gap between critically refracted wave *travel times* from the horizontal and inclined interfaces increases up-dip. This accounts for the different slopes in (c). Of course, this observation does not solve our problem; but you might ask, "What would a travel-time curve look like if we reversed the location

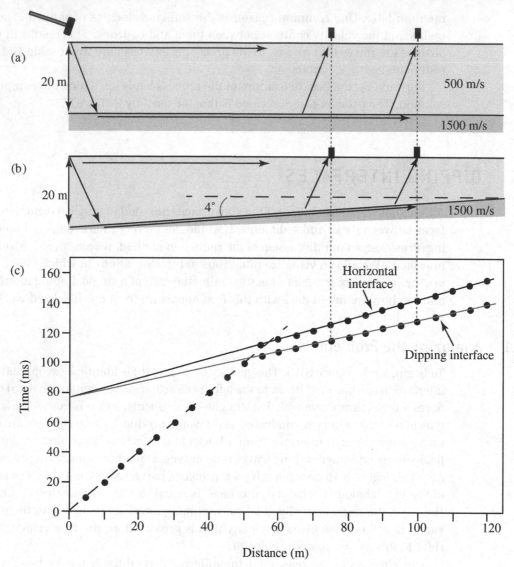

FIGURE 3.15 (a) Horizontal interface at a depth of 20 m with velocities above and below the interface of 500 m/s and 1500 m/s, respectively. (b) A dipping interface with identical depth to that in (a) at the site of the hammer impact and with identical velocities to (a). (c) Travel-time curves for (a) and (b).

of the energy source from the left edge of the diagram to the right edge?" You should be able to predict that it will look different from the horizontal interface case. How will it appear when compared to the curve for the dipping interface that we already examined?

Before attacking this question, we first need to compare the results of an energy source position switch for the case of the horizontal interface. Figure 3.16 illustrates

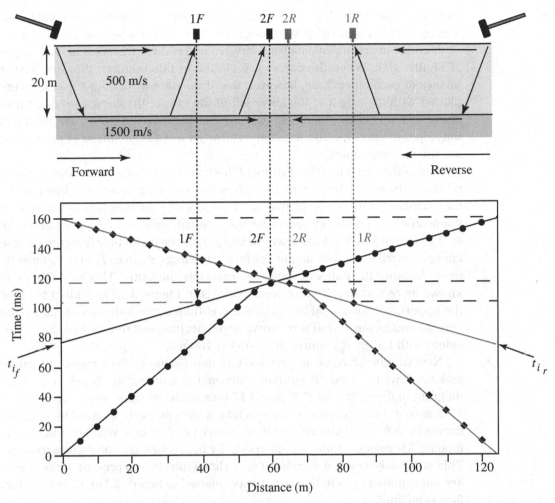

FIGURE 3.16 Correlation of a travel-time curve with wave paths to geophones at equal distances from an energy source for a forward and a reverse traverse.

such a situation. Geophones are distributed along the surface, and an energy source is placed as shown at the left edge of the diagram. After we record travel times from this position, we move the energy source to the other end of the geophone string as shown at the right edge of the diagram and again record times. The first position is called the *forward* traverse because it is the initial position. The second position is termed a *reverse* traverse because the energy source position is reversed relative to the first traverse. The main purpose of Figure 3.16 is to demonstrate that path distances and, hence, travel times are identical from either energy source position to a given geophone interval. For instance, the surface distance from the energy source for the forward traverse to geophone 1F (40 m) is the same as the surface distance from the reverse energy source to geophone 1R. The path distances for both refracted waves are identical, as are their travel times. Such geometry results in travel-time curves for each traverse that are identical.

It is customary to plot data from forward and reverse traverses in the manner of Figure 3.16. When the curves are plotted in this manner, they are mirror images of each other. Note, however, that if the data from both traverses were plotted with the origin at the lower left of the graph, the curves would fit precisely on top of each other. Also observe that because the curves are identical, the slopes of the respective direct and refracted wave segments are the same, as are the intercept times.

One additional important feature of this travel-time curve illustrates a critical relationship to remember when we investigate dipping interfaces. Refer to the time it takes for wave energy that is refracted along the interface to travel from the position of the energy source for the forward traverse to the position of the energy source for the reverse traverse as T_F. T_R is the travel time from the reverse energy source to the position of the forward energy source. T_F and T_R must be equal because the paths of travel are precisely the same. This equivalence is known as *reciprocity*. That reciprocity holds in Figure 3.16 is evident because the uppermost horizontal line on the travel-time curve indicates that both the reverse and forward head wave curve segments intersect the vertical axes (coincident with the energy source locations) at 160 ms.

Now we can return to our previous question: "What would a travel-time curve look like if we reversed the location of the energy source from the left edge of the diagram to the right edge?" Figure 3.17 illustrates the same situation as Figure 3.15 except that a reverse traverse is added. As you likely guessed by now, the curves for a forward and reverse traverse are not the same when an interface is dipping. Therefore, we *always* collect data for both a forward and reverse traverse: This is the only way to determine if an inclined interface is present. If the curves are not symmetric, as in Figure 3.17 (as opposed to Figure 3.16), then the interface is inclined.

The reason that the curves are not symmetric should be clear from Figure 3.17. Geophones 1F and 1R are the same distances from their respective energy sources (40 m) as are geophones 2F and 2R (60 m). Examine the path traveled by the incident and critically refracted waves. If we measure the distance traveled in the V_1 layer to geophone 1F, it is greater than a similar measurement for geophone 1R. This produces a greater travel time as shown on the travel-time curve. This greater travel time for the forward traverse relative to the reverse traverse holds for all source–geophone positions except one. The differential between the forward and reverse times decreases as the source to geophone distance increases. What is the one case for which the times are equal? Recall the requirement of reciprocity. Examine Figure 3.17 and determine to your satisfaction that T_F and T_R, when defined as in our initial discussion of reciprocity, must be equal. Once again these particular path differences are the same, and head wave curve segments intersect the vertical axes at 142 ms for both forward and reverse plots.

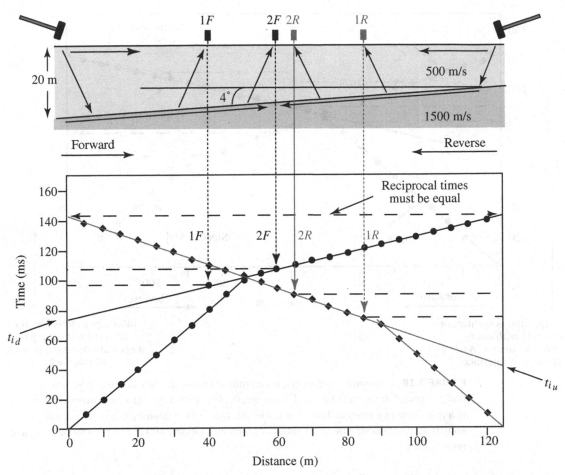

FIGURE 3.17 Correlation of a travel-time curve with geophone positions above a single dipping interface. The purpose of this diagram is to demonstrate the different path distances and arrival times for geophones located at identical offsets for a forward and reverse traverse.

In our discussion of Figure 3.15 we pointed out that the apparent velocity for layer 2 was greater than the true velocity. If we calculate a velocity from the reverse traverse segment, we discover that the apparent velocity is *less* than the true velocity. Actually, if dips are low (<10°), the average of the two apparent velocities is nearly equal to the true velocity.

Before we derive a travel-time equation for the single dipping interface, perhaps a brief review is in order. The main points to keep in mind are summarized in Figure 3.18. Forward and reverse travel-time curves are not symmetric when a dipping interface is present. Velocity V_1 for the layer above the interface is determined in the standard way. V_1 values for the forward and reverse traverses must agree. Reciprocal times must be equal. Inverses of slopes for the head wave arrivals do not give V_2. The average of V_{2F} and V_{2R} approximates V_2 when the dip is low.

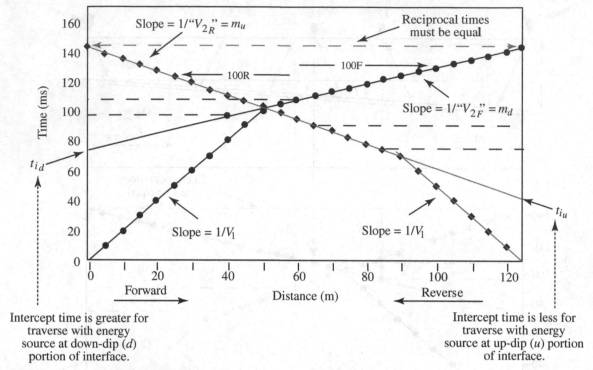

Reciprocal times
must be equal

Slope = $1/$"V_{2R}" $= m_u$

100R

100F

Slope = $1/$"V_{2F}" $= m_d$

t_{i_d}

Slope = $1/V_1$ Slope = $1/V_1$

t_{i_u}

Forward   Distance (m)   Reverse

Intercept time is greater for
traverse with energy
source at down-dip (d)
portion of interface.

Intercept time is less for
traverse with energy
source at up-dip (u) portion
of interface.

FIGURE 3.18 Important features of a travel-time curve for a single dipping interface. Note that reciprocal times must be equal. However, the times from forward and reverse shot points to a geophone at a given distance are not equal. This is illustrated by the arrows labeled 100F and 100R, which designate travel times to geophones located 100 m from the forward (F) and reverse (R) shots.

Slope values are equated to m_u and m_d following the convention for the equation of a straight line: $y = mx + b$ where m designates slope. We can determine the direction of dip of the interface because the intercept time is less for the traverse with its energy source at the up-dip portion of the interface (where the vertical distance to the interface is least). Although this seems like a lot to keep straight, everything is important when we interpret curves related to dipping interfaces.

> To become more adept at working with curves of this type, enter several configurations of velocity, depth, and dip into REFRACT. At first work with only two layers; but when you feel at ease with these curves, attempt some three-layer situations with both interfaces dipping. Following the rules you already know, you should be able to work out similar relationships for this more complicated situation. The convention used in this text for dip angles is that a + angle represents a surface inclined upward toward the right; a − angle represents a surface inclined downward toward the right.

3.5.2 Derivation of a Travel-Time Equation

The derivation of a travel-time equation for the single dipping interface is essentially the same as for the single horizontal interface. The initial equation has a few more terms, and reducing the equation to a more compact form requires a bit more effort. Otherwise everything is familiar.

The symbols employed in the following derivation are identified in Figure 3.19. The quantities h_d and j_d are located at the *down-dip* portion of the interface. In the travel-time equation time$_d$ represents the travel time when the energy source is located in the down-dip position. The intercept time determined from this equation is t_{i_d}, and the slope determined from the derivative of the equation is labeled m_d. Similarly, when the energy source is located at the up-dip portion of the interface, all these symbols take on the "u" subscript as in h_u. You should be aware that this convention is the opposite of that used in most exploration geophysics textbooks. However, we believe that this terminology is easier to keep straight if you relate the subscript to the position of the energy source.

We also note that h is the vertical distance to the interface. This uses h in the same manner as in the case of horizontal interfaces. REFRACT (as well as many other computer programs) refers to this quantity as *thickness*. This is not thickness in the stratigraphic sense but is usually thickness in the sense used by field

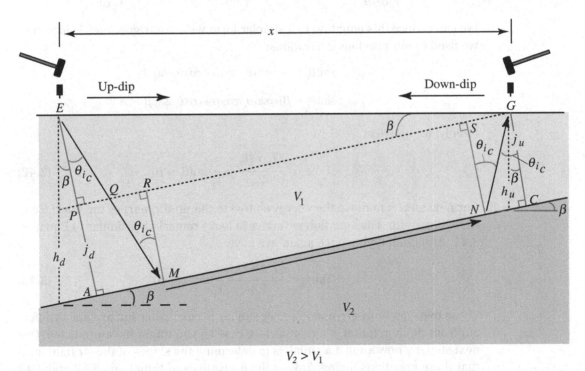

FIGURE 3.19 Diagram illustrating symbols used in derivation of a travel-time equation for a single dipping interface.

explorationists. Thickness as measured perpendicular to layer boundaries always can be computed when the dip is known.

In examining Figure 3.19, recall that θ_{i_c} is measured from the perpendicular to the interface. First we tackle the case when the energy source is located in the down-dip position (we are "shooting" up-dip):

$$\text{Time}_d = \frac{EM}{V_1} + \frac{MN}{V_2} + \frac{NG}{V_1} \tag{3.40}$$

$$EM = \frac{j_d}{\cos\theta_{i_c}}, \ NG = \frac{j_u}{\cos\theta_{i_c}}, \text{ and because } j_u = j_d - EP = j_d - x\sin\beta,$$

$$NG = \frac{j_d - x\sin\beta}{\cos\theta_{i_c}}.$$

Because $MN = PG - AM - NC$, we determine the relevant quantities

$$PG = x\cos\beta, \ AM = j_d\tan\theta_{i_c}, \text{ and } NC = j_u\tan\theta_{i_c} = (j_d - x\sin\beta)\tan\theta_{i_c}.$$

This gives us this basic travel-time equation:

$$\text{Time}_d = \frac{j_d}{V_1\cos\theta_{i_c}} + \frac{x\cos\beta - j_d\tan\theta_{i_c} - (j_d - x\sin\beta)\tan\theta_{i_c}}{V_2} + \frac{j_d - x\sin\beta}{V_1\cos\theta_{i_c}}. \tag{3.41}$$

We can reduce this equation to a simpler form with the trigonometric identities we used in our previous derivations:

$$\sin(\theta_{i_c} - \beta) = \sin\theta_{i_c}\cos\beta - \cos\theta_{i_c}\sin\beta$$

$$\sin(\theta_{i_c} + \beta) = \sin\theta_{i_c}\cos\beta + \cos\theta_{i_c}\sin\beta.$$

Finally, we arrive at

$$\text{Time}_d = \frac{2j_d\cos\theta_{i_c}}{V_1} + \frac{x}{V_1}\sin(\theta_{i_c} + \beta). \tag{3.42}$$

Our next step is to move the energy source to the up-dip part of the interface to shoot down-dip. The equation we arrive at looks remarkably similar to Equation 3.41. After simplifying once again, we have

$$\text{Time}_u = \frac{2j_u\cos\theta_{i_c}}{V_1} + \frac{x}{V_1}\sin(\theta_{i_c} - \beta). \tag{3.43}$$

These two equations are not only very similar to each other but are not as different from the horizontal, single-interface case as you might have predicted. Our next step, by now almost a ritual, is to determine the slopes of the straight lines that these equations define. Taking the derivatives of Equations 3.42 and 3.43 results in

$$\frac{d\text{time}_d}{dx} = \frac{\sin(\theta_{i_c} - \beta)}{V_1} \quad \text{and} \quad \frac{d\text{time}_u}{dx} = \frac{\sin(\theta_{i_c} + \beta)}{V_1}. \tag{3.44}$$

Following the terminology established in Figure 3.18 for slopes and intercept times, we begin by rewriting Equation 3.44 in the form

$$m_d = \frac{\sin(\theta_{i_c} - \beta)}{V_1} \quad \text{and} \quad m_u = \frac{\sin(\theta_{i_c} + \beta)}{V_1}. \tag{3.45}$$

Remembering, as always, that $\sin\theta_{i_c} = V_1/V_2$, we see that we can determine V_2 if we know θ_{i_c}. Of course, we also want to know the dip, β, of the interface. Noting the $+$ and $-$ signs in Equation 3.45, we see that we can solve these slope equations both for β and θ_{i_c}. Therefore, we put these equations in the form

$$\theta_{i_c} - \beta = \sin^{-1}(V_1 m_d) \quad \text{and} \quad \theta_{i_c} + \beta = \sin^{-1}(V_1 m_u). \tag{3.46}$$

Because m_u, m_d, and V_1 are determined directly from the travel-time curve, we solve for θ_{i_c}:

$$\theta_{i_c} = \sin^{-1}(V_1 m_d) + \beta \quad \text{and} \quad \theta_{i_c} = \sin^{-1}(V_1 m_u) - \beta$$

$$2\theta_{i_c} = \sin^{-1}(V_1 m_d) + \sin^{-1}(V_1 m_u)$$

and finally

$$\theta_{i_c} = \frac{\sin^{-1}(V_1 m_d) + \sin^{-1}(V_1 m_u)}{2}. \tag{3.47}$$

Similarly, we determine the relationship for β using

$$\beta = \theta_{i_c} - \sin^{-1}(V_1 m_d) \quad \text{and} \quad \beta = -\theta_{i_c} + \sin^{-1}(V_1 m_u)$$

to arrive at

$$2\beta = \sin^{-1}(V_1 m_u) - \sin^{-1}(V_1 m_d)$$

and

$$\beta = \frac{\sin^{-1}(V_1 m_u) - \sin^{-1}(V_1 m_d)}{2}. \tag{3.48}$$

3.5.3 Determining Thickness

However, we are not yet quite finished. We still need to derive an equation that provides a solution for thickness at both the up-dip and down-dip portions of the interface. At this point we know the critical angle and the velocity of the first layer, so we can express Equations 3.42 and 3.43 in terms of intercept times:

$$t_{i_d} = \frac{2j_d \cos\theta_{i_c}}{V_1} \tag{3.49}$$

and

$$t_{i_u} = \frac{2j_u \cos\theta_{i_c}}{V_1} .$$ (3.50)

Noting from Figure 3.19 that $\cos\beta = j_d/h_d$ and $\cos\beta = j_u/h_u$, we complete our task for the single dipping interface:

$$j_d = \frac{t_{i_d} V_1}{2\cos\theta_{i_c}} ,$$ (3.51)

$$j_u = \frac{t_{i_u} V_1}{2\cos\theta_{i_c}} ,$$ (3.52)

$$h_d = \frac{j_d}{\cos\beta} ,$$ (3.53)

and

$$h_u = \frac{j_u}{\cos\beta} .$$ (3.54)

3.6 MULTIPLE DIPPING INTERFACES

We now could turn to the development of travel-time equations for a situation where more than one interface is present and at least one interface is inclined from the horizontal. The geometrical technique follows the same pattern we used several times already but is much more complex. You would gain relatively little new knowledge by pursuing this in detail, so we simply present the results and encourage you to consult original sources if you are interested in details.

3.6.1 Travel-Time Equation

A clear exposition of the development of travel-time equations for several dipping interfaces is presented by Adachi (1954). Figure 3.20 is based on his work, but the symbols are changed to be consistent with those used in this text. Note that angles θ_i and ϕ_i are measured from the perpendicular to the various interfaces and, therefore, represent angles of incidence and refraction. Angles u_i and d_i are measured from the vertical. These symbols as well as other subscripts follow the same usage as in the case of the single dipping interface.

The equations developed by Adachi are

$$t_d = \frac{x\sin u_1}{V_1} + \sum_{i=1}^{n-1} \frac{h_{di}}{V_i}(\cos d_i + \cos u_i)$$ (3.55)

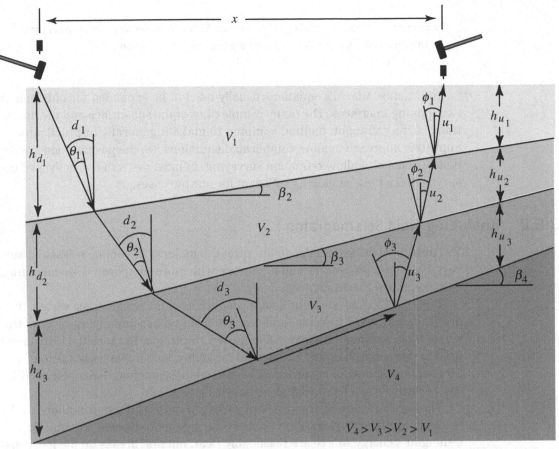

FIGURE 3.20 Diagram illustrating symbols used in travel-time equation for multiple dipping interfaces (after Adachi 1954).

and
$$t_u = \frac{x\sin d_1}{V_1} + \sum_{i=1}^{n-1} \frac{h_{ui}}{V_i}(\cos d_i + \cos u_i) \qquad (3.56)$$

where n = the number of interfaces. We solve these equations by combining the procedures for the single dipping interface and multiple horizontal interfaces. Intercept times are determined for each straight line on the travel-time curve. The same techniques are applied to determine slopes, angle of dip, and critical angle. These quantities then are used to solve for V_2, h_{d_1}, and h_{u_1}. Once we know all quantities related to the shallowest interface, we expand Equations 3.55 and 3.56 with $i = 2$, and the process begins again. As before, this technique is an especially appropriate computer task, and REFRACT is based on these equations.

In practice Adachi's equations usually need to be expanded for only one or two dipping interfaces. The many complexities of the subsurface and the limitations of the refraction method combine to make it generally impracticable to apply this approach to more complicated situations. On the positive side are the typical goals of shallow refraction surveying. In most cases the objectives of the survey make these equations sufficient for our purposes.

3.6.2 Analyzing Field Seismograms

You now have the knowledge to interpret the majority of routine refraction surveys. To review procedures and to practice with more complicated seismograms, let's examine a representative survey.

Figure 3.21 illustrates four field seismograms of the wiggle-trace variety. Two are from a forward traverse, and two are from an accompanying reverse traverse. Field parameters are noted in the figure caption. Because the instrument used in this survey has only 12-channel capability, long traverses requiring 24 or 36 geophones must be recorded in 12-geophone segments. Hence the need for two records for the forward and reverse traverses.

These records are instructive for a variety of reasons. First, conditions at the field site are fairly typical. First breaks on the geophones near the Buffalo-gun ("shotgun") energy source are reasonably clear, but first breaks on the more distant geophones are sometimes difficult to identify. This is common at localities that do not have particularly efficient energy-coupling characteristics (such as dry sand). Computer programs like REFRACT require first break times for every geophone, but these times can be ignored when drawing travel-time curves. You therefore should keep careful records so you know which first break times you have more confidence in. These should guide your interpretation. Once again, remember that obtaining a solution from the travel-time curves is a matter of only a few minutes, so be sure to try several possible fits to your data to determine to what extent your computed subsurface model varies depending on line fits.

Consult the travel-time curve based on our first break picks (Figure 3.22). Segments $1F$ and $1R$ should have the same slopes if the first layer is homogeneous. Their slopes are slightly different but lead to only very slightly different velocity values. Unfortunately, each curve is defined by only one point, so it is difficult to know which line is more reliable. Here is a case where experimenting with interpretations could provide useful information. If layer thicknesses are little affected by the possible extremes in the V_1 values, additional information probably is not warranted. However, if you cannot tolerate the

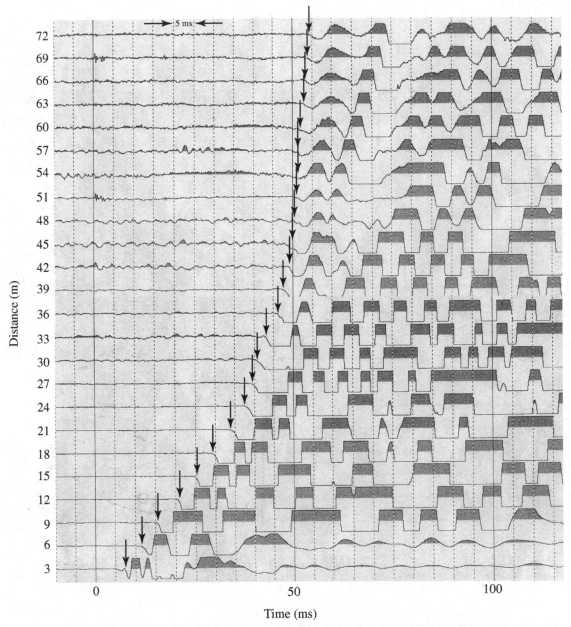

FIGURE 3.21 (a) Field seismogram from the Connecticut Valley, Massachusetts. Forward traverse. First breaks for each trace are indicated by a downward-directed arrow.

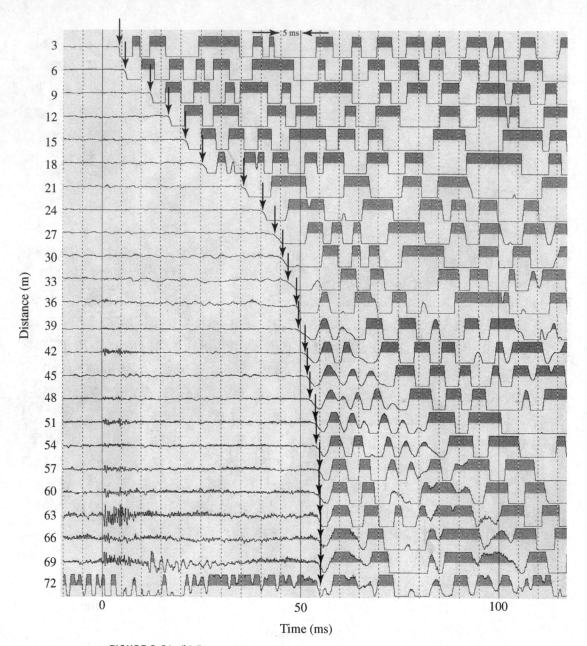

FIGURE 3.21 (b) Reverse traverse.

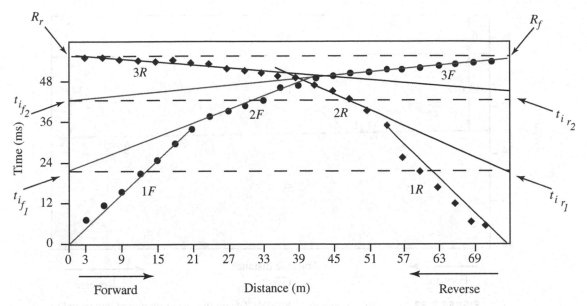

FIGURE 3.22 Travel-time curve based on data from Figure 3.21.

error in thickness interpretations due to this uncertainty, the obvious solution is to place a string of geophones with a smaller interval at the beginning and end of the traverse to record more direct wave arrivals and hence to determine velocities to closer tolerances.

Although some scattering of points exists, segments 2F and 2R are reasonably well defined. Because t_{if1} and t_{ir1} are almost equal, the first interface is essentially horizontal. Note that these two segments must satisfy the principle of reciprocity, even though the intersections of their extensions with the vertical axes are not shown. The data points defining segments 3F and 3R have considerable scatter, but the majority fit on the lines as drawn in Figure 3.22. These two lines satisfy reciprocity as indicated by R_r and R_f and the horizontal dashed line joining these points. However, 3F and 3R have different slopes, and t_{if2} is less than t_{ir2}. This indicates that the second interface is dipping and that the up-dip or shallowest part is located at the energy source position used for the forward traverse. Note, however, that if a third interface is present and is dipping, we cannot use this simple check to determine its direction of dip because the dip of the second interface affects the intercept times for the third interface.

If you examine the structure computed from this positioning of curve segments, you will see that our analysis is correct. By now you should be able to study curves of this complexity and to arrive at the same general conclusions. This is a valuable ability: It enables you to plot data and make a generalized interpretation while still in the field. This is the most appropriate time to try to resolve uncertainties and to gather data for a more complete analysis.

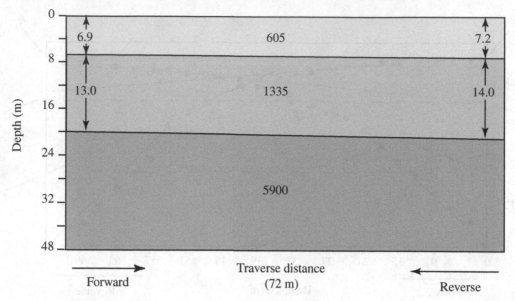

FIGURE 3.23 Structure computed by REFRACT from the curve segments illustrated in Figure 3.22.

You may wish to analyze the seismograms from Figure 3.21 independently. Do not hesitate to substitute your own picks of arrival times or to draw different fits to the data points. In fact, you are encouraged to do so.

3.7 THE NONIDEAL SUBSURFACE

After the previous lengthy introduction to the seismic refraction method, you may think that we have covered most contingencies. Unfortunately, that is nowhere near the true situation. Fortunately, what you have learned is sufficient to cover a great many cases, especially if one is not concerned with detailed subsurface structure but needs only one item of information, such as depth to the water table or bedrock surface. Can you think of some fairly common and simple subsurface configurations that we might add to our repertoire?

One that comes to mind immediately is a lateral change of material. All our models have assumed lateral homogeneity. Another that should occur to you is the assumption of velocity increasing with depth. Although this is true in a general way, especially in shallow subsurface investigations, it often is possible to have a layer with a velocity lower than the one immediately above it. All our derivations also assume planar interfaces, whether horizontal or inclined. What happens if an interface is faulted? Can we deal with cases in which an interface has undulations? Finally, certain combinations of layer thicknesses and velocities can result in critical refractions from a deeper layer arriving before critical refractions from a shallow layer. If this happens, no data points will be available for this shallow layer, and quantities computed from such a travel-time curve will contain errors. We're confident you can think of other situations we should discuss, but these several, either alone or in combination, cover much common subsurface geology. For the sake of time and space, these are the ones we analyze in the following pages.

3.7.1 Hidden Zones: The Low-Velocity Layer

In all our derivations to this point we assumed that each layer has a higher velocity than the layer immediately above it. Although we mentioned that this often is the case, especially in many routine shallow investigations, there is no guarantee that this assumption holds. Most likely you can create reasonable models for which a layer has a lower velocity than the layer above. In such a case with two layers, the consequences are obvious. Critical refraction does not occur; no refracted energy is returned to the surface; first breaks are due to the direct wave; the travel-time curve has one segment only; and we see no evidence that an interface is present.

Three-layer situations provide more combinations for analysis. In the somewhat rare case that a surface layer has a higher velocity than those below, the direct wave traveling through the higher-velocity medium always will arrive before refractions coming from the deeper layers. The basic refraction method will not "see" these layers, and their presence will go undetected unless additional information is available (such as from well logs). Similarly, the deepest layer will not be revealed in a configuration in which $V_1 < V_2 > V_3$. This situation produces a normal travel-time curve for V_1 and V_2, but no critical refraction occurs at the second interface. Although the third layer is undetected, the depth to the first interface will be correct.

Let's now consider the effect of a low-velocity layer sandwiched between layers with higher velocities. Based on the previous discussion, we can understand why the low-velocity layer in the middle of the stack will go undetected. But what is the appearance of a travel-time curve for this arrangement, and what types of errors are introduced into our analysis? The ray paths in Figure 3.24 illustrate that no critical refraction occurs at the first interface because the angle of refraction must be smaller than the angle of incidence (Snell's Law). Critical refraction does occur at the second interface, so a plot of time–distance data for this model produces a travel-time curve with two segments (Figure 3.24). The first segment arises from direct wave arrivals. As wave energy is refracted at the first interface, some energy will arrive at the second interface at the appropriate angle for critical refraction. Equation 3.33 still is valid, and therefore the

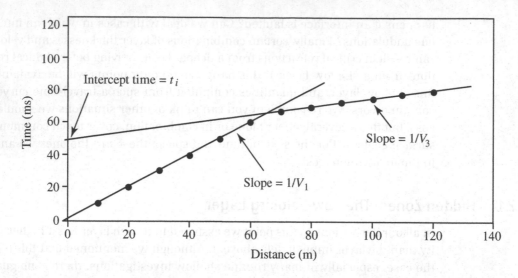

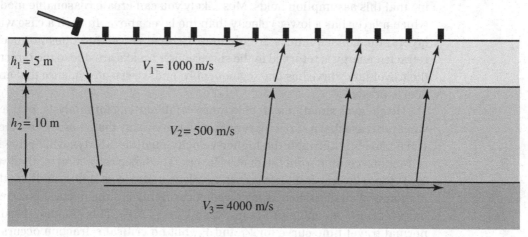

$$V_3 > V_2 < V_1$$

FIGURE 3.24 A plot of time–distance data from Table 3.5 correlated with a diagram of the model used to calculate the data. Note that there is no head wave generated at the V_1–V_2 interface because $V_2 < V_1$. Therefore, there is no indication of the V_2 layer on the travel-time curve.

slope of the second line segment must be equal to $1/V_3$. If the travel-time curve in Figure 3.24 was the only information available to an interpreter, the person would have no evidence for the presence of the low-velocity layer and would compute a two-layer model with a depth to the top of the V_3 layer of 25.5 m. This depth obviously is incorrect because the presence of the V_2 layer did not enter the computation. But what is the magnitude of the error, and is the depth too large or too small?

REFRACT assumes increasing velocity with depth, so we cannot use this program to create a travel-time curve to analyze the problem. However, Table 3.5 provides the necessary information. Table 3.5 contains the model and the time–distance data illustrated in Figure 3.24. The depth to the top of the V_3 layer is the sum of h_1 and h_2, which is 15 m. Thus the depth based on the two-segment curve is greater than the true depth to the top of the V_3 layer. You could continue to investigate this problem by utilizing Table 3.5 as a dynamic table. Enter a model in Table 3.5, type the resulting time–distance data into REFRACT, use REFRACT to compute a model, and compare the result with the original model you entered into Table 3.5. In each case you investigate, you will find that

TABLE 3.5	REFRACTION TIMES FOR A THREE-LAYER CASE WITH A LOW-VELOCITY INTERMEDIATE LAYER

Distance (m)	Direct Wave(ms)	Interface 2 Refraction (ms)
10	10.0	51.9
20	20.0	54.4
30	30.0	56.9
40	40.0	59.4
50	50.0	61.9
60	60.0	64.4
70	70.0	66.9
80	80.0	69.4
90	90.0	71.9
100	100.0	74.4
110	110.0	76.9
120	120.0	79.1

V_1 (m/s)	1000
V_3 (m/s)	4000
h_1 (m)	5
h_2 (m)	10
Interval (m)	10

Intercept time (ms)	49.4

TABLE 3.6	COMPUTATION OF THICKNESS FOR A LOW-VELOCITY SECOND LAYER

V_1 (m/s)	1000	Intercept time (ms)	49.4
V_2 (m/s) (assumed)	500		
V_3 (m)	4000		
h_1 (m) (assumed)	5	h_2 (m)	10.0

the incorrect depth always is greater than the actual depth. The reason for the error is straightforward. A two-segment travel-time curve warrants use of Equation 3.17. The vertical thickness of the first layer, h_1, is directly proportional to t_i. However, the travel time of the critically refracted wave is slowed by passage through V_2, and these arrival times are later than they would be if the entire distance was through V_1 material. Hence the intercept time is greater, and it follows that the calculated depth to the interface is greater than the actual depth.

Unless additional evidence is available, the low-velocity layer will remain undetected. If, however, stream cuts, road cuts, well logs, or other geophysical methods demonstrate the presence of a low-velocity layer in the geologic section, it is possible to approximate the depth to the second interface using a travel-time curve such as the one in Figure 3.24. This travel-time curve provides values for V_1 and V_3. We demonstrated that Equation 3.33 remains valid in the case of the "sandwiched" low-velocity layer. This substantiates that the intercept time, t_i, in Figure 3.24 is correct and is equivalent to t_{i_2} in Equation 3.35. If our independent evidence provides a good estimate for h_1 and V_2, we need only solve Equation 3.35 to determine h_2. Table 3.6 provides a means for testing this assertion. It is a dynamic table and uses the values just discussed to calculate h_2. Values used in the table are the same as those used in Table 3.5 and Figure 3.24.

Explore the sensitivity of h_2 to the magnitudes of V_2 and h_1. Note that V_1, V_3, and the intercept time are related and should not be changed independently. If you want to investigate additional values, use Table 3.5 as a dynamic tool to determine values for a new model. Then insert these into Table 3.6.

Because the presence of a low-velocity layer is a common possibility, every effort should be made to obtain independent information about the subsurface where you are working. The inability of the refraction method to detect low-velocity layers is one of its major shortcomings.

3.7.2 Hidden Zones: The Thin Layer

Unfortunately, a low-velocity layer is not the only circumstance that results in a layer escaping detection by seismic refraction methods. Thin layers or layers that have a low-velocity contrast with a layer below also will escape detection. Let's try to determine why this occurs and what type of error results. A dynamic table (Table 3.7) and REFRACT are useful for our purposes. First enter the following model into REFRACT or into Table 3.7: $V_1 = 500$ m/s, $V_2 = 1400$ m/s, $V_3 = 4500$ m/s, $h_1 = 10$ m, and $h_2 = 15$ m. REFRACT produces a classic three-segment travel-time curve, and a solution to these data using REFRACT yields values that are identical to the starting model. Now change the value of h_2 to 5 m. This time the result is a two-segment curve even though a three-layer model was entered.

TABLE 3.7	REFRACTION TIMES FOR A THIN INTERMEDIATE LAYER		
Distance (m)	Direct Wave (ms)	Interface 1 Refraction (ms)	Interface 2 Refraction (ms)
5	10	40.93	47.65
10	20	44.50	48.76
15	30	48.08	49.87
20	40	51.65	50.99
25	50	55.22	52.10
30	60	58.79	53.21
35	70	62.36	54.32
40	80	65.93	55.43
45	90	69.50	56.54
50	100	73.08	57.65
55	110	76.65	58.76
60	120	80.22	59.87

V_1 (m/s)	**500**
V_2 (m/s)	**1400**
V_3 (m/s)	**4500**
h_1 (m)	**10**
h_2 (m)	**5**
Interval (m)	**5**

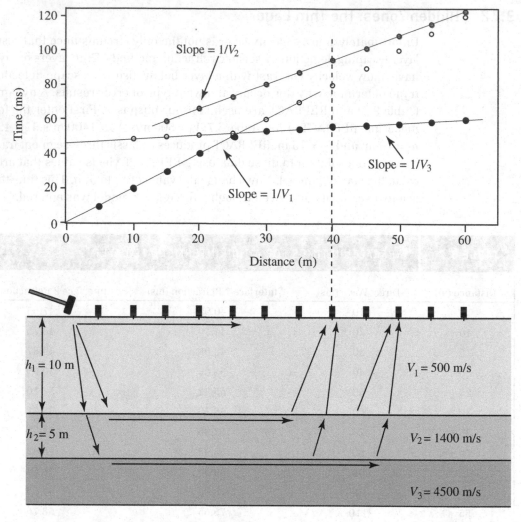

FIGURE 3.25 A thin intermediate layer that is not detected by refraction. The time–distance data from Table 3.7 are correlated with a diagram of the model used to calculate the data. No head waves for the first interface arrive before the direct wave or the head wave from the second interface due to the V_2–V_3 velocity contrast and the thinness of the V_2 layer.

First study Table 3.7 and then inspect Figure 3.25. All head waves from the first interface arrive later than the head waves from the second interface. Usually we can identify only the first energy to arrive at a geophone, so the later-arriving waves will not be identified. Hence we observe a two-layer case when we have a three-layer geologic section. If we use the time–distance data from our second model in REFRACT, we find that the computed vertical thickness to the top of the V_3 layer is 11.7 m. This is less than the true value of 15 m.

The hidden-layer problem due to a thin layer yields a depth value less than the true depth, whereas the low-velocity layer resulted in a depth value greater than the true value. Why? Once again, a two-segment travel-time curve warrants use of Equation 3.17. The vertical thickness of the first layer, h_1, is directly proportional to t_i. However, the travel time of the critically refracted wave (second interface) is speeded by passage through V_2, and these arrival times are earlier than they would be if the entire distance was through V_1 material. Hence the intercept time is less, and it follows that the calculated depth to the interface is less than the actual depth.

Do we have any recourse as we did in the case of the low-velocity layer? If we have no independent information that the V_2 layer may be present, then the layer will be missed, and the computed depth will be accepted as accurate when it actually is too shallow. If, as in the case of the low-velocity layer, independent information indicates that we have a three-layer situation, we can calculate a maximum depth. Because we already have a minimum depth, this process at least provides some limits. We have values for V_1, V_3, and the intercept time for the line through head wave arrivals from the second interface. If independent information indicates that another layer, V_2, is present, we likely have a reasonably good constraint on the value of V_2, and therefore we can draw a line with the correct slope on our curve. This line is adjusted as illustrated in Figure 3.26 until it just touches the intersection of the V_1 and V_3 lines (we are assuming that the V_2 arrivals were just a bit too slow to arrive before the V_3 arrivals that were recorded). This gives us the lowest possible intercept time for the head wave from the first interface.

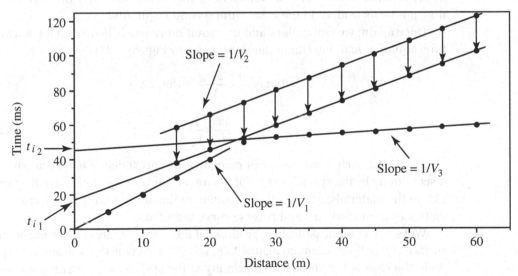

FIGURE 3.26 How to utilize the travel-time curve in Figure 3.25 to calculate a maximum depth to the second interface. The procedure is explained in the text.

You can use this approach in REFRACT. Be sure to draw a V_2 line before drawing the V_3 line because lines are assigned to layers in the order drawn. You then can adjust the slope of the V_2 line until the velocity recorded in a solution is correct.

> Use Table 3.7 and/or REFRACT to investigate a hidden layer due to a low velocity contrast between the first and second layers in a three-layer sequence. Once you obtain a model that produces a two-segment curve, use REFRACT to determine the error in the depth value to the second interface.

3.7.3   Laterally Varying Velocity

Lateral variations in materials at the surface, at least in a minor way, may well be the rule rather than the exception. If these variations are not extreme, surface velocities will not vary significantly, and calculated depths to interfaces will have only small errors. You can gain a sense of the effect of small variations by changing the slope of the direct wave arrivals in REFRACT.

In this section we analyze the effect of an abrupt, vertical change in surface materials (Figure 3.27). The travel-time curve in Figure 3.27 illustrates that a field survey shooting in the direction of the higher-velocity unit sees a curve analogous to that produced by a single horizontal interface. However, a reverse traverse produces a very unusual curve because the first segment has a lower slope than the second segment. It should be clear why the slopes for the first segments of the forward and reverse traverses have the values that they do. However, it may not be as evident in the case of the second segments.

Once again, we can understand the result more easily if we develop a travel-time equation and determine the slope (refer to Figure 3.27):

$$\text{Time} = \frac{y}{V_1} + \frac{x-y}{V_2} \quad \text{(for } x \geq y\text{)} \tag{3.57}$$

$$\frac{dt}{dx} = \frac{1}{V_2}. \tag{3.58}$$

Therefore, in such a case you can recognize the presence of a vertical surface discontinuity by the special pattern of the travel-time curves; determine the velocities of the materials; and locate the position of the discontinuity by the crossover distances of the forward and reverse curve segments.

Will such a unique pattern be produced if the discontinuity is buried beneath a surface layer? If you examine Figure 3.28, you will conclude that the answer is yes. As in the case of the vertical discontinuity at the surface, a reversal in the slope patterns is produced by the sense of the change in velocities between a forward

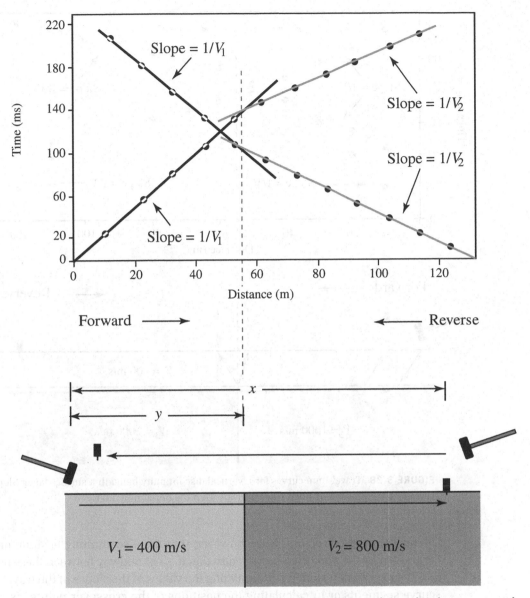

FIGURE 3.27 The effect of a vertical velocity discontinuity on forward and reverse travel-time curves. Note the distinct pattern of slope reversals depending on traverse direction. The position of the vertical discontinuity is marked by the crossover point.

traverse and a reverse traverse. In this case a three-segment curve arises due to the layer above the vertical discontinuity. Once again, the slopes of the travel-time curve segments yield all velocities illustrated in the model. The general position of the discontinuity is indicated by the change in slope of the second and third segments

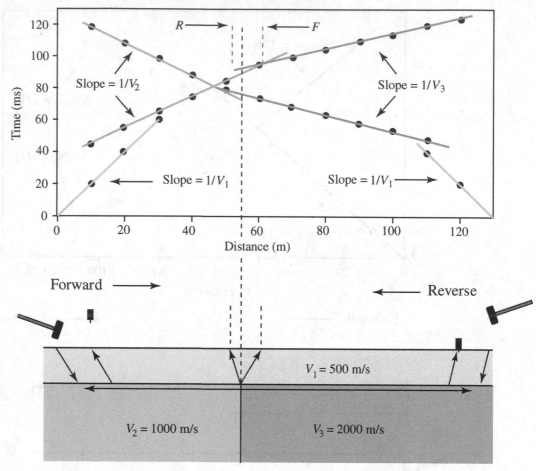

FIGURE 3.28 Travel-time curves for a vertical discontinuity beneath a surface layer. Slope reversals are present for head wave arrivals from the horizontal interface.

of each traverse direction. Note, however, that the discontinuity position only is bracketed by the slope changes and also that it is not midway between these points.

If you are interested in substantiating the values of the slopes of the travel-time curve segments or in calculating the positions of the crossover points, examine Equation 3.8 and rewrite it for the case of the vertical discontinuity. Don't worry about simplifying the equation beyond the form of Equation 3.8.

3.7.4 Interface Discontinuities

In our discussion of interface discontinuities, we confine ourselves to two cases: (1) a sharp, vertical offset in a horizontal discontinuity, and (2) an irregular surface rather than a flat surface. First let's examine the case of the vertical offset, which might represent an erosional step or fault scarp.

Figure 3.29 illustrates ray paths for forward and reverse traverses over a vertical step and resulting travel-time curves. The area of interest is the region surrounding the vertical step because we essentially are dealing with a two-layer situation at significant distances from the step in either direction. First concentrate on the forward traverse. Early arrivals consist of the direct and head waves in the

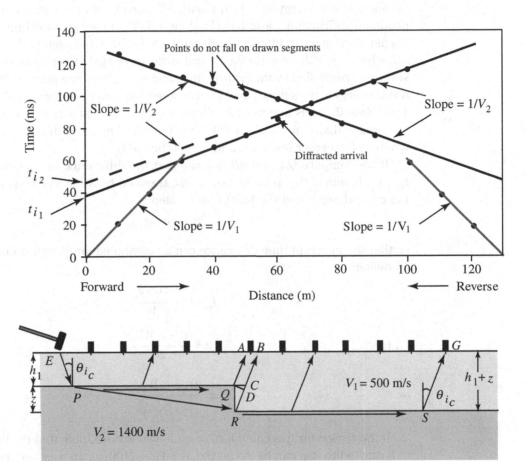

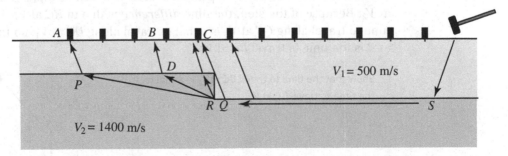

FIGURE 3.29 Ray paths and resulting travel-time curves for forward and reverse traverses over a vertical step. A detailed explanation is given in the text.

classic pattern, producing a two-segment travel-time curve. The last ray path to follow the normal state of affairs is QA. The position of A on the surface is controlled by h_1 and the critical angle, but it is extremely unlikely that a geophone will be located at this exact spot. In Figure 3.29 the last geophone to record head wave energy from the shallower portion of the interface is immediately to the left of A.

Some wave energy follows the path PR, encounters the bottom of the step, produces diffractions, and travels along RS. That energy traveling along the deeper portion of the interface generates waves following such paths as RB and SG, which also will return at the critical angle. Once again the position of B on the surface is controlled by the critical angle and $h_1 + z$. At some point to the right of B this energy arrives first. These arrival times plot along a straight line with slope $1/V_2$. Note that this line must be displaced later because ray paths such as SG are longer than ones such as QA. This later line has an intercept time of t_{i_2}, whereas the earlier line has an intercept time of t_{i_1}.

If we compare QA and RB, we see that the difference in intercept times, $t_{i_2} - t_{i_1}$, is due to the extra distance, RD, traveled by RB at V_1. RD is related to the critical angle and the height of the step z

$$RD = z\cos\theta_{i_c} \tag{3.59}$$

so that the intercept time difference can be computed and z determined by the following:

$$t_{i_2} - t_{i_1} = \frac{z\cos\theta_{i_c}}{V_1} \tag{3.60}$$

and

$$z = \frac{(t_{i_2} - t_{i_1})V_1}{\cos\theta_{i_c}} \tag{3.61}$$

or

$$z = \frac{(t_{i_2} - t_{i_1})V_2 V_1}{(V_2^2 - V_1^2)^{\frac{1}{2}}}. \tag{3.62}$$

If the reason for this computation is not yet obvious, note that the time *delay* at B due to the step can be envisaged as follows. If the step was not present, the time delay difference between positions A and B would be due to travel along QC at V_2. Because of the step, the time *difference* is due to RC at V_1. However, the time to travel along QC at V_2 is equal to travel along DC at V_1, so that the *delay* at B is the time to travel RD at V_1.

Show that the time to travel QC at V_2 is equal to the time to travel DC at V_1.

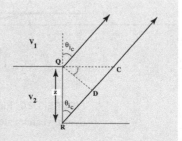

Before proceeding to an analysis of the reverse traverse, we need to point out that the actual arrival pattern is a bit more complex than presented. Diffraction occurs at point Q. These diffracted waves arrive everywhere later than ray QA, but they will arrive earlier at B than ray RB because both are traveling at V_1. At some point to the right of B waves refracted from the deeper portion of the interface will arrive before the diffracted waves. This distance is determined by the velocity ratio and z. Although in Figure 3.29 only one diffracted arrival midway between the early and late refraction segments is illustrated, in practice several such points may be present, making it more difficult to discern the deceptively simple pattern of Figure 3.29. Here we have a case where diffracted waves play a key role (although not a clarifying one) in the form of the travel-time curve.

Naturally, we also collect information from a reverse traverse. In the case illustrated in Figure 3.29, it appears that the step geometry yields two curve segments with slope equal to $1/V_2$ in addition to the direct wave segment. The first of the $1/V_2$ segments is due to head wave arrivals from the deeper portion of the interface. The last such path received at the surface is QC. Wave energy will travel up toward the surface from point R along paths such as RDB and RPA. This energy is refracted upward from the shallower part of the interface, but the angles of refraction do not equal the critical angle. Rays close to the step, such as RDB, will arrive earlier than QC because they follow a somewhat similar path length but with a portion of travel at V_2. Rays such as RPA that travel a considerable distance to the left of R have a portion of their path (RP) at a shallow angle to the interface. The angle of refraction of such rays begins to approximate the critical angle, so that a plot of their arrival times versus distance approximates $1/V_2$.

The offset of the $1/V_2$ segments for the reverse traverse is the opposite of the offset for the forward traverse. Thus the critical criteria for recognizing the presence of the step or fault scarp are the offset $1/V_2$ segments with the offset sense switched for forward and reverse traverses. Remembering these relationships is important—not only for detecting the presence of the step, but for selecting the correct traverse for computing the height of the step. The preceding suggests that under good field conditions the traverse with the shot point located above the shallower portion of the interface is likely to yield superior data.

In all the preceding discussions of the refraction method applied to various subsurface geometries, planar interfaces were assumed. These interfaces produce time–distance data to which straight lines are fit, parameters determined, and geology deduced. Even in the case of a lateral velocity change or an offset interface, the only adjustment to this approach is a single vertical contact. By now you may begin to wonder how one copes with situations when an interface is not planar but possesses an irregular surface with highs and lows—a more realistic geologic scenario. Such situations are considered in the following section.

Before proceeding to the next section, let us call your attention to a valuable resource. Due primarily to space considerations, many variations on the subsurface geometries discussed here are not included. With some careful consideration you should be able to analyze more complicated geometries. What type of travel-time curve would result from a dipping interface with a vertical step? How are time–distance data affected when an interface changes inclination from horizontal to dipping? A good variety of subsurface structures and their associated travel-time curves, as well as analyses of a number of curves, are presented in Mooney (1977, 15-1 to 15-18 and 16-1 to 16-9).

3.8   THE DELAY-TIME METHOD

In this section we take a slightly different approach that ultimately will allow us to map irregular refractor surfaces. Study Figure 3.30(a). Of course, the path $EBCG$ is a familiar one. However, if we concentrate our attention on the interface between V_1 and V_2, which is the refractor surface we are interested in mapping, we might think in terms of the difference in time it takes a wave to travel $EBCG$ as opposed to the time to travel the same distance between source and receiver y in the V_2 layer (path AD). Such a consideration leads to the concept of *delay time*.

Delay time can be defined (Redpath 1973, 10) as the time spent by a wave to travel up or down through the V_1 layer (the slant path) compared to the time the wave would spend if traveling along the projection of the slant path on the refractor (AB or CD in Figure 3.30(a)). For the illustration in Figure 3.30(a) the *total* delay time is

$$T_{EG} = t_R - \frac{y}{V_2} \tag{3.63}$$

where t_R is the total travel time along $EBCG$ or $GCBE$ (recall the requirement of reciprocity). In more advanced analyses this total travel time is referred to as *reciprocal time*. Hence we use the symbol t_R to refer to this quantity. In the development that follows, uppercase T is used to represent delay times, whereas lowercase t is used to refer to travel times that can be measured from time–distance plots.

Because, by definition,

$$T_{EG} = T_E + T_G, \tag{3.64}$$

we can derive an expression for the delay time at a geophone:

$$T_E + T_G = t_R - \frac{y}{V_2}$$

and

$$T_G = t_R - \frac{y}{V_2} - T_E. \tag{3.65}$$

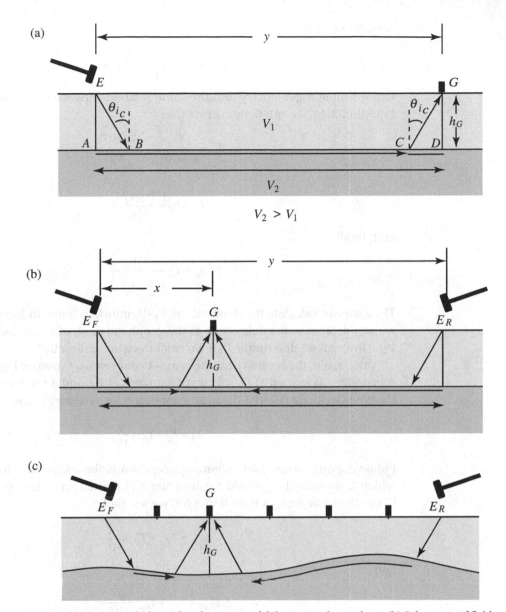

FIGURE 3.30 (a) Symbols used in derivation of delay-time relationships. (b) Schematic of field setup for determining delay times and depths to refractors. (c) Mapping irregular refractors with the delay-time approach.

Also, by definition,

$$T_G = \frac{CG}{V_1} - \frac{CD}{V_2}.$$

(3.66)

Using simple trigonometry and the same relationships we used when deriving Equation 3.14, we can demonstrate that

$$T_G = \frac{h_G}{V_1 \cos\theta_{i_c}} - \frac{h_G \tan\theta_{i_c}}{V_2}$$

$$T_G = h_G \frac{(V_2^2 - V_1^2)^{\frac{1}{2}}}{V_1 V_2}$$

(3.67)

and, finally,

$$h_G = T_G \frac{V_1 V_2}{(V_2^2 - V_1^2)^{\frac{1}{2}}}$$

(3.68)

Thus we can calculate the depth of the V_1–V_2 interface beneath the geophone if we can determine the delay time at the geophone (and, of course, know V_1 and V_2). How can we determine T_G if we can't measure it directly?

Once again, the reverse traverse comes to our rescue. Consider Figure 3.30(b). A geophone is located at G, a forward source is at E_F, and a reverse source is at E_R. We rewrite Equation 3.65 as an expression for reciprocal time:

$$t_R = T_G + T_E + \frac{y}{V_2}.$$

(3.69)

Following our earlier convention, $t_{E_F G}$ represents the travel time from E_F to G, which is measured on a field seismogram and plotted on a time–distance plot. Using the basic form of Equation 3.69, we see that

$$t_{E_F G} = T_{E_F} + T_{G_F} + \frac{x}{V_2}$$

(3.70)

and

$$t_{E_R G} = T_{E_R} + T_{G_R} + \frac{y - x}{V_2}.$$

(3.71)

If we assume that the refractor surface is planar beneath G, then

$$T_{G_F} = T_{G_R}, \quad \text{and}$$

$$t_{E_F G} + t_{E_R G} = T_{E_F} + T_{E_R} + 2T_G + \frac{x}{V_2} + \frac{y - x}{V_2}.$$

(3.72)

Because of the requirement of reciprocity,

$$t_R = T_{E_F} + T_{E_R} + \frac{y}{V_2},$$

and therefore

$$t_{E_FG} + t_{E_RG} = t_R - \frac{y}{V_2} + 2T_G + \frac{y}{V_2}$$

$$t_{E_FG} + t_{E_RG} = t_R + 2T_G$$

and

$$T_G = \frac{t_{E_FG} + t_{E_RG} - t_R}{2}. \qquad (3.73)$$

Because we now have an expression for T_G that is composed entirely of quantities that are obtainable from field observations, we can use Equation 3.68 to calculate h_G. This same procedure can be repeated for each geophone along a typical spread, so that, in theory at least, we can map the form of a refractor's surface by using as many geophones as required data points (Figure 3.30(c)).

Several assumptions are included in the development of the delay-time method that limit its applicability. We assumed (1) that a planar refractor surface lay immediately beneath the geophone and the energy source when we derived expressions for delay times, and (2) that distances along the refractor were equivalent to surface distances x and y. The first assumption effectively limits the application of this method to irregular refractor surfaces only if the relief on the refractor is small in amplitude relative to the average refractor depth (Kearey and Brooks 1984, 124; Redpath 1973, 26). The second assumption limits the method's application to refractor dips of less than 10° (Kearey and Brooks 1984, 125).

Perhaps the best way to visualize application of the delay-time method is to work with a specific example. Table 3.8 (a dynamic table) lists forward and reverse times for an irregular refractor similar to that illustrated in Figure 3.30(c). Each set of forward and reverse times for a specific geophone distance is equivalent to the two energy source-to-geophone times in Equation 3.73. Reciprocal time is the time from the forward source position to the reverse source position. These times and V_1 and V_2 are entered into Table 3.9 (also a dynamic table), which uses Equation 3.73 to compute delay times and Equation 3.68 to calculate depths. Because Table 3.9 is a dynamic table, you can also use it to perform depth calculations for actual field examples. Although Tables 3.8 and 3.9 aid in visualizing the steps in depth computations, we should apply the delay-time approach to an actual travel-time curve.

Figure 3.31(a) is a time–distance plot of the times calculated in Table 3.8, but only the times that would appear as first arrivals are plotted. The only positions where depths can be determined are enclosed in a dashed outline; these are the only geophone positions where travel times are available from both the forward and reverse sources. These few points limit the amount of information about the refractor. Usually we can obtain additional points by increasing the number of geophones, but we must be careful not to use arrivals from a deeper refractor. Strategies for dealing with these problems are covered clearly and in detail in Redpath (1973).

Recall that in the delay-time method the quantity h_G is defined as perpendicular to the refractor surface. Therefore, once depths beneath each geophone

TABLE 3.8 | HEAD WAVE ARRIVAL TIMES FOR AN IRREGULAR REFRACTOR

Position	Distance (m)	Depth (m)	Direct Wave(ms)	Forward Time (ms)	Reverse Time (ms)
Forward Source	0	15	0.00		55.20
Geophone 1	10	14	7.14	21.91	52.14
Geophone 2	20	13	14.29	23.62	49.09
Geophone 3	30	12	21.43	25.33	46.03
Geophone 4	40	13	28.57	28.38	44.33
Geophone 5	50	13	35.71	30.76	41.94
Geophone 6	60	14	42.86	33.82	40.24
Geophone 7	70	15	50.00	36.87	38.53
Geophone 8	80	16	57.14	39.92	36.82
Geophone 9	90	17	64.29	42.98	35.11
Geophone 10	100	19	71.43	46.71	34.08
Geophone 11	110	20	78.57	49.76	32.37
Geophone 12	120	21	85.71	52.82	30.67
Reverse Source	130	21	92.86	55.20	

V_1 (m/s)	**1400**			
V_2 (m/s)	**4200**		Reciprocal time (ms)	55.20
Geophone interval (m)	**10**			

are determined, the refractor can be mapped by drawing arcs with a radius equal to the calculated depths and constructing a smooth curve that is tangent to these arcs (Figure 3.32).

If you have entertained a nagging suspicion that something is slightly amiss, congratulations! Head wave times from an irregular refractor will not lie along a straight line in a time–distance plot, and therefore V_2 cannot be determined accurately. Lacking a good value for V_2, we cannot calculate h_G. How can we circumvent this problem? If we rewrite Equations 3.70 and 3.71, we can arrive at the form

$$t_{E_FG} - t_{E_RG} = T_{E_F} + T_{G_F} + \frac{x}{V_2} - T_{E_R} - T_{G_R} - \frac{y-x}{V_2} \tag{3.74}$$

and

$$t_{E_FG} - t_{E_RG} = T_{E_F} - T_{E_R} + \frac{2x}{V_2} - \frac{y}{V_2}. \tag{3.75}$$

TABLE 3.9	REFRACTOR DEPTHS COMPUTED USING DELAY TIMES			
Position	Forward Time (ms)	Reverse Time (ms)	Delay Time (ms)	Depth (m)
Geophone 1	21.9	52.1	9.4	14
Geophone 2	23.6	49.1	8.8	13
Geophone 3	25.3	46.0	8.1	12
Geophone 4	28.4	44.3	8.8	13
Geophone 5	30.8	42.0	8.8	13
Geophone 6	33.8	40.2	9.4	14
Geophone 7	36.9	38.5	10.1	15
Geophone 8	39.9	36.8	10.8	16
Geophone 9	43.0	35.1	11.5	17
Geophone 10	46.7	34.1	12.8	19
Geophone 11	49.8	32.4	13.5	20
Geophone 12	52.8	30.7	14.1	21

V_1 (m/s)	1400
V_2 (m/s)	4200
Reciprocal time (ms)	55.2

Equation 3.75 demonstrates that travel-time differences plotted against distance yield a line with a slope of $2/V_2$. Figure 3.31(b) is a graph of this type using the times and distances in Table 3.8. Using the calculated slope of 0.476, we arrive at a value for V_2 of 4202 m/s.

The delay-time method represents a somewhat different approach to using typical refraction data and under good conditions provides a much more detailed picture of a refracting interface than the basic approach discussed first in this chapter. However, the delay-time method is fraught with the same difficulties as most commonly employed refraction methods, the most obvious of which is the hidden layer problem.

In the next section we briefly mention other, more advanced refraction methods. Although these may circumvent some of the limitations of the approaches already discussed, they are more complex in design and implementation. They are presented primarily to show additional possibilities.

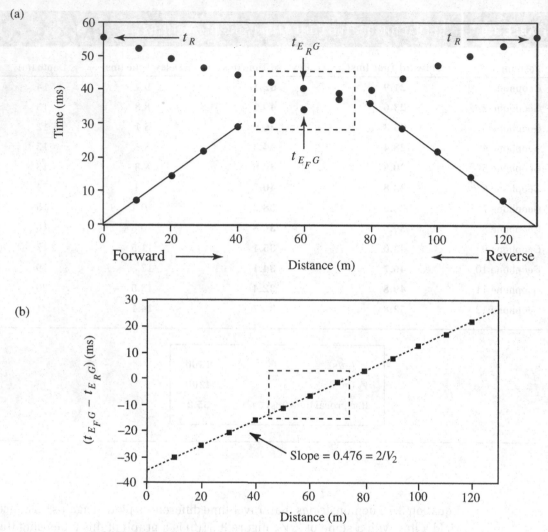

(a)

(b)

FIGURE 3.31 (a) A plot of the direct wave and forward and reverse head wave times listed in Table 3.8. (b) A graph of the difference between forward and reverse travel times for each geophone plotted against distance. The slope of the line fit to these points is used to calculate V_2.

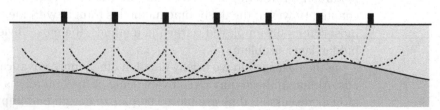

FIGURE 3.32 A refractor surface is constructed from delay-time depths by drawing an arc from each geophone location. The arc's radius is equal to the calculated depth at that position.

3.9   OTHER METHODS

3.9.1   Wave Front Method

The wave front method was developed by Thornburgh (1930) and uses wave front construction to locate points on refracting surfaces. The method is simple to apply in the case of horizontal or dipping planar interfaces but is most useful for delineating irregular refractor surfaces. However, its application in the case of more complex surfaces is tedious. Because of the time involved in constructing these wave fronts, the method has been replaced by other, more powerful methods such as the generalized reciprocal method mentioned in the next section.

Nevertheless, it is instructive to illustrate how the wave front method is applied in a simple case because this shows yet an additional way to view arrival times and to work with seismic data. Consider the planar horizontal interface illustrated in Figure 3.33(a). Refracted energy arriving at the geophone G travels a path $E_F ABG$. The other end of the planar wave front arriving at G is located on the refractor at point P. By using the definition of the critical angle and simple geometry, it is straightforward to show that the time for a wave to travel from B to G at V_1 equals the time for a wave to travel from B to P at V_2:

$$\sin\theta_{i_c} = \frac{V_1}{V_2} = \frac{BG}{BP} \tag{3.76}$$

and

$$\text{Time}_{BG} = \frac{BG}{V_1} = \frac{BP}{V_2} = \text{Time}_{BP} \tag{3.77}$$

A similar wave front associated with point P resulting from a reverse shot at E_R defines point H on the surface. As in the previous derivation,

$$\text{Time}_{CH} = \text{Time}_{CP}. \tag{3.78}$$

The time to travel $E_F ABP$ plus the time to travel $E_R DCP$ must equal the time to travel $E_F ADE_R$, which is reciprocal time t_R. Taking this relationship and using the identities in Equations 3.77 and 3.78, we see that

$$t_{E_F G} + t_{E_R H} = t_R. \tag{3.79}$$

Knowing V_1 and V_2, we calculate θ_{i_c}. If we have a geophone position with a head wave arrival (G), we can construct a planar wave front such as the one at 40 m in Figure 3.33(b). Given a reciprocal time, we need only subtract to find the unknown time, and hence position H, relative to the energy source at E_R. We then construct a second wave front, which locates point P (Figure 3.33(b)). This procedure can be duplicated for every geophone that records an arrival from the refractor of interest. Although this would be a waste of effort for a horizontal surface, two or more points would delimit a planar, dipping refractor, and many points would map in some detail an irregular refractor. In these latter cases the

(a)

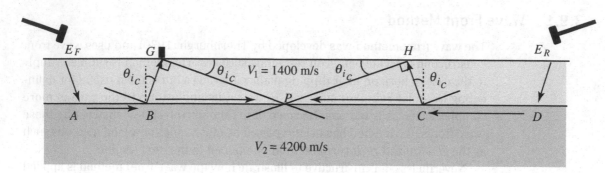

(b)

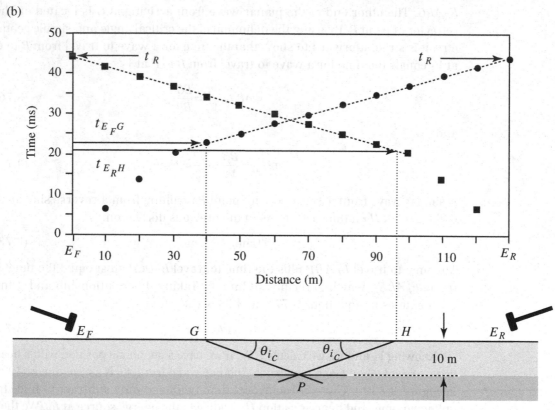

FIGURE 3.33 Diagram illustrating the basic procedure for the wave front method. (a) Relationships used to derive the method. (b) Locating a point on the refractor from the travel-time curve.

wave front construction becomes more complicated and may not be warranted in light of other methods that produce similar results with less effort or superior results with equivalent effort.

3.9.2 Ray Tracing and the Generalized Reciprocal Method

Although computer capabilities were applied for much of the material in this chapter, the processing capabilities of computers barely were challenged. The approaches of ray tracing (Cerveny, Langer, and Psencik 1974) and the generalized reciprocal method (Palmer 1980) use computer capabilities more fully for refraction data from more complicated subsurface situations than have been treated in this text.

In applying the ray tracing method, an investigator proposes a subsurface model based on whatever information is available, including local surface geology, well data, and even refraction data. Ray paths for this model then are calculated by computer and travel times determined. These times are compared with observed times; the model is adjusted with the goal of improving the correspondence between observed and computed times; and new times are computed. This process continues until a realistic model conforms to known geologic information and produces travel times that closely correspond to measured times.

The generalized reciprocal method is a powerful, yet relatively straightforward, approach. It uses reciprocal time and the arrival times at two geophones as does the delay-time method. However, the extensions in analysis that characterize the GRM enable it to deal especially well with irregular refractors because the method can handle dip angles up to 20°. In addition, the GRM can show hidden layers and their probable nature (velocity inversion or thin layer), which helps indicate the nature of depth errors. It also is possible to calculate depths to interfaces below a hidden layer in situations where good control of critical parameters can be achieved.

3.10   FIELD PROCEDURES

3.10.1   Site Selection and Planning Considerations

Once you identify a locality that you will explore using seismic refraction techniques, you should keep a number of considerations in mind. First, you must exhaust all likely sources to determine as much as possible about the local geology, both surface and subsurface, *before* undertaking a refraction survey. Although such advice may seem self-evident, it is amazing how many individuals give this preparation short shrift to get started with seismic data collection. Some may believe that geology can be consulted after geophysical data are in hand, but more effective surveys result from thorough premeasurement preparation.

Assume for a minute that you are exploring for buried valleys cut into bedrock and covered by glacial sands and gravels. If you have 24 geophones available, should you use a cable with 3 m, 5 m, or 10 m take-outs? Your preliminary geologic studies reveal that recorded bedrock depths vary from surface exposures to 70 m. Because you are interested in the areas with greater depths, you might plan to use spread lengths capable of detecting bedrock at depths of from 50 to 70 m. A quick input into REFRACT using a model of 5 m of unsaturated sands and gravels (500 m/s), 65 m of saturated sands and gravels (1400 m/s), and a bedrock velocity of 4500 m/s reveals that the first recorded bedrock refraction is 190 m from the energy source. You want to record a minimum of three or four refractions, so your line length must be at least 210 m; therefore you decide to use a cable with 10 m take-outs. A good rule of thumb for situations with these velocities is to use spread lengths at least three times the target's depth. Although somewhat simplified, this example is intended to illustrate the importance of presurvey planning. In most seismic work, time and funds are limited. You must work as efficiently as possible and be as confident as possible that you are collecting useful data.

Other considerations enter into the placement of geophones. Geophones must be well coupled with the ground. Firm placement of the spike is essential. Often the time and trouble of augering 1 m deep holes, filling them with water to saturate the sediment, and then placing the geophones in the holes yields such improved results that the effort is well worthwhile. If at all possible, lay cable along a line of equal elevation. If geophones are at various elevations, corrections must be made (see the following sections).

Wind and continual traffic must be avoided. Both cause ground motion, which often is of similar or greater amplitude than the waves you are trying to record. Be flexible and return on a calm day or during the night when traffic is at a minimum. Because of some of the inherent uncertainties in refraction surveys, reliable data are essential. Obtaining reliable data under some conditions is next to impossible.

3.10.2 Equipment Considerations

In the majority of what often are termed *shallow* refraction surveys, target depths typically are less than 100 m. Many surveys must operate in populated areas. Results usually influence engineering and/or planning decisions. Contracts tend to be for specific, short-term objectives and cover a limited area of investigation, thus favoring individual and small consulting firms. All these factors directly influence the type of survey that can be conducted, including energy sources, equipment, computing facilities, and time.

At this point it would be a good idea to read once again the material in Chapter 2 addressing energy sources and seismic equipment. When using a signal enhancement seismograph, a hammer source is adequate for many refraction surveys if target depths do not exceed 50 m and saturated sediments are close to the surface. Under more difficult conditions and for greater depths, a minimum of a shotgun or weight-drop source is mandatory. Explosive charges likely would be preferred, but legal, safety, and expense considerations often rule them out.

As demonstrated previously in this section, line lengths for refraction surveys must be at least three times the target's depth. This means that ray paths are long, that much attenuation occurs, and that first break amplitudes are likely to be small. These long path lengths also result in the disappearance of higher frequencies. Even though low-frequency geophones are employed in refraction surveys, the waveform will tend to become smoother and more rounded, making the first break onset even more difficult to recognize. These factors increase the need for careful planning, site selection, and patience in gathering quality data.

If test records contain objectionable noise and the problem cannot be solved by returning to the site at a more propitious time, filters may save the day. Filters are not used as frequently in refraction surveys as they are in reflection surveys; but at sites where noise is a problem, filters may reduce the effect of the noise to some extent. Most modern field seismographs have an oscilloscope display with an option permitting real-time observation of ground motion. Viewing the display while switching filters may help narrow the choice. Another option is to measure dominant noise wavelengths and determine noise frequencies from a test record. Of course, using filters also removes frequency components from signals of interest, which may be more deleterious than noise interference.

3.10.3 Geophone Spread Geometries and Placements

The standard geophone spread is referred to as the *in-line spread* and is illustrated in Figure 3.34(a). The energy source, or shot point, is placed along the same line as the geophone cable and usually is placed at a distance beyond the first geophone that is similar to the take-out interval. The reverse shot is placed at an equivalent position beyond the other end of the spread. This is the arrangement used in many diagrams in this chapter and in REFRACT.

In cases where a single cable length is not sufficient to record arrivals from the target of interest, the geometry illustrated in Figure 3.34(c) is used. The cable is placed at *A* and shot points taken at position 1, then position 2, and then position 3. The cable is moved to *B* and shots taken at positions 1, 2, and 3 once again. Such a plan provides sufficient geophone distribution to record deep arrivals, contains forward and reverse traverses, and has a center shot that is useful in checking variations in velocity and thickness of the first layer along the line of investigation. Of course, this basic plan can be extended to greater lengths if necessary provided sufficient energy is available to record first arrivals at the most distant geophones.

Note that (a) and (c) in Figure 3.34 do not give exact total travel times. Reciprocal times can be estimated from these arrangements, but in cases such as the delay-time method where t_R is included in calculations, Redpath (1973, 33)

(a)   ✳ ■ ■ ■ ■ ■ ■ ■ ■ ■ ■ ■ ✳ ←——— Energy source

(b)   ✳ ■ ■ ■ ■ ■ ■ ■ ■ ■ ■ ■ ✳ ←——— Geophone

(c)   ✳ ■ ■ ■ ■ ■ ■ ■ ■ ■ ■ ■ ■ ✳ ■ ■ ■ ■ ■ ■ ■ ■ ■ ■ ■ ✳
      1   x A 2 B y 3

(d)

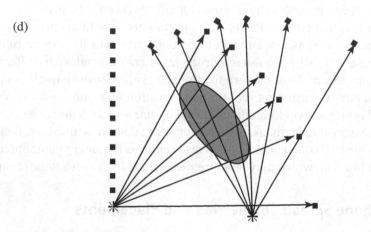

FIGURE 3.34 Geophone spread arrangements. (a) In-line spread. (b) Offset spread. (c) In-line with center shot. (d) Fan shooting.

recommends the offset spread (Figure 3.34(b)). Another viable option, especially where 24 or more geophones are used and the energy source produces minimum damage to the ground (such as a hammer blow), is to disconnect geophone x in Figure 3.34(c) and place shot 1 at that position, then to reconnect geophone x, disconnect geophone y, and place shot 3 at that position.

Several other spread geometries are used and are detailed in advanced texts discussing refraction methods (see the suggested reading). As an example of other possibilities, consider fan shooting (Figure 3.34(d)). Such an arrangement is used to search for a subsurface form with anomalous velocity (the stippled oval in Figure 3.34(d)). Examples include salt domes in earlier, deeper refraction studies (Nettleton 1940) or buried bedrock hills in shallow exploration. A normal refraction line is shot to determine local velocities and depths. This is followed by one or more shots with geophones arranged in a fan shape at equal distances from the shot point. Anomalous increases or decreases in velocity along some shot–geophone lines can help define the general area of the target.

In addition to the geophone spread chosen, placement of the line is important, especially in complex regions. Also, you should always undertake preliminary analysis of data as they are being collected. To illustrate the importance of both of these points, consider the subsurface structure diagrammed in Figure 3.29. Plotting time–distance data in the field would reveal the likely presence of a fault or step. A more complete analysis of the structure might be possible by removing the spread to the left of its present position (in the diagram), taking a shot, and then repeating the process by moving to the right of the diagram.

In the last section of this chapter the "weathering correction" is mentioned. To make this correction reliably, good control of thickness and velocity variations in the surface layer is essential. This can be achieved only by several traverses with relatively small geophone intervals. The main advice here is to plan carefully, continually monitor data, have objectives and data requirements carefully worked out and clearly in mind, and be flexible.

In some cases energy constraints, site access, or other factors may preclude the long spread lengths required to receive refractions from the target. This may not spell disaster depending on the survey requirements. If you have a good grasp of the methods detailed in this chapter, some information always can be salvaged. For example, consider the travel-time curve in Figure 3.35.

Imagine you obtained this curve in an investigation where your goal was to determine bedrock depth. Analysis of the curve suggests a two-layer case with velocities of 800 m/s and 1500 m/s. No bedrock refractions are present, and you

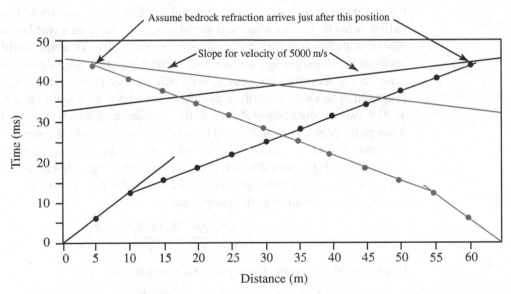

FIGURE 3.35 Travel-time curve illustrating method to determine minimum depth to bedrock when no bedrock refractions are received at geophones. This approach assumes that a bedrock refraction is present just beyond the last position sampled in the survey.

have no time to run a longer spread. Can you deduce anything about bedrock depth? The answer is yes. You can obtain a minimum figure for bedrock depth if you have a reasonably good value for bedrock velocity.

Assume a bedrock refraction immediately after the last geophone in the spread in Figure 3.35. Draw a line with the correct slope for bedrock velocity so that the line barely touches the last point on the curve. Repeat for the reverse traverse. Your solution now will be for a three-layer case, and you will have the minimum depth at which bedrock could be present. In the case of Figure 3.35 the minimum depth is 24 m. This information could be all you require if your assignment was to determine if any bedrock in the area is shallower than 15 m.

> Use REFRACT to create data for a two-layer case (unsaturated sand overlain by saturated sand). Save these data and then read in the data you've saved. Pretend this data set represents field data from a consulting assignment. Assuming a bedrock velocity of 4500 m/s, determine the minimum depth at which bedrock could be present.

Because computer programs such as REFRACT provide easy data input, line fitting, and speedy computations, use them creatively to glean as much information as the data support.

3.10.4 Corrections to Data

If a geophone spread cannot be arranged so all the detectors are at the same elevation, a portion of the observed variations in arrival times will be due to these elevation differences and not to subsurface structure. To understand how corrections for this situation are derived, consult Figure 3.36. Because of the variation in topography along the spread, the energy source and geophones are projected to lie along a common datum. For example, we assume that E is located at A, a vertical distance of Z_E below the surface, and that G is at D. The actual wave path, $EQRG$, is transformed to $APSD$. Travel times in the surface layer are reduced by EB/V_1 and CG/V_1 (denoted by the minus sign in Figure 3.36) and increased by PQ/V_2 and RS/V_2 (denoted by the plus sign). Because $AB = PQ$ and $CD = RS$, we can confine our derivation to triangles EAB and GCD.

The correction time at the energy source is

$$t_{c_E} = \frac{AB}{V_2} - \frac{EB}{V_1} = \frac{Z_E \tan\theta_{i_c}}{V_2} - \frac{Z_E}{V_1 \cos\theta_{i_c}}. \tag{3.80}$$

Using the critical angle, Equation 3.80 simplifies to

$$t_{c_E} = Z_E \frac{\sin^2\theta_{i_c} - 1}{V_1 \cos\theta_{i_c}} \tag{3.81}$$

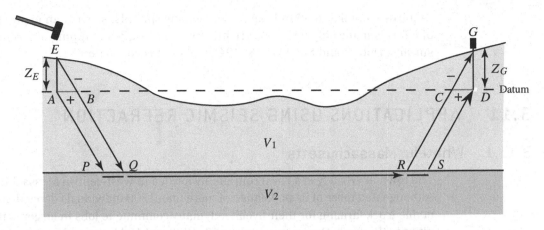

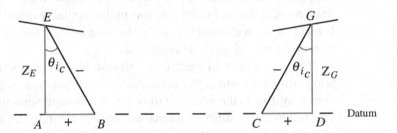

FIGURE 3.36 Symbols and geometry used in the text to derive the elevation correction.

which can be reduced further to

$$t_{c_E} = -Z_E \frac{(V_2^2 - V_1^2)^{\frac{1}{2}}}{V_2 V_1}.$$ (3.82)

The same relationship holds at the geophone position except that Z_G substitutes for Z_E. Thus the total elevation correction is

$$t_{elevation} = -(Z_E + Z_G) \frac{(V_2^2 - V_1^2)^{\frac{1}{2}}}{V_2 V_1}.$$ (3.83)

If a buried shot is used, its depth simply is subtracted from Z_E. If the energy source or geophone is below the datum, the relationship remains the same except that Z_E or Z_G takes on a negative value.

If the surface layer possesses significant variations in thickness, this can introduce unacceptable time variations because this layer's velocity is typically very low. In such cases a correction must be made (often, though misleadingly, referred to as the *weathering correction*). Before such a correction can be introduced, however, good control of surface layer thickness and velocity must be in hand. This requires a number of closely spaced geophone spreads, so such an expenditure of effort is made only when we are interested in detailed mapping of a refractor surface. Dobrin and Savit (1988, 494) explain this correction.

3.11 APPLICATIONS USING SEISMIC REFRACTION

3.11.1 Whately, Massachusetts

The town of Whately is a rural community located in west-central Massachusetts. Although a number of its population of more than 1300 individuals depend on vegetable truck farming for their livelihood, many commute to jobs in larger communities to the south. Over several months in 1983 and 1984, pesticide contamination was discovered in the wells of many homeowners. Because most of the water supply for the community is derived from shallow wells that penetrate an unconfined aquifer, the contamination was understandably viewed with great alarm. The level of concern ultimately resulted in state funding for a study of the contamination problem and possible remedies.

The study was undertaken by a Boston-based engineering firm, which in turn subcontracted some of the subsurface investigations to a geophysical consulting company. Due to the extent of the contamination, it soon became obvious that an alternative water source was mandatory. Fortunately, the local geologic database is good, and some well logs are available. These logs (see Figure 3.37) demonstrate the presence of a sand and gravel layer sandwiched between a thick sequence of glacial-lake clays above and compacted glacial till and/or arkosic bedrock beneath. The thick clay layer forms an impermeable barrier between the deeper sands and gravels and the surface sands that form the contaminated aquifer. The consulting firms reasoned that the deeper sands and gravels might constitute a confined aquifer if areas of sufficient thickness could be discovered. Some prior, but spotty, geophysical investigations suggested the presence of anomalously deep bedrock areas, which might be prime locations for the aquifer thicknesses required.

Seismic refraction and electrical resistivity investigations were initiated in October 1984 and were supplemented by carefully selected drilling sites. Both geophysical methods were unable to differentiate between the clay layer and underlying sands and gravels, but both were able to locate the bedrock surface at all occupied sites. Ultimately, the geophysical survey delineated a buried river valley filled with the lower sequence of sands and gravels to an extent sufficient to constitute a viable water source for the community. This source now is developed and, with sufficient caution, is likely to remain uncontaminated due to the impermeable clay layer.

A typical travel-time curve from the Whately area is presented in Figure 3.38. Although this time–distance plot is from a site where bedrock is much shallower than the 60–70 m depths associated with the buried valley, the curve exhibits

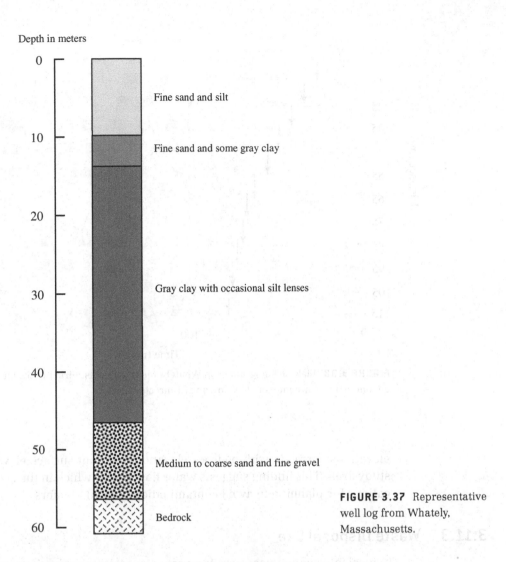

Depth in meters

Fine sand and silt

Fine sand and some gray clay

Gray clay with occasional silt lenses

Medium to coarse sand and fine gravel

Bedrock

FIGURE 3.37 Representative well log from Whately, Massachusetts.

the classic three-layer case typical throughout the area. The first layer is unsaturated sands and alluvial deposits. The second layer consists of saturated sand, gravels, silts, and the lake clays. The third layer typically is bedrock.

3.11.2 Southeastern New Hampshire

A study conducted in New Hampshire in 1974 (Birch 1976) illustrates how seismic refraction studies can be used to map water table elevations as well as bedrock elevations. This study is noteworthy because it was conducted quickly and at low cost and produced useful results. Fieldwork outlined a bedrock trough that could be correlated with a meta-sedimentary unit with relatively low erosional resistance. Analysis of water table elevations led to the conclusion that these

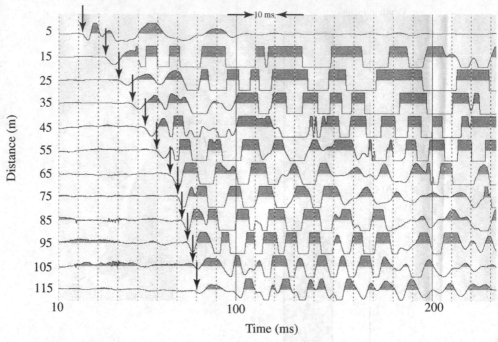

FIGURE 3.38 Field seismogram from Whately, Massachusetts. First breaks for each geophone trace are indicated by downward-directed arrows.

elevations correlate with bedrock elevations and not surface elevations in the study area. This finding suggests water flow paths, which in turn contain implications for planning to avoid contamination of local aquifers.

3.11.3 Waste Disposal Site

Typical information necessary for waste disposal site evaluations includes depth to bedrock, geometry of the bedrock surface, and thickness and velocity characteristics of the overburden. This information often is best gathered by seismic refraction; where quite detailed information is necessary, the GRM should be the method of choice.

Lankston (1990) describes in detail such a study for a municipality considering a waste disposal site at an isolated location near a major river. Figure 3.39 illustrates the information obtained using the GRM. In this study three seismic lines were oriented north–south and three were oriented east–west. Interpretations of depths at line intersections vary by less than 5 percent. Especially striking is the degree of detail obtained concerning the basalt surface as well as the position of lateral velocity variations in the basalt (interpreted by Lankston as layering in the basalt).

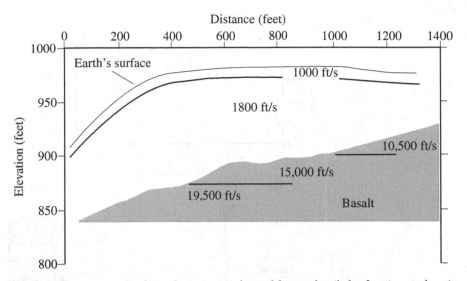

FIGURE 3.39 An example of a geologic section derived from a detailed refraction study using the generalized reciprocal method (GRM). Note that this method provides much detail about the surface of the basalt refractor and is able to determine lateral velocity changes, which are interpreted as layering in the basalt. (Modified from Robert W. Lankston. 1990. High-resolution refraction seismic data acquisition and interpretation. In *Geotechnical and Environmental Geophysics, Volume 1: Review and Tutorial,* ed. Stanley H. Ward, 45–73. Society of Exploration Geophysicists Investigations in Geophysics, no. 5.)

3.11.4   Maricopa Area, Arizona

The area near the Maricopa Mountains in Arizona was one of several locations in the United States proposed as a site for the Superconducting Super Collider and, as such, required that detailed geologic and geophysical information be collected. Gravity, seismic, and electrical methods in conjunction with surface geology and well logs were used to determine the subsurface geology at specific sites within the Maricopa area.

Figure 3.40 illustrates a seismic refraction study from one location at the Maricopa site (Sternberg, Poulton, and Thomas 1990). The refraction travel-time curve (Figure 3.40(a)) is relatively straightforward. The subsurface geology derived from this seismic information is illustrated in Figure 3.40(b). The location of a logged well also is shown. The results of logging cuttings as the well was drilled and the drill rate are diagrammed in Figure 3.40(c). The seismic-based location of the 1000–2130 m/s interface is 35 m at the well location. The lithology log shows this velocity change (sand versus compacted fanglomerate) to be significantly deeper (at about 42 m). However, the drill rate indicates a pronounced change from soft to hard material at 37 m. This likely is the true location of the interface, whereas the lithology log is misleading due to inaccuracies inherent in working with cuttings from loose, soft material traveling to the surface.

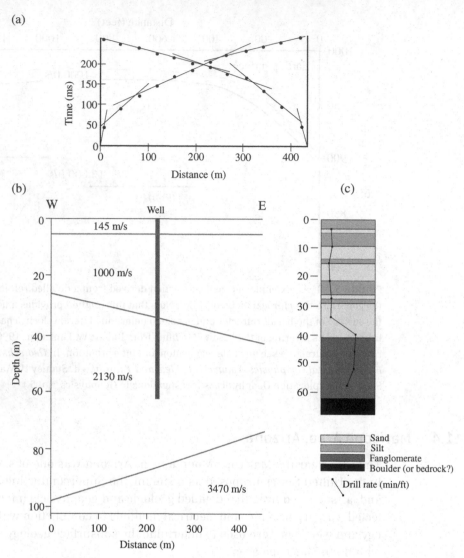

FIGURE 3.40 Seismic investigation of a proposed site for the Superconducting Super Collider. (a) Travel-time curves at site MR1. (b) Interpretation of seismic data. (c). Lithology log and drill rate log at the well location identified in (b). (Modified from Ben K. Sternberg, Mary M. Poulton, and Scott J. Thomas. 1990. Geophysical investigations in support of the Arizona SSC Project. In *Geotechnical and Environmental Geophysics, Volume 3: Geotechnical*, ed. Stanley H. Ward, 211–28. Society of Exploration Geophysicists Investigations in Geophysics, no. 5.)

The well bottomed in very hard material encountered at 62 m. Seismic bedrock is much deeper at this location (approximately 82 m), and it is likely drilling was terminated after striking a large boulder. In general, well information is essential for aiding and constraining geophysical interpretations. In this case the drill rate log provides excellent confirmation of the 1000–2130 m/s interface depth.

However, the lithology log is misleading and provides a cautionary note. One always should use well logs as an additional, but not infallible, source of information. Well logs, especially those from unconsolidated deposits, are only as good as the expertise of the drilling company and the person logging the well.

As we conclude this refraction seismology chapter, we hope you now appreciate the strengths of the method as well as some of its weaknesses. In favorable geologic conditions this approach to exploring the subsurface returns valuable data at reasonable cost. Many of the uncertainties inherent in applying seismic refraction disappear when the method is used in conjunction with other geophysical approaches. Seismic reflection often can discern the presence of low-velocity layers, for example. Your work will best be served if you keep the inherent problems of the method in mind, learn the standard field approaches well enough to design flexible surveys, and always attempt to employ additional geophysical methods. In the next chapter we continue to use seismic techniques via the reflection method. This approach has its own strengths and weaknesses, but many of these tend to complement the refraction approach so that the two, used in conjunction, constitute a formidable pair.

PROBLEMS

3.1 Refer to Figure 3.38. Assuming horizontal interfaces, determine the subsurface geology at this site.

3.2 Figure 3.41 is a field seismograph from a region with horizontal interfaces. Interpret the data.

3.3 Using the information in Figure 3.42, determine as much as possible about the subsurface geology at the site where the information was recorded. Be sure to include comments concerning sources of possible error.

3.4 Calculate the critical distances to the first and second interfaces in the diagram below.

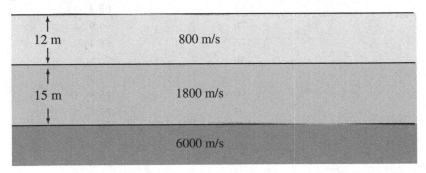

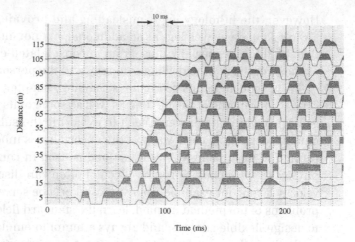

FIGURE 3.41 Field seismogram for Problem 3.2.

3.5 Interpret the following refraction data:

Distance from Shot (m)	Forward Traverse (ms)	Reverse Traverse (ms)
5	7.1	7.1
10	14.3	14.3
15	21.4	18.6
20	28.6	22.0
25	35.7	25.4
30	42.9	28.8
35	48.5	32.2
40	51.0	35.6
45	53.5	39.0
50	56.0	42.3
55	58.4	45.7
60	60.9	49.1
65	63.4	52.5
70	65.9	55.9
75	68.4	59.3
80	70.8	62.7
85	73.3	66.1
90	75.8	69.5
95	77.5	72.8
100	78.3	76.2
105	79.1	77.9
110	80.0	79.0
115	80.8	80.2
120	81.6	81.3

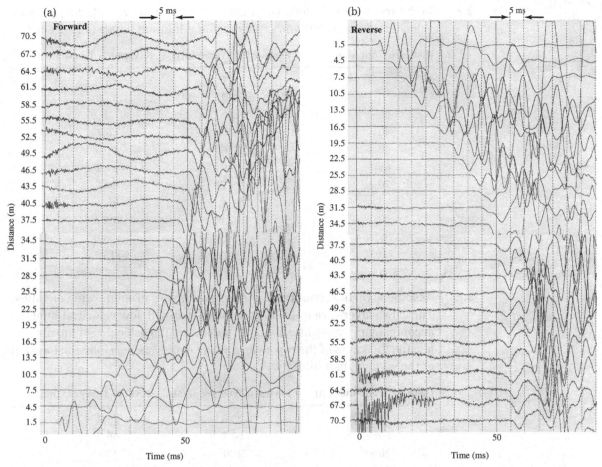

FIGURE 3.42a Forward traverse.

FIGURE 3.42b Reverse traverse.

3.6 Provide a qualitative interpretation for this travel-time curve but be as specific as possible:

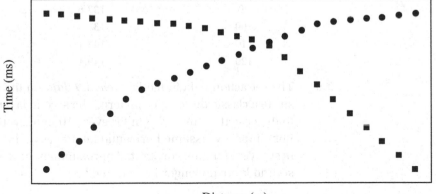

3.7 In the area where the following time–distance data were collected, bedrock velocities average 5000 m/s. What is the minimum depth to bedrock?

Distance from Shot (m)	Forward Traverse (ms)	Reverse Traverse (ms)
10	16.7	16.7
20	33.3	33.3
30	50	50
40	66.7	66.7
50	77.6	77.6
60	83.9	83.9
70	90.1	90.1
80	96.4	96.4
90	102.6	102.6
100	108.9	108.9
110	115.1	115.1
120	121	121

3.8 Assume a vertical contact separates two layers of moderate velocities. These two layers are overlain by a third layer with low velocity. Using the accompanying time–distance data, determine the velocities of the materials, the thickness of the top layer, and, as closely as possible, the location of the vertical discontinuity.

Distance from Shot (m)	Forward Traverse (ms)	Reverse Traverse (ms)
10	33.3	33.3
20	66.7	66.7
30	100	100
40	111.4	119.8
50	114.3	125.1
60	117.3	134.1
70	120.2	137
80	123.1	140
90	133.6	142.9
100	138.9	145.8
110	144.1	148.8
120	149.4	151.7

3.9 The refraction data in file *Problem 3.9 data* on the resources website illustrate a classic three-layer pattern. The second layer in this sequence thins to the east at a rate of 0.5 m for every 10 m of ground distance traversed. For simplicity, assume horizontal interfaces and that the thickness of the first layer remains constant. Approximately how far to the east will the second layer no longer be detected?

3.10 These data are taken from a traverse across a buried vertical discontinuity. Explain why the travel-time curve has the appearance it does. Then determine as much as possible about velocities, the thickness of the overburden, and the location of the vertical discontinuity.

Distance from Shot (m)	Forward Traverse (ms)	Reverse Traverse (ms)
10	12.5	12.5
20	25	25
30	37.5	37.5
40	50	50
50	64.2	62.5
60	69.2	75
70	74.2	80.8
80	79.2	85.8
90	84.2	90.8
100	89.2	96.7
110	94.2	99.2
120	99.2	101.7

3.11 The following refraction data produce a classic two-layer pattern. However, well logs indicate that the typical sequence consists of three layers with a low-velocity layer occupying an intermediate position. The depth to the top of the low-velocity layer remains essentially constant at 3 m. The velocity of the layer is 1200 m/s. Determine the depth to the top of the third layer (bedrock) using only the information on the travel-time curve. Next determine the actual depth to bedrock based on the additional information from well logs. What is the percentage error involved when we use only the information on the travel-time curve?

Distance from Shot (m)	Forward Traverse (ms)
5	2.8
10	5.6
15	8.3
20	11.1
25	13.9
30	16.7
35	19.4
40	22.2
45	25
50	27.8
55	28.7
60	29.5

3.12 Data file *Problem 3.12 data* (located on the resources website) is from a seismic traverse in an area where supplementary information indicates a 2200 m/s layer underlies a 1400 m/s layer. Utilizing this information, what constraints can you place on the depth to bedrock (average velocity = 3900 m/s) in the area?

3.13 The time–distance data presented here were recorded in an area where steep fault scarps in bedrock are overlain by alluvial debris. Deduce as much information as possible from the data.

Distance from Shot (m)	Forward Traverse (ms)	Reverse Traverse (ms)
5	3.6	3.6
10	7.1	7.1
15	10.7	10.7
20	14.3	14.3
25	17.9	17.9
30	21.4	21.4
35	23	25
40	24	28.6
45	24.9	30.4
50	25.8	25.8
55	26.7	26.7
60	27.7	27.7
65	28.6	28.6
70	29.5	29.5
75	30.4	30.4
80	31.4	31.4
85	37.8	32.3
90	38.7	33.2
95	39.7	34.1
100	40.6	35.1
105	41.5	36
110	42.4	36.9
115	43.4	37.9
120	44.3	38.8

3.14 The following time–distance values were recorded from forward and reverse traverses. Because your contract requires detailed information on a bedrock refractor, use the delay-time method to obtain depths and then construct the bedrock surface. Reciprocal time is 40.7 ms.

Distance from Shot (m)	Forward Traverse (ms)	Reverse Traverse (ms)
10	8.3	8.3
20	14.2	16.7
30	16.8	25
40	20.3	30
50	21.3	30.2
60	25.6	33.7
70	28.2	34.7
80	28.4	34.1
90	31.8	36.7
100	34.5	36.9
110	35.5	37.9
120	39.7	40.5

REFERENCES CITED

Adachi, R. 1954. On a proof of fundamental formula concerning refraction method of geophysical prospecting and some remarks. *Kumamoto Journal of Science*, Series A, v. 2, 18–23.

Birch, Francis. S. 1976. A seismic ground-water survey in New Hampshire. *Ground Water*, v. 14, 94–100.

Bonini, W.E., and R.R. Bonini. 1979. Andrija Mohorovicic: Seventy years ago an earthquake shook Zagreb. *EOS*, v. 60, no. 41, 699–701.

Cerveny, V., J. Langer, and I. Psencik. 1974. Computation of geometric spreading of seismic body waves in laterally inhomogeneous media with curved interfaces. *Geophysical Journal of the Royal Astronomical Society*, v. 38, 9–19.

Dobrin, Milton B., and Carl H. Savit. 1988. *Introduction to geophysical prospecting*, 4th ed. New York: McGraw-Hill.

Kearey, P., and M. Brooks. 1984. *An introduction of geophysical exploration*. London: Blackwell Scientific Publications.

Lankston, Robert W. 1989. The seismic refraction method: A viable tool for mapping shallow targets into the 1990s. *Geophysics*, v. 54, 1535–42.

Lankston, Robert W. 1990. High-resolution refraction seismic data acquisition and interpretation. In *Geotechnical and environmental geophysics, volume 1: Review and tutorial*, ed. Stanley H. Ward, 45–73. Society of Exploration Geophysicists Investigations in Geophysics, no. 5.

Mooney, Harold M. 1977. *Handbook of engineering geophysics*. Minneapolis: Bison Instruments.

Nettleton, L.L. 1940. *Geophysical prospecting for oil.* New York: McGraw-Hill.

Palmer, Derecke. 1980. *The generalized reciprocal method of seismic refraction interpreta-tion.* Tulsa, OK: Society of Exploration Geophysicists.

Redpath, Bruce B. 1973. *Seismic refraction exploration for engineering site investigations.* Livermore, CA: Explosive Excavation Research Laboratory, Technical Report E-73-4.

Sternberg, Ben K., Mary M. Poulton, and Scott J. Thomas. 1990. Geophysical investigations in support of the Arizona SSC Project. In *Geotechnical and environmental geophysics, volume 3: Geotechnical,* ed. Stanley H. Ward, 211–28. Society of Exploration Geophysicists Investigations in Geophysics, no. 5.

Thornburgh, H.R. 1930. Wave-front diagrams in seismic interpretation. *Bulletin of the American Association of Petroleum Geologists,* v. 14, 185–200.

SUGGESTED READING

Benjumea, B., T. Teixido, and J.A. Pena. 2001. Application of the CMP refraction method to an archeology study (Los Millares, Almeria, Spain). *Journal of Applied Geophysics,* v. 46, no. 1, 77–84.

Lankston, Robert W. 1990. High-resolution refraction seismic data acquisition and interpreta-tion. In *Geotechnical and environmental geophysics, volume 1: Review and tutorial,* ed. Stanley H. Ward, 45–73. Society of Exploration Geophysicists Investigations in Geophysics, no. 5.

Musgrave, A.W., ed. 1967. *Seismic refraction prospecting.* Tulsa, OK: Society of Exploration Geophysicists.

Palmer, Derecke. 1986. *Refraction seismics.* London: Geophysical Press.

Redpath, Bruce B. 1973. *Seismic refraction exploration for engineering site investigations.* Livermore, CA: Explosive Excavation Research Laboratory, Technical Report E-73-4.

Robinson, Edwin S., and Cahit Coruh. 1988. *Basic exploration geophysics.* New York: John Wiley & Sons.

Rolph, T.C., J. Shaw, E. Derbyshire, and Z.S. An. 1994. Determining paleosol topography using seismic refraction. *Quaternary Research,* v. 42, no. 3, 350–53.

Sain, K., and, K.L. Kaila. 1996. Interpretation of first arrival times in seismic refraction work. *Pure and Applied Geophysics,* v. 147, no. 1, 181–94.

Sheriff, R.E., and L.P. Geldart. 1982. *Exploration seismology, volume 1: History, theory, and data acquisition.* Cambridge, England: Cambridge University Press.

Sundararajan, N., Y. Srinivas, M.N. Chary, et al. 2004. Delineation of structures favorable to groundwater occurrence employing seismic refraction method—A case study from Tiruvuru, Krishna district, Andhra Pradesh. *Proceedings of the Indian Academy of Sciences—Earth and Planetary Sciences,* v. 1113, no. 3, 259–67.

CHAPTER 4

Seismic Exploration:
The Reflection Method

The 1920s witnessed a great deal of fundamental effort in applying seismic theory to exploration for natural resources, particularly petroleum. During this decade developments in equipment and field methods were rapid, so a technique that was relatively unproven at the beginning of the decade was well on its way to assuming the overwhelming prominence that is so evident today. In 1921 the first reflection was received and identified in the field from a shale–limestone contact (Sheriff and Geldart 1982, 6–7), but most early effort concentrated on the refraction method and was directed at the discovery of salt domes. This early dominance of refraction soon declined as the reflection method recorded a number of successes. By the beginning of the 1930s reflection began to replace refraction as the dominant seismic method in petroleum exploration (Sheriff and Geldart 1982, 12). If you are interested in a vivid account of this early history, especially in the details of instrument development and application, we highly recommend the Sheriff and Geldart (1982) account.

You already are aware of some shortcomings of the refraction method: (1) the hidden layer problem, which can be acute when dealing with the geologic sequences encountered in petroleum exploration; (2) the requirement for geophone cable lengths of three to four times the depth of the target horizon—a requirement that obviously becomes untenable when probing to depths on the order of kilometers; and (3) the energy source needed to create recognizable head wave arrivals at distant geophones. In view of these difficulties and due to the dominance of reflection in the petroleum industry, you might expect the same history to hold for shallow seismic exploration. However, we already noted that refraction still is the most widely used technique for seeking information about shallow targets. What is the reason for this different history in the deep and shallow realms?

If you recall the discussion of energy partitioning in Chapter 2, you already may realize that the difficulties caused by the low-velocity layer disappear in reflection work. Also, because reflected waves are received at very short shot–receiver distances (even with distance equal zero), long line lengths are not required. This suggests that the size of the energy source compared to refraction can be much smaller. Nothing in these relationships answers our question, but perhaps a clue lurks in Figure 2.21. Note that reflections always arrive after several other waves. This suggests that reflected energy might be more difficult to recognize than the first arrivals associated with the refraction method. And indeed this is the case. The vast computer resources provided by petroleum exploration companies are used primarily for data processing, which is directed at enhancing reflection arrivals so that they may be recognized and interpreted.

Only within the recent past has the combination of a relatively low-cost, multi-channel, signal enhancement seismograph together with the increased processing capabilities of the personal computer made shallow-target reflection surveys a viable option to refraction. The reason, as before, is that these capabilities provided the impetus to develop field and other techniques to further clarify the arrival of reflected waves from shallow targets. Our strategy in this chapter is somewhat similar to that of the preceding chapter. We begin with geometrical considerations and work toward situations of increasing complexity. After developing these fundamentals, we turn to field techniques and design matters. However, throughout this discussion, we continue to call attention to relationships that ultimately will facilitate our recognition of reflections from shallow interfaces.

4.1 A SINGLE SUBSURFACE INTERFACE

In beginning our analysis of reflected waves, we follow the same general plan as when we discussed refracted waves. First we treat the simple case of a single horizontal interface and develop a travel-time equation for that special case. We graph this equation on a time–distance plot to investigate how the travel-time curve might aid our analysis. We then analyze this equation to see what it might tell us about the subsurface, specifically the depth and velocity to an interface. Finally, we examine seismograms to apply the methods we developed to actual field data.

4.1.1 Using REFLECT

A modeling program such as REFLECT is useful for exploring a number of the parameters involved in understanding patterns of reflections and their relationships to other wave arrivals. REFLECT is similar to REFRACT in that you enter a subsurface model for which arrival times then are calculated. Whereas REFRACT plotted results as data points on a time–distance curve (since this is the basic presentation for refraction analysis), REFLECT produces a simulated field seismogram. This mode of presentation was chosen to begin the process of recognizing reflections on records. It is possible to plot reflections from several interfaces as well as to add arrivals of additional wave types (direct wave, refraction from the first interface, and so on), thereby creating patterns that bear some relationship to those typically encountered on field records or that are useful in elucidating certain strategies employed during fieldwork.

Detailed instructions for using REFLECT are in Appendix B. Although computation times are slightly longer for REFLECT than for REFRACT, we hope the ability to create these simulated seismograms and their usefulness in reinforcing concepts discussed in the text will encourage you to use this program frequently. A number of simplifying assumptions were adopted while developing REFLECT to keep the size of the program manageable and to limit computation time. These simplifications have no bearing on your use of the program for most purposes suggested in this text.

4.1.2 Derivation of a Travel-Time Equation

The derivation of time of travel for a wave reflected from a single horizontal interface is straightforward. Referring to Figure 4.1, we see that

$$\text{Time} = \frac{EA + AG}{V_1},$$

$$EA = AG = \left(\left(\frac{x}{2} \right)^2 + h_1^2 \right)^{\frac{1}{2}},$$

and

$$\text{Time} = \frac{(x^2 + 4h_1^2)^{\frac{1}{2}}}{V_1}. \tag{4.1}$$

If we take the derivative of this equation, we see that we will not be so fortunate as we were in refraction because we don't have a simple way to determine velocity. To understand the form of Equation 4.1 more completely, let's create a table relating reflection travel times to source–geophone distances for a given subsurface model. If we take the values in Table 4.1 and plot them in a standard time–distance graph, we arrive at the travel-time curve in Figure 4.2(a). More points are added for small values of x in the graph than are present in Table 4.1 to define more completely the form of the curve in this region.

The curve form in Figure 4.2(a) is reminiscent of a hyperbola. An equation for a hyperbola that is symmetric about $x = 0$ is

$$\frac{y^2}{a^2} - \frac{x^2}{b^2} = 1. \tag{4.2}$$

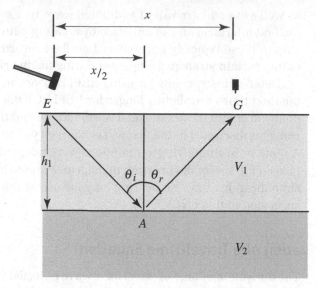

FIGURE 4.1 Diagram illustrating symbols used in derivation of time of travel for reflection.

Equation 4.1 can be rearranged in the form of Equation 4.2, so our initial impression is correct. Of course, the relationship would be even more apparent if we placed geophones along a line symmetrically about $x = 0$ so that the curve in Figure 4.2(a) also extended to the left of the energy source (see Figure 4.20).

> Show that Equation 4.1 can be placed in the form of Equation 4.2. Hint: Square both sides and then divide through by one of the terms that contains only constants.

We can see how the reflections in Table 4.1 would look on a field seismogram by placing the same model into REFLECT (Figure 4.3(a)). The time and distance coordinates for REFLECT plots are arranged to be the same as those on field seismograms generated by most seismographs used in shallow refraction and reflection surveying. To more directly compare the pattern of arriving reflections with the arrangement of standard time–distance curves, we simply rotate the simulated seismogram in Figure 4.3(a) through 90° to arrive at Figure 4.3(b). Although the coordinates are slightly different, the similarity between Figures 4.2(a) and 4.3(b) is evident. Later in our discussion of reflection surveying, we apply corrections to the arrival of reflected pulses to generate a seismic "picture" of the subsurface. In geologic sections depth increases downward, so seismic sections are arranged with time increasing downward. We can obtain a sense of how these appear by flipping Figure 4.3(b) about a horizontal axis to obtain Figure 4.3(c). Note that no corrections have been applied to these data. Also,

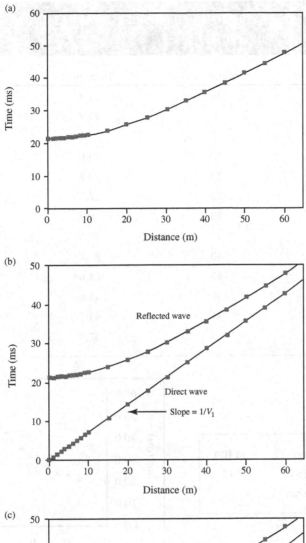

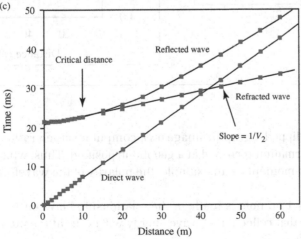

FIGURE 4.2 A plot of reflection time–distance data from Table 4.1. More points are added at short distances from the energy source to establish more clearly the form of the travel-time curve. (a) Reflection arrivals only. (b) Direct wave arrivals added to demonstrate the asymptotic relationship to the hyperbolic form of the reflection arrivals. (c) Head wave arrivals added to (b).

TABLE 4.1

	TRAVEL TIMES FOR REFLECTED WAVES AND NORMAL MOVE-OUT

Geophone	Distance (m)	Time (ms)	NMO (ms)
Source	0	21.4	00.0
1	5	21.7	00.3
2	10	22.6	01.2
3	15	24.0	02.5
4	20	25.8	04.3
5	25	27.9	06.5
6	30	30.3	08.9
7	35	32.9	11.5
8	40	35.7	14.3
9	45	38.6	17.2
10	50	41.6	20.2
11	55	44.7	23.3
12	60	47.9	26.5

Velocity (m/s)	**1400**
Thickness (m)	**15**
Increment (m)	**5**

we've simply flipped a graphic image on a computer screen rather than plotting the digital information recorded at a geophone position. Thus, while Figure 4.3(c) suffices for the moment as an example, the polarity of the waveforms is 180° out of phase.

In Figure 4.1 no note is made of the relationship between V_1 and V_2 for the simple reason that reflection of wave energy will occur at the interface if $V_1 < V_2$

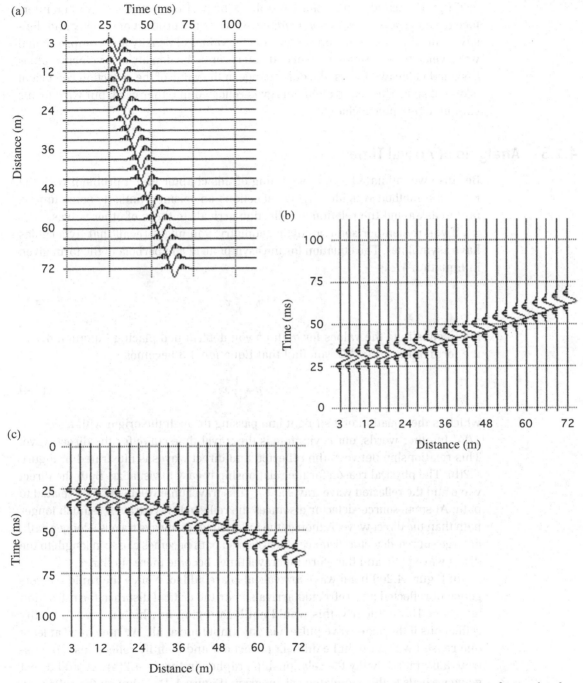

FIGURE 4.3 Three variations in presentation of reflection arrivals on seismogram. (a) A portion of a simulated seismogram produced by REFLECT. (b) Seismogram rotated so axes correspond to conventional time–distance curve. (c) Similar to (b) except time increases downward (as does depth). This is the arrangement used when seismic sections are prepared.

or if $V_1 > V_2$. Actually, the factor controlling the partitioning of energy at an interface is known as *acoustical impedance,* which is the product of velocity and density. Table 2.7 lists reflection coefficients for values of density and compressional wave velocity at normal incidence. If you reverse the density and velocity values assigned to the two layers, the only change in the value of the reflection coefficient is one of sign. Thus equivalent energy is reflected in either case, but waveforms exhibit a 180° phase change.

4.1.3   Analysis of Arrival Times

Because we intimated at the beginning of this chapter that a problem with the reflection method was identifying reflections on seismograms, it seems important to ask about the relation of reflection arrivals to those of other waves.

If you remember some analytic geometry, you might recall that hyperbolas have asymptotes. The equation for the asymptote of a hyperbola of the form given in Equation 4.2 is

$$y = \frac{a}{b} x .$$

(4.3)

If you substitute the values for a and b you determined placing Equation 4.1 in the form of Equation 4.2, you find that Equation 4.3 becomes

$$y = \frac{1}{V_1} x$$

(4.4)

which is the equation of a straight line passing through the origin with a slope of $1/V_1$. In other words, our asymptote is the travel-time curve for the direct wave. This relationship between the reflected and direct waves is illustrated in Figure 4.2(b). The physical reason for the relationship is easy to visualize. Both the direct wave and the reflected wave travel at V_1. Thus travel time will be directly related to path. At small source–detector distances the reflected wave travels a much longer path than the direct wave. Although the reflected wave always travels a longer path, at large source–detector distances the reflected path becomes closer in length to the direct wave path, and therefore the travel times become more similar.

In Figure 4.2(c) head wave arrivals also are added. Notice the rapid convergence of reflected and refracted arrivals near the critical distance for refraction (at about 10 m). Clearly, this would not be a region in which to try to identify reflections if the head wave pulse has any amplitude at all. We now have at least one reason why it might be difficult to sort out and identify reflections. To illustrate a bit more vividly the relationship graphed in Figure 4.2(b), we add direct wave arrivals to this simulated seismogram (Figure 4.4). Although the reflection still is discernible, its pattern is not nearly as clear due to interference with the waveform of the direct wave arrival. Add a bit of noise and a few other complicating factors, and the reflection might no longer be identified.

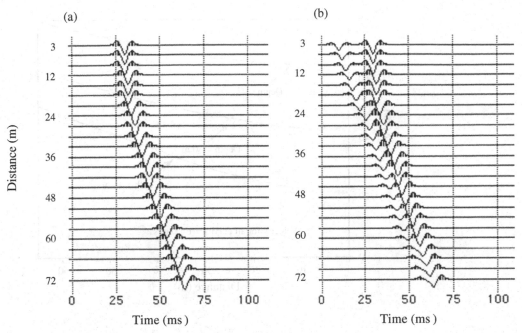

FIGURE 4.4 (a) Reflected arrivals as in Figure 4.3(a). (b) Direct wave pulse added to pattern in (a).

Examine Equation 4.1 once again. You should see that velocity and depth affect the form of the reflection travel-time curve. Because we can expect reflections from various depths traveling at a range of velocities to appear on our field seismograms, it seems wise to determine how these factors affect curve form. Figure 4.5(a) illustrates the effect of depth differences. All curves were calculated using a velocity of 1400 m/s. The degree of curvature is greatest for the shallow reflections, with curvature decreasing with increasing depth. The reason for this change is the different path lengths followed by the reflected energy. In Figure 4.6 path lengths to 12 geophones are illustrated for two different layer thicknesses, one of which is twice the thickness of the other. For constant velocity, the degree of curvature of a time–distance plot will depend on the difference in path lengths. Even though the reflection paths for the thicker layer in Figure 4.6 are almost twice the length of those in the thinner layer, the difference between the longest and shortest path in the thinner layer is almost twice the difference in the thicker layer. Thus we see the reason for the difference in curvature for cases such as h_1 and $2h_1$ (15 m and 30m) in Figure 4.5(a).

Increasing velocity also reduces the curvature of reflection travel-time curves for a constant h_1. This effect is diagrammed in Figure 4.5(b). Take a moment to clarify these relationships in your mind: They are of major importance in reflection surveying.

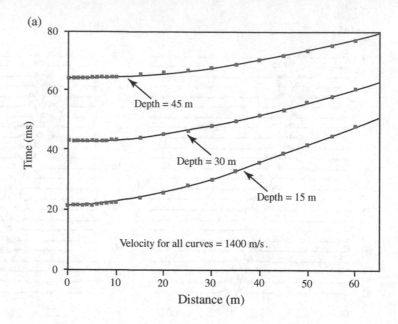

(a)

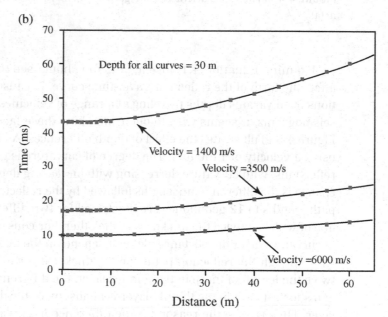

(b)

FIGURE 4.5 Time–distance plots demonstrating the effect of depth and velocity on curvature of reflection travel-time data. (a) Effect of depth with velocity constant. (b) Effect of velocity with depth constant.

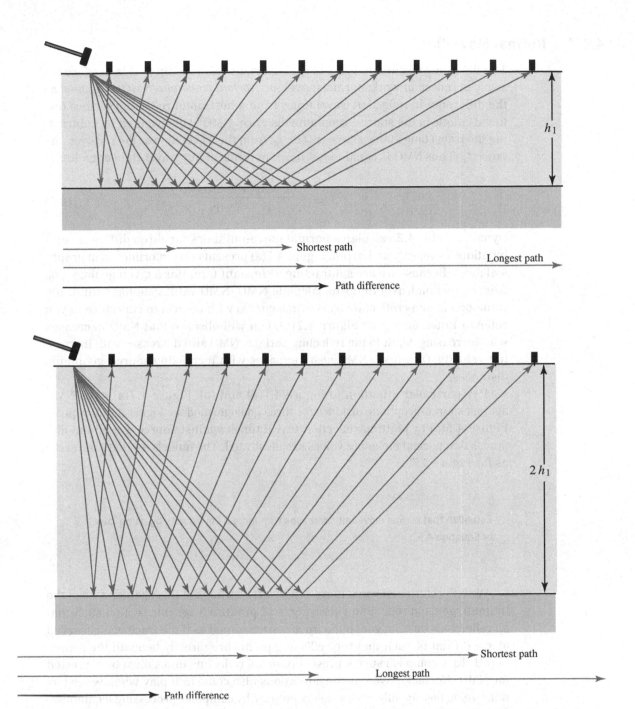

FIGURE 4.6 Reflection paths for two layer thicknesses: h_1 and $2h_1$. Geophone spread is the same in each case. Differences in path lengths for the two cases are evident. The differences between the longest and shortest paths for each case are shown diagrammatically.

4.1.4 Normal Move-Out

The foregoing discussion leads to an important relationship that is used extensively in reflection work: normal move-out. *Normal move-out (NMO)* is defined as the difference in reflection travel times from a horizontal reflecting surface due to variations in the source–geophone distance. NMO is determined by subtracting the travel time at the source ($x = 0$), t_0, from the travel time at the receiver distance t_x. Thus NMO is equal to $t_x - t_0$, or, in terms of our travel-time equation,

$$T_{NMO} = \frac{(x^2 + 4h_1^2)^{\frac{1}{2}}}{V_1} - \frac{2h_1}{V_1} \tag{4.5}$$

Dynamic Table 4.2 calculates normal move-out times for three different combinations of velocity and depth. Figure 4.7(a) presents this information in graphical form. Because we are subtracting a constant term (for a given geology), t_0, from reflection travel times to calculate NMO, NMO values should exhibit the same behavior as reflection travel-time curves with respect to curvature. If you refer to Equation 4.5 or Figure 4.7(a), you will observe that NMO decreases with increasing depth to the reflecting surface. NMO also decreases with increasing velocity. Of course, NMO also increases with increasing source–geophone distances.

Pay particular attention to Figure 4.7(a) and (b). Figure 4.7(a) plots NMO against source–geophone distance for three geologic models. Figure 4.7(b) repeats Figure 4.5(a) by plotting reflection travel times against source–geophone distance. Two normal move-out values are illustrated. The models used are the same as for Figure 4.7(a).

Establish that normal move-out decreases with increasing velocity by using Table 4.2 or Equation 4.5.

We will return to normal move-out in later sections of this chapter. One of the ultimate goals in reflection surveying is to produce a seismic section such that all reflections on each geophone trace are presented as if the trace was recorded at $x = 0$ (that is, such that the reflecting points are directly beneath the source point). To accomplish such a presentation, all reflecting times must be corrected for NMO. Normal move-out considerations also come into play when we determine velocities and other parameters entirely by computer processing techniques. When undertaking these discussions, we also derive an alternative equation for calculating NMO. Before proceeding to these more esoteric analyses, however, it might be prudent to first investigate how to determine velocity and interface depth for the simple two-layer case.

Geophone	Distance	NMO – 1	NMO – 2	NMO – 3
Source	0	00.0	00.0	00.0
1	3	00.1	00.1	00.0
2	6	00.4	00.2	00.1
3	9	00.9	00.5	00.3
4	12	01.7	00.8	00.6
5	15	02.5	01.3	00.9
6	18	03.6	01.9	01.3
7	21	04.7	02.5	01.7
8	24	06.0	03.3	02.2
9	27	07.4	04.1	02.8
10	30	08.9	05.1	03.5
11	33	10.4	06.1	04.2
12	36	12.0	07.1	05.0
13	39	13.7	08.3	05.8
14	42	15.4	09.5	06.7
15	45	17.2	10.7	07.6
16	48	19.0	12.0	08.6
17	51	20.8	13.4	09.6
18	54	22.7	14.8	10.7
19	57	24.6	16.3	11.8
20	60	26.5	17.8	13.0
21	63	28.4	19.3	14.2
22	66	30.4	20.9	15.4
23	69	32.3	22.5	16.7
24	72	34.3	24.1	18.0

Increment	3

Velocity 1	1400	Thickness 1	15
Velocity 2	1400	Thickness 2	30
Velocity 3	1400	Thickness 3	45

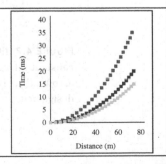

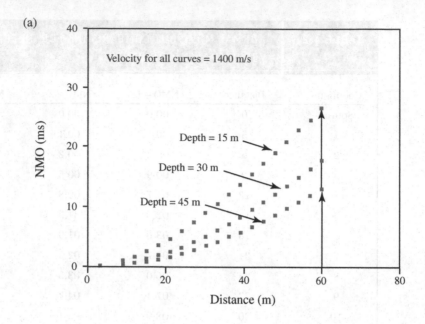

(a)

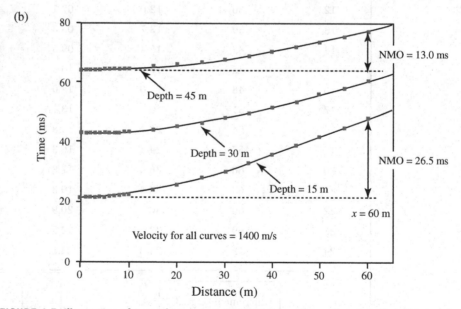

(b)

FIGURE 4.7 Illustration of normal move-out. (a) NMO curves for various interface depths at constant velocity. NMO for shallowest and deepest interface at $x = 60$ m shown by vertical arrows. (b) A repetition of reflection times in Figure 4.5(a) with NMO values at $x = 60$ m designated.

4.1.5　Determining Velocity and Thickness

The most straightforward method for determining velocity and thickness when dealing with a single interface was first proposed by Green (1938). This method, referred to as the $x^2 - t^2$ method, is appropriate for working with the type of data typically available when we explore shallow targets.

If we square both sides of the basic reflection travel-time equation,

$$\text{Time} = \frac{(x^2 + 4h_1^2)^{\frac{1}{2}}}{V_1}, \tag{4.1}$$

we arrive at an equation that can be written in the form of the equation for a straight line, $y = mx + b$:

$$t^2 = \frac{x^2 + 4h_1^2}{V_1^2} = \frac{1}{V_1^2}x^2 + \frac{4h_1^2}{V_1^2}. \tag{4.6}$$

Thus, on an $x^2 - t^2$ graph, the data points plot as a straight line with slope equal to the inverse of V_1^2 (see Figure 4.8(a)). This line is projected back to $x = 0$, which gives us t_0^2 (see Figure 4.8(a)). Velocity and t_0 now are known. If we examine the form of Equation 4.6 at $x = 0$, we see that

$$t_0^2 = \frac{4h_1^2}{V_1^2}$$

$$t_0 = \frac{2h_1}{V_1}$$

and

$$h_1 = \frac{t_0 V_1}{2}. \tag{4.7}$$

If our initial conditions of constant velocity and a horizontal interface are met, the $x^2 - t^2$ method is straightforward to apply. Because of the requirement that data be squared, it is easiest to use some type of computer aid if possible rather than simply a hand calculator. The dynamic Table 4.3 illustrates how a spreadsheet program can fulfill this requirement. Source–geophone distances and reflection arrival times are typed in. Ideally, the x^2 and t^2 values are graphed by the spreadsheet program, which provides the slope and t_0^2 values. Alternatively, a separate graphing program can be used. These two quantities then are placed back in the spreadsheet template for final determination of the velocity and thickness of the layer about the reflecting horizon. The time–distance values in Table 4.3 are taken from the simulated seismogram illustrated in Figure 4.8(b). The x^2 and t^2 values in Figure 4.8(a) are from Table 4.3.

(a)

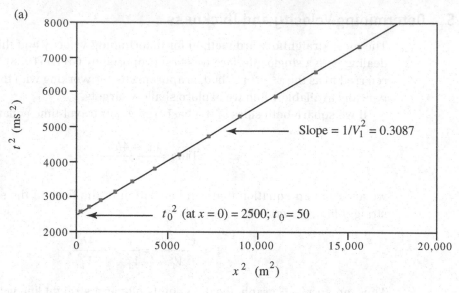

(b)

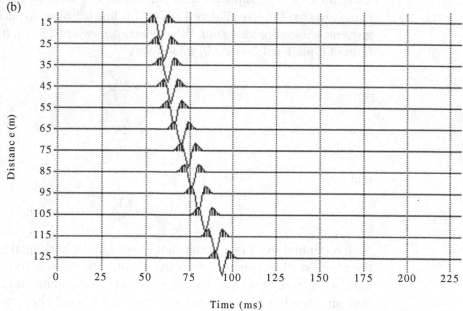

FIGURE 4.8 An example of the $x^2 - t^2$ method. (a) An $x^2 - t^2$ graph. (b) The seismogram from which the data points for (a) were taken.

Another alternative is to use a program, such as REFLECT, designed expressly for the $x^2 - t^2$ method. As always, it is a good idea to work through a method once or twice by hand (using Table 4.3 to provide the squared data) before relying entirely on computer automation.

TABLE 4.3	$x^2 - t^2$ METHOD		

Distance (m)	Time (ms)	x^2	t^2
15	**50.7**	225	2570.49
25	**51.9**	625	2693.61
35	**53.6**	1225	2872.96
45	**55.9**	2025	3124.81
55	**58.6**	3025	3433.96
65	**61.7**	4225	3806.89
75	**65.1**	5625	4238.01
85	**68.8**	7225	4733.44
95	**72.7**	9025	5285.29
105	**76.8**	11,025	5898.24
115	**81.1**	13,225	6577.21
125	**85.6**	15,625	7327.36

Slope	**0.3086**
t^2 at $x = 0$	**2500**

Velocity (m/s)	1800
Thickness (m)	45

Interval (m):	**10**
Shot offset (m):	**15**

4.1.6 Applying the $x^2 - t^2$ Method to a Field Seismogram

Now let's apply the $x^2 - t^2$ method to real field data. Figure 4.9 illustrates a seismogram recorded on the floodplain of the Connecticut River in Massachusetts during a survey attempting to map the bedrock surface. The survey was designed especially to record reflections. Details of survey design will be presented in a later section of this chapter. Although there are suggestions of earlier reflections, the first distinct reflection is denoted by arrows. For this discussion we will assume this is the first reflection on the record and, therefore, will treat the data as in a two-layer case.

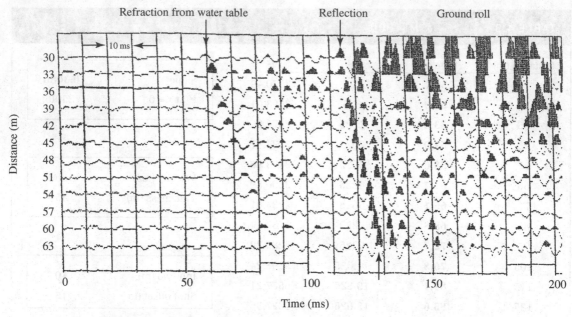

FIGURE 4.9 Field seismogram from the Connecticut Valley, Massachusetts. The reflection selected for analysis is marked by arrows. Other wave arrivals of interest are noted and will be discussed in detail later in this chapter.

We also will postpone a detailed discussion of how to recognize reflections on records. However, two good criteria are normal move-out and waveform similarity. If our source–geophone interval is not large relative to the reflector's depth, all paths will be similar in length, and the energy at each geophone should be similar. The reflection on the record exhibits only slight NMO but does possess good waveform similarity.

Usually it is difficult to identify the arrival of the onset of the reflected energy, and the seismogram in Figure 4.9 is no exception. In such a situation the first trough or peak of the reflection that can be identified from trace to trace is selected for analysis. Clearly the rule here is consistency. To practice identifying reflection arrival times, record time–distance data much as you did in analyzing refraction seismograms. Using REFLECT, Table 4.3, or some other means, determine the velocity and thickness of material above the reflecting horizon.

An $x^2 - t^2$ graph constructed from the reflections in Figure 4.9 is presented in Figure 4.10. The computed thickness of 53 m agrees well with bedrock depths in the area determined from refraction studies. Refraction profiles also identify an unsaturated layer that averages 7 m in thickness (velocity = 350 m/s) and saturated sediment velocities of 1100 m/s. These figures suggest an average velocity to bedrock of 1000 m/s, which is close to that calculated from the $x^2 - t^2$ graph.

Finally, to satisfy our curiosity concerning the amount of NMO that should be exhibited by reflection from an interface at 53 m, a velocity of 955 m/s, and

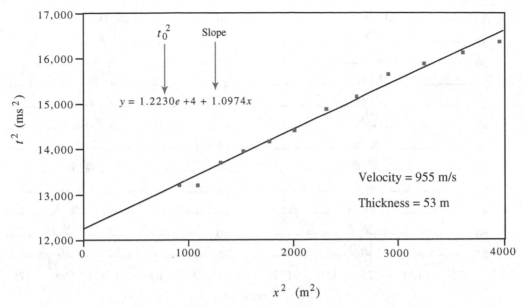

FIGURE 4.10 An $x^2 - t^2$ plot for the reflection identified in Figure 4.9.

source–geophone offsets used in our survey, we place these values into REFLECT. The simulated seismogram is illustrated in Figure 4.11(a). The curvature of the reflection arrivals is essentially the same as in the field seismogram. Although this anticipates a bit our discussion of field survey design, we might ask if we would be able to identify this reflection in the absence of obvious NMO if the amplitudes were not so great. We can adjust the shot offset value in REFLECT, but this only moves the pattern in Figure 4.11(a) earlier in time. Increasing the geophone interval from 3 m to 6 m does result in significantly greater NMO values that are easily seen in Figure 4.11(b). Perhaps the best solution is to keep a 3 m interval and to use 24 geophones or two 12-geophone arrays. The main lesson from this brief discussion (which will be expanded later) is that survey design is especially critical in providing an optimum situation in which to identify reflections.

4.2 MULTIPLE HORIZONTAL INTERFACES

When we developed the refraction method, the step from the single interface to multiple interfaces was not particularly difficult. Although the trigonometric complexity increased somewhat, the equations were of similar form, and the required unknown quantities could be obtained from travel-time curves. If we worked our way through the equations, obtaining information from layer 1, then layer 2, and so forth, both velocities and thicknesses for every layer could be calculated. This

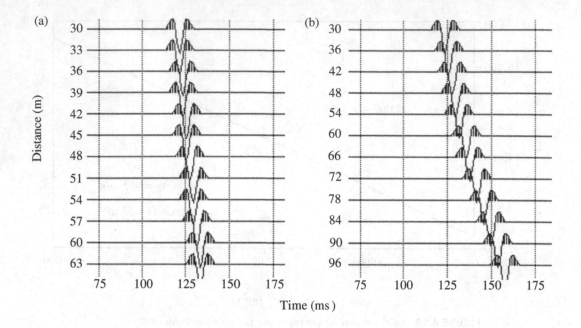

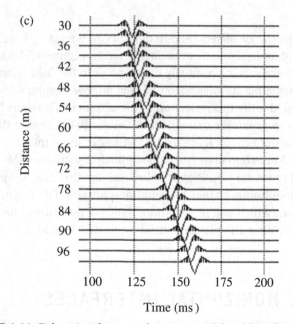

FIGURE 4.11 Enhancing the normal move-out exhibited by reflections from an interface by adjusting the geophone interval. The geophone interval for (a) is 3 m; the interval for (b) is 6 m. (c) 24 traces with an interval of 3 m.

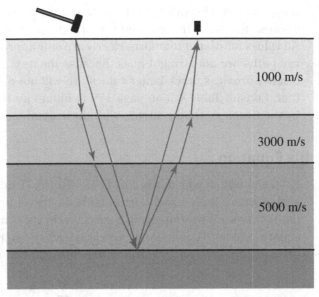

FIGURE 4.12 Actual reflection ray paths in a multilayered sequence compared with straight-line ray paths.

capability to determine all velocities and thicknesses in the refraction method ultimately depends on the fact that we are dealing with critically refracted rays, and hence know the angle of refraction (90°) at each interface.

Unfortunately, the same jump from the single interface to multiple interfaces in the reflection method involves more complexities than were encountered in the refraction approach. If we consider Figure 4.12 for a moment, it should be obvious that we cannot apply the straight-line ray path that was used to derive Equation 4.1 to a reflection from, for example, the third interface. Refraction occurs at the first and second interfaces so that the actual path of the reflected ray is quite different than the straight-line path. In addition, each source–receiver reflection path from a reflecting horizon is different than all others. Of course, given a subsurface model and an angle of inclination for a ray at the energy source, we can calculate the precise path of that ray. However, when we do not know the subsurface and have only time–distance data for a given reflection, there simply are too many variables to permit the derivation of a unique equation that does not involve some type of approximation. What are our options?

One option is to neglect refraction at interface boundaries and to assume that the straight-line path is a permissible approximation, especially if source–receiver distances are kept small. When using this option (which we will refer to as the *Green method*), we simply create an $x^2 - t^2$ plot for each reflection event on a seismogram. Because each line through the squared time–distance values results in a velocity value, an h value for the thickness of material to a reflecting interface, and a t_0 value, we can obtain velocity and thickness values for each layer in the sequence. The values for velocity and thickness of the first layer are correct because the ray paths involved are straight lines (this is the one-interface case). All values for deeper interfaces clearly are only approximations because the actual ray paths are not straight lines. Because the next method (the *Dix method*) we discuss provides much better values, we will not discuss the Green method further. Consult Table 4.5 on page 179 to obtain an idea of how values computed by the Green method compare with those obtained via the Dix method.

4.2.1 The Dix Equation

A second option was derived by C. Hewitt Dix (1955). Dix demonstrated that in a case where there are n horizontal beds, travel times can be related to actual paths traversed by employing a special velocity value known as the root-mean-square velocity, V_{rms}. In the general case where there are n horizontal beds and Δt_i is the *one-way vertical travel time* through bed i, the Dix equation states

$$V_{rms}^2 \approx \frac{\sum_{i=1}^{n} V_i^2 \Delta t_i}{\sum_{i=1}^{n} \Delta t_i}. \tag{4.8}$$

For example, if we want to determine the rms velocity to the second in a series of reflecting interfaces, we expand Equation 4.8 to arrive at

$$V_{rms} = \left(\frac{V_2^2 \Delta t_2 + V_1^2 \Delta t_1}{\Delta t_2 + \Delta t_1} \right)^{\frac{1}{2}}. \tag{4.9}$$

Careful inspection of Equation 4.9 should lead you to an understanding of why this velocity is termed the root-mean-square velocity.

Although we soon will see that the Dix equation provides very good results, note that it is an approximation and assumes that source–receiver distances are kept small relative to the distances to reflecting interfaces. In addition to Dix's article, good discussions relating to the derivation of this equation are in Sheriff and Geldart (1982, 86–87) and Robinson and Coruh (1988, 102–5).

4.2.2 Determining Velocities

Dix's analysis allows us to continue to use an $x^2 - t^2$ plot to determine layer velocities and thicknesses. Because travel times are related to actual paths by the rms velocity, Equation 4.6 still is valid if velocity terms are replaced by V_{rms}:

$$t^2 = \frac{1}{V_{rms}^2} x^2 + t_0^2. \tag{4.10}$$

Recall that we still are working with an approximation. In a following section we will take a more detailed look at the nature of this approximation and try to see how good it is. For the moment little has changed with respect to the $x^2 - t^2$ plot except that now slope $= 1/V_{rms}^2$.

Let's consider an actual interpretation problem. For simplicity, we consider the seismogram in Figure 4.13(a). Three reflection events are present. If we prepare an $x^2 - t^2$ plot from this information, we arrive at the three lines in Figure 4.13(b). Lines were fit to data points using a least squares approach, and the equations of these fits are given to the right of each line. Beginning at the shallowest reflection, the slope of each line is 0.11113, 0.21007, and 0.13243. These slopes result in the rms velocities listed along the left portion of each line. Because the ray paths connected with the first reflecting horizon are straight lines, $V_{rms_1} = V_1$. Of course, we also seek V_2 and V_3. Can we obtain these from the rms velocities?

We begin with Equation 4.8. Given n reflecting horizons, let $V_{n_{rms}}$ represent the rms velocity to the nth reflector and let $V_{n-1_{rms}}$ represent the rms velocity to the $(n-1)$th reflector. Then

$$V_{n_{rms}}^2 = \frac{\sum_{i=1}^{n} V_i^2 \Delta t_i}{\sum_{i=1}^{n} \Delta t_i} \tag{4.11}$$

and

$$V_{n_{rms}}^2 \sum_{i=1}^{n} \Delta t_i = \sum_{i=1}^{n} V_i^2 \Delta t_i. \tag{4.12}$$

Next we expand the summation on the right of the equal sign in Equation 4.12 and write an equation for $V_{n-1_{rms}}$:

$$V_{n_{rms}}^2 \sum_{i=1}^{n} \Delta t_i = \sum_{i=1}^{n-1} V_i^2 \Delta t_i + V_n^2 \Delta t_n \tag{4.13}$$

$$V_{n-1_{rms}}^2 = \frac{\sum_{i=1}^{n-1} V_i^2 \Delta t_i}{\sum_{i=1}^{n-1} \Delta t_i} \tag{4.14}$$

and

$$V_{n-1_{rms}}^2 \sum_{i=1}^{n-1} \Delta t_i = \sum_{i=1}^{n-1} V_i^2 \Delta t_i \tag{4.15}$$

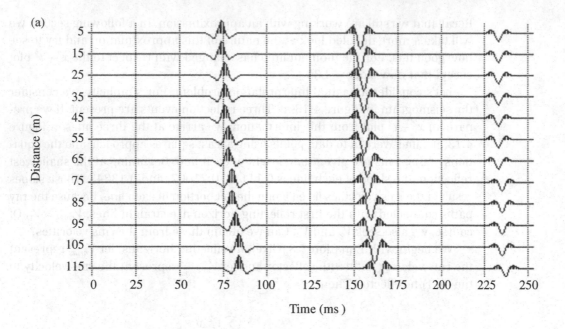

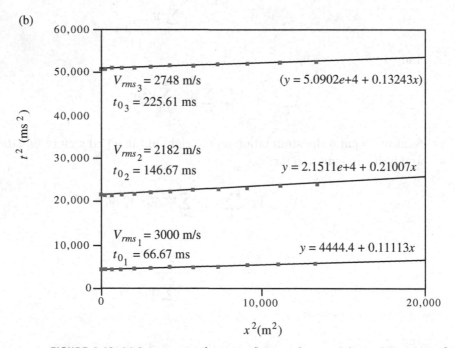

FIGURE 4.13 (a) Seismogram showing reflections from multilayered model. (b) $x^2 - t^2$ plot of time–distance values taken from (a). RMS velocities noted. Results of Dix analysis given in text.

Finally, we substitute the equivalency established in Equation 4.15 into Equation 4.13, divide by Δt_n, and solve for $V_n{}^2$:

$$V_n^2 = \frac{V_{n_{rms}}{}^2 \sum\limits_{i=1}^{n} \Delta t_i - V_{n-1_{rms}}{}^2 \sum\limits_{i=1}^{n-1} \Delta t_i}{\Delta t_n} . \qquad (4.16)$$

All that is left is to recast Equation 4.16 into a form that uses quantities directly obtainable from an $x^2 - t^2$ diagram. Remembering that Δt_i is the *one-way verti-cal travel time* through bed i and that t_{0_n} is the two-way vertical travel time from reflector n (at $x^2 = 0$ on an $x^2 - t^2$ diagram), we can write

$$\sum_{i=1}^{n} \Delta t_i = \Delta t_1 + \Delta t_2 + \Delta t_3 \ldots \ldots$$

$$= \frac{t_{0_1} + (t_{0_2} - t_{0_1}) + (t_{0_3} - t_{0_2}) \ldots \ldots}{2}$$

$$= {t_{0_n}}\Big/{2} \qquad (4.17)$$

and also

$$\Delta t_n = {(t_{0_n} - t_{0_{n-1}})}\Big/{2} . \qquad (4.18)$$

Therefore, we finally arrive at

$$V_n^2 = \frac{V_{n_{rms}}{}^2 \dfrac{t_{0_n}}{2} - V_{n-1_{rms}}{}^2 \dfrac{t_{0_{n-1}}}{2}}{\dfrac{(t_{0_n} - t_{0_{n-1}})}{2}} \qquad (4.19)$$

or

$$V_n^2 = \frac{V_{n_{rms}}{}^2 t_{0_n} - V_{n-1_{rms}}{}^2 t_{0_{n-1}}}{t_{0_n} - t_{0_{n-1}}} . \qquad (4.20)$$

Equation 4.20 allows us to determine the velocity of any unit for which we have reflections from the interfaces bounding the unit on its upper and lower surfaces. For this reason V_n is referred to as an *interval* velocity. Equation 4.20, in addition to Equation 4.8, often is referred to as the *Dix equation*. Now let's return to Figure 4.13(b) and apply Equation 4.20 to the task of determining interval velocities.

RMS velocities from the three reflecting horizons and square roots of the t^2 values at $x^2 = 0$ for the three lines also are noted in Figure 4.13(b). These are the t_0 values that represent two-way vertical travel times from each reflector. In determining interval velocities we simply select the interval of interest and substitute the appropriate quantities in Equation 4.20. For illustration, let's assume we want to determine V_3. We write

$$V_3^2 = \frac{V_{3_{rms}}^2 t_{0_3} - V_{2_{rms}}^2 t_{0_2}}{t_{0_3} - t_{0_2}},$$

$$V_3^2 = \frac{(2748 \text{ m/s})^2 (0.22561 \text{ s}) - (2182 \text{ m/s})^2 (0.14667 \text{ s})}{(0.22561 \text{ s} - 0.14667 \text{ s})},$$

and

$$V_3 = 3569 \text{ m/s}.$$

In many presentations of determining interval velocities the subscripts L for *lower* and U for *upper* are substituted for n and $n-1$. Here *lower* is synonymous with *deeper* and *upper* is synonymous with *shallower*. Equation 4.20 then becomes

$$V_L^2 = \frac{V_{L_{rms}}^2 t_{0_L} - V_{U_{rms}}^2 t_{0_U}}{t_{0_L} - t_{0_U}}. \qquad (4.21)$$

REFLECT is one of the computer programs supplied with this text, and it uses this terminology. Although the computations in determining interval velocities are straightforward using a hand calculator, REFLECT speeds the process and tends to minimize errors.

Calculate the interval velocity for the second unit from the data in Figure 4.13(b).

4.2.3 Determining Thicknesses

Once we have interval velocities, calculating thicknesses is trivial. Because travel time equals path distance divided by velocity, vertical thickness (h_n) is equal to velocity times the one-way vertical travel time:

$$h_n = V_n \left(\frac{t_{0_n} - t_{0_{n-1}}}{2} \right). \qquad (4.22)$$

If we use this equation to find the thickness of the third unit in our three-unit sequence of Figure 4.13, we have

$$h_3 = V_3 \left(\frac{t_{0_3} - t_{0_2}}{2} \right)$$

and

$$h_3 = 3569 \text{ m/s} \left(\frac{0.22561 \text{ s} - 0.14667 \text{ s}}{2} \right) = 141 \text{ m}.$$

4.2.4   Further Discussion of the Dix Method

In the past the Dix method was used extensively in reflection analysis. Most large exploration firms now use computer power to automate the determination of velocities, and the Dix method has fallen into disuse in these circles. Although we will discuss these more sophisticated procedures in a following section, many small operations still utilize the Dix approach. For this reason, it seems prudent to take a slightly more detailed look at the assumptions of the method and how its results compare with known values.

We already called attention to the role of refraction at layer boundaries in shaping the form of the reflection ray path. This is emphasized once again in Figure 4.14. Actually, a valuable procedure in analyzing reflection travel times from such a multilayered situation is to prepare a computer program or a spreadsheet template to calculate these times for a given subsurface model. The dynamic

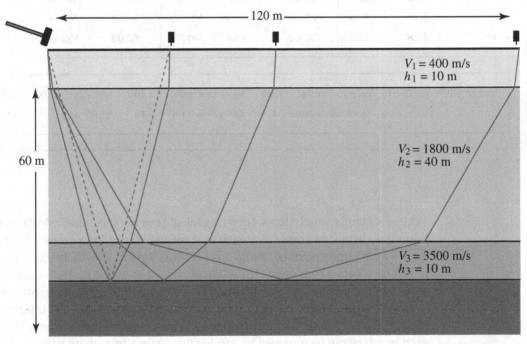

FIGURE 4.14 Reflection ray paths for source–receiver offsets of 30 m, 60 m, and 120 m for the model parameters noted in the diagram. A straight-line ray path is shown for the 30 m offset.

TABLE 4.4

Theta 1 (θ)	X1 (m)	T1 (ms)	Theta 2 (θ)	X2 (m)	T2 (ms)	Theta 3 (θ)	X3 (m)	T3 (ms)
0.50	0.17	50.00	2.25	3.32	94.48	4.38	4.85	100.21
0.90	0.31	50.01	4.05	5.98	94.56	7.90	8.76	100.33
1.30	0.45	50.01	5.86	8.66	94.69	11.45	12.71	100.52
1.70	0.59	50.02	7.67	11.37	94.87	15.04	16.74	100.78
2.10	0.73	50.03	9.49	14.11	95.09	18.70	20.87	101.13
2.50	0.87	50.05	11.32	16.88	95.37	22.43	25.14	101.56
2.90	1.01	50.06	13.16	19.71	95.71	26.27	29.59	102.08
3.30	1.15	50.08	15.01	22.60	96.10	30.24	34.26	102.71
3.70	1.29	50.10	16.88	25.57	96.55	34.37	39.25	103.47
4.10	1.43	50.13	18.76	28.61	97.07	38.72	44.65	104.39
4.50	1.57	50.15	20.67	31.76	97.66	43.34	50.63	105.51
4.90	1.71	50.18	22.60	35.02	98.32	48.35	57.51	106.92
5.30	1.85	50.21	24.56	38.41	99.08	53.91	65.85	108.78
5.70	2.00	50.25	26.54	41.96	99.93	60.33	77.06	111.47
6.10	2.14	50.28	28.56	45.68	100.89	68.38	96.14	116.39
6.50	2.28	50.32	30.62	49.62	101.97	82.03	192.48	143.18

Increment (m)	**0.40**	Velocity 1 (m/s)	**400**	Velocity 2 (m/s)	**1800**	Velocity 3 (m/s)	**3500**
		Thickness 1 (m)	**10**	Thickness 2 (m)	**40**	Thickness 3 (m)	**10**

Table 4.4 supplies such times. Given a geologic model, this table calculates reflection time–distance values from each of three interfaces for a range of incidence angles at the first interface. *Theta* 1 represents the angle of incidence at the first interface; *Theta* 2 represents the angle of refraction at the first interface and the angle of incidence at the second interface; and so on. X1 and T1 represent horizontal distance and time values for rays reflected from the first interface. X2, X3, T2, and T3 follow the same pattern. A wide range of time–distance data from each interface can be obtained by varying the *Theta* 1 increment value (only a portion of the actual table is illustrated in the text).

Time–distance values of reflections that undergo refraction at overlying interface(s) no longer lie along a hyperbolic curve as do time–distance values in the case of a single reflecting horizon. It follows that in an $x^2 - t^2$ plot, data points no longer will lie along a straight line. These points actually define a curve that is slightly convex toward greater t^2 values. Recall that Dix's analysis assumed small x values (source–receiver offsets). Actually, in determining root-mean-square velocity values, we should be drawing a straight line that is tangent to the curve defined by points near $x^2 = 0$. Because the $x^2 - t^2$ curve is nearly straight in this region, however, for all practical purposes the tangent is the same as the line that best fits the data points.

Another approach to this discussion is summarized in Sheriff and Geldart (1983, 19). They state, "Note that the Dix formula implies that the travel paths to the $(n–1)$th and nth reflectors are essentially identical except for the additional travel between the two reflectors." Refer to Figure 4.14. Assume that the geophone interval is 2.5 m for the 30 m spread, 5 m for the 60 m spread, and 10 m for the 120 m spread. You should be able to picture that geophones at 27.5 and 30 m meet this criterion, that those at 55 and 60 m come close, and that those at 110 and 120 m likely do not.

To more fully appreciate the practical implications of the preceding discussion and to review some generalizations, let's formulate a model and undertake a Dix analysis on three quite different variations in geophone spread length. The model we will use is defined in Figure 4.14. Because the total depth to the third reflecting surface is 60 m, let's investigate reflection time–distance values in the ranges 0–30 m, 0–60 m, and 0–120 m. Ray paths to the maximum of each of these ranges are illustrated in Figure 4.14 to show how the shape of the path varies with increasing source–receiver distances.

Figure 4.15 illustrates three separate $x^2 - t^2$ diagrams. Each contains lines fit by a least squares analysis to reflection time–distance values from each of the reflectors in Figure 4.14. Each diagram corresponds to one of the ranges selected for analysis, and therefore they are labeled 30 m, 60 m, and 120 m. Equations for each of the lines are included, and the diagram illustrates that line orientations differ in each case.

Results of a Dix analysis are summarized in Table 4.5. Green method results are included also for informational purposes. They clearly are not nearly as reliable as the Dix values. As expected, results for the 30 m spread correspond well to model parameters. The 60 m values are acceptably close. At 120 m, however, we begin to see a percentage of error that is unacceptable given the perfect data we are working with. Remember that these results are only for this one model. Other subsurface configurations (thicknesses and velocity contrasts) would produce different results. In some cases the divergences would be greater and in some cases less than are presented here. However, we would observe the same trends from small offsets to large offsets.

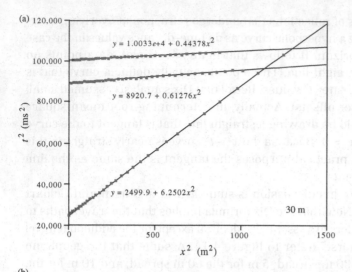

(a)

$y = 1.0033e+4 + 0.44378x^2$

$y = 8920.7 + 0.61276x^2$

$y = 2499.9 + 6.2502x^2$

30 m

x^2 (m^2)

t^2 (ms^2)

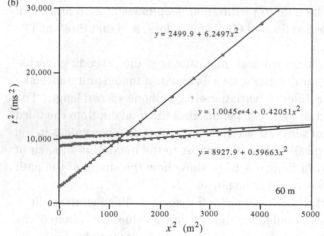

(b)

$y = 2499.9 + 6.2497x^2$

$y = 1.0045e+4 + 0.42051x^2$

$y = 8927.9 + 0.59663x^2$

60 m

x^2 (m^2)

t^2 (ms^2)

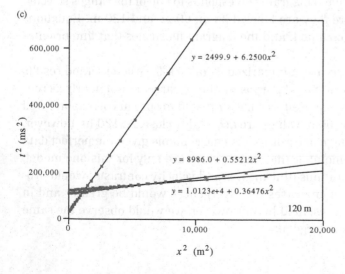

(c)

$y = 2499.9 + 6.2500x^2$

$y = 8986.0 + 0.55212x^2$

$y = 1.0123e+4 + 0.36476x^2$

120 m

x^2 (m^2)

t^2 (ms^2)

FIGURE 4.15 Comparison of $x^2 - t^2$ plots for the model defined in Table 4.5 and illustrated in Figure 4.14. (a) Data points for various source–geophone distances up to a maximum of 30 m. (b) Data points for various source–geophone distances up to a maximum of 60 m. (c) Data points for various source–geophone distances up to a maximum of 120 m.

TABLE 4.5

Parameters	Model	Green 30 m	60 m	120 m	Dix 30 m	60 m	120 m
Velocity 1 (m/s)	400	400	400	400	400	400	400
Velocity 2 (m/s)	1800	2250	2293	2411	1812	1839	1912
Velocity 3 (m/s)	3500	5254	5585	6529	3542	3736	4234
Thickness 1 (m)	10	10	10	10	10	10	10
Thickness 2 (m)	40	50	51	54	40	41	43
Thickness 3 (m)	10	15	16	19	10	11	12

It also is instructive to examine the details in the variation of these data points. Figure 4.16 depicts 30 m, 60 m, and 120 m plots for data from the third and deepest reflecting horizon. In the 30 m case, Figure 4.16(a), the points define a nearly straight line because source–receiver offsets are small relative to the depth of the interface. This line produces the best results (Table 4.5). In Figure 4.16(b) the points clearly do not lie on a straight line. This divergence from a straight line is most marked for the 120 m case, Figure 4.16(c), and this line yields the poorest results. Because the least squares fit gives all points equal weight, points with large source–receiver values are included and lead to increased error. Note that if we examine the lines going from the 30 m to the 60 m to the 120 m spreads, the slopes are decreasing and t_0^2 values are increasing, which leads to the increased velocities and thickness cataloged in Table 4.5.

Examine Figure 4.16(c) again. Even in this case, where some of the larger values clearly are unacceptable, we could arrive at much better values by drawing a tangent to the data points at small x^2 values or by fitting a line to data points in a way that gives greater weight to points at small x^2 values. Such a line is identified in Figure 4.16(c). It has a greater slope and lower t_0^2 values, which are precisely what is required to produce a solution that corresponds more closely with our initial model.

This discussion should have made you more aware of the need to keep source–receiver offsets small relative to the depths at which exploration targets are assumed to lie. Ideally, spreads should not exceed target depths. If this is not possible, lines drawn on $x^2 - t^2$ plots should give greater weight to points at small x^2 values. Admittedly, the quality of average reflection data and the difficulty of accurately picking the beginning of the reflection pulse often will generate a considerable spread in data points. Nevertheless, if you employ the Dix method for resolving reflection data, you should follow these guidelines.

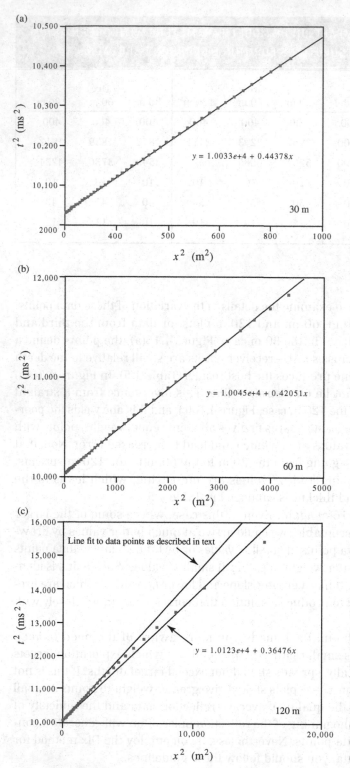

FIGURE 4.16 A more detailed presentation of selected material from Figure 4.15. Data points are squared time–distance values for reflections from the third and deepest interface defined in Table 4.5.
(a) 30 m geophone spread.
(b) 60 m geophone spread.
(c) 120 m geophone spread.

If long spread lengths are necessary, it is possible to gain more accuracy by incorporating additional higher-order terms in the Dix approximation. Shah and Levin (1973) develop this extension and include an error analysis.

4.2.5 Analyzing a Field Seismogram Containing Multiple Reflections

Let's apply a Dix analysis to a field seismogram recorded in an area where bedrock depths tend to average from 50 m to 100 m. The water table in the area is fairly close to the surface, and there is a thick sequence of glacial lake clays overlain and underlain by sands and gravels. Although there are several possible reflections on the seismogram, only two are selected for analysis. These are indicated on Figure 4.17.

After compiling time–distance values for these two reflections, perform a Dix analysis. The most convenient approach is to use REFLECT for each group of data to obtain rms velocities, interval velocities, and thicknesses.

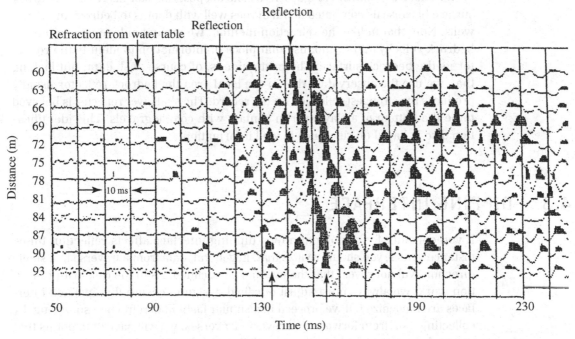

FIGURE 4.17 Field seismogram from the Connecticut Valley, Massachusetts. Reflections selected for analysis are marked by arrows.

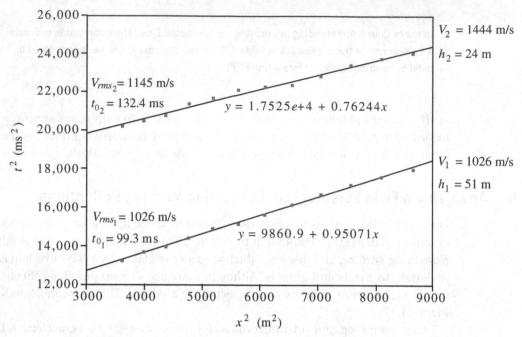

FIGURE 4.18 An $x^2 - t^2$ plot and results of Dix analysis for reflections identified in Figure 4.17.

Solutions are shown in Figure 4.18. The deepest reflector likely is the bedrock surface because its computed depth agrees well with depths to bedrock in nearby wells. Note that unlike the refraction method, we are not able to determine the bedrock velocity and cannot use that piece of information to identify the reflector as the bedrock surface. The best guide is, of course, well data; but lacking that, we use the observation that in the local area the bedrock reflector usually produces the strongest reflection on the record. The shallower reflector is believed to be the bottom of a clay layer in contact with coarse gravels. This identification also is based on reference to well information.

4.3   DIPPING INTERFACE

If you consider for a moment how a dipping interface affects reflection time–distance values, you should realize that, in general, data from a traverse in a single direction will not reveal that the interface is inclined. As in the case with refraction work, we always must adjust our field design to ensure that inclined interfaces are recognized. If we proceed in a similar fashion to refraction surveying, by collecting data from forward and reverse traverses, we will have data points that define different slopes and therefore different velocities. This would indicate an inclined interface for essentially the same reasoning discussed in Chapter 3.

In reflection work it is convenient, for a number of reasons that should become more obvious as we work through the remainder of this chapter, to use geophone spreads distributed equally on either side of a center shot (a split-spread arrangement). This is the geometry we will use to develop a travel-time equation for a single dipping interface, which of course will hold also when the dip equals zero.

4.3.1 Derivation of a Travel-Time Equation

Figure 4.19 diagrams a reflecting horizon that is inclined upward to the right. To maintain the same conventions as we elected when developing equations for the refraction method, the dip angle is taken to be positive when the slope is upward toward the right. The vertical distance EP to the interface beneath the shot point is thickness and is denoted by h; the perpendicular to the interface, EM, is denoted by j. Because we will be using a center shot, x is taken to be positive to the right and negative to the left. Thus, for a positive angle, $+x$ values increase up-dip. The derivation that follows is based on a presentation by Slotnick (1959, 48–52).

In developing reflection equations, a useful device is the concept of the *image point*. Simply imagine extending the line EM, which is perpendicular to the reflecting horizon, an equal distance on the other side of the interface. This gives us the line MQ, which is also of length j. The image point is at Q (see Figure 4.19). If we now draw a line from Q to G, which is the geophone position, we have a particularly useful geometric construct. Because triangles EMN and MQN are congruent, EN is equal to QN. Therefore, the path QNG is equal in length to the actual path traversed by the reflected wave, ENG.

This relationship makes it straightforward to develop our travel-time equation by using triangle EQG. Before we move on to this objective, however, note another useful application of the image point. Once Q is located, if we draw a line QG to some point on the surface, we need only draw a line from the energy source to the point where QG crosses the reflecting horizon N to construct the ray that will be reflected from the interface to point G. This procedure makes the task of constructing a ray path to emerge at a selected surface position a trivial exercise. Try constructing such a ray path without using an image point!

Path distance is most easily derived by using the Law of Cosines:

$$a^2 = b^2 + c^2 - 2bc\,(\cos A) \tag{4.23}$$

where a, b, and c are the sides of the triangle and A is the angle opposite side a. Thus, for the triangle in Figure 4.19,

$$(QG)^2 = (EQ)^2 + (EG)^2 - 2(EQ)(EG)\cos(90 - \beta) \tag{4.24}$$

$$(QG)^2 = (2j)^2 + (x)^2 - 2(2j)(x)\cos(90 - \beta)$$

$$(QG)^2 = 4j^2 + x^2 - (4jx)\cos(90 - \beta)$$

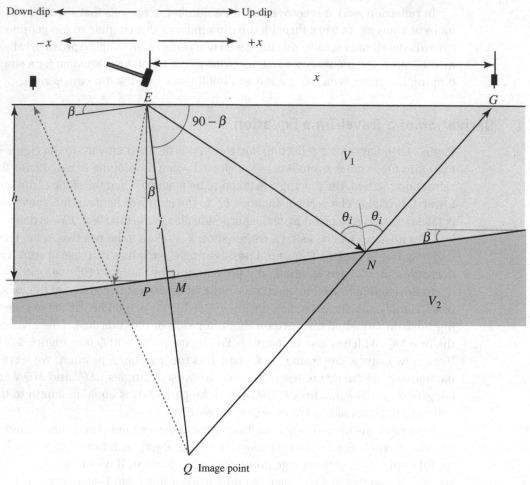

FIGURE 4.19 Diagram illustrating symbols used in derivation of a travel-time equation for a reflected ray. Dip convention is similar to that used in refraction analysis. β is positive as illustrated.

and

$$(QG)^2 = 4j^2 + x^2 - (4jx)\sin\beta. \qquad (4.25)$$

The travel-time equation thus becomes

$$t^2 = \frac{4j^2 + x^2 - 4jx\sin\beta}{V^2} \qquad (4.26)$$

or

$$t_x = \frac{(4j^2 + x^2 - 4jx\sin\beta)^{\frac{1}{2}}}{V} \qquad (4.27)$$

Geophone	Distance (m)	Time (ms)
12	−36	49.09
11	−33	47.91
10	−30	46.79
9	−27	45.73
8	−24	44.74
7	−21	43.81
6	−18	42.96
5	−15	42.18
4	−12	41.49
3	−9	40.88
2	−6	40.36
1	−3	39.94
Source	0	39.61
1	3	39.38
2	6	39.25
3	9	39.23
4	12	39.30
5	15	39.48
6	18	39.76
7	21	40.13
8	24	40.60
9	27	41.17
10	30	41.82
11	33	42.55
12	36	43.36

Velocity (m/s)	**1500**
Thickness (m)	**30**
Dip (°)	**8**
Increment (x)(m)	**3**

t_0 (ms)	39.61
t_{min} (ms)	39.23
x_{min} (m)	8.27

Equation 4.26 can be shown to have the form of a hyperbola as was the case for our horizontal interface equation. However, you would think that there should be some difference in form or position of the hyperbola, due to the dipping horizon. One way to investigate the difference is to use dynamic Table 4.6 to create time–distance data and then to graph these data. Note that all values in Table 4.6 follow the conventions established at the beginning of this section. Thickness is the vertical distance from the energy source to the reflecting horizon. The positive x values are oriented toward the right and increase up-dip because the dip angle in the table is positive.

Table 4.6 was used to produce time–distance values for a horizontal interface by setting β equal to zero and for an interface dipping at 8°. Travel-time curves for each set of values and a diagram of the model associated with each are illustrated in Figure 4.20. As you already realize, the travel-time curve for the horizontal reflecting horizon is a hyperbola with an axis of symmetry that is parallel to the time axis and passes through the point $x = 0$. If we refer to the minimum time on this hyperbola as t_{min} and the x position at which time is a minimum as x_{min}, then clearly $t_0 = t_{min}$ and $x_0 = x_{min}$.

If we now investigate Figure 4.20(b), it is obvious that these relationships no longer hold. The shortest path and, therefore, the one of least time at constant velocity is no longer the vertical path located beneath the energy source. That this statement is correct can be proven by looking at the vertical line that passes through the image point. Because this line is perpendicular to the ground surface, it is the shortest of all the lines emanating from the image point and contacting the ground surface. Therefore, the ray path constructed using this vertical line must be the shortest ray path. Based on the geometry of constructing the image point, this ray path always must be located up-dip from the energy source location, and thus the least-time position (the position of the hyperbola symmetry line), which has coordinates x_{min}, t_{min}, is always offset up-dip from x_0. Figure 4.20 helps clarify these relationships.

Now the reason for a split-spread arrangement becomes less murky. If the reflection arrivals are relatively conspicuous, their hyperbolic form should immediately indicate if the reflections are from a dipping surface, and, if so, the direction of dip. Our next task is to determine how to calculate dip amount, thickness, and velocity. First let's develop a few more relationships that we can use to arrive at these unknown values.

4.3.2 Determining Dip, Thickness, and Velocity

If we differentiate Equation 4.26 and set the result equal to zero, we can find the value of x (x_{min}) for which the time term is a minimum. Accordingly,

$$\frac{dV^2 t^2}{dx} = 2tV^2 \frac{dt}{dx} = 2x - 4j\sin\beta$$

$$\frac{dt}{dx} = \frac{x - 2j\sin\beta}{tV^2} = 0 \tag{4.28}$$

and

$$x_{min} = 2j\sin\beta. \tag{4.29}$$

We now can find an expression for t_{min} by substituting Equation 4.29 into Equation 4.26.

$$t_{min}^{\;2} = \frac{4j^2 + (2j\sin\beta)^2 - 4j(2j\sin\beta)\sin\beta}{V^2}$$

$$t_{min}^{\;2} = \frac{4j^2\cos^2\beta}{V^2} \tag{4.30}$$

$$t_{min} = \frac{2j\cos\beta}{V} \tag{4.31}$$

Next we also derive an expression for t_0, which because $x = 0$ at t_0, also is easily produced from Equation 4.26.

$$t_0 = \frac{2j}{V} \tag{4.32}$$

If we have good data, we should be able to acquire values for t_0 and t_{min} from a travel-time curve. Realizing this and examining Equations 4.31 and 4.32, we see that

$$\frac{t_{min}}{t_0} = \frac{\dfrac{2j\cos\beta}{V}}{\dfrac{2j}{V}} \tag{4.33}$$

$$\frac{t_{min}}{t_0} = \cos\beta \tag{4.34}$$

or

$$\beta = \cos^{-1}\left(\frac{t_{min}}{t_0}\right). \tag{4.35}$$

Based on this derivation, the sign of β must be determined from the dip direction, which should be clearly indicated on the travel-time curve.

Because we now know β and can determine x_{min} in the same fashion we obtained t_{min}, Equation 4.29 supplies j.

$$j = \frac{x_{min}}{2\sin\beta} \tag{4.36}$$

(a)

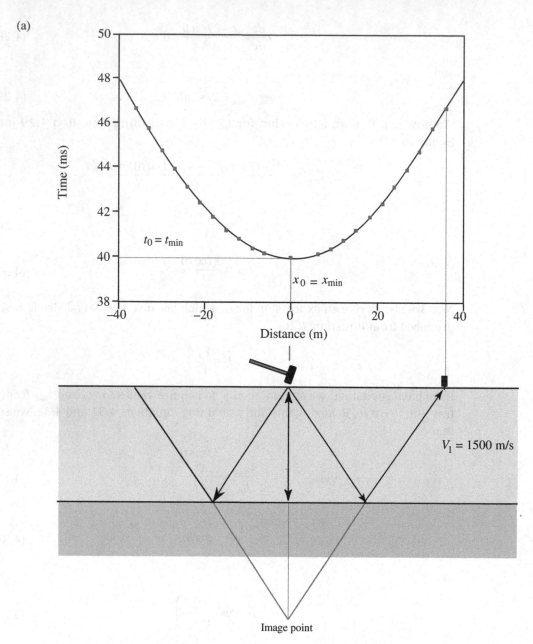

FIGURE 4.20 Diagram illustrating differences in travel-time curves for reflections from (a) a horizontal interface and (b) an inclined interface. All parameters are the same except for the dip of the interface. Model values are the same as in Table 4.6.

(b)

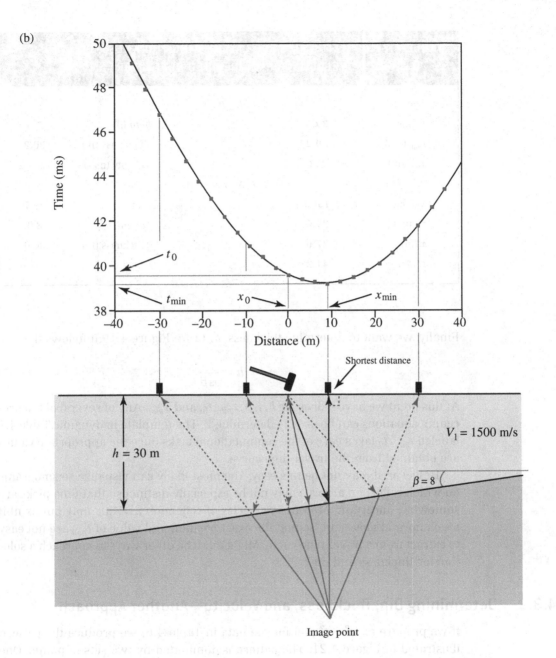

Image point

TABLE 4.7	TEMPLATES FOR CALCULATING THICKNESS, DIP, AND VELOCITY FOR AN INCLINED REFLECTOR			

t_{min}/t_0

t_0 (ms)	**39.6**		beta (β)	7.9
t_{min} (ms)	**39.2**		Thickness (m)	30.2
x_{min} (m)	**8.3**		Velocity (m/s)	1511.2

$x^2 - t^2$ Average Data

Velocity (m/s)	**1500**		i (m)	29.7
t_0 (ms)	**39.6**		beta (β)	8.0
x (m)	**27.0**		Thickness (m)	30.0
t (ms)	**41.2**			

Finally, we want to determine thickness, h. From Figure 4.19 it follows that

$$h = \frac{j}{\cos\beta}. \tag{4.37}$$

At this point we have values for $h, j, \beta, t_{min}, t_0$, and x_{min}. Any of several of the preceding equations can be used to determine V. The template in dynamic Table 4.7 labeled t_{min}/t_0 takes care of the computational tasks once the appropriate values are obtained from the travel-time curve.

However, it may not be this easy. You most likely can visualize seismograms on which reflection arrivals may not be especially distinct so that time picks are somewhat uncertain. Moreover, a series of reflection arrivals may not exhibit much normal move-out, so that the exact position and value of t_{min} are not easy to extract from a travel-time curve. Might there be other ways to approach a solution for thickness and dip?

4.3.3 Determining Dip, Thickness, and Velocity—Another Approach

If we prepare an $x^2 - t^2$ plot for the data in Table 4.6, we produce the pattern illustrated in Figure 4.21. The pattern is dominated by two sets of points. One set represents down-dip times and the other up-dip times. If we try to fit a straight line to these points, the result is a line that appears to bisect the data points. The equation for this line is labeled *actual data*. If we use the slope of this line to determine a velocity in the conventional manner, the result is 1499 m/s compared to the 1500 m/s used in Table 4.6 to generate the time–distance values. If we solve for thickness as in the case of a horizontal layer, the result is 29.70 m. This is essentially the same as the perpendicular thickness of 29.71 computed

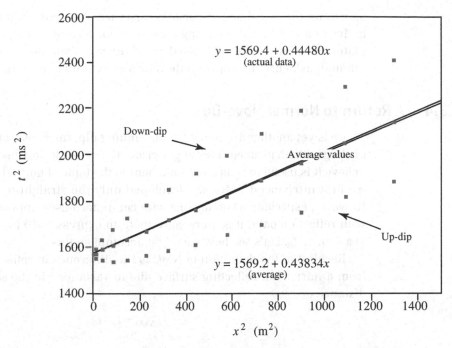

FIGURE 4.21 An $x^2 - t^2$ graph for the reflection data in Table 4.6. One line is fit to the data points. The other line is fit to the average of time values at corresponding distances from the energy source.

using the dip of 8° and thickness of 30 m that generated the data. Clearly, there seems to be some merit in this approach, but why does it work? After all, the data are for an inclined reflecting horizon.

Because the line we produced in Figure 4.21 appears to bisect time values for equivalent distances from the source, let's investigate the result of taking an average by summing our travel-time Equation 4.26 for an inclined interface and substituting $+x$ and $-x$ values:

$$t^2 + t^2 = \frac{4j^2 + (-x)^2 - 4j(-x)\sin\beta}{V^2} + \frac{4j^2 + (+x)^2 - 4j(+x)\sin\beta}{V^2} \quad (4.38)$$

$$2t^2 = \frac{x^2 + 4j^2 + x^2 + 4j^2}{V^2}$$

$$t^2 = \frac{x^2 + 4j^2}{V^2}. \quad (4.39)$$

Equation 4.39 is the same as Equation 4.6 except that j replaces h, which tells us that we can use an average of our time values at equivalent distances from the source to determine velocity and the value of j. Once we have these values, we can substitute them and one of our time–distance data pairs into Equation 4.26 to arrive at a value for β. Knowing β we solve for h using Equation 4.37. The template in dynamic Table 4.7 labeled $x^2 - t^2$ *Average Data* takes care of the computational task once the appropriate values are obtained from the $x^2 - t^2$ plot.

4.3.4 A Return to Normal Move-Out

There is yet another approach to determining dip, thickness, and velocity when dealing with an inclined reflecting horizon. Before exploring this final method, we believe it is useful to return for a moment to the topic of normal move-out. When we first introduced NMO, we developed only the straightforward definition. However, especially when dealing with computer-based approaches to working with reflection data, it is more convenient to express NMO in terms of the time at $x = 0$, or t_0. Let's see how we might accomplish this.

Recall that the definition of NMO is the difference in reflection travel times from a horizontal reflecting surface due to variations in the source–geophone distance or

$$T_{NMO} = t_x - t_0$$

$$T_{NMO} = \frac{(x^2 + 4h_1^2)^{\frac{1}{2}}}{V_1} - \frac{2h_1}{V_1}. \tag{4.5}$$

Now let's examine our basic travel-time equation once again so that we can express as much as possible in terms of t_0. For convenience in development we will represent V_1 by V and h_1 by h.

$$t_x = \frac{(x^2 + 4h^2)^{\frac{1}{2}}}{V} \tag{4.1}$$

$$t_x^2 = \frac{x^2}{V^2} + \frac{4h^2}{V^2} = \frac{x^2}{V^2} + t_0^2 \tag{4.40}$$

$$t_x^2 = \frac{x^2 + V^2 t_0^2}{V^2} = \left(\frac{x^2}{V^2 t_0^2} + 1 \right) t_0^2 \tag{4.41}$$

$$t_x = t_0 \left(1 + \frac{x^2}{V^2 t_0^2} \right)^{\frac{1}{2}} \tag{4.42}$$

Remembering that our goal is to derive an equation that expresses NMO in terms of t_0, we see that Equation 4.42 presents a problem because a t_0 term is part of an expression that is raised to the one-half power. Reaching backward in time, we will use a relationship learned in high school mathematics to expand

$$\left(1 + \frac{x^2}{V^2 t_0^2}\right)^{\frac{1}{2}}.$$

According to the generalized binomial theorem,

$$(1+z)^a = 1 + az + \frac{a(a-1)}{2 \cdot 1} z^2 + \frac{a(a-1)\dots(a-n+1)}{n!} z^n \dots$$

So if we set

$$a = \frac{1}{2} \quad \text{and} \quad z = \left(\frac{x^2}{V^2 t_0^2}\right),$$

Equation 4.42 can be expressed as

$$t_x = t_0 \left(1 + \frac{x^2}{2V^2 t_0^2} - \frac{x^4}{8V^4 t_0^4} + \frac{x^6}{16V^6 t_0^6} \dots\right). \tag{4.43}$$

Examining the variables within the parentheses, we see that we can ignore all but the first two terms *if*

$$\frac{x}{V t_0} \lll 1.$$

If we do so, then our expression reduces to

$$t_x = t_0 + \frac{x^2}{2 t_0 V^2}. \tag{4.44}$$

Recalling our objective to express T_{NMO} in terms of t_0, we see that because

$$T_{NMO} = t_x - t_0,$$

then

$$T_{NMO} = \frac{x^2}{2 t_0 V^2}. \tag{4.45}$$

However, we must keep in mind the conditions under which Equation 4.45 is valid. Equation 4.7 tells us that our former qualification

$$\frac{x}{V t_0} \lll 1$$

is equivalent to saying

$$\frac{x}{2h} \lll 1.$$

Thus for Equation 4.44 to be a useful approximation, horizontal distance divided by twice the thickness must be much less than one. Where this is not so, a much better approximation results from using the first three terms in Equation 4.44 rather than just the first two. Therefore, we have

$$T_{NMO} = \frac{x^2}{2t_0 V^2} - \frac{x^4}{8t_0^3 V^4}. \qquad (4.46)$$

Table 4.8 lists values computed using Equation 4.45 (column labeled "First Two Terms") and Equation 4.46 (column labeled "First Three Terms"). These can be compared with actual NMO values calculated using Equation 4.5. The column labeled "Condition" lists the value of $x/t_0 V$. Because this is a dynamic table, you can investigate values for which Equation 4.45 provides a good approximation for NMO values and for which the use of Equation 4.46 is advised.

Figure 4.22(a) illustrates the error involved in using Equation 4.45 for various thicknesses. Figure 4.22(b) is the same except Equation 4.46 was used. Note that the maximum vertical scale value in Figure 4.22(a) is 0.6 ms whereas that in Figure 4.22(b) is only 0.1 ms. Of course, all the errors in Figure 4.22(b) are much less than those in Figure 4.22(a). With the exception of a thickness value of 30 m (the one used in Table 4.8), all the errors are less than 0.10 ms. If you reexamine some of the field seismograms you've worked with in previous sections, you will see that such errors are acceptable when we are fortunate to pick events with a precision of 0.50 ms. However, the error for a thickness of 30 m is 0.55 ms at a source–geophone distance of 36 m. This error increases to 6.32 ms at a distance of 72 m. Such an error clearly is unacceptable. This analysis emphasizes that we cannot always use Equation 4.45 when dealing with the depths of reflectors typically encountered in shallow exploration surveys. Once again we see the importance of attempting to keep geophone spreads relatively short when reflecting depths are not great.

> The preceding paragraphs discussed the effect of source–geophone distance on the error involved in using Equation 4.45 to calculate NMO. Determine how velocity affects this approximation and how these errors compare with those due to source–geophone distances.

The previous discussion considered only NMO calculations for a single horizontal interface. What level of complexities will be introduced when multiple interfaces are introduced? If you consider the problem for a moment, it might occur to you that we will be able to use, in some fashion, V_{rms}. You certainly are correct. Taner and Koehler (1969) demonstrated that Equation 4.45, or Equation 4.46 for that matter, is valid for multiple horizontal interfaces if we substitute V_{rms} for the velocity of layer 1:

Geophone	Distance (m)	Travel Time (ms)	NMO (ms)	Condition	First Two Terms	First Three Terms
Source	0	40.00	0.00	0.00	0.00	0.00
1	3	40.05	0.05	0.05	0.05	0.05
2	6	40.20	0.20	0.10	0.20	0.20
3	9	40.45	0.45	0.15	0.45	0.45
4	12	40.79	0.79	0.20	0.80	0.79
5	15	41.23	1.23	0.25	1.25	1.23
6	18	41.76	1.76	0.30	1.80	1.76
7	21	42.38	2.38	0.35	2.45	2.37
8	24	43.08	3.08	0.40	3.20	3.07
9	27	43.86	3.86	0.45	4.05	3.84
10	30	44.72	4.72	0.50	5.00	4.69
11	33	45.65	5.65	0.55	6.05	5.59
12	36	46.65	6.65	0.60	7.20	6.55

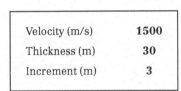

Velocity (m/s)	**1500**
Thickness (m)	**30**
Increment (m)	**3**

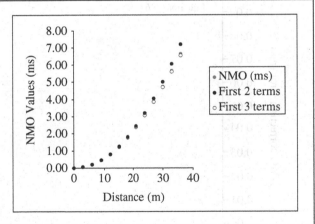

$$T_{NMO} = \frac{x^2}{2t_0 V_{rms}}. \tag{4.47}$$

In a following section when we discuss data processing techniques, we will investigate the accuracy of this approximation using a dynamic table. Now, however, it is time to return to our initial objective in this major section: determining dip, thickness, and velocity when an inclined reflecting interface is present.

(a)

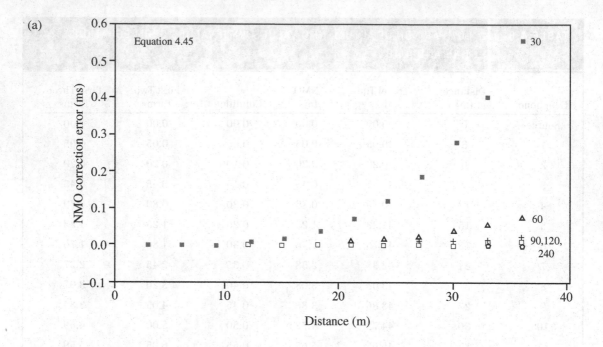

(b)

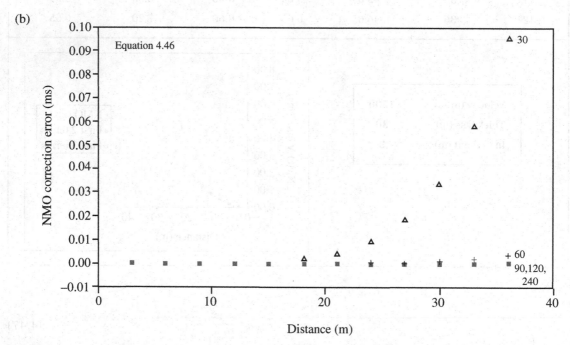

FIGURE 4.22 Plots of differences in NMO values using Equation 4.5 and an approximation using Equations 4.45 and 4.46. Thicknesses used were 30, 60, 90, 120, and 240 m.

4.3.5 Determining Dip, Thickness, and Velocity—Yet Another Approach

This "yet another approach" essentially uses the displacement of the hyperbolic form of the travel-time curve when dealing with an inclined interface. Figure 4.23(a) graphically illustrates normal move-out, and Figure 4.23(b) illustrates a new relationship, dip move-out. *Dip move-out (DMO)* is the difference in travel time to geophones at equal distances from the source position ($x = 0$) in a split-spread geometry:

$$T_{DMO} = t_{+x} - t_{-x}.$$

(4.48)

Because dip move-out is related to the dip of the reflecting horizon, there should be a way to use this time to determine dip. We begin with the basic travel-time equation for reflections from a horizontal interface, rearrange this equation to incorporate t_0, and then expand the resulting equation using the generalized binomial theorem as we did in our recent discussion of normal move-out. In the following development, please recall the distinction we make between the quantities h and j. Thus

$$t_x = \frac{(4j^2 + x^2 - 4jx\sin\beta)^{\frac{1}{2}}}{V}$$

(4.27)

$$t_x^2 = \frac{x^2}{V^2} + \frac{4j^2}{V^2} - \frac{4jx\sin\beta}{V^2}$$

(4.49)

$$t_x^2 = t_0^2 + \frac{x^2}{V^2} - \frac{4jx\sin\beta}{V^2}$$

(4.50)

$$t_x^2 = t_0^2 \left(1 + \frac{x^2 - 4jx\sin\beta}{4j^2} \right)$$

(4.51)

and

$$t_x = t_0 \left(1 + \frac{x^2 - 4jx\sin\beta}{4j^2} \right)^{\frac{1}{2}}.$$

(4.52)

Expanding Equation 4.52, we have

$$t_x = t_0 \left(1 + \left(\frac{x^2 - 4jx\sin\beta}{8j^2} \right) - \frac{1}{8} \left(\frac{x^2 - 4jx\sin\beta}{4j^2} \right)^2 + \dots \right).$$

(4.53)

Once again, we take only the first two terms in the expansion, and, remembering that $t_0 = 2j/V$, we arrive at

$$t_x = t_0 + \frac{x^2 - 4jx\sin\beta}{4jV}.$$

(4.54)

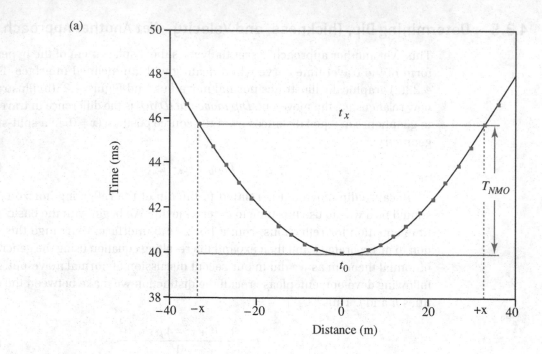

(a)

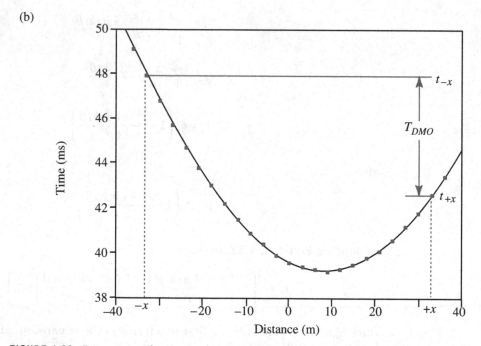

(b)

FIGURE 4.23 Comparison of normal move-out and dip move-out definitions on time–distance curves. Dip move-out (DMO) is the difference in travel time to geophones at equal distances from the source position ($x = 0$) in a split-spread geometry.

When we developed the $x^2 - t^2$ *Average Data* method we summed reflection travel-time equations for $+x$ and $-x$ cases. However, recalling that we are developing a method to take advantage of the availability of dip move-out, which we defined in Equation 4.48, we see that

$$T_{DMO} = t_{+x} - t_{-x} = \left(t_0 + \frac{(+x)^2 - 4j(+x)\sin\beta}{4jV} \right) - \left(t_0 + \frac{(-x)^2 - 4j(-x)\sin\beta}{4jV} \right) \quad (4.55)$$

$$T_{DMO} = -\frac{2x\sin\beta}{V} \quad (4.56)$$

and

$$\beta = \sin^{-1}\left(-\frac{VT_{DMO}}{2x} \right). \quad (4.57)$$

We determine T_{DMO} for a selected x value on our travel-time curve and V from an $x^2 - t^2$ plot. This gives us β from Equation 4.57, and j, and then h, from one of several equations presented previously.

Use Table 4.6 to generate time–distance data for inclined reflecting horizons where thicknesses are 30 m and 300 m. Determine β using Equation 4.57. Are problems in accuracy present for cases in which thickness is not much greater than offset distance? If so, must we abandon the dip move-out approach to determining β, or is there an alternative?

In concluding an analysis of methods to determine horizon dips, it is important to realize that the true dip of a plane cannot be determined from a seismic line unless we are fortunate to orient the survey perpendicular to the strike of the plane. Often, if true dips are desired, we arrange survey lines perpendicular to one another to provide the information necessary to determine the true dip. If dips are low (less than 10°), then the standard methods found in structural geology textbooks may be applied. These demonstrate how true dips can be determined from apparent dips acquired along two different orientations. If dips are not low, then more complicated calculations are needed because the line along which reflection points occur is not parallel to the survey line. An excellent development of the problem with understandable solutions is presented in Slotnick (1959, 127–62).

4.4 ACQUIRING AND RECOGNIZING REFLECTIONS FROM SHALLOW INTERFACES

At the beginning of this chapter we mentioned the seemingly anomalous situation of reflection methods being dominant in petroleum exploration and refraction methods being dominant in shallow exploration. We suggested a few reasons why this might be so, among which was the problem of recognizing many reflection arrivals and the vast data processing required to solve this difficulty and to meet other requirements when working with reflections. As you work through this chapter, perhaps you also appreciate that velocity data may not be as straightforward to acquire or as accurate as those produced by refraction work. In many situations, therefore, requirements favor the information supplied by refraction work.

We also pointed out some of the limitations of refraction surveys and several advantages of working with reflections. Surely the possibility of producing seismic sections that reveal great detail about the shallow subsurface is a powerful incentive for mastering reflection work in shallow exploration.

You should recall the two criteria for recognizing reflections on seismograms: (1) similarity of waveforms and (2) exhibition of normal move-out. However, to apply these and other tests we first must produce a record on which reflections are sufficiently clearly developed that these tests are possible. Several strategies are available: equipment design, data processing, and survey design. In most cases all are used in varying proportions to produce meaningful records. As such, each will be discussed in sections of this chapter to follow. Here we concentrate on a fundamental concept of survey design that often results in data suitable for interpretation with a minimum of computer processing.

4.4.1 The Optimum Window

Let's accept Figure 4.24(a) as a simulated seismogram from a typical shallow geologic section in the Connecticut Valley of Massachusetts. The model used is located in the upper right corner of Figure 4.24(a). We assume that our instruments are configured to maximize reflection arrivals. Waveforms for the several wave types illustrated are very simplified (see Appendix B), but the locations are accurate. We ignore the reflection from the water table, which is at a depth of 5 m. The water table refraction is shown because it is often clearly developed on shallow reflection records. At the moment we especially are concerned with the relative positions of the reflections from the interfaces at 30 m and 50 m and the air wave and ground roll (surface wave).

To make this simulated seismogram more realistic, imagine a fair amount of noise spread about the record so as to render the individual pulses less clearly defined. Also extend the ground roll significantly toward increased time (see Figure 4.26(a) for the extent of ground roll interference on an actual record). It should be fairly obvious that the reflections will be all but impossible to discern.

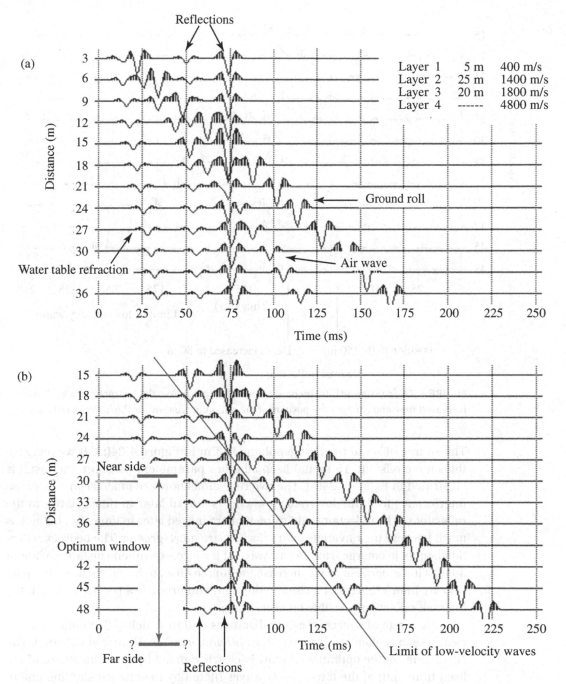

FIGURE 4.24 An illustration of the importance of source–receiver distance in reflection surveying. (a) Typical reflection record showing positions of low-velocity waves (ground roll and air wave), refraction from the water table, and reflections from two interfaces. (b) An increased source–geophone distance moves low-velocity waves toward greater time at a rate faster than reflections, which become easier to discern. Position of the *optimum window* is indicated.

(c)

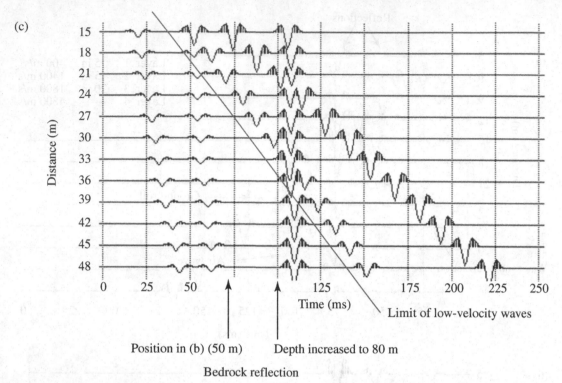

Position in (b) (50 m)   Depth increased to 80 m

Bedrock reflection

FIGURE 4.24 (continued) (c) Increased bedrock depth moves the bedrock reflector toward increased time and changes the position of the optimum window for the bedrock reflection.

The source offset to the first geophone is 3 m in Figure 4.24(a). If we increase the source offset to 15 m and keep all other parameters constant, the result is illustrated in Figure 4.24(b). The two reflections now are much clearer because interference from the low-velocity waves is moved later in time relative to the reflections. The reflections and refraction are moved later in time also, but not as much because their average path velocities are much greater. The bedrock reflection begins to emerge from the air wave at a source–geophone distance of about 30 m. This is because as we increase the source offset to 30 m, we move the low-velocity interference still farther to the right and produce a record on which the bedrock reflection is visible on every trace.

This range of source–receiver distances within which reflections are most clearly seen is often referred to as the *optimum window* (Hunter et al. 1984). The "near side" of the optimum window is clearly defined by the coincidence of the least time limit of the low-velocity waves (noted by the shaded slanting line in Figure 4.24(b)) and by the reflections of interest. However, the upper limit or the "far side" of the optimum window is much less obvious. Several considerations influence the location of the maximum source–receiver distance. As source–receiver distances increase, reflections from shallow interfaces may begin to interfere with reflections from deeper interfaces such as the bedrock reflection. Figure 4.25 illustrates such a case. The model used is slightly different from that in Figure 4.24. Ground roll is not included although the limit of the low-velocity waves is indicated. The reflection from the 10 m deep interface does not interfere with the bedrock reflection in Figure 4.25(a) for which the source offset is 15 m and the geophone interval is 3 m. However, in Figure 4.25(b), where the initial source offset is 30 m, this same reflection begins to merge with the bedrock reflection on the 63 m geophone trace. If we are interested primarily in the bedrock reflection, the optimum window extends from approximately 39 m to 63 m. However, if we hope to separate reflections from all three interfaces, the far-side limit shrinks drastically to a maximum of 45 m.

If the water table refraction is particularly well developed, it may interfere with shallow reflections as the source–receiver distance is increased. Compare the position of this refraction with the first reflection in Figure 4.24(a) and (b). Also, as source–receiver distances increase, amplitude and phase changes in the incoming pulse may make it difficult to correlate from trace to trace on a given record. Pullan and Hunter (1985) document this effect quite vividly. They conclude that phase changes should not be a problem in tracing bedrock reflections as long as source–receiver distances do not exceed bedrock depth. We developed a restriction of this same order when investigating the Dix method and when developing an expression for NMO in terms of t_0. By now you should understand that appropriate source–receiver distances are quite sensitive to the depth of the horizon of interest. You will need to keep in mind what these limits are and the conditions under which they apply.

Finally, note that the position of the optimum window may vary in a survey if the depths of the interfaces being studied change. The optimum window selected in Figure 4.24(b) would no longer apply in Figure 4.24(c) because the bedrock surface has increased in depth from 50 m to 80 m.

After you feel you have a good grasp of the optimum window concept and the near-side controls, study Figure 4.26. Illustrated in this figure are two field seismograms from different sites in the Connecticut Valley of Massachusetts. The air wave and ground roll are present on both records, but the air wave is more clearly defined on Figure 4.26(a). The (a) and (b) parts of Figures 4.24 and 4.26 correlate reasonably well, although in both cases the limit of the low-velocity waves is present at greater times on the actual field records. This correlation demonstrates that modeling programs such as REFLECT can provide useful results when we use them knowing their limitations.

How, then, is the optimum window determined? In the best case well logs exist for the exploration area. These should enable the production of useful models that can reduce initial field efforts considerably. In many cases, however, we simply employ a split spread and continue to expand its length by adding geophone cable to each end. Eventually, assuming conditions are satisfactory for recording reflections, reflections will emerge from the low-velocity waves. Spread length

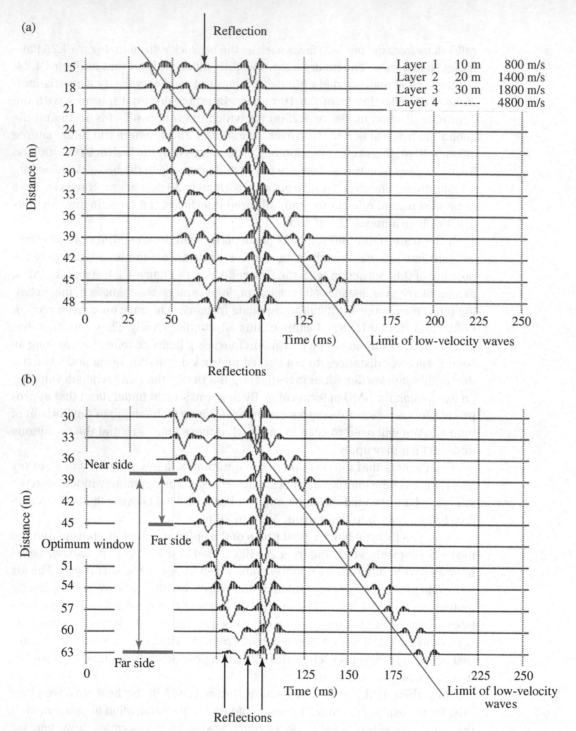

FIGURE 4.25 Factors influencing the position of the far side of the optimum window. (a) Three reflections from the designated model. (b) Source–receiver distance increased. Position of near side of optimum window clearly delimited. Position of far side less clear. At the larger source–geophone distances, the shallow reflection interferes with the bedrock reflection.

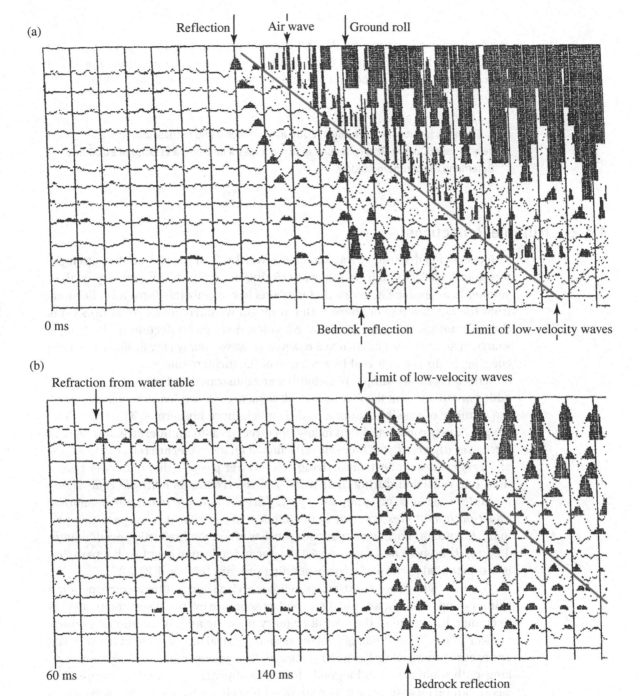

FIGURE 4.26 Field seismograms from the Connecticut Valley, Massachusetts. The geophone interval for both records is 3 m. (a) Source offset is 30 m. The record contains two prominent reflections: a well-defined air wave and obvious ground roll. Compare with Figure 4.24(b). (b) This record is from a different site than (a). The source offset is 60 m. The air wave is present but is difficult to discern. The water table refraction is just visible at about 75 ms on the first trace.

can be increased as time, funds, and conditions permit to develop distinct NMO and to watch for interfering reflections and phase changes. Care should be taken in acquiring this initial record to define the best possible optimum window.

> A 120 m thick unit with a velocity of 2000 m/s overlies a 250 m thick unit with a velocity of 4000 m/s. This latter unit in turn overlies a unit with a velocity of 6000 m/s. Determine the optimum window.

4.4.2   Multiple Reflections

The preceding discussion was directed toward a field procedure that attempts to separate reflections from other interfering wave types. Our primary goal, of course, is to produce a record that contains in recognizable form all reflections from the depth range of interest. One problem we have not as yet addressed is that of multiple reflections. We've considered simple reflection paths from a source to an interface and then to a receiver. However, many equally likely but more complex paths are followed by a portion of the incident energy.

Three relatively simple possibilities are illustrated in Figure 4.27(a). If we added another unit in the relatively thick sequence labeled "saturated sediments," the number of possible paths would be much more numerous. The reflections we previously considered that follow the direct path to an interface and back to a receiver (that consist of only one reflection at an interface) are referred to as *primary reflections*. Reflections arriving at a receiver position that underwent more than one reflection during their travel are referred to as *multiple reflections*. Multiple reflections, or multiples, typically are divided into two main categories. *Short-path multiples*, as their name implies, are similar in total path to a primary, but a short portion of their path consists of a second reflection event (Figure 4.27(a)). *Long-path multiples* consist of a second reflection event with a path that is a substantial portion of a corresponding primary (Figure 4.27(a)).

As you might guess, short-path multiples arrive shortly after a corresponding primary and, as such, are not discerned as separate events. Their main effect is to change the shape of the overall primary pulse by adding additional undulations or a "tail." This prolonging of the primary arrival may itself tend to overwrite or obscure other primaries arriving close in time to our initial primary. However, long-path multiples stand a good chance of appearing as distinct events that easily can be misinterpreted as primaries. If such a mistake occurs, subsurface reconstructions will be incorrect. How serious a problem are long-path multiples, and what strategy might we develop to deal with them?

In attacking this problem, first let's return to our discussion of energy partitioning in Chapter 2 and apply this information to a rather simple subsurface model that is common in shallow surveying (Figure 4.27(b)). This exercise is

(a)

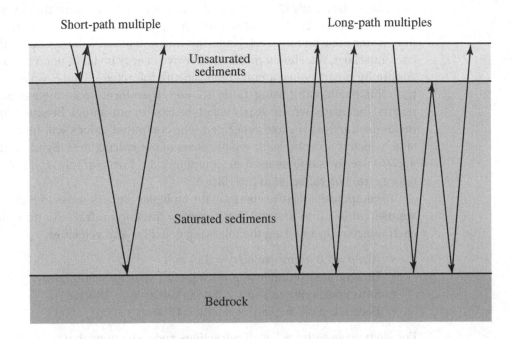

(b)

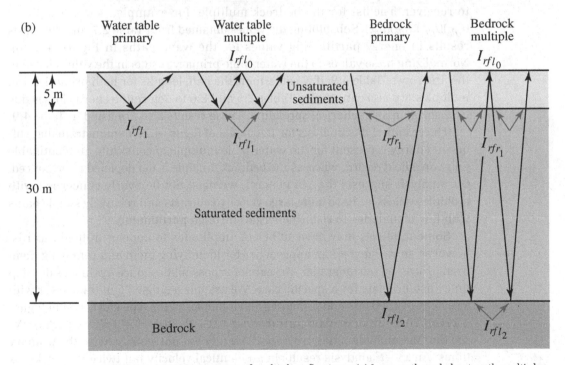

FIGURE 4.27 Various types of multiple reflections. (a) Long-path and short-path multiples. (b) Symbols used in text discussion of losses due to reflection and refraction at interfaces for primary and multiple reflections.

intended to be mainly illustrative and does not pretend to cover the range of multiple arrivals that might occur in more complicated situations. Here we address only water table and bedrock primaries and one possible long-path multiple for each interface. You should recall that P-wave energy incident upon a velocity discontinuity is partitioned among reflected and refracted P- and S-waves. We compute this partitioning using Table 2.7 and, therefore, make the assumption of normal incidence, which clearly is not the case in our model. Because on normal incidence no S-waves are generated, our computed values will be on the high side, which is acceptable for the purposes of our calculations. Symbols in Figure 4.27(b) are the same as used in Equation 2.31. For example, I_{rfl_1} refers to the energy fraction reflected at interface 1.

To compute the relative energy of the multiples and primaries in Figure 4.27(b), we use Table 2.7 to determine the energy fractions refracted and reflected at each interface by entering the following densities and velocities:

- Air: $\rho = 1.0$ g/cm^3, velocity = 335 m/s.
- Unsaturated sediments: $\rho = 1.8$ g/cm^3, velocity = 500 m/s.
- Saturated sediments: $\rho = 2.0$ g/cm^3, velocity = 1800 m/s.
- Bedrock: $\rho = 2.6$ g/cm^3, velocity = 4800 m/s.

For each wave path we list all refractions and reflections that occur from source to receiver. The list for the bedrock multiple, for example, is $(I_{rfr_1})(I_{rfl_2})(I_{rfr_1})$ $(I_{rfl_0})(I_{rfr_1})(I_{rfl_2})(I_{rfr_1})$. Substituting the values obtained from Table 2.7 into these lists results in energy partitioning values for the wave paths in Figure 4.27(b). Normalizing these values to the water table primary results in the values listed in the first row of Table 4.9. If we determine the path lengths for both primaries and multiples, we also can employ Tables 2.5 and 2.6 to determine energy losses due to absorption and spherical spreading. These results also are listed in Table 4.9.

The results of this analysis for the model of Figure 4.27 demonstrate that sufficient energy is present for the water table multiple to constitute an identifiable pulse on a field record, whereas the bedrock multiple is too depleted to be noticed. Our analysis suggests that, in general, we need not be overly concerned with multiples unless we have quite large velocity contrasts and relatively simple paths with few boundaries to reduce energies through partitioning.

Some multiples may have sufficient amplitudes to appear on field records, however, so we must adopt approaches for identifying them and removing them from processed seismograms. To search for possibilities, once again let's develop time–distance data for a specific case. We assume a single 75 m thick layer with a velocity of 1800 m/s and primaries and multiples of the type illustrated in Figure 4.27(b). Time–distance data for this model are illustrated in Figure 4.28(a). We see that the multiple times are *approximately* but not exactly twice the primary times. An $x^2 - t^2$ analysis results in an identical velocity but twice the thickness for the multiple (Figure 4.28(b)). This results from the fact that the path length of the multiple for a given source–receiver distance is identical to that which would

TABLE 4.9	RELATIVE ENERGY COMPARISONS AMONG PRIMARY REFLECTIONS AND MULTIPLES			
	Water Table Primary	Water Table Multiple	Bedrock Primary	Bedrock Multiple
Energy partitioning	1.000	0.359	0.353	0.045
Plus absorption	1.000	0.323	0.189	0.013
Plus spherical spreading	1.000	0.314	0.160	0.009

be followed by an imaginary primary to an imaginary interface at exactly twice the thickness of the actual interface. Finally, if we plot normal move-out values for both the primary and the multiple, it is obvious that the multiple possesses less NMO (Figure 4.28(c)). This difference in NMO also is evident from Figure 4.28(a).

Thus reflection events of smaller amplitude and twice the time of other reflection events are good candidates for multiples. If analysis produces closely similar velocities to the event of greater amplitude and thickness values that are twice as great, a multiple event is indicated. Because of the difference in NMO values, the amplitudes of multiples can be further reduced during processing as described in a later section. Multiples constitute a serious problem in the deep reflection surveys utilized by petroleum exploration firms, and much effort is expended in identifying and eliminating them. They are a less serious consideration in shallow work, but one nonetheless should be alert for such a possibility.

4.4.3 Diffractions

In Chapter 2 we noted that when a wave encounters a sudden change in curvature of an interface, diffraction occurs (see Figure 2.19). Recall that effects of diffraction are most pronounced when sharp changes in curvature of interfaces have radii that are similar to those of seismic wavelengths. In reflection work we try to employ energy sources that produce a high-frequency content in the source signal and use high-frequency geophones. As 100 Hz frequencies are associated with 15 m wavelengths, we should expect to encounter diffracted energy in our work.

In Chapter 3 when we discussed refraction travel-time curves associated with discontinuous reflectors, we learned that diffracted waves could occur as first arrivals in some situations but that, in general, it would not be an easy matter to confidently recognize these as diffractions on field records. In reflection work, however, as we continuously record geophone response, we stand a much better chance of recognizing diffraction arrivals. Therefore, let's determine what pattern diffractions should illustrate on field seismograms.

(a)

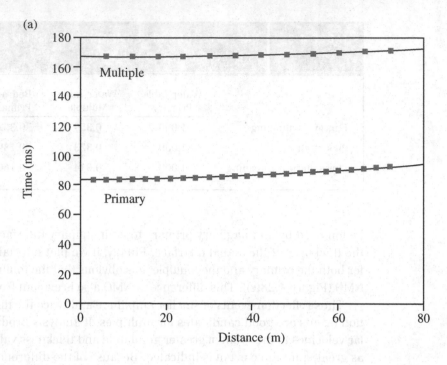

(b)

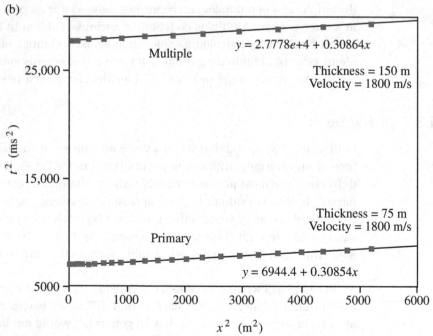

FIGURE 4.28 Comparison of primary reflection and multiple for a single layer with a velocity of 1800 m/s and a thickness of 75 m. (a) Time–distance plot for primary and multiple. (b) $x^2 - t^2$ plot for primary and multiple.

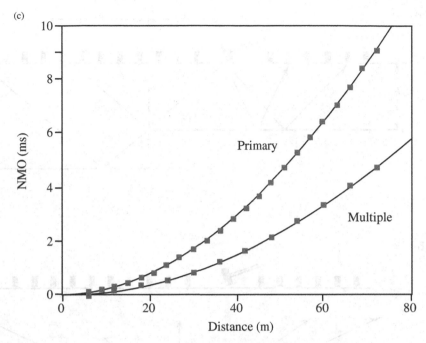

(c)

FIGURE 4.28 *(continued)* (c) NMO plot for primary and multiple.

Figure 4.29(a) illustrates the general case of a step in a reflecting surface with the energy source offset from the vertical projection of the step. For the particular geometry illustrated, reflections will arrive in the normal manner with the exceptions that (1) there is a finite limit to the right of the step for reflections from the shallower portion of the interface, and (2) reflections from the deeper portion of the interface begin to arrive at some distance after this limit. After energy arrives at the upper edge of the step, diffractions will begin to spread outward toward all geophones in the spread. Diffractions also will be generated from the lower edge of the step, but for simplicity we do not consider them in this analysis.

Our previous reflection equations obviously are valid for the two sets of reflections involved. It is a trivial matter to calculate the last point at which reflections from the upper interface arrive and the first point at which reflections from the lower interface arrive. Travel times t_t for diffractions consist of two components: (1) the travel time t_d of a diffracted wave from the edge of the step to a given horizontal distance x_g, and (2) the time t_s for the energy from the source to reach the step edge. The path distances are straightforward to determine because we can apply right triangles in both cases. For diffraction travel time

$$t_d = \frac{(x_g^2 + h_1^2)^{\frac{1}{2}}}{V_1}. \tag{4.58}$$

(a)

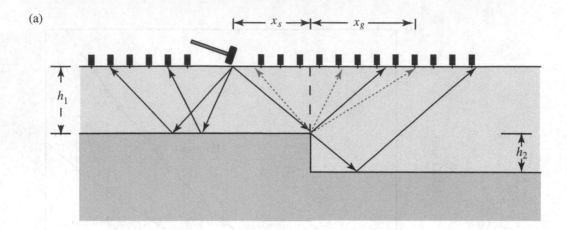

(b)

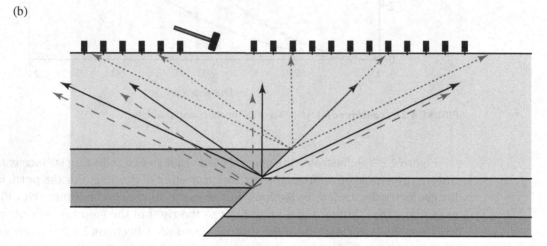

FIGURE 4.29 Diffraction at an edge. (a) Reflection and diffraction paths from a step in a horizontal interface. Symbols are explained in the text. (b) Some diffraction paths from three edges in a faulted horizontal unit.

Similarly, for travel time from source to edge

$$t_s = \frac{(x_s^2 + h_1^2)^{\frac{1}{2}}}{V_1} \tag{4.59}$$

and, finally,

$$t_t = \frac{(x_s^2 + h_1^2)^{\frac{1}{2}}}{V_1} + \frac{(x_g^2 + h_1^2)^{\frac{1}{2}}}{V_1}. \tag{4.60}$$

The diffraction times determined by Equation 4.58 plot as a hyperbola, as do the total times determined by Equation 4.60 because the first term in this equation simply adds a constant time to each diffraction time (compare diffraction times in Figure 4.30(a) and (c)). Figure 4.30(a) shows a typical travel-time plot for the model in Figure 4.29(a). The reflection arrivals produce their characteristic hyperbolic curve symmetric about the source with the exception that part of the curve is displaced later in time by the presence of the step in the reflector. The diffractions also produce a hyperbolic curve as mentioned. This curve illustrates greater NMO than the reflections even though both diffractions and reflections travel entirely in the V_1 medium. Note that the minimum diffraction time is located directly above the position of the step's edge.

If you are interested in working with diffraction arrival patterns, experiment with Table 4.10. Column 3 lists diffraction times from an edge to geophones located at the horizontal distances from the vertical projection of the edge listed in column 2. Column 5 lists total travel times. If the shot is not directly above the edge, the correct horizontal positions for geophones for the total travel times are listed in column 4. The shot position is determined relative to the edge position but is always at $x = 0$.

Figure 4.30(b) illustrates a situation similar to that in Figure 4.29(b), which illustrates beds displaced by a high-angle reverse fault. Diffractions from the uppermost three edges are plotted as well as reflections from the uppermost two horizons on the upthrown block. Reflections from the downthrown block are not shown to avoid clutter. Note that diffractions provide some information about the position and direction of inclination of the fault because diffractions from each edge lie vertically above the edge. Also note the overlap of reflection and diffraction arrivals, especially the reflections from the second interface. On a field seismogram these overlaps would create a good deal of interference, thus obscuring reflections of interest.

Figure 4.30(c) illustrates the special case when the shot point is directly over the edge. This is included because of the symmetric relation between reflections and diffractions. It perhaps is easier to observe the difference in normal move-out in this case. If we correct all times in Figure 4.30(c) for normal move-out, the result is the pattern in Figure 4.30(d). Due to the NMO difference between reflections and diffractions, the correction flattens the reflections but only reduces the amplitude of the diffraction hyperbola. If we take Figure 4.30(d) and flip it so that time increases downward and draw a wavelet at each arrival point, we create the simulated seismic section in Figure 4.30(e).

Design a model similar to that in Figure 4.29(a). Place the step at somewhat greater depth and reduce its amount of relief. Place the shot point fairly far to the right of the step. Using Table 4.1 and Table 4.10, gather appropriate travel times for reflections from both portions of the offset interface and for diffractions from the upper edge of the step. Be sure not to show reflections at positions where they cannot occur. Finally, graph the data.

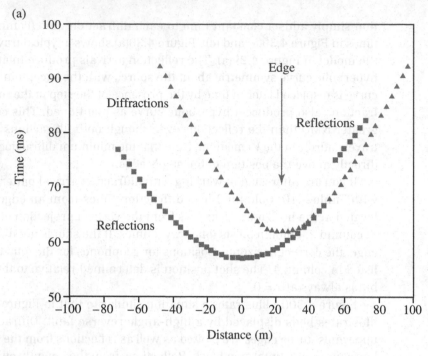

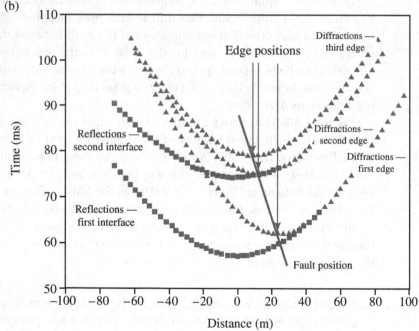

FIGURE 4.30 Diffraction and reflection patterns over a vertical step and an inclined fault. See Figure 4.29 for symbols and models. (a) Diffractions and reflections for the general case where the energy source is displaced from the diffracting edge: $h_1 = 40$ m, $h_2 = 5$ m, and $x_s = -24$ m. (b) Some diffractions and reflections from a model similar to that in Figure 4.29 (b). Parameters similar to (a) except additional edges are present at 52 m and 55 m and are offset in the $-x$ direction as indicated.

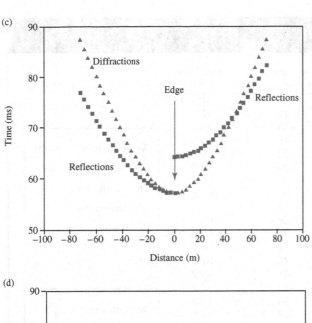

(c)

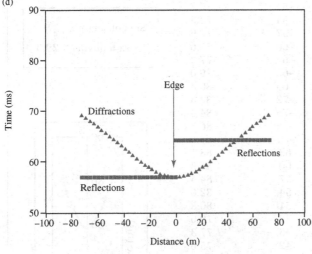

(d)

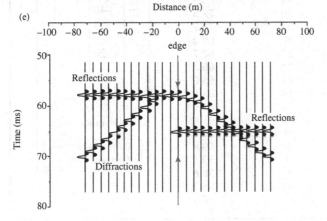

(e)

FIGURE 4.30 *(continued)*
(c) Diffractions and reflections
from a vertical step where the
energy source is located directly
over the step: $h_1 = 40$ m, $h_2 = 5$ m.
(d) Arrival times in (c) corrected
for NMO of reflection times.
(e) Simulated seismic section
based on times in (d).

TABLE 4.10 | DIFFRACTION TRAVEL TIMES

Geophone	Distance (m)	Diffraction Time (ms)	Offset Correction (m)	Total Time (ms)
Source	0	28.6	−9	57.9
1	−3	28.7	−12	57.9
2	−6	28.9	−15	58.2
3	−9	29.3	−18	58.6
4	−12	29.8	−21	59.1
5	−15	30.5	−24	59.8
6	−18	31.3	−27	60.6
7	−21	32.3	−30	61.6
8	−24	33.3	−33	62.6
9	−27	34.5	−36	63.8
10	−30	35.7	−39	65.0
11	−33	37.0	−42	66.3
12	−36	38.4	−45	67.7
13	−39	39.9	−48	69.2
14	−42	41.4	−51	70.7
15	−45	43.0	−54	72.3
16	−48	44.6	−57	73.9
17	−51	46.3	−60	75.6
18	−54	48.0	−63	77.3
19	−57	49.7	−66	79.0
20	−60	51.5	−69	80.8
21	−63	53.3	−72	82.6
22	−66	55.1	−75	84.4
23	−69	57.0	−78	86.3
24	−72	58.8	−81	88.1
24	72	58.8	63	88.1
23	69	57.0	60	86.3
22	66	55.1	57	84.4
21	63	53.3	54	82.6
20	60	51.5	51	80.8
19	57	49.7	48	79.0
18	54	48.0	45	77.3
17	51	46.3	42	75.6
16	48	44.6	39	73.9
15	45	43.0	36	72.3
14	42	41.4	33	70.7
13	39	39.9	30	69.2
12	36	38.4	27	67.7
11	33	37.0	24	66.3
10	30	35.7	21	65.0
9	27	34.5	18	63.8
8	24	33.3	15	62.6
7	21	32.3	12	61.6
6	18	31.3	9	60.6
5	15	30.5	6	59.8
4	12	29.8	3	59.1
3	9	29.3	0	58.6
2	6	28.9	−3	58.2
1	3	28.7	−6	57.9

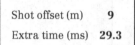

Velocity (m/s)	**1400**
Thickness (m)	**40**
Increment 1 (m)	**−3**
Increment 2 (m)	**−3**

Shot offset (m)	**9**
Extra time (ms)	**29.3**

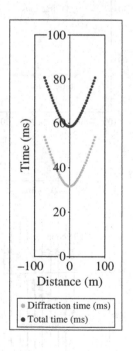

Diffraction time (ms)
Total time (ms)

This section emphasizes the misleading effect diffractions might have in suggesting the presence of structures or interfaces that simply do not exist. This is in addition to the other deleterious effects noted previously. Diffraction effects are especially common in seismic sections acquired during petroleum exploration efforts, and a great deal of effort and funds have been expended to deal constructively with this problem. Most of this occurs during data processing and will be mentioned briefly in a following section. Diffraction effects generally are not as obvious a problem in shallow work, largely because short geophone spreads often are sufficient to solve the problem at hand. In these cases diffraction patterns are not sufficiently well developed to be recognized as such. However, many diffractions likely are present on records and function mainly as interference, the origin of which cannot be ascertained. When continuous profiling is employed for shallow investigations, diffractions may appear as clearly developed as on many deeper records. For example, in Figure 4.36 study carefully the wavelet patterns in and around the reflections produced by bedrock hills. Certainly many patterns reminiscent of those in Figure 4.30 are in evidence. Other good illustrations of diffraction patterns on seismic sections can be found in Dobrin and Savit (1988, 310), Kearey and Brooks (1984, 90–93), and Robinson and Coruh (1989, 185). Now let's turn to a general discussion of how one goes about creating a record such as that in Figure 4.36.

4.5 COMMON FIELD PROCEDURES

In this section we are interested primarily in three fundamental field operations using geophone spreads that are common in shallow reflection surveys. Before beginning that discussion, however, a few comments are in order concerning hardware considerations. Much of this material was covered in Chapter 2, and you are urged to read once again the sections addressing energy sources, geophones, and seismographs.

4.5.1 Equipment Considerations

In reflection work we constantly are concerned with enhancing the high-frequency component of the seismic signal. One reason is to remove lower-frequency noise to improve our chance to observe reflections. The other reason is to improve resolution so we may observe detailed structures in the shallow subsurface.

As an example of the importance of frequency in vertical resolution, consider Figure 4.31. This diagram illustrates two thin reflectors spaced 7.5 m apart. We assume average velocity is 1500 m/s. Also illustrated are two seismic pulses with wavelengths λ of 10 m and 20 m, which have frequencies of 150 and 75 Hz respectively. These are considered to be sine curves for simplicity. The signal

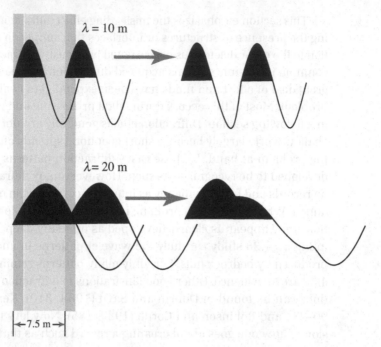

$\lambda = 10$ m

$\lambda = 20$ m

←7.5 m→

FIGURE 4.31 Diagrammatic explanation for increased vertical resolution achieved from higher-frequency (shorter-wavelength) pulse. The 10 m wavelength pulse delineates both reflectors, but the 20 m pulse effectively records only a single reflection event.

received at a given geophone from such a reflection will be the composite waveform illustrated. Clearly, the 150 Hz signal delineates both thin beds, but the 75 Hz signal does not. Although Figure 4.31 is diagrammatic only and thus is simplified (for instance, reflections from the bottom of each bed are not considered), it does emphasize the role of signal frequency in vertical resolution. Many seismologists maintain that the limit of vertical resolution is $\lambda/4$ (Sheriff and Geldart 1982, 119; Sheriff 1989, 330–31); but in shallow work the more likely limit is, at best, $\lambda/2$.

Does frequency affect horizontal resolution, and, if so, in what way? In considering this question, we must recall that wave fronts actually are spherical and not the planar wave fronts we have assumed for our derivations. The radius R_1 of the area on the reflector primarily responsible for producing the signal (the first Fresnel zone) is determined by

$$R_1 = \tfrac{1}{2} V_1 \left(\frac{t_0}{f} \right)^{\tfrac{1}{2}} = (\tfrac{1}{2} \lambda h_1)^{\tfrac{1}{2}} \qquad (4.61)$$

where V_1 is velocity, t_0 is travel time, h_1 is thickness, and f is frequency (Sheriff and Geldart 1982, 119–21). If we assume a velocity of 1500 m/s and a thickness of 30 m, then a frequency of 100 Hz has a Fresnel zone with radius equal to 15 m, and a 50 Hz signal leads to a radius of 21 m. We achieve better horizontal resolution with higher frequencies as was the case in vertical resolution. Of course, geophone spacing also determines in what detail we sample the subsurface. For horizontal reflecting surfaces, the area of reflection is located halfway between receiver and source. If our geophone spread has a geophone interval of 10 m, we are sampling the subsurface at intervals of 5 m, whereas an interval of 3 m results in a 1.5 m spacing of samples (see Figure 4.32).

To create and receive these high frequencies, we use energy sources that tend to produce a high-frequency spectrum with substantial energy values. As noted in Chapter 2, this often tends to be a shotgun source for shallow work. High-frequency geophones, typically 100 Hz, are desirable as is a seismograph with filters that can remove lower frequencies. These typically are set at 100 or 200 Hz. In good conditions when we are interested only in mapping the bedrock surface, such equipment demands may not be necessary. We have had good results with bedrock reflections from an average depth of 50 m using a hammer source, 14 Hz geophones, and 100 Hz low-cut filters. However, we can obtain better reflections and more detail in the material above bedrock when using a 10-gauge shotgun source, 100 Hz geophones, and 200 Hz low-cut filters.

Although not directly related to frequency considerations, a signal enhancement seismograph is essential. When using high-frequency geophones and low-cut filters, a substantial portion of the seismic energy is removed. It often is mandatory to fire two or more shots, thereby summing the incoming signal, to produce a record with discernible reflections. Digital recording capabilities also are necessary for all but the most basic reflection survey in order to process the recorded data by computer. The following section discusses typical geophone spreads for reflection work. Two of the three rely on digital processing to greater or lesser degrees, and the remaining one can benefit from some processing.

4.5.2 Geophone Spreads

In this section we cover three fundamental approaches to the acquisition of seismic reflection data. One is the standard split spread we already mentioned. The others, used primarily when seismic sections are desired, are referred to as the *common offset* method and the *common depth point* method.

4.5.2.1 SPLIT SPREAD
When we use reflection surveying primarily to determine subsurface characteristics at one location, split spreads with a center shot arrangement normally are employed. The first step is to deploy a geophone string and to record data beginning with a small shot offset. This serves two purposes. First it supplies data that can be used for refraction determination of the near-surface velocities to be used in static corrections. It also supplies some information about the optimum window. Most likely the shot offset will have to be increased until the optimum window is clearly delineated.

This is a critical phase of the survey and must not be given short shrift. The major subsurface target of the exploration work needs to be clearly defined because this affects the optimum offset choice. Also, we need to analyze noise factors and to observe how low-cut filters affect the quality of the desired reflections. For example, a 200 Hz low-cut filter may remove a substantial portion of noise but reduce the amplitude of target reflections to an undetectable level. Once we have the correct shot offset and the best filter combination, we start the general survey.

Typically in shallow reflection surveys a group of 12 (or 24) geophones is strung along a preselected survey line (GL_1 in Figure 4.32). If possible the geophones are placed in augered holes to improve ground coupling and to reduce interference from the air wave. One or more shots are taken at the preselected shot offset (E_1 in Figure 4.32). The geophone string then is moved to the right of the shot point GR_1 and more shots are taken. This process and geometry result in records containing the best possible reflections of interest, permit dip determinations if dips are present, and produce (we hope) good NMO to aid in delineating reflections and for velocity and depth determinations. If the survey purpose is to supplement refraction work or to obtain information that was not possible with refraction (too large an energy source required or other reasons), the survey might end at this stage.

Typically more information is desired, so additional lines are deployed. The geophone string GR_1 is left in place, and the shot is moved to E_2. The string now functions as GL_2 and records information from additional points on the reflector. The next step is to move GL_2 to GR_2 and so forth. This process of accumulating data along a line is termed *profiling*. Note that in the case illustrated in Figure 4.32 not every point on the reflector is sampled. If this is unacceptable, we can institute a slightly different procedure to ensure complete coverage. A seismic

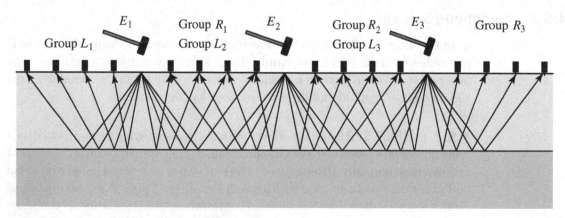

FIGURE 4.32 A simple form of seismic profiling with single coverage, center-shot spreads. A 4-geophone spread is shown for simplicity, although shallow exploration spreads typically contain 12 or 24 geophones.

section can be produced from these data by correcting for statics and removing NMO as discussed in the following section on data analysis. However, if a seismic section and continuous coverage are desired, we usually would select either the common offset or common depth point method.

4.5.2.2 COMMON OFFSET

The common offset method (also termed *optimum offset*) was developed by Hunter et al. (1984) and is remarkably simple to implement and actually quite efficient. As already described, we first must delineate the optimum window and determine the optimum offset. Once we know this value, the geophones are distributed and the first shot (E_1 in Figure 4.33) is taken. The channel on the seismograph corresponding to geophone G_1 is then "frozen" so it will not be affected by subsequent shots. The remaining channels are cleared to remove all data, and the next shot is taken at E_2. Channel G_2 is frozen, the display is cleared, and shot E_3 is executed. This process continues until all 12 (or 24) channels contain data. The data are transferred to disk or tape via digital recording capabilities, all channels are cleared, and the process continues.

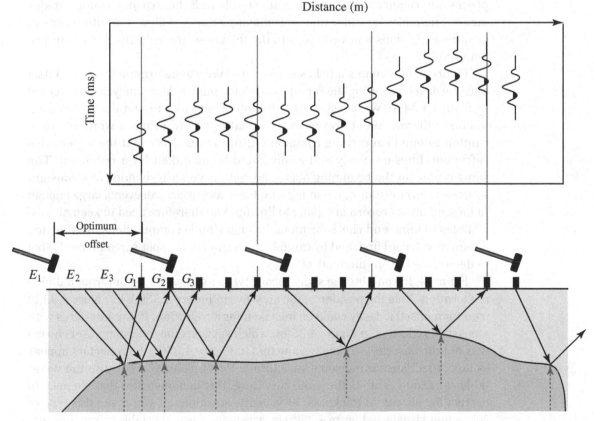

FIGURE 4.33 An example of common offset profiling. Dashed-line arrows emphasize position of reflecting points relative to position of receiving geophones.

Although only every fourth shot position is illustrated in Figure 4.33, note that common offset procedures produce continuous subsurface coverage. Also, and most important, is the lack of normal move-out in the field seismogram because all source–receiver distances are the same. Thus NMO corrections are not required to produce a seismic section, and the raw data give a reasonable indication of the reflector geometry. Close examination of Figure 4.33 does, however, reveal two important relationships to keep in mind. First, the "relief" of the reflector as portrayed on the seismogram will be enhanced or depressed depending on the velocity of the overlying material. Of course, this can be corrected, as we will see when discussing data processing operations. Second (because shot points in the diagram are to the left of geophones), the undulations on the reflector are shifted to the right relative to the reflecting point but by unequal amounts depending on the local orientation of the reflecting surface. Therefore, there is a "stretching" distortion in the horizontal plane as well. In most cases where relief on reflectors is modest and/or interfaces have only slight dips and for most survey purposes, this horizontal distortion is not appreciable and is acceptable.

Although common offset records normally should be digitally processed, the processing requirements can be quite simple and the expense is not a major factor. The major consideration is removing effects of elevation differences in geophone locations and variations in the thickness and velocity of near-surface material.

Figure 4.34 contains a field seismogram taken to determine optimum offset and one developed using the common offset technique. After analyzing the record in Figure 4.34(a), we selected an optimum offset of 42 m and then proceeded to obtain the common offset record. This latter record brings a word or two of caution to mind concerning common offset records. Note that the water table refraction lines up nicely and easily could be mistaken for a reflection. The same is true for the beginning of ground roll. As you are shooting at a constant source–receiver distance, assuming a uniform subsurface, all events large enough to be seen on the record are going to line up. You therefore need to keep all possibilities in mind and check for them through simple computations, by shooting a refraction record first, and by careful observation of the split-spread records shot to determine the optimum offset.

Figure 4.35 contains the same record types as Figure 4.34 but from a different location. Note the position of the air wave in Figure 4.35(a). In Figure 4.35(b) (common offset) it easily could be mistaken for a reflection. It also interferes with a possible reflection in Figure 4.35(b), which suggests the optimum offset chosen was not the best choice. Finally, note that in Figure 4.35(b) the reflectors appear to be curved, suggesting some structure or relief. However, because the water table refraction exhibits the same curvature, this demonstrates that we need to correct for elevation, thickness, or velocity variations. If you make this type of correction visually to Figure 4.35(b) by arranging the water table refractions into a straight line, the other reflectors become more or less horizontal also.

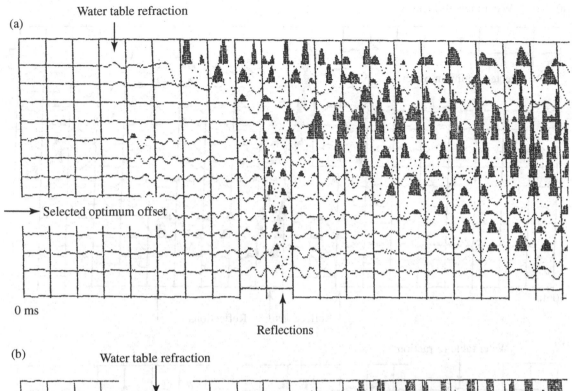

(a)

Water table refraction

Selected optimum offset

0 ms

Reflections

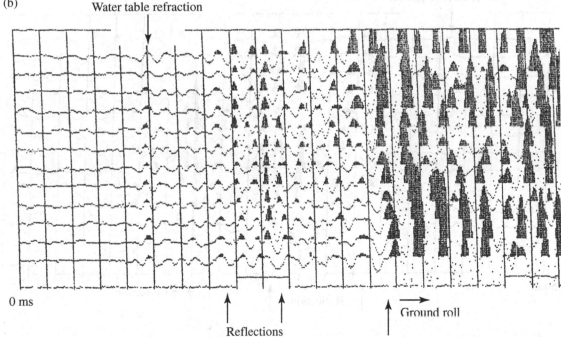

(b)

Water table refraction

0 ms

Reflections

Ground roll

FIGURE 4.34 (a) Field seismogram recorded with an 18 m shot offset and a 3 m geophone interval. (b) Field seismogram recorded using the common offset method. Offset to each geophone is 42 m. Note that the water table refraction easily could be mistaken for a reflection.

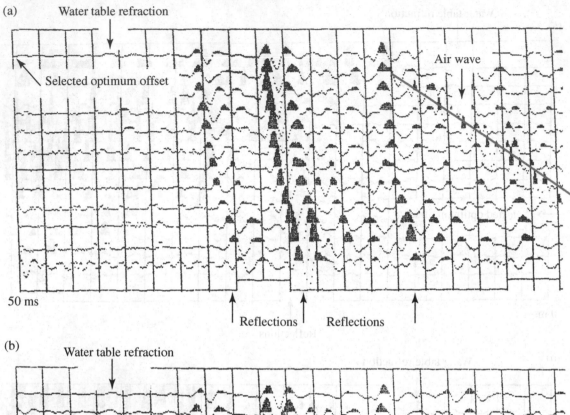

(a) Water table refraction

Selected optimum offset

Air wave

50 ms

Reflections   Reflections

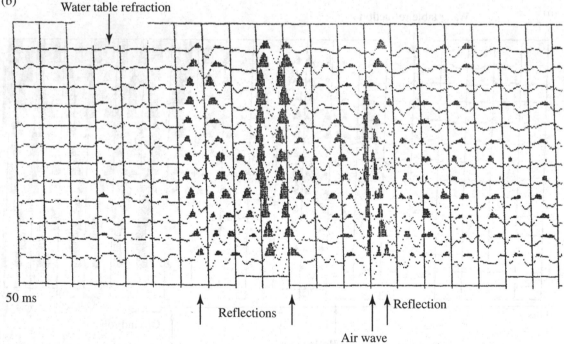

(b) Water table refraction

50 ms

Reflections   Reflection

Air wave

FIGURE 4.35 (a) Field seismogram recorded with a 60 m offset and a 3 m geophone interval. (b) Common offset record shot with a 60 m offset. Note how the air wave lines up and appears to be a reflection.

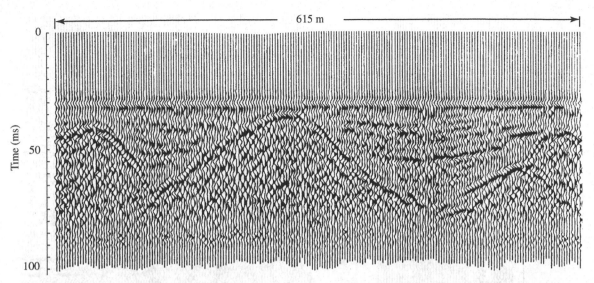

FIGURE 4.36 Seismic section recorded near Shawville, Quebec. Shot offset 30.5 m. Geophone interval 3 m. Maximum relief of the bedrock hills is approximately 40 m. Notice good reflections within unconsolidated glacial sediments deposited between the bedrock highs. (Modified from J.A. Hunter, S.E. Pullan, R.A. Burns, R.M. Gagne, and R.L. Good. 1984. Shallow seismic reflection mapping of the overburden–bedrock interface with the engineering seismograph—some simple techniques. *Geophysics,* v. 49, 1381–85.)

As an example of what can be achieved with the common offset method, examine Figure 4.36. This remarkable record of bedrock highs and lows was produced using a 7.25 kg hammer, 50 Hz geophones, 300 Hz low-cut filters, a 12-channel signal enhancement seismograph, and an Apple II computer for digital processing. Although this section was produced at what might be termed an "optimum site" in terms of available structure, bedrock depths, and good energy coupling, it does emphasize the capabilities of the common offset reflection method at localities for which the reflection potential is moderate to good.

The seismic section illustrated in Figure 4.37 also exhibits several strong reflectors. As we might expect, the bedrock reflector is clearly delineated. Compact till also has a high velocity, and the contrast between till and overlying sands and gravels would be expected to produce a good reflection, which it does. Note also the strong reflections due to contacts between clay beds and sands and gravels.

4.5.2.3 COMMON DEPTH POINT
Much of what we have said about field and equipment considerations also applies to this method. The main objective of common depth point (CDP) investigations is to sample each subsurface point several times. If the information on each record is then summed, after suitable pre-processing to be described in the next section, true reflection arrivals will be

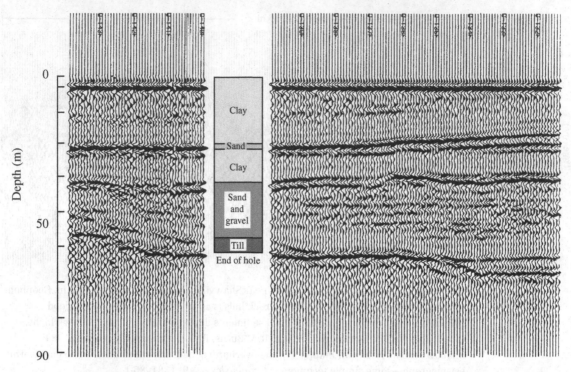

FIGURE 4.37 Seismic section recorded near Val Gagné, Ontario. Notice the excellent reflections from bedrock, till, and clay–sand interfaces. Survey parameters: shot offset = 15 m, 300 Hz low-cut filters, 100 Hz geophones, 12-gauge Buffalo gun source. (Courtesy James Hunter and Susan Pullan, Geological Survey of Canada.)

enhanced and various unwanted signals will tend to be reduced or eliminated, thereby producing superior records. A key component of CDP work is extensive computer processing of digital records for velocity analysis, removal of NMO, summing of traces from the same subsurface point, and various other crucial corrections.

To understand the basic design of the CDP method, let's examine Figure 4.38(b). For simplicity we'll assume we are operating with a string of 6 geophones, although the actual minimum we would use would be 12. The first shot point is E_1, and the geophone string for this shot is labeled $G_{11} \ldots G_{61}$ (geophones 1–6 and shot 1). The field record corresponding to shot E_1 is so designated in Figure 4.38(a). The shot point then is moved to E_2, and the geophone string position becomes $G_{12} \ldots G_{62}$. Another field record is made, the shot point is moved again, and so on. There are a number of procedures for speeding fieldwork so that the entire string of six geophones is not moved for every shot. One obvious option is to move the geophone nearest the shot next to the most distant geophone after each shot. However, the main procedural point to grasp is the manner in which the shot is moved along the line.

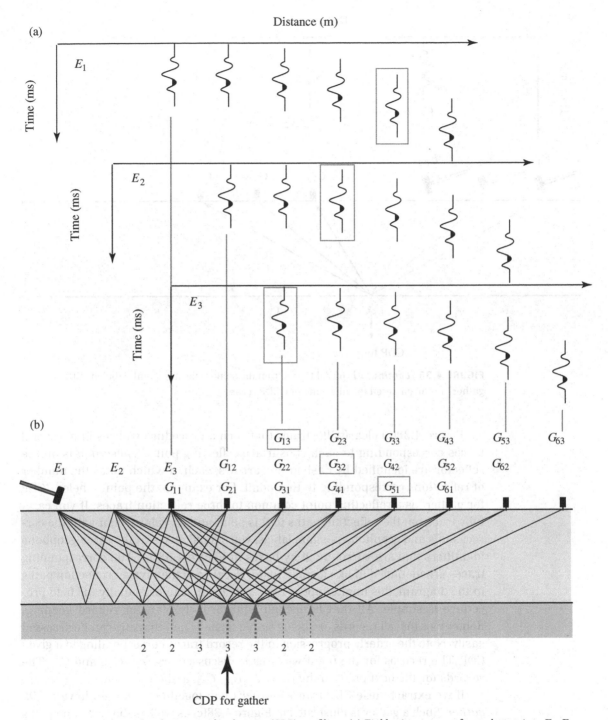

FIGURE 4.38 An example of common depth point (CDP) profiling. (a) Field seismograms from shot points E_1, E_2, and E_3. Traces from a common point on the reflector are outlined. (b) Geophone positions and shot points for CDP profiling. Dashed-line arrows designate positions on the reflector from which more than one trace is recorded.

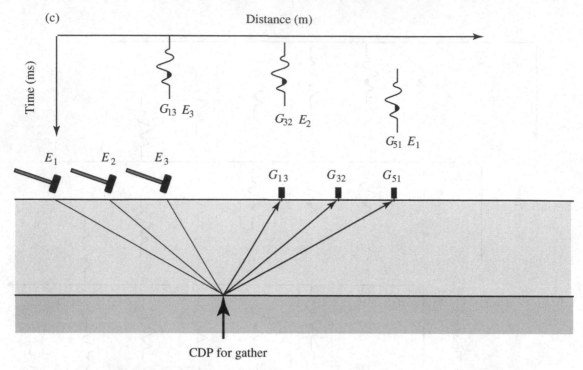

FIGURE 4.38 (*continued*) (c) All traces from the same reflecting point (labeled "CDP for gather" in (b)) gathered together into one CDP record.

Figure 4.38(b) clearly illustrates that such a procedure will result in several traces corresponding to each subsurface reflecting point. These points on the reflector are identified by dashed-line arrows, each of which notes the number of reflections corresponding to that point. For example, the point labeled "CDP for gather" is a reflecting point common to three reflection traces. If you carefully trace out the reflection paths to this point and correlate them with the correct shots and geophone position labels, you should arrive at the three geophone identifiers that are surrounded by boxes in Figure 4.38(b). The corresponding traces are identified in like fashion in (a). Because of the myriad reflection paths in the diagram, this might appear to be very complicated. Actually the field procedure is simple. All that is needed is an orderly plan for record keeping. Remember that all records are being stored digitally and that any can be accessed easily. Note the orderly progression in the record traces corresponding to a given CDP. The records for the point we already discussed are G_{51}, G_{32}, and G_{13}. The records for the next point to the right are G_{61}, G_{42}, and G_{23}.

If we extract these CDP traces (*gather* them together), we then have a *CDP gather*. Such a gather is illustrated in Figure 4.38(c) as well as the reflection paths corresponding to these traces. In an actual survey, more geophones would be used, and we would have more traces per CDP. In our illustration we have threefold coverage. In most shallow exploration work the minimum coverage would be sixfold. As long as we keep the distance between shot points and the geophone interval the same, and if n represents the number of geophones, our coverage is equal to $n/2$-fold.

It is easy to see why a CDP gather is ideal for determining the NMO associated with a given reflection because all traces on the record are from the same subsurface point. Moreover, after we correct for NMO, we can sum these traces to produce an enhanced reflection signal. In this simple case any other undulations on these traces are unlikely to undergo substantial enhancement when the traces are summed. Therefore, the CDP method is extremely effective in improving the amplitude of reflections while at the same time reducing unwanted noise.

Now it is time to discuss the various digital processing techniques required to transform raw reflection data into a seismic section. Unlike split-spread and common offset profiling, which provide only single-fold coverage, common depth point profiling demands digital processing from the onset. Because of the computer power required to perform this task, the CDP method has received less use in shallow reflection work to date than the other methods. Personal computers now are capable of performing CDP processing, and as their power increases and cost decreases, it is likely that CDP will become the method of choice in shallow exploration as it is in the exploration endeavors performed by petroleum firms.

4.6 COMPUTER PROCESSING OF REFLECTION DATA

As mentioned previously, enormous amounts of computer time on very high-speed computers are regularly utilized by petroleum exploration companies to process seismic reflection data. Such time, expense, and expertise normally are not available to the small (sometimes one-person) companies active in solving geotechnical problems using shallow reflection surveys. Although this rapidly is changing, our goal in this section is to concentrate on processing that is essential to meet the normal objectives of the shallow survey while briefly introducing techniques that may be applied to shallow reflection data and that normally are utilized when targets are located at greater depths.

4.6.1 The Static Correction

The first step in processing reflection data is to apply both a topographic correction and what is called the *weathering correction*. The reasons and strategy are similar to those used in refraction work and were discussed in Chapter 3. In perhaps the majority of cases, a low-velocity layer that varies in thickness is present at the surface. Because the velocity of this layer is substantially lower than the layer underlying it, travel times are affected unequally to an extent that cannot be ignored.

A generalized situation is presented in Figure 4.39(a). In shallow reflection surveys the low-velocity layer often consists of unsaturated sediments that have low velocities relative to their saturated counterparts situated beneath the water table. Hence the term "weathering correction" is not entirely appropriate here, but we retain it because of its widespread usage in the petroleum industry. The symbols h_{w_g} and h_{w_s} refer to the thickness of the low-velocity layer beneath the geophone and shot point, and e_g, e_s, and e_d refer to the elevations of the geophone, shot point, and datum respectively.

In making these corrections we assume vertical paths through the low-velocity layer. Due to the velocity contrasts this is not a bad assumption; but in situations where relief is excessive or the velocity contrast is not large, it might be wise to sample the error introduced by assuming vertical paths. Is this assumption acceptable for the investigation diagrammed in Figure 4.39(b)? Assume our maximum source–geophone offset is 40 m and the thickness of the 1500 m/s layer is 40 m. We use Snell's Law and the given velocities to calculate the angles of incidence and refraction for the ray paths illustrated. Using these angles and the thickness of the low-velocity layer, we find that the actual ray path and th`e vertical path differ by 0.01 m. At 400 m/s this translates into a time difference of 0.025 ms. If you examine the time scale on any of the field seismographs illustrated in Chapters 3 or 4, you should see that this time difference is negligible.

The first step is to remove the effect of the low-velocity layer. Let t_w represent the weathering correction time. In the case illustrated in Figure 4.39(a), the correction represents the time of travel at 400 m/s less the time of travel at 1500 m/s (essentially we calculate the delay due to this material possessing a velocity of 400 m/s rather than a velocity of 1500 m/s). In mathematical form

$$t_w = \frac{h_w}{V_w} - \frac{h_w}{V_1} \tag{4.62}$$

where h_w represents either h_{w_g} or h_{w_s} depending on whether we are correcting for the geophone or the shot.

Once the weathering correction is applied, we effectively change the 400 m/s velocity to 1500 m/s. The topographic correction now is applied simply by dividing the difference in elevation between the geophone (or shot) and the datum by, in our example, 1500 m/s. Therefore, the elevation correction is

$$t_e = \frac{e_g - e_d}{V_1}. \tag{4.63}$$

These corrections are simple to apply if the necessary information is available. If reflection spreads do not have source–geophone distances sufficiently small to sample refractions from the base of the low-velocity layer, then special refraction surveys are undertaken. If possible, the datum should be chosen as the lowest point in the survey because statics always reduce arrival times. Note that the static correction is applied to the entire trace. Every point on a trace

(a)

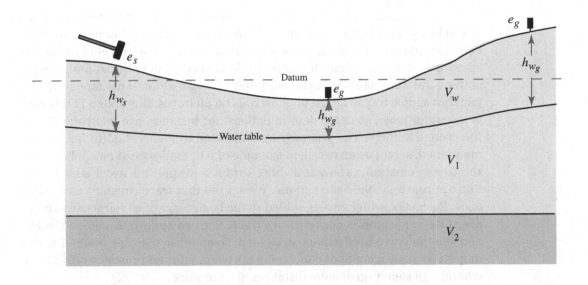

(b)

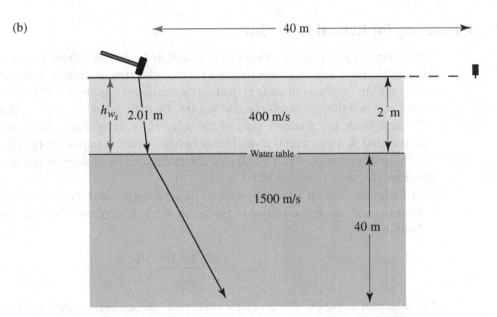

FIGURE 4.39 Static corrections. (a) Parameters used in elevation and weathering corrections. (b) A model demonstrating the validity of the vertical path assumption.

from a given geophone is moved the same amount in time depending on static correction for that geophone position.

In some situations a simpler approach may suffice. Assume the base of our low-velocity layer is the water table. If we have common offset data, we typically can identify the water table refraction arrival. Also, if relief along our spreads is not great, then the water table is very close to horizontal. Simply shifting the water table refractions so they all begin at the same time is an elegant but simple way to adjust later arrivals on all traces. Our datum in this case is the water table, so depth determinations, for instance, are relative to it, not the ground surface. The seismogram illustrated in Figure 4.35(b), as already mentioned, serves as an excellent example of why statics must be applied. This record also emphasizes how straightforward it is to apply the water table refraction approach to static corrections. We set the first trace constant as a reference. Each succeeding trace is shifted so the beginnings of all refraction arrivals are at the same time as our reference trace. You can apply this method visually and see the curved reflections straighten. Remember that this method works only given our assumptions concerning the water table and *common offset data,* wherein all source–geophone distances are the same.

In contrast to static corrections, *dynamic* corrections move different parts of a given trace by different amounts in time. It is to these that we now turn.

4.6.2 Correcting for Normal Move-Out

If our primary goal in a reflection survey is simply to identify reflectors and determine thicknesses and velocities within a very limited area, we might work entirely with split-spread data, identify reflections visually, and apply $x^2 - t^2$ analyses to arrive at a solution for subsurface structure. Usually, however, if we undertake reflection work, our goal is a view of the subsurface in time and distance coordinates (that is, a seismic section). In this case we attempt to plot every reflection arrival at its correct t_0—as if it traveled a vertical path to a reflector and back. In other words, we correct for NMO.

Of course, we can take split-spread data, determine thicknesses and rms velocities, and use these values in Equation 4.5 to determine NMO values (see Equation 4.64):

$$T_{NMO} = \frac{(x^2 + 4h^2)^{\frac{1}{2}}}{V_{rms}} - \frac{2h}{V_{rms}}. \tag{4.64}$$

We then can use these NMO values to correct various reflections on all traces. This would be done via computer, but still would demand a tremendous amount of time and effort from the person undertaking such an analysis. Imagine all the steps involved and the number of maneuvers in a relatively simple case where each 12-geophone spread contains six obvious reflections and where eight spreads were recorded. It doesn't take much of that kind of labor to search for a better way.

One approach is to record only common offset data that do not require NMO corrections. In such a case our section does not display vertical path arrivals, but we still produce an acceptable picture of the subsurface if we keep our sampling interval small (for example, geophone intervals of 1–3 m). However, we still require velocity information to obtain depths for our sections, so a somewhat more automated approach remains desirable.

Once we begin to acquire CDP data in any quantity, some automation is mandatory. In addition, the approaches described work only with reflection arrivals that are clear and obvious. As long as we are automating, why not inquire if other advantages might also accrue through computer processing—such as identifying weak reflections? Although a number of strategies exist for automated velocity determinations, we concentrate here on outlining three separate designs, which for convenience we label "Velocity Analysis A," "Velocity Analysis B," and "Velocity Analysis C." All involve computer processing and have elements in common, but "B" is somewhat more sophisticated than "A," and "C" typically employs much more sophisticated algorithms and demands considerably more processing power than either "A" or "B."

4.6.2.1 VELOCITY ANALYSIS A
A typical computer program used to facilitate this analysis uses an equation such as Equation 4.65 to compute the NMO correction:

$$T_{NMO} = \frac{x^2}{2t_0 V_{st}^2} - \frac{x^4}{8t_0^3 V_{st}^4}.$$

(4.65)

Of course we know the offsets x used in our field design, but we do not know velocities or t_0. Remembering that we do not want to employ the $x^2 - t^2$ method but wish to somehow automate this determination, we rely on the computer's speed to solve the problem essentially by a trial and error method. Here we refer to the velocities we select for trials as *stacking velocities* or V_{st}. These are similar in value to root-mean-square velocities (determined via Dix's method) but are arrived at through the trial and error process described next.

Let's examine a very simplified outline of how this proceeds using a dynamic table, Table 4.11, as an aid. Assume we have a 24-trace record that contains two identifiable reflections. Figure 4.40(a) illustrates a time–distance curve for two reflections (time is oriented to increase downward as in a seismic section). To reduce clutter, we present only points on this curve rather than wavelets. Our goal is to find values of V_{st} and t_0 for each reflecting horizon such that NMO is completely removed for arrivals from that horizon; all points plot at the same time.

Our computer program will allow us to enter maximum and minimum values and increment sizes for both velocity and t_0. In its simplest implementation, the program takes these values, calculates NMO corrections for one combination, adjusts all digitized data points on all traces by the computed amount, and plots the results (Figure 4.40(b)). Another pair of values is selected, corrections once

TABLE 4.11 USING V_{st} AND t_0 TO AUTOMATE VELOCITY ANALYSIS

Distance (m)	Reflection Times (ms)	NMO Correction 1 (ms)	Corrected Time 1 (ms)	NMO Correction 2 (ms)	Corrected Time 2 (ms)	NMO Correction 3 (ms)	Corrected Time 3 (ms)
0	64.3	00.0	64.3	00.0	64.3	00.0	64.3
3	64.3	00.0	64.3	00.1	64.3	00.0	64.3
6	64.4	00.1	64.3	00.3	64.1	00.1	64.3
9	64.6	00.3	64.3	00.6	64.0	00.2	64.4
12	64.9	00.6	64.3	01.1	63.7	00.3	64.5
15	65.2	00.9	64.3	01.7	63.4	00.5	64.6
18	65.6	01.3	64.3	02.5	63.1	00.8	64.8
21	66.0	01.7	64.3	03.3	62.7	01.0	65.0
24	66.5	02.2	64.3	04.3	62.2	01.4	65.2
27	67.1	02.8	64.3	05.4	61.7	01.7	65.4
30	67.8	03.5	64.3	06.6	61.1	02.1	65.6
33	68.5	04.2	64.3	07.9	60.6	02.6	65.9
36	69.2	04.9	64.3	09.3	59.9	03.0	66.2
39	70.1	05.8	64.3	10.7	59.3	03.5	66.5
42	70.9	06.6	64.3	12.3	58.7	04.1	66.8
45	71.9	07.5	64.3	13.8	58.1	04.7	67.2
48	72.9	08.5	64.4	15.4	57.4	05.3	67.6
51	73.9	09.5	64.4	17.0	56.8	05.9	68.0
54	75.0	10.5	64.4	18.7	56.3	06.6	68.4
57	76.1	11.6	64.5	20.3	55.8	07.3	68.8
60	77.3	12.7	64.6	21.9	55.4	08.1	69.2
63	78.5	13.8	64.7	23.5	55.0	08.8	69.7
66	79.7	15.0	64.8	25.0	54.8	09.6	70.1
69	81.0	16.1	64.9	26.4	54.6	10.4	70.6
72	82.3	17.3	65.0	27.7	54.7	11.2	71.1

Model values:	Increment (m)	**3**
	Thickness (m)	**45**
	Velocity (m/s)	**1400**
Correction velocities: (m/s)	Velocity 1	**1400**
	Velocity 2	**1000**
	Velocity 3	**1800**
Correction t at $x = 0$: (m/s)	t_0-1	**64.3**
	t_0-2	**64.3**
	t_0-3	**64.3**

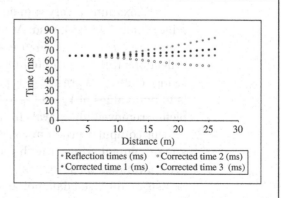

again computed, and results plotted (Figure 4.40(c)). The process continues until all combinations of t_0 and V_{st} that originally were designated are utilized and results plotted. You can gain a feeling for the process by entering trial values in Table 4.11 and observing the results of the corrections. Although you can enter only three pairs of trial values at a time, you can observe how quickly the computations proceed. Therefore, attempting a much larger number of trials would not be insignificant in processing time but would be much more efficient than attempting to employ $x^2 - t^2$ analyses.

Although Figure 4.40(d) illustrates a successful correction for the earlier of the two reflections on the record, the points at 60 m and more do not line up nearly as well as the points at distances less than 60 m. Can you explain the discrepancy, knowing that the reflecting horizon is at 45 m and that the velocity to this first interface is 1400 m/s?

If we entered reasonable ranges for t_0 and V_{st} before running our correction program, we should find some traces for which the corrections are successful (Figure 4.40(d) and (e)). Using these values, we can produce a final seismic section in which all reflections are corrected for NMO and also have the information necessary to determine interval velocities and thicknesses.

Assume that the digital sampling interval on our seismic traces is 0.5 ms. Further assume that a reflection arrives at 64.0 ms at a shot–geophone offset of 3 m. Velocities must be between 500 and 3000 m/s. Use Table 4.11 to determine realistic values for the minimum value, maximum value, and trial increment for the zero-offset time.

Although this approach is not elegant, it works reasonably well and does not require more than modern personal computer capabilities. It is adequate for much shallow exploration work, is not difficult or expensive to implement, and is equally useful for routine data or for CDP data. One drawback is the potentially large number of trial results that need to be scanned to pick the trial values that successfully remove NMO from a reflection. Its major drawback, however, is that we work only with reflections that can be identified in order to be checked for NMO removal. Methods "B" and "C" do not possess this drawback, but as you might surmise, they impose much more extensive computational demands.

4.6.2.2 VELOCITY ANALYSIS B A portion of this approach is similar to "A" in that many trials are executed using combinations of t_0 and V_{st}. For each set of trial values all traces are shifted to correct for NMO. After this adjustment the computational algorithm moves through all traces at small time increments. At

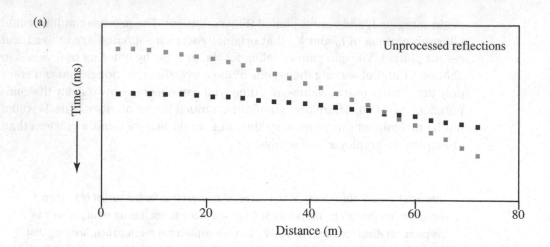

(a)

Unprocessed reflections

Time (ms)

Distance (m)

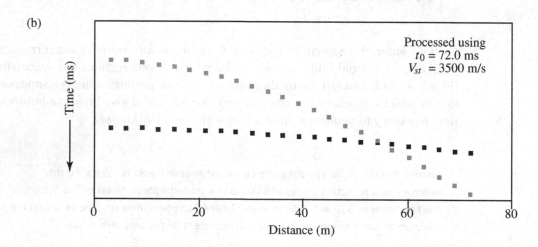

(b)

Processed using
$t_0 = 72.0$ ms
$V_{st} = 3500$ m/s

Time (ms)

Distance (m)

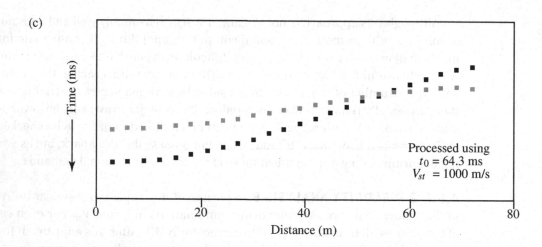

(c)

Time (ms)

Processed using
$t_0 = 64.3$ ms
$V_{st} = 1000$ m/s

Distance (m)

(d)

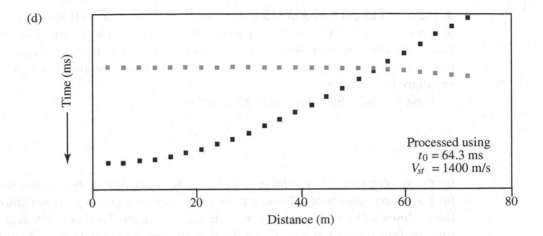

Processed using
$t_0 = 64.3$ ms
$V_{st} = 1400$ m/s

(e)

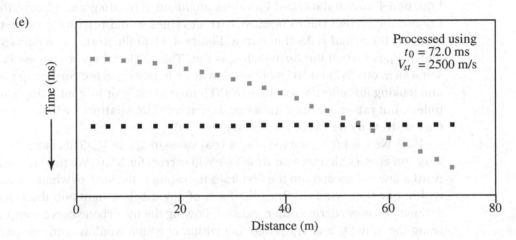

Processed using
$t_0 = 72.0$ ms
$V_{st} = 2500$ m/s

FIGURE 4.40 Velocity analysis of reflections processed using various values of t_0 and V_{st}.
(a) Original reflections. (b) and (c) Unsuccessful corrections. (d) Early reflection successfully
processed and stacking velocity determined. (e) Later reflection successfully processed and
stacking velocity determined.

each time increment the amplitudes of all traces are added together. If the trial
values are correct for a given reflection, after the NMO correction, all troughs
and peaks for that reflection will line up on the record. When added together,
these will result in a signal that is much greater than for other trial values that
are not correct for this reflection. If we plot amplitudes against all trial values
for a record on which several reflections are present, we should see several large
amplitude peaks. The t_0 and V_{st} associated with these peaks give us the information we need to produce a seismic section.

4.6.2.3 VELOCITY ANALYSIS C In this approach we assume that each point on a selected trace is the onset of a possible reflection and then test this assumption on all other traces. Of course, most tests will fail, but some will succeed. In this brief account we concentrate on how such tests are made and how success or failure is determined.

First we utilize Equation 4.42 and substitute V_{st}

$$t_x = t_0 \left(1 + \frac{x^2}{V_{st}^2 t_0^2} \right)^{\frac{1}{2}} \tag{4.66}$$

for the velocity term. If we select a t_0 and a V_{st}, we can calculate reflection times for the source–geophone offsets that we used in our field survey. As we know, these times will plot along a hyperbolic curve. Figure 4.41(a) illustrates a time–distance curve for an actual reflection and two curves determined using Equation 4.66 and the t_0 and V_{st} values identified in the diagram. Clearly, these are not the correct values because correct values would produce a curve that plots on the actual reflection curve. Figure 4.41(b) illustrates two curves for which V_{st} is correct but for which t_0 is not. This trial and error process is the same as in our "A" and "B" methods; however, here we are not correcting traces and looking for reflections with no NMO (either visually or by computing amplitudes), but rather we are comparing actual reflection patterns with computed curves. How do we make these comparisons?

First we compute a curve using a trial value of t_0 and V_{st}. This curve begins at t_0 and crosses all traces on which we will correct for NMO. We then move forward a few milliseconds on the first trace to "capture" the form of whatever exists within our time window. Once the form of any displacements on the trace is defined, we move across other traces, following the hyperbolic curve computed using the initial t_0 and V_{st}, remaining within our time window, and comparing curve forms from trace to trace. If an actual reflection lies within the window and follows the computed curve, the comparison algorithm will be successful and will return a high value. If not, a low value will be returned. In Figure 4.42(a) our time window cuts across the actual reflection, so comparison of waveforms yields a low value. In Figure 4.42(b) two combinations using a constant V_{st} and two values of t_0 fail; but one set of trial values, as indicated on the diagram, defines a window that parallels the NMO of the true reflection arrivals. This set of trial values would yield a very high comparison success value. After all initially selected trial values of t_0 and V_{st} are used, a table of comparison success values is created. These comparison values are analyzed to determine successful t_0 and V_{st} trials that most likely indicate reflections (or multiples).

4.6.3 Stacking CDP Gathers

At this point we have applied static corrections, undertaken velocity analysis, and computed NMO corrections for our traces. Although we noted in our discus-

(a)

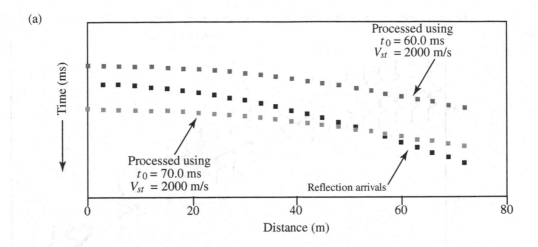

(b)

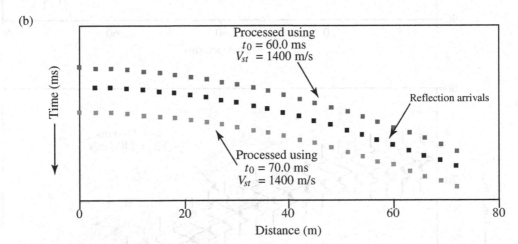

FIGURE 4.41 Hyperbolic curves computed for various values of t_0 and V_{st} compared with the time–distance curve for an actual reflection.

sions of these two corrections that they are applied to field data collected using a variety of field designs, most modern processing strategies use common depth point data. Recall that we gathered together traces representing reflections from the same depth point. Now that static and NMO corrections have been applied to these gathers, waveforms representing reflection arrivals should line up across the record if the reflecting surface is horizontal (Figure 4.43). If these traces are summed together (or *stacked*), reflection arrivals will be enhanced, whereas noise and other arrivals (such as refractions) in general should be degraded.

The stacked traces now can be presented in their correct position (for vertical travel paths), each at the position of their common reflecting point, to create

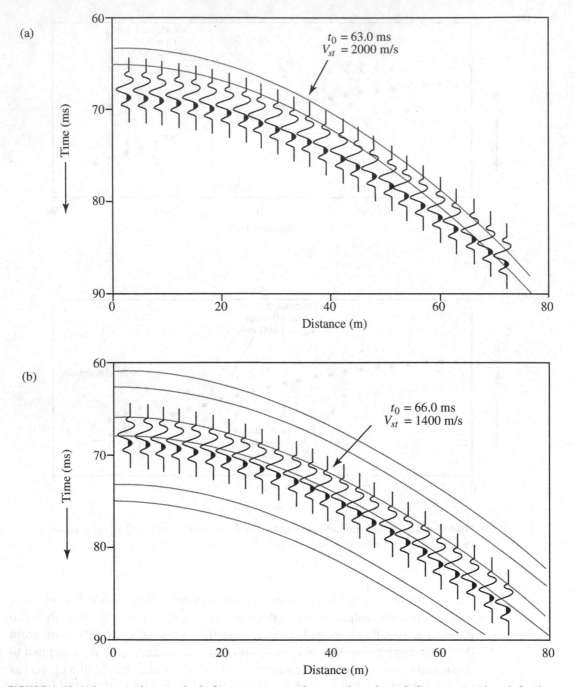

FIGURE 4.42 Velocity analysis method of comparing waveforms within a hyperbolic curve window defined by trial values of t_0 and V_{st}. (a) Trial values define a curve that cuts across reflection arrivals and for which comparison of waveforms within the window would return a low value. (b) Windows defined using the correct V_{st} and three values of t_0, one of which is correct. Comparison of waveform within the correct window clearly would meet with success and result in a high value.

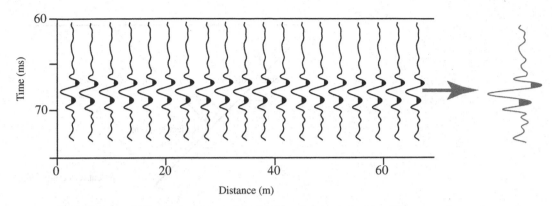

FIGURE 4.43 CDP gathers corrected for NMO and then stacked to yield a waveform of enhanced amplitude.

a seismic section. If diffractions are minimal and interfaces have only low dips or are horizontal, for many routine shallow investigations processing may end at this stage. After all, we have a reasonable representation of the subsurface in time and have the values required to compute interval velocities and thicknesses. However, a number of additional processing stages can improve the data still further. The time and cost of applying these additional techniques often is not justified for shallow work of limited extent and duration, but usually is justified when deeper targets with greater rewards are sought.

4.6.4   Migration

In our discussion of the production of seismic sections we assumed that reflecting points were directly beneath the zero-offset position. However, if the reflecting horizon is inclined, this will not be the case. Consider Figure 4.44(a). Here the reflector dips to the right, and the actual zero-offset path is indicated by the solid black line, with arrowheads at each end. When plotted in the conventional manner, the true reflecting point assumes the position identified as "apparent reflecting point." Although this clearly is not a trivial matter, its seriousness is underscored by Figure 4.44(b). First consider the reflecting surface inclined to the right. The apparent reflecting points related to this reflector define a line (gray in the diagram) that is displaced upward and to the right compared to the real reflector and is inclined at a different angle. If we add a reflector dipping to the left, as would be the case if a syncline is present, this surface also is displaced. The displacement of one point is illustrated by an arc with an arrowhead attached. The combination of the two apparent displacements creates a pattern of two crossing lines that possibly could be mistaken for an anticline and at least obscures the true structure.

We need to discover how to move these apparent reflecting points to their true positions. This process is referred to as *migration*. Modern migration techniques are computer-based and are extremely sophisticated. Because of this, and

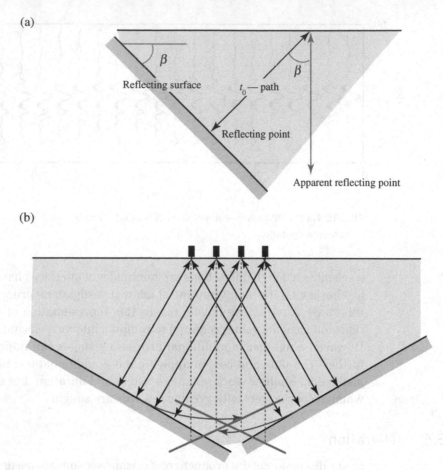

(a)

β

Reflecting surface

t_0 — path

β

Reflecting point

Apparent reflecting point

(b)

FIGURE 4.44 Displacement of true reflecting points when zero-offset traces are plotted directly beneath a receiver. (a) Diagram illustrating that displacement of reflecting point is about a circular arc and equal in degrees to dip of reflecting horizon. (b) Diagram illustrating same displacements as in (a) but demonstrating how false images can be produced when reflecting points are not plotted in their correct position.

because migration is not normally attempted in shallow work, we do not pursue the theory and operation of migration in this text. Several of the references in the suggested reading at the end of this chapter contain excellent summaries of this topic. Our goal in the next few paragraphs is to demonstrate two straight-forward approaches to migration that formed the basis of techniques used to migrate sections before computers assumed this task.

Consult Figure 4.44(a) once again. Both lines with arrowheads at each end are the same length. Given a surface receiving point and a reflection time, we know that the reflecting point must be located somewhere on a circular arc with radius of length $V_1(t_0/2)$. If we know the dip β of the reflecting surface, simple geometry demonstrates that the displacement is through the angle β.

Consider Figure 4.45(a). If we take each apparent reflecting point and move it along a circular arc, we see that a line tangent to all these arcs defines the true position of the reflector. Although such a procedure is applicable only in the simplest of cases, it does emphasize that true reflecting surfaces can be reconstructed.

Figure 4.45(b) illustrates how the apparent dip of a reflector can be determined using the apparent reflecting points. Once the dip is known, the position of the actual point is determined by calculating values for m and y. In this illustration lines with arrowheads at each end follow the same convention as in Figure 4.44. Because the vertical zero-offset path and the actual zero-offset path both have

(a)

(b)

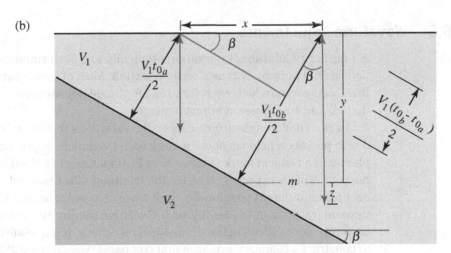

FIGURE 4.45 Illustrations of simple manual (noncomputer) methods for migrating reflecting points. (a) Migrating about circular arcs. (b) Geometric method to compute horizontal and vertical displacement of reflecting point.

the same length, we know the lengths of both reflection paths illustrated. We can determine from the diagram that the difference in path lengths is related to the dip of the reflector by Equation 4.67:

$$\sin \beta = \frac{V_1(t_{0_b} - t_{0_a})/2}{x}.$$ (4.67)

Given β, it is trivial to compute y, m, and z. Thus, knowing the position of the apparent reflecting point and having calculated m and z, we establish the position of the actual reflecting point.

In addition to placing inclined reflectors in their correct spatial positions and orientations, migration typically attempts to remove diffractions. We already examined diffractions and realize how they often obscure reflections. They are relatively easy to identify visually because of their typical hyperbolic form on seismic sections. Diffraction migration takes advantage of the special geometry of diffractions by employing computer algorithms that search for diffraction patterns and then remove them in a way that affects the surrounding waveforms as little as possible.

We are confident you realize by now that migrating a seismic section is possible, although exceedingly complex in practice, and that a migrated section is much superior to an unmigrated one for interpreting subsurface relationships. It also likely is evident that migration is costly in terms of computer processing time and in developing programs to perform the task. Thus it is attempted only where the value to be gained outweighs the cost of the process. For those of you inclined to pursue migration in greater detail, a particularly readable summary with examples is presented by Waters (1981).

4.6.5 Waveform Adjustments

A number of additional operations typically are performed on seismic traces before final presentation as a seismic section. Most of these may be grouped into three categories, which we refer to as amplitude adjustments, frequency adjustments, and transmission adjustments.

In the shallow subsurface especially, variations in transmission efficiencies due to pockets of heterogeneous materials and variations in ground coupling of geophones can result in marked differences in amplitudes of reflections from the same horizon. This is especially true for the common offset method because the shot occupies a different position for each trace. These variations may make reflections more difficult to identify or the seismic section harder to interpret as it becomes more difficult to trace individual reflections. It is a relatively simple matter to construct a computer program that compares waveforms within a narrow time window from trace to trace and then normalizes these trace-to-trace amplitudes. Also, arrivals from deep reflectors or interfaces with moderate reflection coefficients tend to have reduced amplitudes. It is again a straightforward matter to specify time windows in which amplitudes on all traces should be increased or decreased by a specified amount. Although a relatively crude approach, this often is remarkably effective in increasing the readability of shallow sections.

Frequency adjustments involve filtering the digital seismic record. You likely recall our discussion of filtering the incoming seismic signal to remove unwanted frequencies so that reflections will not be obscured as amplifiers became saturated with high-amplitude, low-frequency signals. Much of this effort was directed at removing or at least reducing ground roll with a large low-frequency component to preserve higher-frequency reflection signals. Of course, once we employ filters before recording data, the part of the signal removed is lost forever. As seismographs typically used for shallow exploration become more sophisticated and as digital memory limitations vanish, virtually all of the incoming signal can be preserved. We then apply digital filters to remove unwanted frequencies and thereby enhance signals of interest. The beauty of this approach is that no information is lost in the filtering process because the original digital record always is available. Thus various filter combinations can be applied to achieve the best possible results for a particular geologic situation.

Transmission adjustments attempt to remove certain effects that altered the shape of wavelets as they moved through the subsurface. Two simple examples emphasize the goals of this processing task. Recall that higher-frequency components of a seismic signal are removed at a greater rate than low-frequency components as the signal is transmitted through the Earth. One type of transmission adjustment, therefore, is to attempt to restore the lost higher-frequency components. Because in frequency filtering we attempted to remove components, this attempt to restore components typically is referred to as *inverse filtering*. Another example involves short-path multiples. In our discussion of multiple reflections, we noted that the main effect of short-path multiples is to lengthen the seismic wavelet. If lengthened wavelets appear at a number of positions on a trace, weaker reflections could be hidden or difficult to identify with confidence. Filtering these extended wavelets to remove all but the true reflection pulse clearly is a valuable operation. Yet another adjustment attempts to remove long-path multiples, which, when strong, also obscure reflections occurring late on the trace.

4.6.6 Seismic Sections: Time Sections and Depth Sections

After some or all of the previous corrections and adjustments to seismic data occur, the final seismic section is plotted and interpreted. The section can be presented with time increasing downward, which is referred to as a *time section* (Figure 4.46). In shallow exploration work time sections are the norm because the goals of the survey often do not warrant developing a *depth section*. A time section provides a general view of subsurface relationships and, when combined with a depth scale based on a velocity analysis, may suffice. In surveys where the main interest lies within the saturated sediment column above bedrock, the processing required to convert time sections to depth sections may not be warranted because velocities within the saturated sediments tend to be quite similar, and hence the spacing between reflectors on the seismic sections remains constantly related to actual thicknesses.

In situations where velocities vary with depth or may vary laterally, depth sections always should be prepared. This, of course, requires detailed knowledge of velocity values, both vertically and along the bearing of the seismic line. However, because computer-based velocity analysis is used to prepare the seismic section by removing NMO effects to stack CDP gathers, there is little excuse

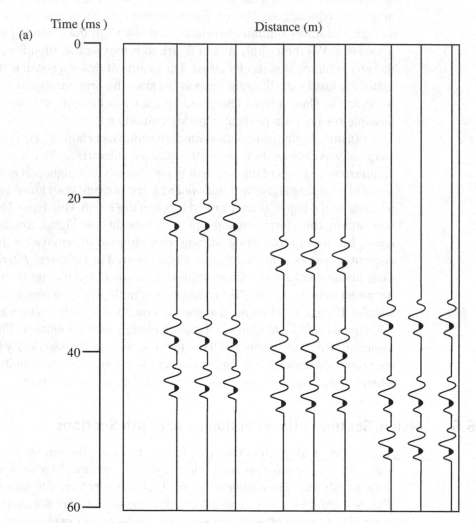

FIGURE 4.46 An illustration of time versus depth sections. (a) A time section in which thicknesses of all units remain constant but velocities of the upper two units vary laterally. (b) A depth section with all traces corrected according to well-defined velocity functions.

for not applying the next step and preparing depth sections. Figure 4.46 attempts to demonstrate the value of depth sections compared to time sections. Figure 4.46(a) is a schematic of several traces with well-defined reflection arrivals. There appear to be three strong reflectors. If we interpret these patterns based on the positions of the reflectors in time, it appears that the upper unit thickens to the right as the first reflections arrive progressively later moving from left to right. It also appears that the second unit thins to the right as the time interval between the earliest and next earliest decreases to the right. Based on the differences in time between the first and second reflection arrivals and the second and third arrivals, the second unit appears considerably thicker than the third unit. Reflector

(b)

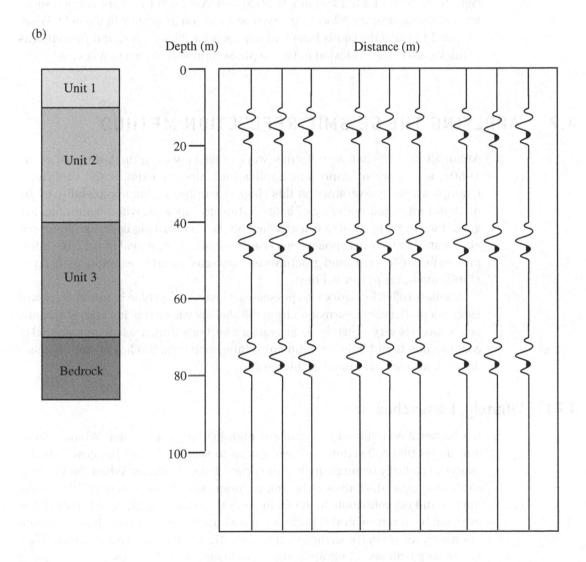

patterns also suggest that the base of the third reflector is significantly deeper to the right.

None of this is true. In fact, unit boundaries are horizontal and unit thicknesses are constant across the entire section. Unit two and unit three are equal in thickness. Depth to bedrock remains constant. Each unit possesses a different velocity, which is greater than that of the overlying unit. The reason for the displacements of the reflections in time from left to right on the time section is a variation in the velocities of the upper two units. Velocities in the first unit range from 500 m/s for the first three traces to 400 m/s for the next three traces to 300 m/s for the three traces farthest to the right. Velocities in the second unit increase from left to right from 2000 m/s to 2500 m/s to 3000 m/s. Although these are simple, simulated seismograms, the effects portrayed occur all too frequently in the field. When we need to make decisions based on depths, unit thicknesses, and the patterns of thicknesses, it is critical to have a depth section with which to work.

4.7 APPLYING THE SEISMIC REFLECTION METHOD

Although *routine* shallow reflection work dates only from the latter half of the 1980s, a number of important applications already exist in the literature. Examples already examined in this chapter emphasize the applicability of the method to detailed mapping of bedrock topography and, when conditions are good, the stratigraphy of unconsolidated materials overlying bedrock. These relatively straightforward applications clearly contain a great deal of valuable information for engineering and groundwater concerns. Another example of this type of information is presented next.

Shallow reflection work also possesses considerable value in terms of hazard assessment. Detailed information from the shallow subsurface in zones of recently active faults is very valuable, as are seismic sections from areas with potential of collapse due to sinkhole formation or mining activities. Studies of each of these hazards also are examined in this section.

4.7.1 Whately, Massachusetts

In Chapter 3 we outlined the aquifer contamination problem that Whately faced and noted that refraction data were used to secure alternative groundwater sources primarily through depth-to-bedrock determinations. When the Whately study commenced, shallow reflection techniques were not in general use, so the method did not contribute to the main study's conclusions. Since that time, however, we have run several reflection lines within Whately. These lines, although not nearly as extensive as those in Figure 4.37, are similar in appearance. They nicely map bedrock topography and contain distinct reflections in the sediment column. One of these reflections correlates with the base of the clay that delineates the top of the confined aquifer. A study now is under way to map the regional aquifer geometry to ensure a well-planned development of this resource.

4.7.2   Meers Fault, Oklahoma

The Meers fault, located in southwestern Oklahoma, has produced significant surface displacements within the last 2000 years. Although a recognizable scarp marks the trend of the fault, no historical record of activity exists (Lawson et al. 1979). Detailed information from the shallow subsurface might contribute information useful in establishing the degree of hazard the Meers fault poses. For this reason and to investigate the potential of shallow reflection surveys in detailing shallow structures, a reflection survey was undertaken by Miller, Steeples, and Myers (1990, 18–25). The summary that follows is taken from the results of this work.

The survey used the common depth point approach with a 24-channel seismograph, 100 Hz geophones, and analog low-cut filters. A typical geophone spread used geophone intervals of 2 m with a shot offset of 20 m. Data reduction consisted dominantly of static corrections, muting ground roll and the air wave, velocity analysis, NMO corrections, frequency filtering, CDP stacking, and automatic gain control. Figure 4.47 illustrates a typical final record and structural interpretation.

In addition to providing detailed information about the deformation zone surrounding the Meers fault, this survey demonstrated that shallow reflection work can reveal a great deal of structural complexity (see the structural interpretation in Figure 4.47). The survey recorded reflections from depths up to 200 m, achieved a minimum bed resolution of 6 m, and demonstrated that the detection limit for faulting is approximately 2 m. Structural conclusions include demonstrating that the major fault is a high-angle reverse fault, that secondary faulting is present within a depth range of 30–200 m, and that the majority of secondary deformation is within 100 m of the Meers fault scarp. One obvious application of such detailed, shallow structural information is to help select sites for trenching in hopes of determining the displacement history of the fault and the present hazard associated with it.

4.7.3   Cavity Detection

The ability to locate and outline subsurface cavities resulting from dissolution or mining activities promises substantial mitigation of these hazards. Shallow seismic reflection surveys possess this ability, assuming that prominent reflectors are present.

Steeples, Knapp, and McElwee (1986, 295–301) were able to recognize sinkholes produced by dissolution of the Hutchinson Salt along Interstate Highway 70 in Russell County, Kansas. Subsidence associated with sinkhole formation tended to be gradual and caused the highway to be twice rebuilt at sinkhole sites.

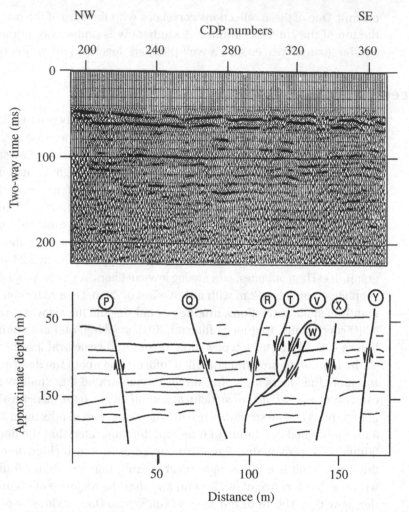

FIGURE 4.47 CDP stack and a line drawing interpretation of the stacked section for a seismic line across the Meers fault. (From Richard D. Miller, Don W. Steeples, and Paul B. Myers. 1990. Shallow seismic reflection survey across the Meers fault, Oklahoma. *Geological Society of America Bulletin*, v. 102, 18–25.)

However, a catastrophic collapse at a saltwater disposal well associated with a nearby oil field heightened public concern. To better define this hazard, a seismic reflection survey was conducted in the area. The survey demonstrated that it was possible to recognize sinkholes by displaced strata. As reflections were secured from the top and bottom of the salt unit, changes in thickness due to solution could be determined. This determination delineated areas where continued sinkhole development and/or growth could be expected. These areas now can be monitored more closely to reduce hazard. Another area of sinkhole formation in Kansas due to the same process is illustrated in Figure 4.48. Because

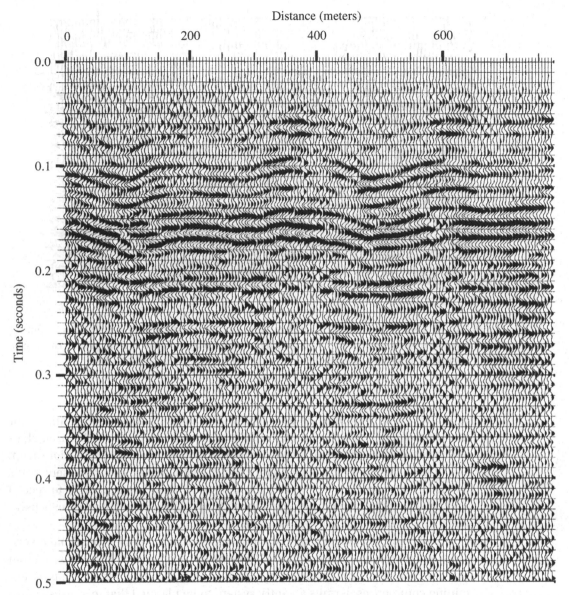

FIGURE 4.48 Seismic reflection record section illustrating disruption of strata due to sinkhole formation: Reno County, Kansas. Disturbed stratification is evident for section times less than 170 ms. Sinkhole formation is due to salt dissolution. The base of the salt unit is at 210 ms. (Courtesy Richard D. Miller and Don W. Steeples, Kansas Geological Survey.)

dissolution at this site also is beneath a major highway, the reflection record is invaluable in defining areas of greatest hazard.

Collapse of roofs in shallow coal mines constitutes an acute hazard in many coal-mining regions. Unfortunately, the location of many abandoned mines is

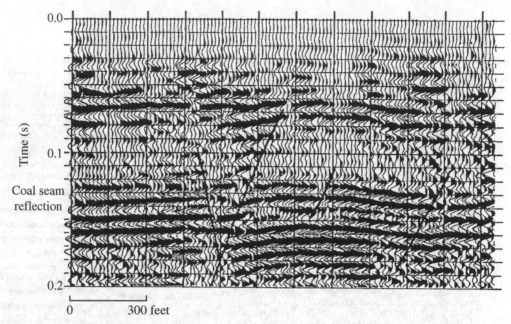

FIGURE 4.49 CDP seismic section illustrating locations of faults cutting coal seams. (From Lawrence M. Gochioco and Steven A. Cotten. 1989. Locating faults in underground coal mines using high-resolution seismic reflection techniques. *Geophysics,* v. 54, 1521–27.)

unknown, and therefore the degree of the hazard cannot be ascertained nor hazard reduction efforts undertaken. Branham and Steeples (1988) used shallow reflection techniques to locate abandoned mines by tracing reflections from a coal unit. Areas where reflections from the coal are absent represent voids formed from removal of the coal by mining activities. The capability to record coal reflections from depths as shallow as 10 m promises a relatively inexpensive approach to quantifying the mine hazard.

In addition to cavities, faults cutting coal seams obviously constitute a serious hazard. Gochioco and Cotten (1989) demonstrated that shallow reflection work is capable of finding such faults. Knowing the location of these faults before mining commences permits an entry system to be planned that increases safety. Figure 4.49 illustrates how these faults appear on a CDP seismic section.

4.7.4 Other Applications

Other informative studies include those by Slaine et al. (1990), who mapped overburden stratigraphy and bedrock relief in the vicinity of a proposed hazardous waste facility; Miller et al. (1990), who delineated an intra-alluvial aquifer's geometry, offset relations of the aquifer and surrounding sediments, and bedrock structural relationships; and Pullan and Hunter (1990), who illus-

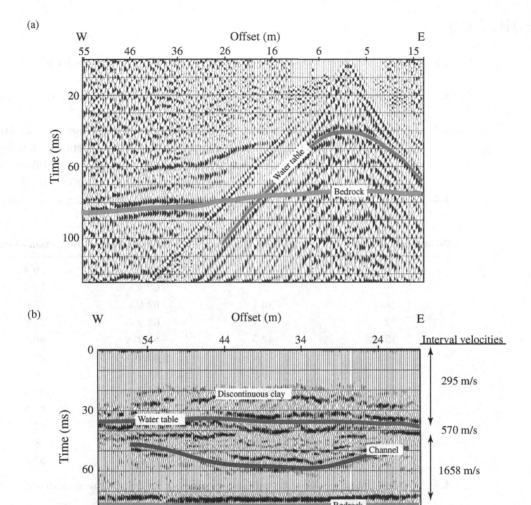

(a)

W   Offset (m)   E

55   46   36   26   16   6   5   15

Time (ms)

20

60

100

Water table

Bedrock

(b)

W   Offset (m)   E

54   44   34   24

Interval velocities

Time (ms)

0

30

60

90

Discontinuous clay

Water table

Channel

Bedrock

295 m/s

570 m/s

1658 m/s

FIGURE 4.50 Common midpoint seismic reflection data collected near Clay Center, Kansas, on July 31, 2004. Data were collected using two 72-channel Geometrics StrataView seismographs using a 0.5 m geophone interval and a 1 m source interval. The source was a .223-caliber rifle. (a) Raw field data. (b) Processed and interpreted section. (Courtesy Don W. Steeples, University of Kansas.)

trated the utility of the shallow seismic reflection technique in mapping a number of buried bedrock valleys.

In closing this chapter we thought it appropriate to illustrate (see Figure 4.50) what is becoming routinely possible in shallow reflection exploration using signal enhancement seismographs with 72 (or more) channels supplemented by processing on personal computers.

PROBLEMS

4.1 Refer to Figure 4.51(a). Determine the thickness and velocity of material above the first reflecting horizon.

4.2 A prominent reflection is indicated in Figure 4.51(b). What are the velocity and thickness of material above this reflecting horizon?

4.3 Figure 4.52 illustrates a center-shot, wiggle-trace seismogram. The shot offset is 34 m, and the geophone interval is 48 m. Three prominent reflections are indicated by arrows. Assuming these are the only reflections present, determine rms velocities, interval velocities, and unit thicknesses.

4.4 Using the time–distance values given below, interpret the subsurface geology as completely as possible.

Distance from Shot (m)	Reflection #1 (ms)	Reflection #2 (ms)	Reflection #3 (ms)
3	21.4	62.3	79.4
6	25	62.4	79.5
9	30.1	62.6	79.6
12	36.1	62.9	79.9
15	42.5	63.2	80.1
18	49.2	63.6	80.5
21	56.2	64.1	80.9
24	63.3	64.7	81.3
27	70.4	65.4	81.8
30	77.6	66.1	82.4
33	84.9	66.9	83
36	92.2	67.7	83.7

4.5 Using the time–distance values given below, interpret the subsurface geology as completely as possible.

Distance from Shot (m)	Reflection #1 (ms)	Reflection #2 (ms)	Reflection #3 (ms)
3	40.5	102.5	116.8
13	47.7	103.1	117.2
23	61	104.4	118
33	77.2	106.3	119.2
43	94.9	108.9	120.9
53	113.3	112.1	123
63	132.2	115.8	125.5
73	151.4	120	128.4
83	170.8	124.6	131.6
93	190.3	129.7	135.1
103	209.9	135.1	138.9
113	229.5	140.8	143

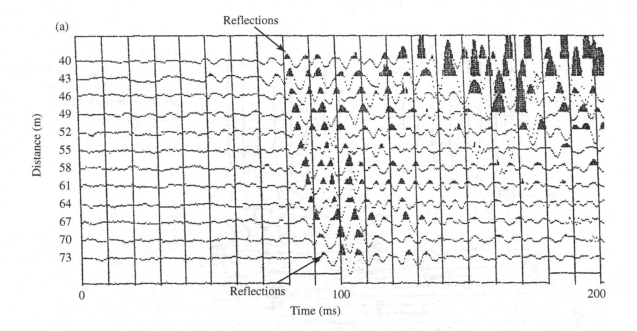

(a)

Reflections

Distance (m): 40, 43, 46, 49, 52, 55, 58, 61, 64, 67, 70, 73

Reflections

Time (ms): 0, 100, 200

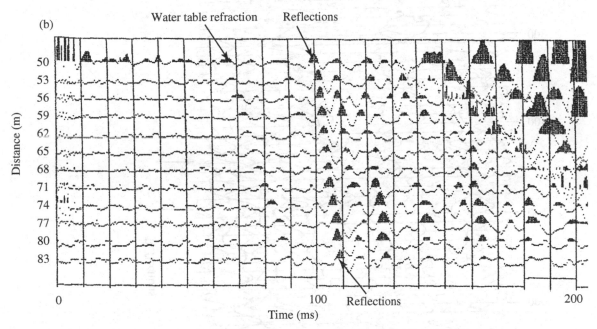

(b)

Water table refraction   Reflections

Distance (m): 50, 53, 56, 59, 62, 65, 68, 71, 74, 77, 80, 83

Time (ms): 0, 100, 200

Reflections

FIGURE 4.51 (a), (b) Field seismograms with reflections for analysis.

(a)

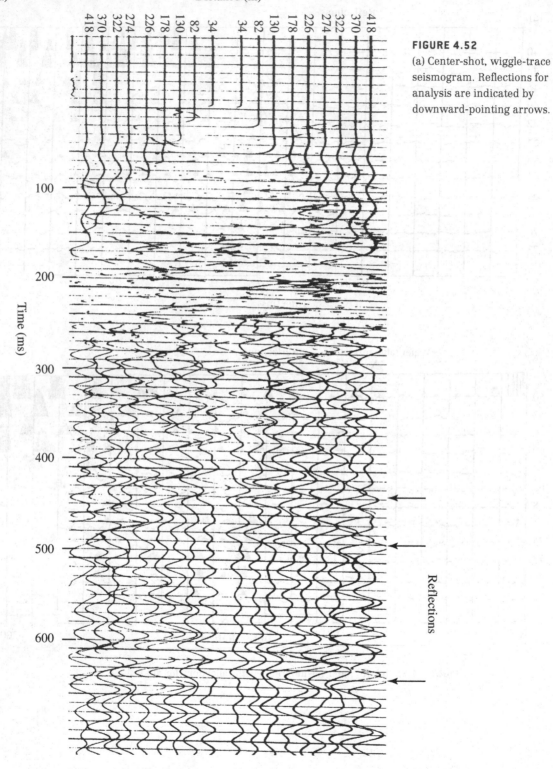

Distance (m)

418 370 322 274 226 178 130 82 34 34 82 130 178 226 274 322 370 418

Time (ms)

100

200

300

400

500

600

FIGURE 4.52
(a) Center-shot, wiggle-trace
seismogram. Reflections for
analysis are indicated by
downward-pointing arrows.

Reflections

Reflections

(b)

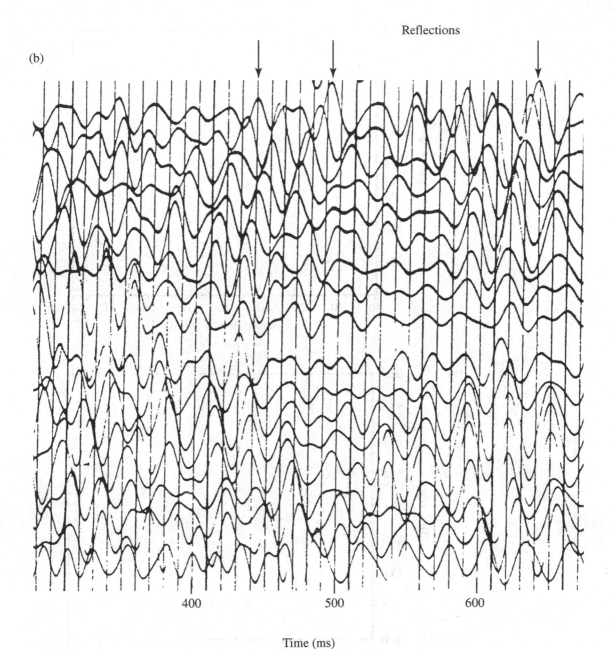

Time (ms)

FIGURE 4.52 *(continued)* (b) Enlarged view of (a) to aid in picking reflection times.

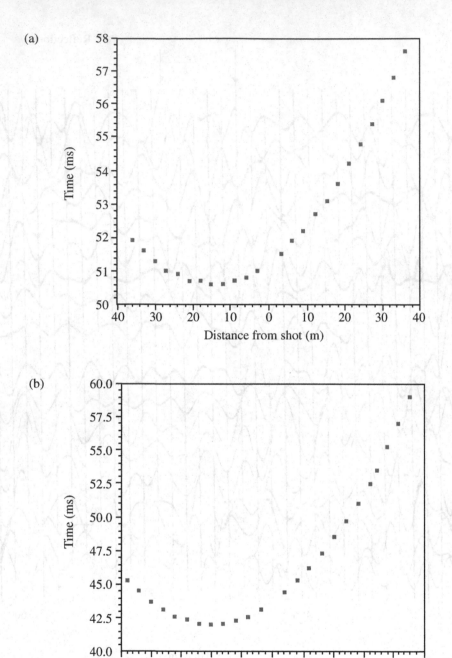

FIGURE 4.53 Reflection time–distance data. (a) Data for Problem 4.6. (b) Data for Problem 4.10.

4.6 Interpret the time–distance data illustrated in Figure 4.53(a) on page 258.

4.7 A target horizon in a reflection survey is believed to be at an approximate depth of 35 m. Explain why the maximum shot–geophone distance should not exceed 35 m. Include two different reasons in your explanation.

4.8 Use both the t_{min}/t_0 and the $x^2 - t^2$ methods to interpret the following time–distance data from a dipping reflector:

Distance from Shot (m)	Reflection Time (ms)
36	22.1
33	21.44
30	20.8
27	20.21
24	19.66
21	19.15
18	18.68
15	18.27
12	17.92
9	17.62
6	17.38
3	17.21
0	Shot
3	17.06
6	17.09
9	17.18
12	17.34
15	17.56
18	17.84
21	18.19
24	18.59
27	19.04
30	19.54
33	20.08
36	20.67

4.9 The reflection in Figure 4.54 (page 260) is very obvious. Determine the thickness and velocity of the material above the reflecting horizon.

4.10 Using dip move-out, interpret Figure 4.53(b) on page 258.

4.11 An 80 m thick sequence of saturated sands with a velocity of 1350 m/s overlies a 40 m thick sequence of till with velocity of 2800 m/s. The till is in contact with bedrock, which has a velocity of 5500 m/s. Determine the optimum window when the target is the bedrock reflection.

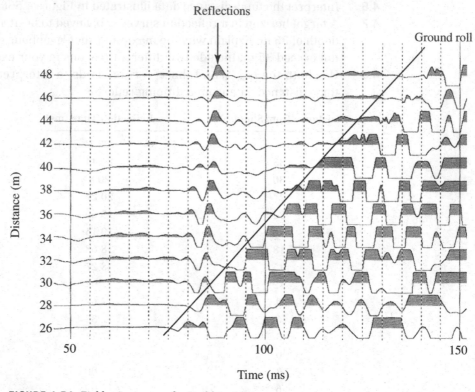

FIGURE 4.54 Field seismogram for Problem 4.9.

4.12 Calculate the depth to bedrock and the velocity of the sediments above bedrock using the field record in Figure 4.26(b).

4.13 Construct a travel-time curve for the model illustrated below. Show reflection times for all geophones and diffractions from the upper edge of the step.

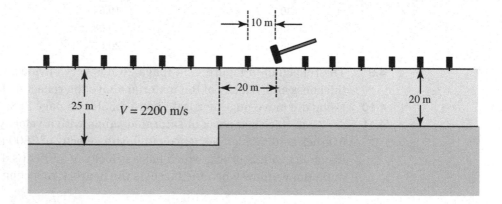

4.14 Calculate elevation and weathering corrections for the three geophone locations in the sketch shown here.

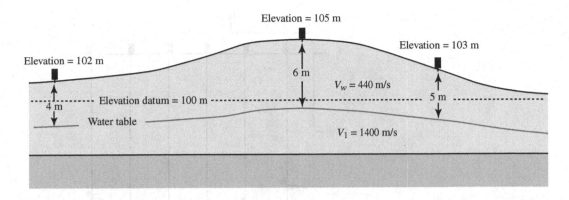

4.15 Determine V_{st} and t_0 for each of the reflections in the accompanying simulated seismogram. The correct arrival times are listed below. Use Table 4.11 on page 234.

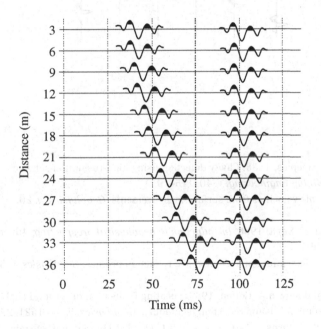

Distance (m)	Time (ms)
3	33.71
6	34.8
9	36.55
12	38.87
15	41.67
18	44.85
21	48.33
24	52.07
27	56
30	60.09
33	64.31
36	68.64

Distance (m)	Time (ms)
3	93.35
6	93.42
9	93.53
12	93.68
15	93.87
18	94.1
21	94.38
24	94.69
27	95.05
30	95.45
33	95.89
36	96.37

4.16 Perform migration on the traces illustrated in the accompanying seismic section to portray subsurface relationships in their true form.

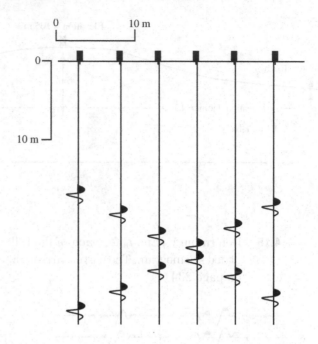

REFERENCES CITED

Branham, K.L., and D.W. Steeples. 1988. Cavity detection using high-resolution seismic reflection methods. *Mining Engineering*, v. 40, 115–19.

Dix, C. Hewitt. 1955. Seismic velocities from surface measurements. *Geophysics*, v. 20, 68–86.

Dobrin, Milton B., and Carl H. Savit. 1988. *Introduction to geophysical prospecting*, 4th ed. New York: McGraw-Hill.

Green, C.H. 1938. Velocity determinations by means of reflection profiles. *Geophysics*, v. 3, 295–305.

Gochioco, Lawrence M., and Steven A. Cotten. 1989. Locating faults in underground coal mines using high-resolution seismic reflection techniques. *Geophysics*, v. 54, 1521–27.

Hunter, J.A., S.E. Pullan, R.A. Burns, R.M. Gagne, and R.L. Good. 1984. Shallow seismic reflection mapping of the overburden–bedrock interface with the engineering seismo-graph—some simple techniques. *Geophysics*, v. 49, 1381–85.

Kearey, P., and M. Brooks. 1984. *An introduction to geophysical exploration*. London: Blackwell Scientific Publications.

Lawson, J.E. Jr., R.L. DuBois, P.H. Foster, and K.V. Luza. 1979. *Earthquake map of Oklahoma*. Oklahoma Geological Survey Map GM-19, scale 1/750,000.

Miller, Richard D., Don W. Steeples, Robert W. Hill Jr., and Bobby L. Gaddis. 1990. Identifying intra-alluvial and bedrock structures shallower than 30 meters using seismic reflection techniques. In *Geotechnical and environmental geophysics, volume 3: Geotechnical,* ed. Stanley H. Ward, 89–97. Society of Exploration Geophysicists Investigations in Geophysics, v. 5.

Miller, Richard D., Don W. Steeples, and Paul B. Myers. 1990. Shallow seismic reflection survey across the Meers fault, Oklahoma. *Geological Society of America Bulletin*, v. 102, 18–25.

Pullan, S.E., and J.A. Hunter. 1985. Seismic model studies of the overburden–bedrock reflection. *Geophysics,* v. 50, 1684–88.

Pullan, S.E., and J.A. Hunter. 1990. Delineation of buried bedrock valleys using the optimum offset shallow seismic reflection technique. In *Geotechnical and environmental geophysics, volume 3: Geotechnical,* ed. Stanley H. Ward, 75–87. Society of Exploration Geophysicists Investigations in Geophysics, v. 5.

Robinson, Edwin S., and Cahit Coruh. 1988. *Basic exploration geophysics.* New York: John Wiley & Sons.

Shah, Pravin M., and F.K. Levin. 1973. Gross properties of time–distance curves. *Geophysics,* v. 38, 643–56.

Sheriff, Robert E. 1989. *Geophysical methods.* Englewood Cliffs, NJ: Prentice Hall.

Sheriff, R.E., and L.P. Geldart. 1982. *Exploration seismology, volume 1: History, theory, and data acquisition.* Cambridge, England: Cambridge University Press.

Sheriff, R.E., and L.P. Geldart. 1983. *Exploration seismology, volume 2: Data processing and interpretation.* Cambridge, England: Cambridge University Press.

Slaine, David D., Peeter E. Pehme, James A. Hunter, Susan E. Pullan, and John P. Greenhouse. 1990. Mapping overburden stratigraphy at a proposed hazardous waste facility using shallow seismic reflection methods. In *Geotechnical and environmental geophysics, volume 2: Environmental and groundwater,* ed. Stanley H. Ward, 273–80. Society of Exploration Geophysicists Investigations in Geophysics, v. 5.

Slotnick, M.M. 1959. *Lessons in seismic computing.* Tulsa, OK: Society of Exploration Geophysicists.

Steeples, Don W., Ralph W. Knapp, and Carl D. McElwee. 1986. Seismic reflection investigations of sinkholes beneath Interstate Highway 70 in Kansas. *Geophysics*, v. 51, 295–301.

Taner, M. Turhan, and Fulton Koehler. 1969. Velocity spectra—digital computer derivation and applications of velocity functions. *Geophysics,* v. 34, 859–81.

Waters, Kenneth H. 1981. *Reflection seismology*, 2nd ed. New York: John Wiley & Sons.

SUGGESTED READING

Chon, G.H., G.H. Liang, D.R. Xu, Q.S. Zeng, S.H. Fu, X.R. Wei, Z.L. He, and G.G. Fu. 2004. Application of a shallow seismic reflection method to the exploration of a gold deposit. *Journal of Geophysics and Engineering*, v. 1, no. 1, 12–16.

Hunter, J.A., S.E. Pullan, R.A. Burns, R.M. Gagne, and R.L. Good. 1984. Shallow seismic reflection mapping of the overburden–bedrock interface with the engineering seismograph—some simple techniques. *Geophysics,* v. 49, 1381–85.

Juhlin, C. 1995. Imaging of fracture zones in the Finnsjon area, Central Sweden, using the seismic reflection method. *Geophysics*, v. 60, no. 1, 66–75.

Karastathis, V.K., S. Papamarinopoulus, and R.E. Jones. 2001. 2-D velocity structure of the buried ancient canal of Xerxes: An application of seismic methods in archeology. *Journal of Applied Geophysics*, v. 47, no. 1, 29–43.

Knapp, R.W., and D.W. Steeples. 1986. High-resolution common-depth-point seismic reflection profiling: Instrumentation. *Geophysics,* v. 51, 276–82.

Knapp, R.W., and D.W. Steeples. 1986. High-resolution common-depth-point seismic reflection profiling: Field acquisition parameter design. *Geophysics,* v. 51, 283–94.

Miller, R.D., W.E. Doll, C. Merey, and W.E. Black. 1994. Applications of shallow high-resolution seismic reflection to environmental problems. *Environmental Geosciences,* v. 1, no. 1, 32–39.

Sheriff, R.E., and L.P. Geldart. 1982. *Exploration seismology, volume 1: History, theory, and data acquisition.* Cambridge, England: Cambridge University Press.

Sheriff, R.E., and L.P. Geldart. 1983. *Exploration seismology, volume 2: Data processing and interpretation.* Cambridge, England: Cambridge University Press.

Steeples, Don W., and Richard D. Miller. 1990. Seismic reflection methods applied to engineering, environmental, and groundwater problems. In *Geotechnical and environmental geophysics, volume 1: Review and tutorial,* ed. Stanley H. Ward, 1–30. Society of Exploration Geophysicists Investigations in Geophysics, v. 5.

Waters, Kenneth H. 1981. *Reflection seismology,* 2nd ed. New York: John Wiley & Sons.

CHAPTER 5

Electrical Resistivity

5.1   INTRODUCTION

As we move from seismic topics to a discussion of electrical methods for exploring the shallow subsurface, two striking differences are the diversity of electrical methods and their relative complexity in deriving quantitative interpretations from field data as compared to seismic methods.

Although a wide range of methods can be included under the heading of this chapter, these methods group naturally into two categories: (1) those in which current is applied to the Earth and (2) natural energy sources.

5.1.1   Applied Currents

In *electrical resistivity* (ER) methods direct current or low-frequency alternating current is applied at the ground surface, and the potential difference is measured between two points. Variations in resistance to current flow at depth cause distinctive variations in the potential difference measurements, which provide information on subsurface structure and materials.

When the current used in electrical resistivity surveying is switched off, the voltage between the two points used to determine potential difference does not decline immediately to zero. Rather, a small residual voltage slowly decreases. A similar effect occurs when the current is switched on, suggesting that the current flow has electrically polarized the ground. The character of this *induced polarization* (IP) also reveals information about the subsurface.

A third method using applied currents is *electromagnetic* (EM) surveying. In EM surveying a primary electromagnetic field is produced by passing alternating current through a coil. Conducting bodies beneath the surface generate secondary electromagnetic fields that are detected by a receiver coil. Differences between the primary and secondary fields once again provide information about the presence and characteristics of subsurface conducting bodies.

5.1.2 Natural Currents

The flow of charged particles in the ionosphere due to solar emissions is responsible for alternating currents that flow through the upper regions of the Earth. This natural current flow (referred to as *telluric* currents) is altered by the varying conductivity properties of rocks. The telluric method takes advantage of these natural current variations by measuring potential differences at the surface and interpreting these differences in terms of subsurface materials. The *magnetotelluric* method is similar to telluric surveying but measures the magnetic field as well as the electrical field.

The *self-potential* or *spontaneous potential* (SP) approach uses natural electrochemical activity. If an ore body is in contact with solutions possessing different compositions, resultant chemical reactions cause a flow of ions. This leads to potential differences and thus to a flow of current. A simplified view is a system behaving as a battery. Because current is naturally produced, we need employ only two probes to measure potential differences at points on the surface.

5.1.3 A Brief History

The history of electrical prospecting essentially dates from the 1830s, when Robert W. Fox experimented with natural currents associated with sulfide ore deposits at Cornwall, England. Most of the work during the 1800s dealt only with these types of currents. In the early part of the second decade of the 1900s Conrad Schlumberger in France and Frank Wenner in the United States applied current to the ground and measured the resulting potential difference. These investigations established the direct current resistivity method.

In 1914 Schlumberger found a rich ore deposit in Serbia using the self-potential method, and in 1917 the electromagnetic method was introduced by H.R. Conklin. Telluric currents were first studied by O.H. Gish and W.J. Rooney in the United States in the early 1920s. This work was continued and expanded on by the Schlumberger group beginning in 1934. Virtually all this early work quite naturally was associated with ore deposits and their exploration. One of the earliest noncommercial applications was mapping of high-resistivity bedrock by I.B. Crosby and E.G. Leonardon in 1928 during an investigation of a proposed dam site.

With the exception of the magnetotelluric method, essentially all electrical methods had been investigated in some fashion by 1930. Since then, of course, progress has consisted of instrument refinement, development of a solid theoretical basis, and improvement in interpretative methods—primarily, if not entirely, due to enhanced computer capabilities. An especially complete review of the history of electrical prospecting is presented by Van Nostrand and Cook (1966), from which the comments here were selected. As we suggested for a similar history of seismic reflection, reading such a complete history of a particular geophysical method, especially after completing the chapter addressing the method, is satisfying, provides a useful perspective, and, in a sense, helps review concepts.

5.1.4 Chapter Goals

Introducing electrical methods in a text with our objectives presents a number of problems. First, as you now are aware, several electrical methods exist, and portions of each contain unique theoretical aspects as well as operational procedures. Second, all methods are complex theoretically, and quantitative interpretations are not nearly as easy to achieve compared to the seismic methods with which you are familiar.

Covering electrical methods thoroughly, therefore, would take more room and time than we devoted to seismic methods even though electrical methods are not as widely utilized. A superficial coverage of each method would result primarily in confusion, not mastery. In view of these concerns and constraints, we will concentrate on one method only: electrical resistivity. Electrical resistivity is probably the most common method applied to shallow subsurface investigations, especially in groundwater studies. At the end of this chapter we provide an outline of the major additional methods in terms of procedure and applications. We hope that mastery of one method and knowledge of the major applications of the others provide a reasonable solution in discussing electrical methods in an introductory text.

5.2   BASIC ELECTRICITY

The electrical resistivity method measures potential differences at points on the Earth's surface that are produced by directing current flow through the subsurface. This leads to the determination of resistivity distribution in the subsurface and to an interpretation of Earth materials. Because we are working with resistance, current flow, and potentials, a brief review of basic electrical concepts seems warranted.

Figure 5.1 illustrates a basic electrical circuit containing a battery, connecting wires, and a resistor. The battery maintains a potential difference between two points: its positive terminal and its negative terminal. The battery thereby functions as a power source in moving charges through the circuit, much as a pump moves water through pipes. The convention adopted in this text is to define current flow as the movement of positive charges. To accomplish this flow, the battery must move positive charges from a low potential at the negative terminal to high potential at the positive terminal. The work done in this potential change requires that a force be applied. This force is known as electromotive force or *emf,* and the unit of emf is the *volt*.

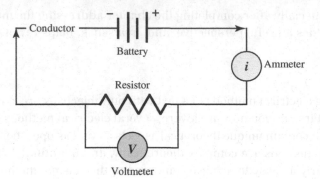

A 9-volt battery maintains a potential difference of 9 volts between its terminals and thus has a certain potential for doing work. As noted, the movement of charges through the conducting wire is termed *current*. Specifically,

$$i = \frac{q}{t} \tag{5.1}$$

where i is current in amperes, q is charge in coulombs, and t is time in seconds.

Another important concept in electrical resistivity surveying is the current density j. *Current density* is defined as the current divided by the cross-sectional area of the material through which it is flowing:

$$j = \frac{i}{A}. \tag{5.2}$$

Because current is defined as the movement of charges across a given cross-sectional area in a unit of time, maintaining a constant current and reducing the cross-sectional area through which it flows must cause reduced spacing of charges and, therefore, an increased current density. Figure 5.2 illustrates this relationship diagrammatically using arrows to represent movement of charges.

You recognize, of course, that copper wire, wood, aluminum, and rubber possess varying resistances to the flow of current. Copper has very low resistance, whereas rubber has an extremely high resistance. Resistance is quantified in the following way: one ohm of resistance allows a current of one ampere to flow when one volt of emf is applied.

Ohm's Law, first presented by German physicist Georg Simon Ohm, states that current is directly proportional to voltage V and inversely proportional to resistance R:

$$i = \frac{V}{R}. \tag{5.3}$$

Consider Figure 5.1. If the battery supplies 9 volts and the resistor has a value of 10 ohms, the current measured by the ammeter will be 0.9 ampere. If resistance is increasing, it will take an increasing voltage to maintain the same current.

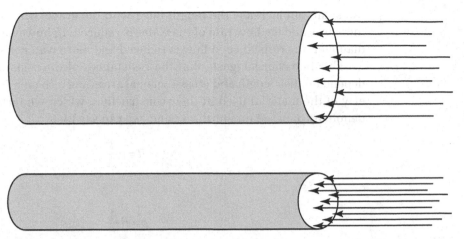

FIGURE 5.2 Diagram illustrating concept of current density j in wires with different cross-sectional area. Current flow is represented by arrows.

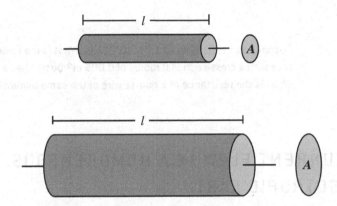

FIGURE 5.3 Two resistors of different lengths l and cross-sectional area A.

Because various geologic materials can be expected to have different resistances to current flow, it might seem fairly straightforward to measure current and voltage to calculate resistance and determine the material in the subsurface. One immediate complication is that resistance depends not only on the material but also on its dimensions. Consider Figure 5.3. The sketches illustrate two resistors with different lengths l and cross-sectional areas A. If these resistors are constructed from the same material, it seems intuitively obvious that they will not have the same resistance to current flow. Remembering that current flow is the movement of charged particles in a given unit of time, we can use a water analogy. Consider an open pipe in which one section is filled with gravel. A pump creates a pressure difference in the pipe and water flows. The gravel causes a resistance to flow relative to the open portion of the pipe. If we keep everything the same but increase the length filled with the gravel, the resistance to flow increases and the flow rate of the water is reduced. If, however, we increase the diameter, the resistance to flow is reduced and more water flows.

This behavior suggests that the resistances of the resistors in Figure 5.3 depend on their length and cross-sectional areas and also on a fundamental property of the material used in their construction, which we term *resistivity* and denote by ρ. Based on our discussion, we can say that

$$R = \rho \frac{l}{A} \tag{5.4}$$

or

$$\rho = R \frac{A}{l}. \tag{5.5}$$

Resistivity units are *resistance·length*, which commonly is denoted by *ohm·m (Ω·m)*. Conductance is the inverse of resistance, and conductivity is the inverse of resistivity.

Copper has a resistivity of 1.7×10^{-8} Ω·m. What is the resistance of 20 m of copper wire with a cross-sectional radius of 0.005 m? Quartz has a resistivity of 1×10^{16} Ω·m. What is the resistance of a quartz wire of the same dimensions?

5.3   CURRENT FLOW IN A HOMOGENEOUS, ISOTROPIC EARTH

5.3.1   Point Current Source

Because the resistivity method consists of applying current and measuring potentials, we begin by considering the potential at a point P_1 when current is applied at a point source C_1. We place the return current electrode at a very great distance and assume material of uniform resistivity, ρ. Because air has infinite resistivity, no current flows upward. Thus current flows radially outward through the Earth equally in all directions so as to define a hemispherical surface (Figure 5.4). Because current distribution is equal everywhere on this surface that is at a distance r from the current electrode C_1, the potential also is equal. These surfaces are known as *equipotential* surfaces. If we define a very thin shell of thickness dr and employ Equations 5.3 and 5.4, we can define the potential difference across the shell to be

$$dV = i(R) = i\left(\rho \frac{l}{A}\right) = i\left(\rho \frac{dr}{2\pi r^2}\right). \tag{5.6}$$

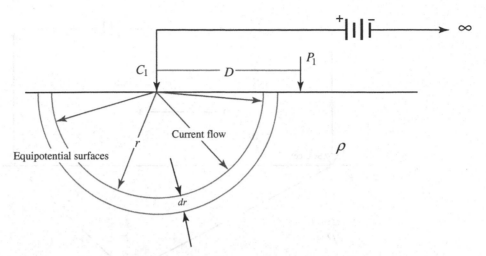

FIGURE 5.4 Diagram illustrating symbols and configuration used to determine potential at P_1 for a single point source of current C_1. The current sink, C_2, is at infinity. The two equipotential surfaces shown are separated by the distance dr.

We now use Equation 5.6 to determine the potential at P_1. In determining the potential at a point, we compare it to the potential at a point infinitely far away, which by convention is arbitrarily defined to equal zero. The most direct way to determine V is to integrate Equation 5.6 over its distance D to the current electrode to infinity, or

$$V = \int_D^\infty dV = \frac{i\rho}{2\pi} \int_D^\infty \frac{dr}{r^2} = \frac{i\rho}{2\pi D} \qquad (5.7)$$

(Van Nostrand and Cook 1966, 28). Equation 5.7 is the fundamental equation in our electrical prospecting discussions. Before using it to develop more practical relationships, let's examine what information we can glean from it. Assuming a resistivity and current, we can map the potential at any point in our section view of Figure 5.4. If we plot enough points, we can draw lines through the points possessing the same potential. This defines the equipotential surfaces. Because current flow must be perpendicular to the equipotential surfaces, we can determine the direction of current flow. Of course, in this case we already know what the result will be; but in the next step the pattern will not be so evident.

5.3.2 Two Current Electrodes

Our next step is an attempt to determine the current flow in a homogeneous, isotropic Earth when we have two current electrodes. In this case the current must flow from the positive current electrode (the source) to the negative current electrode (the sink). The path of the current is not as obvious as in our

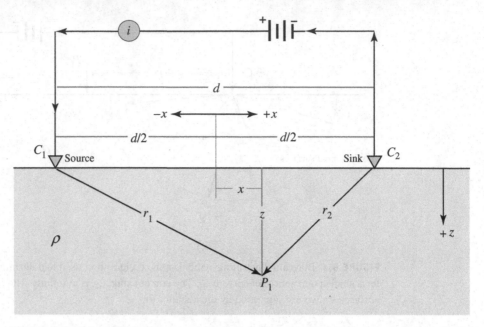

FIGURE 5.5 Diagram illustrating symbols and configuration used to determine potential at P_1 for a current source C_1 and sink C_2.

previous discussion, so we once again determine the potential at a point. From this information we determine the equipotential surfaces and then the current flow. For simplicity we present the derivation only for the case when the current electrodes and our potential points lie in the same plane (Figure 5.5). The more general case is quite similar and can be examined in Van Nostrand and Cook (1966, 30–31).

The potential at point P_1 is determined by using Equation 5.7. The effects of the source at C_1 (+) and the sink at C_2 (–) are both considered, and therefore,

$$V_{P_1} = \frac{i\rho}{2\pi r_1} + \left(-\frac{i\rho}{2\pi r_2} \right). \tag{5.8}$$

Expressing r_1 and r_2 in terms of the x–z coordinate system illustrated in Figure 5.5, we rewrite Equation 5.8 as

$$V_{P_1} = \frac{i\rho}{2\pi} \left\{ \frac{1}{\left[\left(\frac{d}{2} + x \right)^2 + z^2 \right]^{\frac{1}{2}}} - \frac{1}{\left[\left(\frac{d}{2} - x \right)^2 + z^2 \right]^{\frac{1}{2}}} \right\}. \tag{5.9}$$

Now we need only calculate V_{P_1} at many points in the x–z plane, draw contours through points of equal potential to define the equipotential surfaces, and, finally, draw current flow lines perpendicular to the equipotential surfaces.

Horizontal Position (m)

Depth (m)	−10	−9	−8	−7	−6	−5	−4	−3	−2	−1	0	1	2	3	4	5	6	7	8	9	10
0	1.06	1.42	2.04	3.32	7.23	####	7.07	2.98	1.52	0.66	0.00	−0.66	−1.52	−2.98	−7.07	####	−7.23	−3.32	−2.04	−1.42	−1.06
1	1.03	1.36	1.91	2.90	4.91	7.17	4.75	2.57	1.39	0.62	0.00	−0.62	−1.39	−2.57	−4.75	−7.17	−4.91	−2.90	−1.91	−1.36	−1.03
2	0.95	1.22	1.60	2.16	2.85	3.20	2.70	1.85	1.11	0.52	0.00	−0.52	−1.11	−1.85	−2.70	−3.20	−2.85	−2.16	−1.60	−1.22	−0.95
3	0.84	1.04	1.28	1.56	1.82	1.89	1.58	1.28	0.83	0.41	0.00	−0.41	−0.83	−1.28	−1.68	−1.89	−1.82	−1.56	−1.28	−1.04	−0.84
4	0.73	0.86	1.01	1.15	1.25	1.25	1.12	0.89	0.60	0.30	0.00	−0.30	−0.60	−0.89	−1.12	−1.25	−1.25	−1.15	−1.01	−0.86	−0.73
5	0.62	0.71	0.79	0.87	0.90	0.88	0.79	0.63	0.44	0.22	0.00	−0.22	−0.44	−0.63	−0.79	−0.88	−0.90	−0.87	−0.79	−0.71	−0.62
6	0.53	0.58	0.63	0.67	0.67	0.64	0.57	0.46	0.32	0.17	0.00	−0.17	−0.32	−0.46	−0.57	−0.64	−0.67	−0.67	−0.63	−0.58	−0.53
7	0.44	0.48	0.51	0.52	0.52	0.48	0.43	0.34	0.24	0.12	0.00	−0.12	−0.24	−0.34	−0.43	−0.48	−0.52	−0.52	−0.51	−0.48	−0.44
8	0.38	0.40	0.41	0.41	0.40	0.37	0.33	0.26	0.18	0.09	0.00	−0.09	−0.18	−0.26	−0.33	−0.37	−0.40	−0.41	−0.41	−0.40	−0.38
9	0.32	0.33	0.34	0.33	0.32	0.29	0.25	0.20	0.14	0.07	0.00	−0.07	−0.14	−0.20	−0.25	−0.29	−0.32	−0.33	−0.34	−0.33	−0.32
10	0.27	0.28	0.28	0.27	0.26	0.23	0.20	0.16	0.11	0.06	0.00	−0.06	−0.11	−0.16	−0.20	−0.23	−0.26	−0.27	−0.28	−0.28	−0.27
11	0.23	0.23	0.23	0.22	0.21	0.19	0.16	0.13	0.09	0.04	0.00	−0.04	−0.09	−0.13	−0.16	−0.19	−0.21	−0.22	−0.23	−0.23	−0.23
12	0.20	0.20	0.19	0.19	0.17	0.15	0.13	0.10	0.07	0.04	0.00	−0.04	−0.07	−0.10	−0.13	−0.15	−0.17	−0.19	−0.19	−0.20	−0.20
13	0.17	0.17	0.16	0.16	0.14	0.13	0.11	0.08	0.06	0.03	0.00	−0.03	−0.06	−0.08	−0.11	−0.13	−0.14	−0.16	−0.16	−0.17	−0.17
14	0.15	0.14	0.14	0.13	0.12	0.11	0.09	0.07	0.05	0.02	0.00	−0.02	−0.05	−0.07	−0.09	−0.11	−0.12	−0.13	−0.14	−0.14	−0.15
15	0.13	0.12	0.12	0.11	0.10	0.09	0.07	0.06	0.04	0.02	0.00	−0.02	−0.04	−0.06	−0.07	−0.09	−0.10	−0.11	−0.12	−0.12	−0.13
16	0.11	0.11	0.10	0.10	0.09	0.08	0.06	0.05	0.03	0.02	0.00	−0.02	−0.03	−0.05	−0.06	−0.08	−0.09	−0.10	−0.10	−0.11	−0.11
17	0.10	0.09	0.09	0.08	0.07	0.06	0.05	0.04	0.03	0.01	0.00	−0.01	−0.03	−0.04	−0.05	−0.06	−0.07	−0.08	−0.09	−0.09	−0.10
18	0.09	0.08	0.08	0.07	0.06	0.06	0.05	0.04	0.02	0.01	0.00	−0.01	−0.02	−0.04	−0.05	−0.06	−0.06	−0.07	−0.08	−0.08	−0.09
19	0.08	0.07	0.07	0.06	0.06	0.05	0.04	0.03	0.02	0.01	0.00	−0.01	−0.02	−0.03	−0.04	−0.05	−0.06	−0.06	−0.07	−0.07	−0.08
20	0.07	0.06	0.06	0.05	0.05	0.04	0.03	0.03	0.02	0.01	0.00	−0.01	−0.02	−0.03	−0.03	−0.04	−0.05	−0.05	−0.06	−0.06	−0.07

Resistivity (Ω-m)	50		x-increment (m)	1
Current (amperes)	1		z-increment (m)	1
Electrode separation (m)	10		(Potential values are in volts)	

Table 5.1 is a dynamic table based on Equation 5.9. Parameters in the table are related to Figure 5.5. If you examine the values of the potential at the points in the table, you should be able to sense the distribution of the equipotential surfaces. Figure 5.6 is a simple contour plot of the values in Table 5.1 with symbols also keyed to Figure 5.5. If you have a computer contouring package available to you, the spreadsheet template on disk named *Table 5.1 contour* is exactly the same as Table 5.1 but is configured to make it simple to transfer this information to a contouring program. Experiment!

The patterns of equipotential surfaces likely are fairly similar to what you would predict. Note that the current electrodes are displaced slightly from the centers of the hemispheres. Many contour lines close to the current electrodes were omitted for clarity, as you can tell if you examine contour line values and the potential values in Table 5.1.

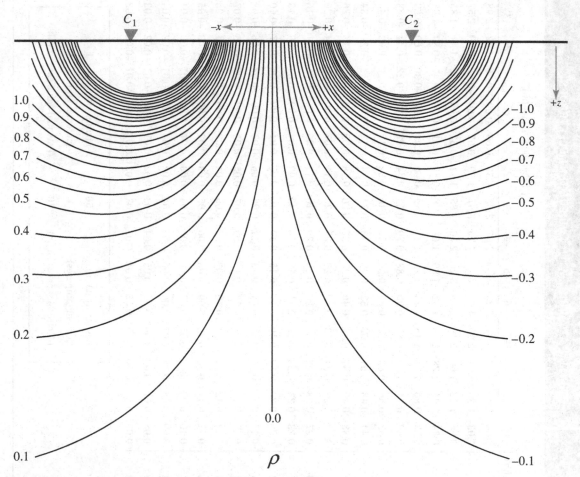

FIGURE 5.6 Contours based on potential values in Table 5.1. Contours represent positions of equipotential surfaces about a current source C_1 and sink C_2. Many contours near the current electrodes are not shown due to the close spacing of contours in this region.

We know that the current flow lines are perpendicular to the equipotential lines, but we do not know how the current is distributed. The mathematical analysis to determine the current distribution is fairly complicated but results in a simple equation that provides current distribution as a fraction of the total current (Van Nostrand and Cook 1966, 30–34). Along a vertical plane midway between the two current electrodes, the fraction of the total current i_f penetrating to depth z for an electrode separation of d is given by

$$i_f = \frac{2}{\pi} \tan^{-1}\left(\frac{2z}{d}\right).$$ (5.10)

The dynamic Table 5.2 uses Equation 5.10, so we can investigate current distribution for various current electrode separations. If you try several values for current electrode separation, you should observe that the values in columns two

TABLE 5.2 PERCENTAGE OF CURRENT PENETRATING A HOMOGENEOUS, ISOTROPIC EARTH

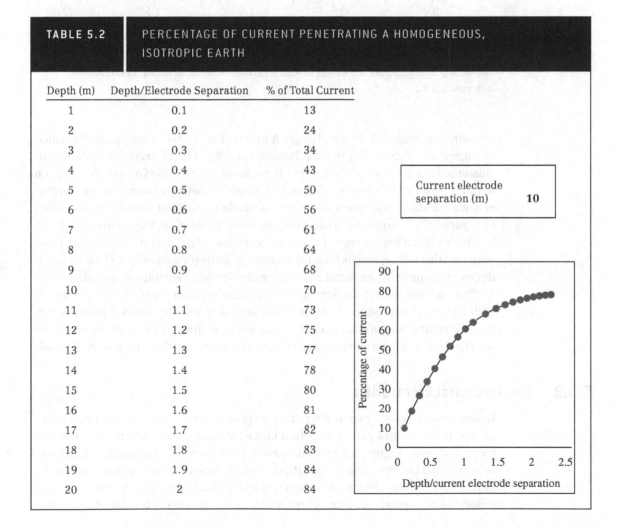

Depth (m)	Depth/Electrode Separation	% of Total Current
1	0.1	13
2	0.2	24
3	0.3	34
4	0.4	43
5	0.5	50
6	0.6	56
7	0.7	61
8	0.8	64
9	0.9	68
10	1	70
11	1.1	73
12	1.2	75
13	1.3	77
14	1.4	78
15	1.5	80
16	1.6	81
17	1.7	82
18	1.8	83
19	1.9	84
20	2	84

Current electrode separation (m) **10**

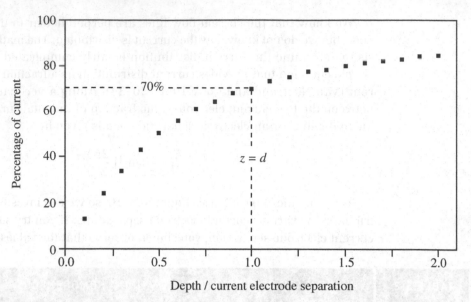

FIGURE 5.7 Plot of results in Table 5.2. Depth is z and current electrode separation is d. The data points illustrate the extent to which current penetrates into a homogeneous, isotropic Earth.

and three in Table 5.2 do not change. A graph of these relationships is illustrated in Figure 5.7. Examining these relationships allows us to arrive at certain conclusions. Fifty percent of the current is confined above a horizontal plane with a depth of one-half the current electrode separation. Seventy percent of the current is confined above a depth equal to the electrode separation (see the dashed lines in Figure 5.7). Clearly, the greater the electrode separation, the greater the depth to which a given percentage of current penetrates. Make a firm mental note, however, that this relationship is for a homogeneous, isotropic subsurface. Once we introduce a discontinuity, we must reexamine the current distribution question.

Now we can take the information that Equation 5.10 supplies and combine it with Figure 5.6 to produce the current flow lines that are diagrammed in Figure 5.8. We will return several times to these current flow lines, their spacing, and their distribution; so be sure to grasp what they represent and how they were derived.

5.3.3 Two Potential Electrodes

In electrical resistivity surveying our goal is to measure the potential difference between two points just as we often make this measurement in electrical circuits. Examine Figure 5.9 and compare it to Figure 5.5. Figure 5.9 illustrates two potential electrodes, P_1 and P_2, that are located on the surface, as are the current electrodes. Because we already derived an equation to determine the potential at a point due to a source and a sink, we obtain the potential differ-

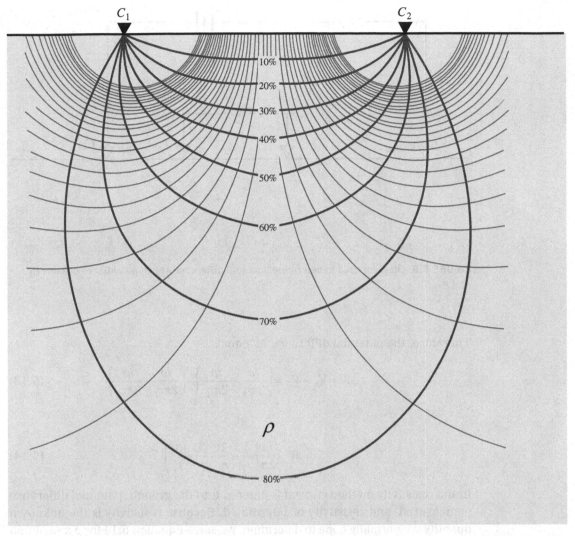

FIGURE 5.8 Equipotential surfaces and current lines of flow. Labels indicate percentage of total current that penetrates to the depth of the line.

ence by determining the potential at one potential electrode P_1 and subtracting from it the potential at P_2. Using Equation 5.8 we determine that

$$V_{P_1} = \frac{i\rho}{2\pi r_1} - \frac{i\rho}{2\pi r_2} \qquad (5.11)$$

and

$$V_{P_2} = \frac{i\rho}{2\pi r_3} - \frac{i\rho}{2\pi r_4}. \qquad (5.12)$$

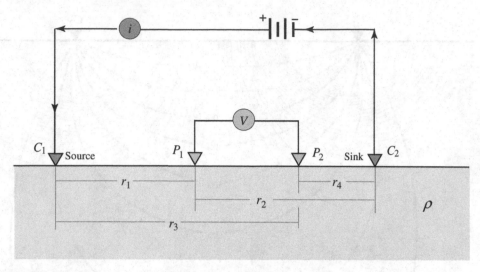

FIGURE 5.9 Diagram used to determine potential difference at two potential electrodes P_1 and P_2.

Therefore, the potential difference ΔV equals

$$\Delta V = V_{P_1} - V_{P_2} = \left(\frac{i\rho}{2\pi r_1} - \frac{i\rho}{2\pi r_2} \right) - \left(\frac{i\rho}{2\pi r_3} - \frac{i\rho}{2\pi r_4} \right) \qquad (5.13)$$

or

$$\Delta V = \frac{i\rho}{2\pi} \left(\frac{1}{r_1} - \frac{1}{r_2} - \frac{1}{r_3} + \frac{1}{r_4} \right). \qquad (5.14)$$

In the resistivity method current is entered into the ground, potential difference is measured, and resistivity is determined. Because resistivity is the unknown quantity we normally hope to determine, we solve Equation 5.14 for ρ and obtain

$$\rho = \frac{2\pi \Delta V}{i} \left(\frac{1}{\dfrac{1}{r_1} - \dfrac{1}{r_2} - \dfrac{1}{r_3} + \dfrac{1}{r_4}} \right). \qquad (5.15)$$

Perhaps we should test our understanding of Equation 5.15 by applying it to a known situation. Let's assume we can place potential electrodes anywhere along the surface as illustrated in Figure 5.9. Further, we will use the values in Table 5.1 for our test. Figure 5.10(a) presents one possible measurement, and Figure 5.10(b) another. Substituting the values in Figure 5.10(a) produces Equation 5.16, which results in a resistivity value of 50 ohm·m. A glance at the model values used to produce Table 5.1 confirms that the resistivity was 50 ohm·m.

(a)

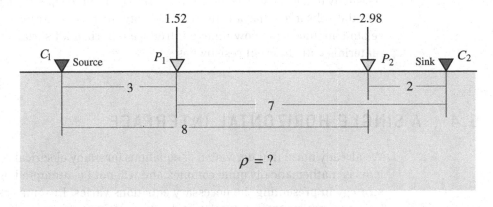

(b)

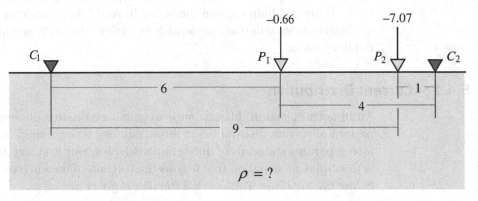

FIGURE 5.10 Measuring potential difference to determine ρ. Potential values for the indicated spacings are taken from Table 5.1. Solutions are presented in the text (Equations 5.16 and 5.17).

$$\rho = \frac{2\pi(4.5 \text{ volts})}{1 \text{ ampere}}\left(\frac{1}{\dfrac{1}{3\text{ m}} - \dfrac{1}{7\text{ m}} - \dfrac{1}{8\text{ m}} + \dfrac{1}{2\text{ m}}}\right) = 50 \text{ ohm·m} \qquad (5.16)$$

Equation 5.17 uses the values illustrated in Figure 5.10(b). As expected, evaluation of the equation gives 50 ohm·m as the resistivity.

$$\rho = \frac{2\pi(6.41 \text{ volts})}{1 \text{ ampere}}\left(\frac{1}{\dfrac{1}{6\text{ m}} - \dfrac{1}{4\text{ m}} - \dfrac{1}{9\text{ m}} + \dfrac{1}{1\text{ m}}}\right) = 50 \text{ ohm·m} \qquad (5.17)$$

These calculations confirm that if we produce a current, measure the electrode spacings, and determine the potential difference, we can arrive at a value for the resistivity of the subsurface materials. However, until this point we have assumed that the subsurface has a constant resistivity. Just as in our development of the seismic method, it is now time to introduce a horizontal surface that separates materials with different resistivities.

5.4 A SINGLE HORIZONTAL INTERFACE

We already noted that derivation of equations for many electrical resistivity problems is mathematically quite complex and will not be attempted in this text. Our strategy in presenting the necessary equations varies. In some cases we simply introduce an important equation and note a reference you may consult if you are interested and have the mathematical background to follow the derivation. In other cases we present a simplified case or present only part of the derivation if we feel this will help explain the concept. And finally, we may present only a qualitative explanation or couple such an explanation with an equation or a partial derivation.

5.4.1 Current Distribution

An important goal in this section is to gain a qualitative understanding for the pattern of current distribution in the subsurface when a single horizontal interface separates materials of different resistivities. Our first step toward this goal is to employ an equation that tells us the fraction of the current that penetrates below the interface. This current fraction is given by

$$i_F = \frac{2\rho_1}{\pi\rho_2}(1+k)\sum_{n=0}^{\infty} k^n \left[\frac{\pi}{2} - \tan^{-1}\left(\frac{2(2n+1)z}{3a} \right) \right] \tag{5.18}$$

where

$$k = \frac{\rho_2 - \rho_1}{\rho_2 + \rho_1}$$

(Van Nostrand and Cook 1996, 89).

Although this equation is relatively complex, none of the quantities should surprise us: current fraction i_F, electrode spacing a, depth of interface z, and resistivities of the materials above and below the interface ρ_1 and ρ_2. Note that in this equation electrode spacing a refers to the spacing between each electrode in the array (Figure 5.11(a)), whereas previously we usually referred to the spacing between current electrodes. The relationships that we can deduce from this equation form the basis for a qualitative interpretation of field measurements, so let's take some time to gather information.

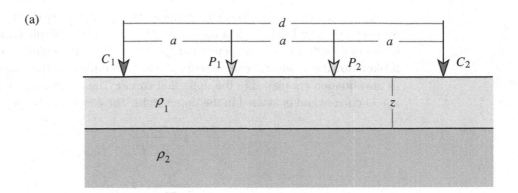

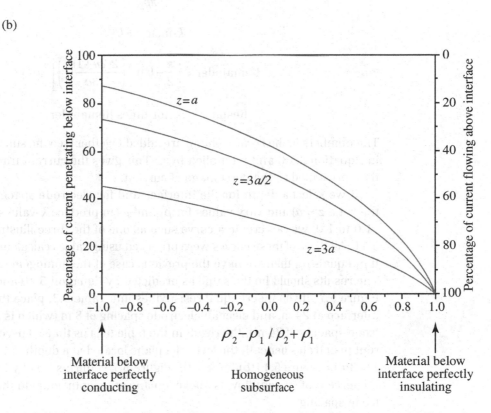

FIGURE 5.11 Current distribution when a single horizontal interface is present. (a) Electrode configuration. (b) Graph of current distribution. Current distribution is controlled by the relationship between interface depth z, electrode separation a, and the two resistivities ρ_1 and ρ_2.

Equation 5.18 is sufficiently straightforward to permit entry into a spreadsheet for evaluation, and Table 5.3 is the result. The appearance of this table is slightly different than those we've seen previously because the equation being evaluated is broken down into several components due to its complexity. The major variables in the equation are placed in the upper left corner. The result is given as percentage of current and is located in the upper right. The components are

$$k = \frac{\rho_2 - \rho_1}{\rho_2 + \rho_1}$$

$$C = \frac{2\rho_1}{\pi\rho_2}(1+k),$$

$$k\text{-factor} = k^n$$

$$\text{Remainder} = \left[\frac{\pi}{2} - \tan^{-1}\left(\frac{2(2n+1)z}{3a}\right)\right], \text{ and}$$

$$\text{Result} = k\text{-factor times Remainder}$$

The numbers in the *result* column are added together (see the summation symbol in Equation 5.18) and multiplied by C. This gives the current fraction, which is then presented as the percentage of current.

If we select a depth for the interface and for electrode spacing (perhaps in the ratio $z = a$) and vary values for ρ_1 and ρ_2 to produce k-values ranging from -1.0 to 1.0, we can create a curve such as one of the three illustrated in Figure 5.11. All three of these curves were produced using values calculated by Table 5.3. If ρ_1 equals ρ_2, then we have the previous case of the homogeneous subsurface. The results should be the same as predicted by Equation 5.10 and illustrated in Figure 5.7. If you insert equal resistivities into Table 5.3, place the depth of the interface at 24 m, and elect an electrode spacing of 8 m (which is a current electrode spacing of 24 m), the result in the table tells us that 30 percent of the current penetrates beneath the level of a plane located at a depth of 24 m (see curve $z = 3a$ in Figure 5.11(b) for $k = 0$). Figure 5.7 confirms this by illustrating that 70 percent of current flow is above a plane at a depth equal to the current electrode spacing.

Now let's examine the situation when $\rho_2 > \rho_1$. The material below the interface has a higher resistance to current flow (it is a poor conductor relative to the layer above). K-values will be positive for this case. The three curves in Figure 5.11(b) all have the same form, so we can generalize our comments. If we select any positive k-value, we see that substantially *less* current penetrates below the level of the interface compared to the homogeneous case (k-value = 0). In other words, current flow tends to avoid a poor conductor in favor of a good conductor. Perhaps a water analogy will make this phenomenon clearer. Visualize a homogeneous sequence of coarse sand with water flowing through. If at some

TABLE 5.3	PERCENTAGE OF CURRENT PENETRATING BELOW A SINGLE HORIZONTAL INTERFACE

k	−0.40		
C	0.89		
n	k-factor	Remainder	Result
0.00	1.00	0.98	0.98
1.00	−0.40	0.46	−0.19
2.00	0.16	0.29	0.05
3.00	−0.06	0.21	−0.01
4.00	0.03	0.17	0.00
5.00	−0.01	0.14	0.00
6.00	0.00	0.11	0.00
7.00	0.00	0.10	0.00
8.00	0.00	0.09	0.00
9.00	0.00	0.08	0.00
10.00	0.00	0.07	0.00
		Sum of results	0.83
		Current fraction	0.74

Electrode spacing a (m)	8	
Depth of interface (m)	8	
ρ_1	70	Percentage of current 74
ρ_2	30	
(Resistivities are in $\Omega \cdot$m)		

depth z we introduce a relatively low-permeability layer, less water will flow below that depth level than in the homogeneous case. Of course, the extreme case is when the layer below the interface is perfectly insulating (or completely impermeable), and no current (or water) penetrates below that level.

If $\rho_2 < \rho_1$ the material below the horizontal interface is less resistant to current flow than the material above the interface and is, therefore, a better conductor. As Figure 5.11(b) demonstrates, more current will penetrate below the interface than in the homogeneous case. Using the water analogy once again, we replace our homogeneous material with a coarse gravel that has very high permeability. More water now will flow below the level of the interface and through the gravel. Note that a substantial percentage of the current flows through the material above the interface even if the material below the interface possesses very low resistivities (is almost perfectly conducting).

In our previous discussion of current flow in a homogeneous, isotropic subsurface we learned that as current electrode spacing is increased, the percentage of current flowing at depth is increased. For example, only 30 percent of the current penetrates below a depth z when the electrode spacing is equal to one-third the depth ($z = 3a$). However, 50 percent of the current penetrates deeper than z when the electrode separation is increased to two-thirds the depth ($z = 3a/2$), and 63 percent of the current penetrates below z when the electrode spacing equals the depth ($z = a$). Remember that we are using electrode spacing a as diagrammed in Figure 5.11(a). Thus, when the electrode spacing is equal to the depth of our plane of reference ($z = a$), the current electrode separation is three times the depth ($3z = 3a = d$).

Figure 5.11(b) indicates, as expected, that current penetration increases as electrode separation increases. However, the percentage of current penetrating below an interface is controlled by the relative magnitudes of ρ_2 and ρ_1 as well as by electrode separation. This relationship plays an important role in data acquisition design, so it is a good idea to thoroughly understand the implications of Figure 5.11.

5.4.2 Current Flow Lines and Current Density

The previous discussion gives us sufficient information to qualitatively assess current flow lines and, more important, current density distribution when a horizontal interface is present. As a first step in this process, we must investigate what happens to the orientation of flow lines and equipotentials when they cross a boundary separating regions of differing conductivities or resistivities. Hubbert (1940, 844–46) demonstrated that the flow lines follow a tangent relationship such that

$$\frac{\tan \theta_1}{\tan \theta_2} = \frac{\rho_2}{\rho_1} \tag{5.19}$$

where θ and ρ are as defined in Figure 5.12(a). If the resistivity ρ_2 of the deeper material is greater, the flow lines bend in toward the normal to the interface (Figure 5.12(b)) and as a consequence are more widely spaced. However, if the reverse is true as in Figure 5.12(c), the flow lines bend away from the normal, become oriented more parallel to the interface, and are closer together.

If we take this relationship and the information we developed to produce Figure 5.11, we can assess the effect of the horizontal interface. Figure 5.13(a) illustrates the pattern of current flow lines for a homogeneous subsurface ($\rho_2 = \rho_1$). If we increase the value of ρ_2 (Figure 5.13(b)), more current flows above the interface,

(a)

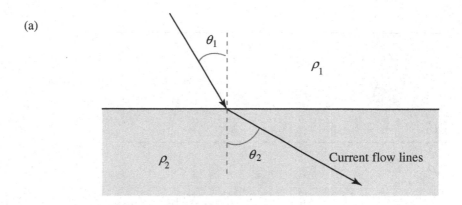

(b)

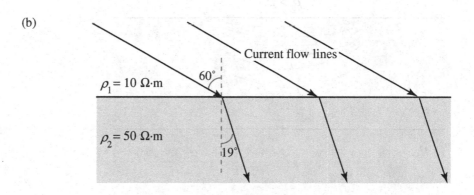

(c)

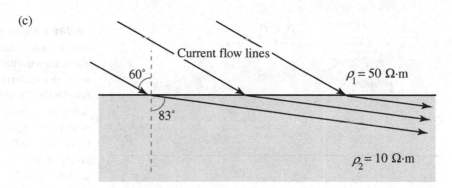

FIGURE 5.12 Refraction of current flow lines at a boundary separating materials of differing resistivities. (a) Symbols used in Equation 5.19. (b) Refraction when $\rho_1 < \rho_2$. (c) Refraction when $\rho_1 > \rho_2$.

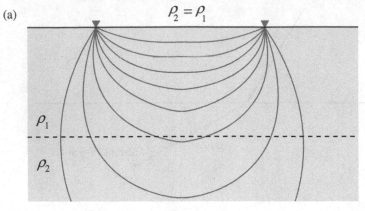

(a) $\rho_2 = \rho_1$

ρ_1

ρ_2

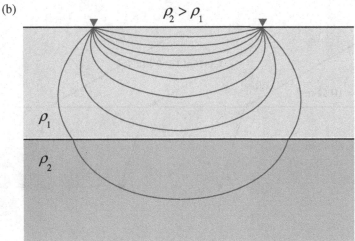

(b) $\rho_2 > \rho_1$

ρ_1

ρ_2

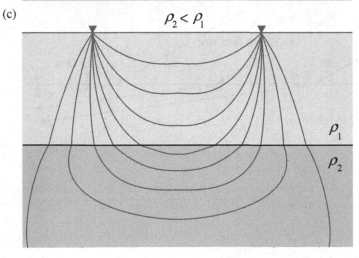

(c) $\rho_2 < \rho_1$

ρ_1

ρ_2

FIGURE 5.13 Qualitative distribution of current flow lines when a horizontal interface separates materials of differing resistivities. (a) Homogeneous subsurface. Dashed line marks position of horizontal interface in (b) and (c). (b) When material with greater resistivity is located beneath the interface, the current flow lines are more closely spaced than in (a). (c) When material with greater resistivity is located above the interface, the current flow lines are less closely spaced than in (a).

the current flow lines are spaced more closely, and the current density is greater in the region above the interface relative to the case of the homogeneous subsurface. If $\rho_2 < \rho_1$, the opposite effect takes place. A greater percentage of current flows beneath the interface (Figure 5.13(c)), the current flow lines are spaced more widely in the material above the interface, and the current density there is reduced.

Our real purpose is to determine, qualitatively at the moment, how the presence of the interface affects ΔV readings at the surface. We are just about ready to take this step, but first we need to introduce the notion of *apparent resistivity*.

5.4.3 Apparent Resistivity

When we derived Equation 5.15, we assumed a homogeneous, isotropic subsurface. As demonstrated previously, any combination of electrode spacings and current results in a potential difference that provides the correct value for the resistivity of the subsurface (as of course should be the case if our equation is correct). Once the subsurface is nonhomogeneous, the value determined for the resistivity is extremely unlikely to equal the resistivity of the material in which the electrodes are inserted. Equation 5.15 thus defines a different quantity, which is termed the *apparent resistivity*, ρ_a. Inasmuch as nonhomogeneity is the rule, we now write

$$\rho_a = \frac{2\pi\Delta V}{i}\left(\frac{1}{\dfrac{1}{r_1}-\dfrac{1}{r_2}-\dfrac{1}{r_3}+\dfrac{1}{r_4}}\right). \tag{5.20}$$

What does this equation tell us? How do we interpret apparent resistivity values in terms of subsurface geology?

If we examine Equations 5.2, 5.3, and 5.5, it becomes straightforward to demonstrate that the potential is proportional to current density:

$$V \propto j. \tag{5.21}$$

When we measure the potential difference between the potential electrodes, the values are proportional to the current density in the cylinder of material near the surface that extends between these two electrodes. If we examine Equation 5.20 and remember the relationship expressed in Equation 5.21, we conclude that, all else being equal, variations in current density near the surface will result in variations in apparent resistivity (Mooney 1958). Let's return to Figure 5.13.

In the case of Figure 5.13(a) we already know that Equation 5.20 will yield a value of $\rho_a = \rho_1$. In the case represented by Figure 5.13(b) the current density is increased in the upper layer relative to the homogeneous case, so ρ_a will be greater than ρ_1. Conversely, in Figure 5.13(c) the current density is decreased due to the lower-resistivity material at depth, and therefore the value of ρ_a will be less than ρ_1. Notice that the electrode spacing remains constant in this example.

5.4.4 Qualitative Development of the Resistivity Pattern over a Horizontal Interface

By now you should have a good grasp of how the horizontal interface affects current flow, current density, and the resistivity that we measure at the surface (the apparent resistivity). But how do we use this information to explore the subsurface as we did using seismic methods—to determine the depth to the interface and the nature of the geologic materials above and below the interface? As a first step we continue our qualitative approach, but shortly we will develop quantitative methods.

If a horizontal interface is deep relative to electrode spacing a or current electrode spacing d as defined in Figure 5.11(a), it is unlikely that the presence of the interface will have much effect on the current density between the potential electrodes. Therefore, ρ_a will equal or be very close to ρ_1. This relationship is diagrammed in Figure 5.14, where dashed black lines represent current flow if no interface is present and solid gray lines represent current flow due to the interface. In Figure 5.14(a) both sets of lines coincide. If we simply move this electrode array about on the surface, all readings will be the same assuming the ρ_1 layer is laterally homogeneous. However, if we expand the electrode spacing, more current will flow at depth, and the current density will be affected by the ρ_2 layer. The apparent resistivity value will be greater than ρ_1 but still will be substantially less than ρ_2 (Figure 5.14(b)). If electrode spacing is increased still more, the effect of the higher-resistivity material below the interface is even greater, as illustrated in Figure 5.14(c). The value of apparent resistivity continues to increase and gradually approaches the ρ_2 value while becoming increasingly greater than the ρ_1 value.

In many resistivity investigations electrode spacing is plotted against apparent resistivity. If we sketch a curve based on our previous analysis, it should have a form similar to that in Figure 5.15(a). Of course, the actual form will depend on the resistivity contrast, the depth of the interface, and the electrode spacings. At the moment it is important only that you understand why the general shape of the curve is as illustrated. If we reverse the situation so that the material below the interface has a lower resistivity than the material above, the curve assumes the form illustrated in Figure 5.15(b). Various curves for other subsurface situations are illustrated in Mooney (1958), and we encourage you to study them.

Following the form of the previous discussion, describe why the curve in Figure 5.15(b) appears as it does.

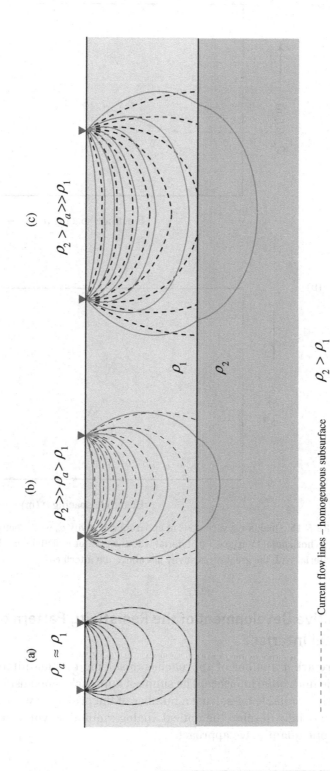

FIGURE 5.14 Effect on apparent resistivity as electrode spacing is increased. Dashed black lines represent current flow line distribution for homogeneous subsurface. Solid gray lines represent actual current flow lines due to horizontal interface. As the distance between current electrodes is increased, ρ_a approaches the value of ρ_2.

(a)

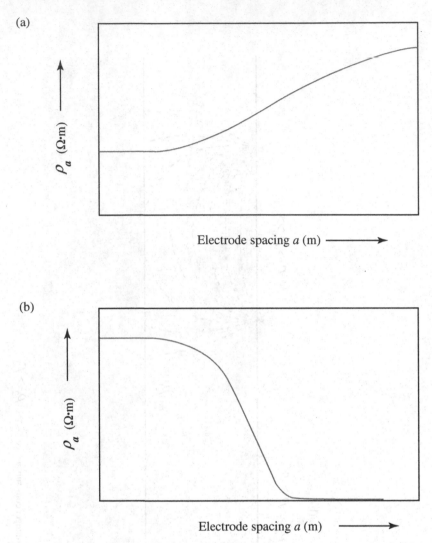

Electrode spacing a (m) $\longrightarrow$

(b)

Electrode spacing a (m) $\longrightarrow$

FIGURE 5.15 Qualitative variations in apparent resistivity ρ_a with electrode spacing (a) for a single horizontal interface. (a) Material with the greater resistivity lies below the interface. (b) Material with the greater resistivity lies above the interface.

5.4.5 Quantitative Development of the Resistivity Pattern over a Horizontal Interface

There are a number of approaches to solving the quantitative problem of the single horizontal interface. The simplest approach is one using an optical analog, which is suitable for solving a number of the problems we are interested in. We will partially develop the optical analog method so you can gain some insight into one quantitative approach.

As a first step in the optical analog approach, we assume a planar boundary that behaves as a semitransparent mirror. If we place a light source at point C_1 and an observation point at P_1, part of the light will travel directly to P_1 and part will be reflected from the mirror (Figure 5.16). The total light reaching P_1 will be the sum of the two. The amount reflected depends on a property of the mirror referred to as the *reflection coefficient* (k) and is equal to the reflection coefficient times the intensity of the light source. In subsequent development of this approach it is more direct to construct an image of point C_1 on the other side of

(a)

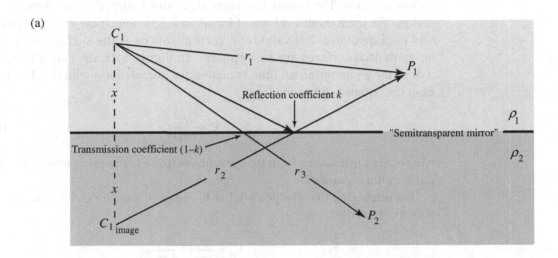

(b)

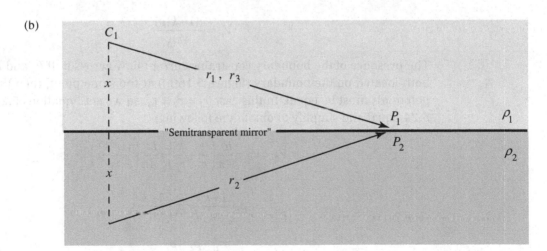

FIGURE 5.16 Determination of potential at points P_1 and P_2 using an optical analog for a case when a single interface separates two regions of different resistivities. (a) General case. (b) Case when P_1 and P_2 are at the same position on the boundary.

the mirror. This is the same approach we used in deriving equations for reflected rays in Chapter 4. Recall that C_1 and its image are equidistant from the boundary (mirror) and lie on a line that is perpendicular to the boundary. Recall also that the distance r_2 is equal to the reflected path. If our observation point is at P_2 rather than P_1, the amount of light reaching P_2 is only that which is transmitted through the mirror. This amount is due to the intensity of the light source, the *transmission coefficient* $(1-k)$, and the distance of P_2 from C_1.

To develop the electrical case, we place a current source at C_1 with intensity i and develop equations for the potential at P_1 and P_2 following the optical analog as just outlined. The planar boundary separates materials with resistivities ρ_1 and ρ_2. We use a modified form of Equation 5.7 to determine the potential at P_1 and P_2. Equation 5.7 is valid for a current source at the surface because the equipotential surfaces are hemispheres. In Figure 5.16 the current source is within the ρ_1 material, and the equipotential surfaces are spheres. For such a case the equation takes the form

$$V = \frac{i\rho}{4\pi D} \tag{5.22}$$

where D is the distance from the current electrode to the point where the potential is being measured.

Remembering that the potential at P_1 is due to a direct and a reflected component, we write

$$V_{P_1} = \frac{i\rho_1}{4\pi r_1} + \frac{ik\rho_1}{4\pi r_2}. \tag{5.23}$$

The potential at P_2 (due to transmission) is

$$V_{P_2} = \frac{i(1-k)\rho_2}{4\pi r_3}. \tag{5.24}$$

The presence of the boundary constrains current flow across it. If P_1 and P_2 are both located on the boundary (Figure 5.16(b)) at the same point, then the two potentials must be equal. In this case $r_1 = r_2 = r_3$, so we set Equations 5.23 and 5.24 equal and simplify to obtain the following:

$$V_{P_1} = V_{P_2} = \frac{i\rho_1}{4\pi}\left(\frac{1}{r_1} + \frac{k}{r_2}\right) = \frac{i\rho_2}{4\pi}\left(\frac{(1-k)}{r_3}\right)$$

$$\rho_1\left(\frac{1+k}{r_1}\right) = \rho_2\left(\frac{1-k}{r_1}\right)$$

$$k = \frac{\rho_2 - \rho_1}{\rho_2 + \rho_1}. \tag{5.25}$$

This derivation is useful for two reasons. First, we now have sufficient information to map the equipotentials for this two-medium boundary problem. Also, we know how to determine the reflection coefficient. You may recall that this same factor appeared in Equation 5.18 when we first began our qualitative consideration of the horizontal interface problem. Although we accepted Equation 5.18 without explaining its derivation, our present inquiry is beginning to shed light on various components of that equation.

We will not pursue this case further. Instead we now turn to the actual problem at hand—that of the horizontal interface. If you inspect Figure 5.17 and compare it to Figure 5.16, you will see that we must add another surface to consider both the effect of the Earth's surface and the buried interface. To derive the potential equations we must consider both surfaces, but we proceed essentially as before.

A current source C_1 of intensity i is placed within the ρ_1 layer. The point P_1 at which we want to know the potential is placed a distance r from C_1 at the same depth. The placement of C_1 and P_1 is for convenience; they will be placed on the surface later in the derivation. The upper boundary (the Earth's surface) separates ρ_0 from ρ_1. Our object of interest, the horizontal interface, serves as the lower boundary and separates ρ_2 from ρ_1. The reflection coefficient for the upper boundary is designated by $k_{1,0}$ and follows the form of Equation 5.25 (using ρ_1 and ρ_0), as does the reflection coefficient for the lower boundary $k_{1,2}$. Image sources of C_1 are referred to as such with a number in parentheses indicating the layer in which they reside.

Referring to Figure 5.17(a), we proceed as in the previous example. The potential at P_1 due to C_1 is simply

$$^{primary}V_{P_1} = \frac{\rho_1 i}{4\pi r}. \tag{5.26}$$

If we consider a single reflection from the upper boundary (equivalent to placing an image of C_1 in the upper material), the intensity of the current is i times the reflection coefficient $k_{1,0}$, and the contribution to the potential at P_1 is

$$^{upper}_{first}V_{P_1} = \frac{\rho_1 i k_{0,1}}{4\pi(r^2 + 4m^2)^{\frac{1}{2}}}. \tag{5.27}$$

The qualifiers associated with the potential symbol V signify that the image is in the upper material and is the first image constructed there. It is demonstrated in Figure 5.17 that we also can have a single reflection from the lower boundary. This contribution to the potential is

$$^{lower}_{first}V_{P_1} = \frac{\rho_1 i k_{1,2}}{4\pi(r^2 + 4(z-m)^2)^{\frac{1}{2}}}. \tag{5.28}$$

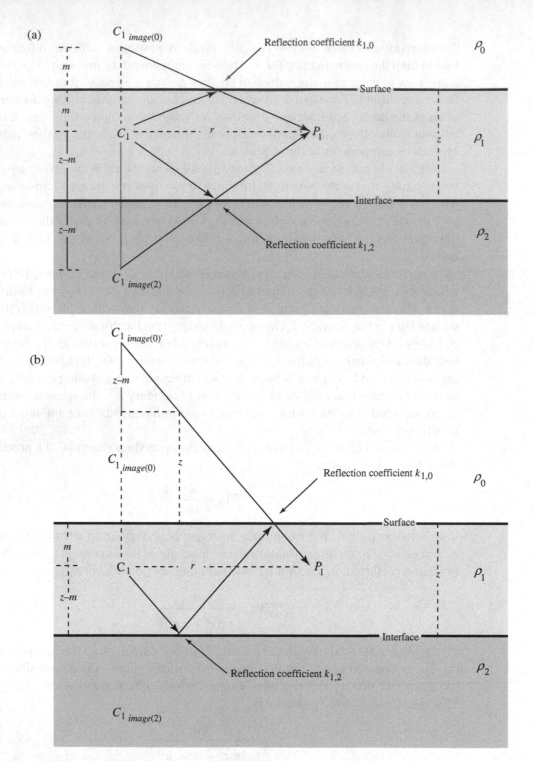

FIGURE 5.17 Determination of potential at point P_1 using an optical analog for the case of a horizontal interface beneath the surface. (a) First images in upper and lower materials. (b) Second image in upper material.

Reverting to the optical analog for a moment, we can consider a ray emanating from C_1 reflecting from the lower boundary, reflecting from the upper boundary, and then arriving at P_1. As illustrated in Figure 5.17(b), this is equivalent to placing a second image in the upper layer ($C_{1_{image(0)'}}$), which is located at a distance $2z$ from C_1. Because this path involves reflections from both boundaries, both reflections contribute to a reduction in intensity. Current density, therefore, is multiplied by the product of both reflection coefficients. The equation for the potential due to the second image in the upper material is

$$\substack{upper \\ second} V_{P_1} = \frac{\rho_1 i k_{0,1} k_{1,2}}{4\pi(r^2 + 4z^2)^{\frac{1}{2}}}.$$

(5.29)

We can continue this process and consider as a next step the path that is similar to that in Figure 5.17(b) but reflects twice off the lower boundary and twice off the upper boundary. The equation for this path is

$$\substack{upper \\ third} V_{P_1} = \frac{\rho_1 i k_{1,0}^2 k_{1,2}^2}{4\pi(r^2 + 16z^2)^{\frac{1}{2}}}.$$

(5.30)

Note that now each reflection coefficient is squared. We can continue this process producing more images in the upper medium that contribute less to the potential at P_1. If you examine Equations 5.29 and 5.30, you should note the similarity. Equations incorporating more of this group of images would continue to have this form. This suggests that these equations can be expressed as the infinite series

$$\substack{upper \\ even} V_{P_1} = \sum_{n=1}^{\infty} \frac{\rho_1 i k_{1,0}^n k_{1,2}^n}{4\pi[r^2 + (2nz)^2]^{\frac{1}{2}}}$$

(5.31)

(Keller and Frischknecht 1966, 109). Up to this point we have four equations that represent contributions to the potential at P_1 (5.26, 5.27, 5.28, and 5.31). In addition to these three other infinite series must be included. In arriving at Equation 5.31 we considered only an even number of reflections that could be related to an image in the upper medium. Other combinations are (1) an odd number of reflections with images in the upper medium, (2) an even number of reflections with images in the lower medium, and (3) an odd number of reflections with images in the lower medium. The total potential at P_1 is the sum of all of these.

The next step is to place C_1 and P_1 on the surface. In this case, because air has infinite resistivity, $k_{1,0} = 1$ and $m = 0$. These substitutions are made in all the equations we added together to determine the potential at P_1. We rearrange these to arrive at

$$V_{P_1} = \frac{\rho_1 i}{2\pi r}\left[1 + 2\sum_{n=1}^{\infty} \frac{k_{1,2}^n}{\left[1 + (2nz/r)^2\right]^{\frac{1}{2}}}\right]$$

(5.32)

(Keller and Frischknecht 1966, 112). However, we ultimately are interested in determining the potential difference due to a source and a sink in a configuration such as that illustrated in Figure 5.11(a). Equation 5.32 provides the potential at P_1 due to the source at C_1, which are a distance a apart. Deriving an equation for the potential at P_2 due to C_1 is trivial because the only difference is the separation of the two points, which is now $2a$. Thus the equation for the potential difference is

$$\Delta V = V_{P_1} - V_{P_2} = \frac{\rho_1 i}{4\pi a}\left[1 + 4\sum_{n=1}^{\infty}\frac{k_{1,2}^n}{\left[1 + (2nz/a)^2\right]^{\frac{1}{2}}} - 2\sum_{n=1}^{\infty}\frac{k_{1,2}^n}{\left[1 + (nz/a)^2\right]^{\frac{1}{2}}}\right]. \quad (5.33)$$

Remember that our ultimate goal is to develop an equation similar to Equation 5.20 that provides values of apparent resistivity for various electrode spacings in the case of the horizontal interface. If we use the electrode spacings in Figure 5.11(a), we can rewrite Equation 5.20 as

$$\rho_a = \frac{2\pi\Delta V}{i}\left(\frac{1}{\frac{1}{a} - \frac{1}{2a} - \frac{1}{2a} + \frac{1}{a}}\right) \quad (5.34)$$

and

$$\rho_a = \frac{2\pi a \Delta V}{i}. \quad (5.35)$$

Equation 5.33 does not consider the contribution of the electrode at C_2, but this simply doubles the potential difference ΔV. We make this adjustment and then rewrite Equation 5.35 using the expression for ΔV given in Equation 5.33 to obtain

$$\rho_a = \left(\frac{2\pi a}{i}\right)\left\{\frac{\rho_1 i}{2\pi a}\left[1 + 4\sum_{n=1}^{\infty}\frac{k_{1,2}^n}{\left[1 + (2nz/a)^2\right]^{\frac{1}{2}}} - 2\sum_{n=1}^{\infty}\frac{k_{1,2}^n}{\left[1 + (nz/a)^2\right]^{\frac{1}{2}}}\right]\right\}$$

and

$$\rho_a = \rho_1\left[1 + 4\sum_{n=1}^{\infty}\frac{k_{1,2}^n}{\left[1 + (2nz/a)^2\right]^{\frac{1}{2}}} - 2\sum_{n=1}^{\infty}\frac{k_{1,2}^n}{\left[1 + (nz/a)^2\right]^{\frac{1}{2}}}\right]. \quad (5.36)$$

The form of Equation 5.36 is valid only for the electrode configuration in Figure 5.11(a). Other electrode configurations require modifications of Equation 5.36. These other configurations will be considered in a later section. It's been a long path to arrive at this point, and a number of steps have been neglected to keep the development to a minimum. We hope this gives you sufficient insight into the optical analog approach to obtaining a quantitative solution to the single interface problem. If you are interested in a complete and thorough analysis, consult Keller and Frischknecht (1966).

Let's place Equation 5.36 into a dynamic table (Table 5.4) to investigate the variation of apparent resistivity with electrode spacing. This table is designed much like Table 5.3. The major variables are in the upper left corner. The result is apparent resistivity and is located in the upper right. The components are

$$k = \frac{\rho_2 - \rho_1}{\rho_2 + \rho_1},$$

k-factor $= k^n,$

$$\text{Divisor 1} = \left[1 + (2nz/a)^2\right]^{\frac{1}{2}},$$

$$\text{Divisor 2} = \left[1 + (nz/a)^2\right]^{\frac{1}{2}},$$

Sum 1 = k-factor/Divisor 1,

Sum 2 = k-factor/Divisor 2, and

$\rho_a = \rho_1[1 + 4(\text{Sum 1 column total}) - 2(\text{Sum 2 column total})].$

Finally, we have a quantitative approach to determining the variation of apparent resistivity with electrode spacing as defined in Figure 5.11(a) (all electrodes spaced at equal intervals). If we select the subsurface model as it appears in Table 5.4 and vary the electrode spacing, we can generate the curve in Figure 5.18(a). The general form of the curve should come as no surprise: It is similar to our qualitative prediction (Figure 5.15(a)). What may be a surprise, however, is the magnitude of the electrode spacing a necessary for the apparent resistivity to approach the resistivity of the second layer. As electrode spacing is increased, more current penetrates to a greater depth, so, as you already know, we "sample" material at increasingly greater depths. However, the shallow material always is included in the measurement. Therefore, for the measured apparent resistivity to approach the resistivity of the second layer, a great volume of the subsurface must be sampled. As you might imagine, this seriously limits the resistivity method because it requires sufficient space in which to place electrodes.

5.4.6 Using RESIST

Although we could continue to use Table 5.4 and a graphing application to construct apparent resistivity curves, it certainly is easier and more elegant to use a computer program written especially for this purpose. RESIST is such a program. It accepts resistivities and thicknesses for up to 10 layers, calculates apparent

TABLE 5.4		VARIATION OF APPARENT RESISTIVITY WITH ELECTRODE SPACING FOR A SINGLE HORIZONTAL INTERFACE			

k	0.82				
n	k-factor	Divisor 1	Divisor 2	Sum 1	Sum 2
1.00	0.82	40.01	20.02	0.02	0.04
2.00	0.67	80.01	40.01	0.01	0.02
3.00	0.55	120.00	60.01	0.00	0.01
4.00	0.45	160.00	80.01	0.00	0.01
5.00	0.37	200.00	100.00	0.00	0.00
6.00	0.30	240.00	120.00	0.00	0.00
7.00	0.25	280.00	140.00	0.00	0.00
8.00	0.20	320.00	160.00	0.00	0.00
9.00	0.16	360.00	180.00	0.00	0.00
10.00	0.13	400.00	200.00	0.00	0.00
			Column totals	0.04	0.08

Electrode spacing a (m)	**1**
Depth of interface (m)	**20**
ρ_1	**10**
ρ_2	**100**
(Resistivities are in $\Omega \cdot$m)	

Apparent resistivity ($\Omega \cdot$m)	10

resistivities, and plots apparent resistivity against electrode spacing. Henceforth, for convenience, whenever we mention electrode spacing (or *a-spacing* as is sometimes the practice), we mean the spacing between individual electrodes, which is identified by the letter a in Figure 5.11(a). Later in this chapter other variations will be discussed and labeled differently.

Figure 5.18(b) illustrates a curve drawn by RESIST with the same input parameters that appear in Table 5.4. RESIST uses a different mathematical approach for solving the single horizontal interface problem, but the results are the same as those the optical analog produces. Actually, the results of Equation 5.36 (which is used in Table 5.4) are plotted as small, filled circles in Figure 5.18(b). The line passing through these points is the RESIST curve from Figure 5.18(c), which is overlaid on these data points to emphasize that both approaches produce identical results.

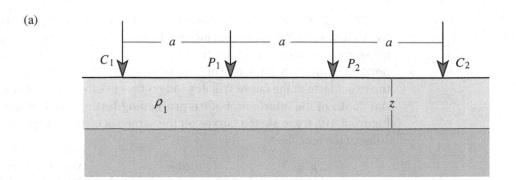

(a)

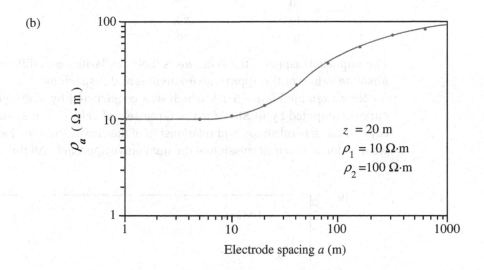

(b)

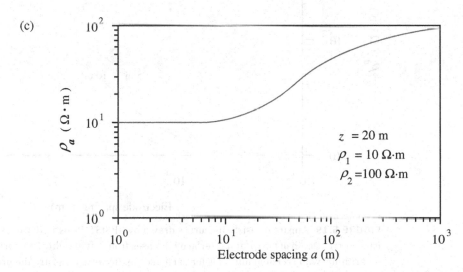

(c)

FIGURE 5.18 Apparent resistivity plotted against electrode spacing. (a) Geometry and parameters used. (b) Points calculated with Table 5.4. (c) Curve computed using RESIST.

Although we have not yet discussed procedures to interpret apparent resistivity curves in terms of subsurface geology, it is not too early to begin to assemble some basic properties of such curves. Apparent resistivity curves for the single horizontal interface have one of the two forms presented in Figure 5.15. Clearly, the exact form of the curve will depend on the resistivity values of the layers and the depth of the interface. Before proceeding further (and without examining Figure 5.19), try to sketch curves on the same set of axes for the following three subsurface models:

ρ_1 ($\Omega \cdot$m)	ρ_2 ($\Omega \cdot$m)	Depth to Interface
10	200	5
10	200	10
10	200	20

The important aspect of these curves is their similarities and differences, not the absolute values of the apparent resistivities and a-spacings.

Now examine Figure 5.19, which was constructed by overlaying the three curves computed by RESIST for the three sets of values. You should be able to explain the general shape and relationship of the curves from the basic principles we developed in our discussion of the qualitative approach. All three curves have

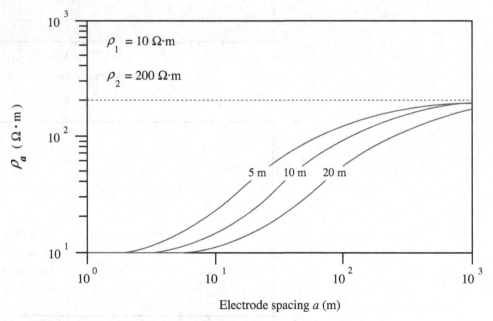

FIGURE 5.19 Apparent resistivity curves drawn by RESIST. Resistivity values are constant for each curve. Depth of horizontal interface increases from left to right. The horizontal scale is logarithmic with six equal intervals for each decade. Because these are the preferred sampling intervals for field measurements, they are displayed on all curves calculated by RESIST. These values are 1.0, 1.47, 2.15, 3.16, 4.64, 6.81, 10.0, and so on.

similar shapes, but curves for the deeper interfaces are progressively offset to the right. This makes sense if we remember the principle of current penetration, electrode spacing, and current density as affected by poor conductors. The 5 m curve begins to rise earlier because the effect of the shallow interface is felt even at small electrode spacings. In the case of the 20 m curve an electrode spacing of almost 10 m (current electrode spacing of 30 m) is necessary to reveal the presence of a higher-resistivity layer at depth. The 5 m curve begins to flatten noticeably at an a-spacing in the vicinity of 100 m, but the 20 m curve still is rising steeply at this point. It should be obvious that a 5 m thick layer constitutes only a small fraction of the volume of material included in a measurement with an a-spacing of 200 m. Therefore, the apparent resistivity for this a-spacing is quite close to the 200 ohm·m true resistivity of the deeper layer. This is not true when the interface is at 20 m, and so the ρ_a value is only about 80 ohm·m.

Sketch a set of three curves following the layout of Figure 5.19. Use the same interface depths but reverse the resistivities so the resistivity of the shallow layer is 200 Ω·m and that of the deeper layer is 10 Ω·m. Be sure you can defend the relative positions of the curves.

Before we proceed to the next step, try to sketch three curves for which the interface remains the same, the resistivity of the shallow layer is constant at 10 Ω·m, but the resistivity of the deeper layer takes on values of 100 Ω·m, 200 Ω·m, and 500 Ω·m.

As a final exercise before we continue our general analysis by considering the effect of multiple interfaces, let's investigate one further set of relationships for the single interface case. In the preceding example resistivities remained constant and only the depth of the interface varied. Now let's hold constant the resistivity of the layer above the interface and the interface depth while we vary the resistivity of the layer below the interface. Figure 5.20 provides a comparison for five apparent resistivity curves drawn under the stated constraints. Dashed lines are included in the figure to facilitate a visual comparison of the latter portion of each curve and the true resistivity of the deeper layer for which the curve was computed. The form and relationship of the curves should come as no surprise. As the resistivity of the deeper layer increases, larger electrode spacings are required before the slope of the curve begins to decrease and approach the resistivity value of that deeper layer.

The form of the curves in Figures 5.19 and 5.20 suggests that we should be able to devise a scheme to interpret apparent resistivity data in terms of resistivities and interface depths. We soon will see that this indeed is the case; but before we consider interpretations of field data, it seems prudent to consider more complicated models.

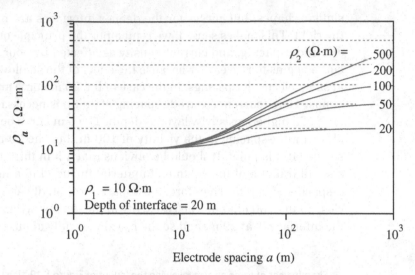

FIGURE 5.20 Apparent resistivity curves for constant depth to an interface. Resistivity of the layer above the interface also is constant. Dashed lines are drawn at resistivity values assigned to the layer beneath the interface. Pay special attention to the different curve forms for the various resistivities assigned to the deeper layer.

5.5 MULTIPLE HORIZONTAL INTERFACES

Based on our development of a quantitative solution to the single interface problem, it should be evident that we could continue in this direction for investigating the effect of more than one interface. However, the complexity of the image theory approach becomes extreme, so that in practice more advanced mathematical solutions usually are utilized. For our purposes it is sufficient to develop apparent resistivity curves using RESIST and to qualitatively analyze these curves based on the principles we discussed previously. To keep this discussion relatively brief, we consider only two horizontal interfaces (three layers). You are encouraged to use RESIST to conduct your own analysis of the three-interface (four layers) case.

Because we are dealing with three layer resistivities in the two-interface case, four possible curve types exist: $\rho_1 > \rho_2 > \rho_3$ (Q), $\rho_1 > \rho_2 < \rho_3$ (H), $\rho_1 < \rho_2 > \rho_3$ (K), and $\rho_1 < \rho_2 < \rho_3$ (A). The letters used to refer to the various curves are those typically used in electrical resistivity literature. We use them here only for convenience in referring to specific curve types. The four curves are illustrated in Figure 5.21. Interface depths are the same for all curves and are defined in Figure 5.21(a). Resistivities used are 1, 10, and 100 ohm·m. The ρ_1, ρ_2, and ρ_3 values for each curve are self-evident. Curves for $\rho_1 > \rho_2 > \rho_3$ (Q) and $\rho_1 < \rho_2 < \rho_3$ (A) are

(a)

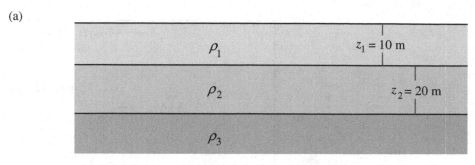

(b)

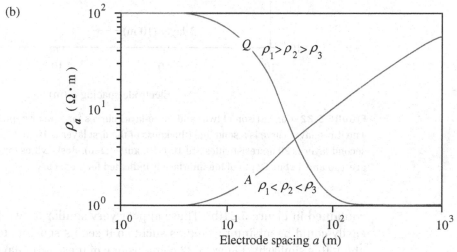

(c)

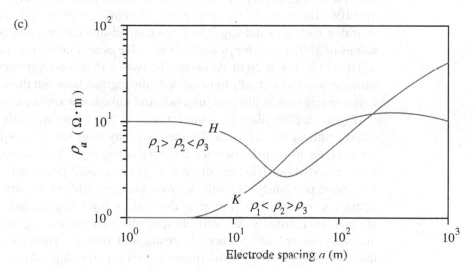

FIGURE 5.21 The four basic apparent resistivity curve types for two horizontal interfaces. Resistivities used to compute the curves are 1, 10, and 100 Ω·m. Thicknesses for each case are as defined in (a).

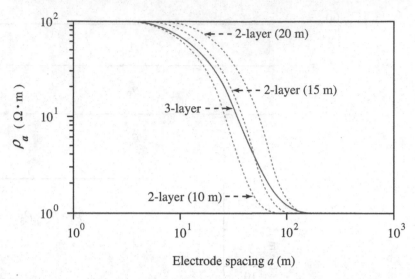

FIGURE 5.22 Comparison of two- and three-layer curves for similar depths and resistivities. The three-layer curve is a solid line (thickness of the first layer is 10 m; thickness of the second layer is 20 m; resistivities are 100, 10, and 1 $\Omega \cdot$m). Resistivities for the two-layer cases are 100 and 1 $\Omega \cdot$m. Depth of the interface is indicated for each curve.

contained in Figure 5.21(b). These appear very similar to two-layer curves and easily could be misinterpreted as such, so it seems sensible to inquire further about this similarity. Figure 5.22 contains one of these three-layer curves (Q with $\rho_1 = 100$ ohm·m, $\rho_2 = 10$ ohm·m, $\rho_3 = 1$ ohm·m, $z_1 = 10$ m, and $z_2 = 20$ m), which is drawn with the solid line. The three dashed-line curves were computed using values of 100 ohm·m for ρ_1 and 1 ohm·m for ρ_2 and horizontal interface depths of 10 m, 15 m, and 20 m. Although the two- and three-layer curves are not the same, it could be virtually impossible to distinguish between them with field data that are acquired at discrete intervals and almost always contain some "noise."

Figure 5.21(c) illustrates the remaining two curve types (H and K). These clearly indicate the presence of a low-resistivity layer within a sequence of higher resistivities or the presence of a high-resistivity layer lying between layers with lower resistivities. The form of curve K can be explained using the principles we developed previously. At small electrode spacings current density is affected only by the low resistivity (1 ohm·m) of the shallow layer, which leads to small values of apparent resistivity. As electrode spacings increase, the higher resistivity (100 ohm·m) of layer 2 affects current density, and apparent resistivities rise. As electrode spacings continue to increase, a greater percentage of current flows in the deepest layer, which has a lower resistivity (10 ohm·m) than the intermediate layer. This also affects current density distributions so that apparent resistivities begin to decrease. Note that even though the thickness of the second layer is substantial (20 m), apparent resistivities never exceed 13 ohm·m even though the resistivity of the layer is 100 ohm·m. Be sure you understand why curve H in Figure 5.21(c) has the particular form that it does.

> How many basic curve types exist for the four-layer case? How many of these are unique in the sense that they clearly arise from four layers?

Although complexities and, therefore, identification and interpretation difficulties increase as the number of interfaces increases, the basic principles determining apparent resistivity values versus electrode spacing remain consistent. We will return to this subject during our discussion of interpreting field values; but we turn now to a different subsurface model, that of the vertical contact.

5.6 VERTICAL CONTACT

A rapid lateral change in resistivity can be modeled as a vertical contact. Such geometry often is taken to represent a vertical fault, but sudden changes in composition due to depositional processes or even a pronounced elevation gradient (vertical cliff) also fit this model. The equations for this situation are no more complicated than for multiple horizontal interfaces. The method of images works well for the vertical contact. If you return to the section where we first considered the quantitative approach to a single horizontal interface and rotate Figure 5.16 90°, you should appreciate the similarity of the vertical contact problem to the single horizontal interface problem.

Complications arise, however, as electrode spacing is increased (or the entire spread is moved) because eventually one or more electrodes cross the contact (which is not a consideration when the interface is horizontal). This changes the boundary conditions, and a different potential equation must be applied. We need five equations to satisfy all possible variations in electrode placements as illustrated in Figure 5.23(a). We've chosen not to derive or present the relevant equations here; rather, we concentrate on the apparent resistivity patterns associated with measurements taken during traverses that cross a vertical contact. These equations and their derivation are present in Van Nostrand and Cook (1966, 52–54) or Telford, Geldart, and Sheriff (1990, 556–57).

When discussing apparent resistivity patterns associated with horizontal interfaces, we increased electrode spacings to achieve increased current penetration and, therefore, changes in the observed apparent resistivity. However, two different approaches are feasible when a vertical contact is present. When dealing with such an abrupt lateral change, we can detect the contact either by fixing the center of the electrode spread and systematically increasing electrode spacing (*expanding-spread traversing*) or by holding electrode spacing constant and systematically moving the entire spread along a traverse line (*constant-spread traversing*). Expanding-spread traversing also is referred to as *depth profiling, electrical drilling*, or *resistivity sounding*, whereas constant-spread traversing often is called *electrical mapping* or *resistivity profiling*.

5.6.1   Constant-Spread Traverse

Let's first examine a constant-spread traverse across a vertical contact. Figure 5.23(a) diagrams the essential elements of the traverse. The distance from the spread center to the contact is x, which is positive to the right of the contact and negative to the left. All electrodes are equally spaced a distance a apart. The entire spread is moved at equal increments along a line perpendicular to the strike of the contact. Table 5.5 contains results for a given selection of resistivities, electrode spacing, and traversing increment. Casual examination of the apparent resistivities in the table reveals that the apparent resistivities do not vary smoothly as was the case for horizontal interfaces. A plot of Table 5.5 values (Figure 5.23(c)) confirms this impression. Actually, there are four points on the curve where the value and direction of slope suddenly change.

In trying to develop a sense of why these slope reversals occur, it helps to plot electrode positions relative to the contact on a map view. Figure 5.23(b) illustrates several important arrangements. Note that the traverse typically is along a single line. The positions are staggered in the diagram for clarity. As you read through the following discussion, refer frequently to Figure 5.23(a) and (b). As expected, when the electrode spread is to the far left ($x = -5$) or far right ($x = 5$), apparent resistivities approximate the actual resistivities. As the spread is moved closer to the contact ($x = -2$), apparent resistivities decrease relatively quickly because the spread is approaching a lower-resistivity medium. Slope reversals occur at $x = -1.5, -0.5, 0.5$, and 1.5. These values represent spread positions where an electrode crosses from one side of the contact to the other. Therefore, we should be able to locate the contact quite closely by noting the positions of spread electrodes when reversals take place. Remember that Figure 5.23 is idealized, and it is very unlikely that field values of a and spread-center positions will locate the contact as closely as in our model.

Why is the change in apparent resistivity values much more pronounced between $x = -3$ to $x = -2$ than from $x = 3$ to $x = 2$? Hint: Note that the negative x values are in the higher-resistivity medium. Use the current density distribution model or a water flow analogy to arrive at a qualitative explanation.

We strongly suggest that, if possible, you work with Table 5.5 to produce curves of your own for various values of the quantities in boldface. For best results use x increments that are smaller than the value you select for electrode spacing.

TABLE 5.5	CONSTANT-SPREAD SURVEY ACROSS A VERTICAL CONTACT

x (m) (Spread Center to Vertical Contact)	Apparent Resistivity ($\Omega \cdot$m)
−5.0	99.48
−4.8	99.39
−4.5	99.28
−4.3	99.14
−4.0	98.96
−3.8	98.72
−3.5	98.41
−3.3	97.98
−3.0	97.37
−2.8	96.48
−2.5	95.13
−2.3	92.94
−2.0	89.09
−1.8	81.49
−1.5	63.18
−1.3	64.29
−1.0	65.91
−0.8	68.44
−0.5	72.73
−0.3	64.55
0.0	55.00
0.3	40.55
0.5	12.73
0.8	13.16
1.0	13.41
1.3	13.57
1.5	13.68
1.8	11.85
2.0	11.09
2.3	10.71
2.5	10.49
2.8	10.35
3.0	10.26
3.3	10.20
3.5	10.16
3.8	10.13
4.0	10.10
4.3	10.09
4.5	10.07
4.8	10.06
5.0	10.05

ρ_1	**100.0**
ρ_2	**10.0**
Electrode spacing a (m)	**1.0**
x increment (m)	**0.25**

k-factor	−0.82

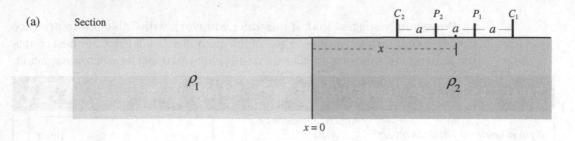

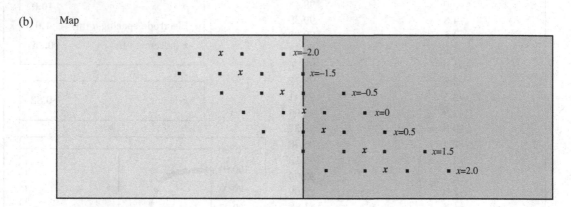

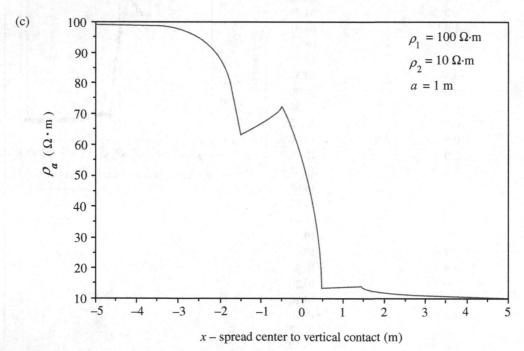

FIGURE 5.23 Apparent resistivity for a constant-spread traverse oriented perpendicular to a vertical contact. (a) Section view illustrating parameters used in Table 5.5. (b) Map view designating critical positions of electrode spreads. (c) Graph of values in Table 5.5.

The equations used are not defined at $|x| = a$ and $|x| = 3a/2$. If you select values that result in these equalities, the words "not defined" will appear instead of an apparent resistivity. To avoid this, we suggest selecting an a-spacing such as 2.01 instead of 2.0.

Of course, when we are working in the field, it would be extremely fortuitous for a traverse to be oriented perpendicular to the trace of a vertical contact. Fortunately, the main difference in an apparent resistivity curve for a traverse at an angle to the contact is a reduction in the size of the slope reversals. These slope reversal points, or *cusps*, become less pronounced as the angle between the traverse and contact trace decreases. For representative curves consult Van Nostrand and Cook (1966, 120). If a traverse is parallel to the contact, apparent resistivity values remain constant. Their magnitude depends on the distance of the traverse line from the contact.

5.6.2   Expanding-Spread Traverse

In an expanding-spread traverse the electrode center is fixed and electrode spacings are systematically increased. Similarly to the previous example, we link the values in a dynamic table (Table 5.6) to a specific diagram (Figure 5.24). In this example the spread center is at $x = -2.0$, and the traverse is perpendicular to the contact. The results in Table 5.6 are plotted in Figure 5.24(c), which reveals two sharp slope discontinuities. Various expanding electrode spreads are illustrated in Figure 5.24(b). Once again the slope discontinuities can be correlated with electrodes crossing the contact. Comparison of the $a = 1.0$ and $a = 2.0$ spread geometries suggests that the slope discontinuity for an a-spacing of 1.25 correlates with a current electrode moving across the vertical contact. For an a-spacing of 4.0 m, a potential electrode crosses the contact that corresponds to the second slope discontinuity. Although perhaps not as dramatic as the resistivity curve obtained by constant-spread traversing, the expanding-spread curve certainly is sufficiently distinctive both to establish the presence of a rapid lateral change in resistivity and to establish reasonably accurately the location of the change.

A traverse oriented parallel to a vertical contact and located close to the contact produces a curve that is remarkably similar to a two-layer curve. How can we explain this behavior when a constant-spread traverse taken in the same orientation produces constant apparent resistivity values? This seeming contradiction is easily explained if we remember that equipotential surfaces in a homogeneous, isotropic material are hemispherical. When the electrodes are close to the contact but are spaced very close together relative to the contact distance, the effect of the material on the opposite side of the contact will be minimal. The situation is analogous to that illustrated in Figure 5.14(a). However, as electrode spacing is increased, a greater percentage of current flow not only penetrates more deeply but spreads laterally as well. If a material with different resistivity is encountered in this lateral direction, current density will be affected. Thus apparent resistivity values will change also.

TABLE 5.6 | EXPANDING-SPREAD SURVEY ACROSS A VERTICAL CONTACT

ρ_1	100.0
ρ_2	10.0
Electrode spacing a (m)	0.25
Spread center to vertical contact (x) in m	−2.1

k-factor	−0.82

Electrode Spacing a (m)	Apparent Resistivity (Ω·m)
1.00	89.30
1.25	73.36
1.50	63.84
1.75	64.87
2.00	65.87
2.25	66.84
2.50	67.79
2.75	68.69
3.00	69.56
3.25	70.39
3.50	71.19
3.75	71.94
4.00	72.67
4.25	71.81
4.50	70.93
4.75	70.16
5.00	69.48
5.25	68.86
5.50	68.31
5.75	67.79
6.00	67.33

If we use the configuration in Figure 5.24(a), the resistivity to the right of the contact is low. Assume we place our electrodes as shown but rotate them 90° so they are parallel to the contact. An initial reading for a small a-spacing would be close to 100 ohm·m. As we increase electrode spacing, apparent resistivity will decrease as more current flows in the lower-resistivity material and

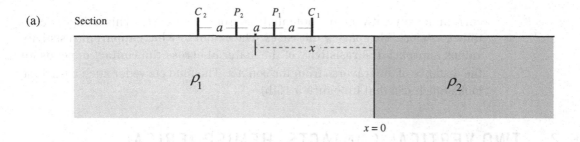

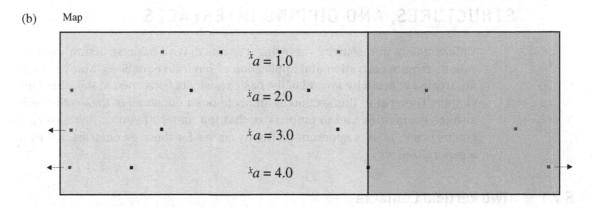

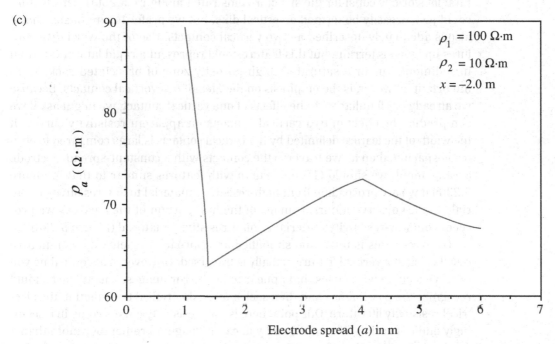

FIGURE 5.24 Apparent resistivity for an expanding-spread traverse oriented perpendicular to a vertical contact. (a) Section view illustrating parameters used in Table 5.6. (b) Map view illustrating electrode positions for various a-spacings. (c) Graph of values in Table 5.6.

current density between the potential electrodes decreases. This behavior continues as electrode spacing increases. The extent to which apparent resistivity values approach the resistivity of the material across the contact depends on the distance of the traverse from the contact. The final curve for such a traverse looks much like that in Figure 5.15(b).

5.7 TWO VERTICAL CONTACTS, HEMISPHERICAL STRUCTURES, AND DIPPING INTERFACES

Unfortunately, in resistivity surveying it seems that all basic structural configurations require derivation and application of separate equations. Many of these are treated in detail by several of the basic resistivity texts cited at the end of this chapter. Our goal in this section is simply to point out some of these other subsurface geometries and to emphasize that the material you've just absorbed equips you to predict apparent resistivity curves for these geometries, if only in a generalized way.

5.7.1 Two Vertical Contacts

First let's briefly consider the structure illustrated in Figure 5.25(a). This feature could just as easily be termed a vertical dike, but we prefer to emphasize that it also is adequately described as two vertical contacts. Use of the word *dike* conjures up igneous terrain, but this feature could represent a rapid lateral variation in sediment type or a saturated, high-porosity zone of brecciated rock. More important, however, is the emphasis on the idea of *two* vertical contacts. Because we already are familiar with the effects of one vertical contact, we might ask if we can predict the effect of two vertical contacts on apparent resistivity curves. If the width of the feature delimited by the vertical contacts is large compared to electrode spacing, then as we traverse the contacts with a constant-spread electrode arrangement, we should (1) see a curve with features similar to that in Figure 5.23(c) if we are proceeding from high-resistivity material to low-resistivity material, and (2) observe a mirror image of the first portion of the curve as we proceed from low-resistivity material to high-resistivity material (Figure 5.25(b)).

Of course, this is much too simplified an approach to yield any quantitative results if such a vertical feature actually is traversed. However, if at some time you observe a curve with cusps that appear to be mirror images, you at least should recognize the possibilities and then seek a more theoretical treatment in the electrical resistivity literature. Our point here is that resistivity curves come in a seemingly endless variety of shapes, and you cannot begin to remember more than a few basic features. But if you grasp the essential features and behavior of a few curve types (single horizontal interface, multiple horizontal interfaces, vertical contact), you certainly will be more effective in the field when required to make decisions concerning electrode spacing, traverse type, and traverse direction.

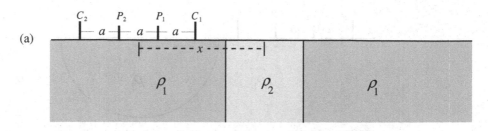

(a)

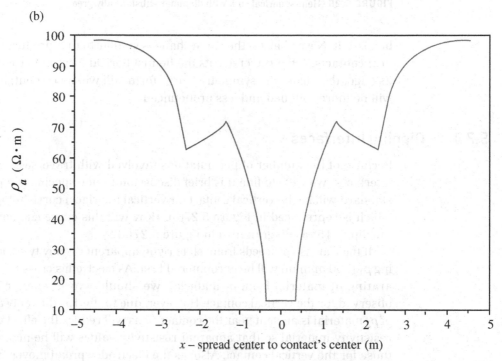

(b)

FIGURE 5.25 Constant-spread traverse across two vertical contacts. (a) Diagram of relationships. (b) Qualitatively derived sketch of likely curve shape.

5.7.2 Hemispherical Structures

A number of subsurface configurations are better approximated by a hemispherical form than vertical contacts (Figure 5.26). An abandoned river channel filled with gravel incised into floodplain deposits of silts and clays and a sinkhole in limestone filled with dissimilar material are two common examples. Let's assume we are sufficiently fortunate to cross the center of such a form with a constant-spread traverse and an a-spacing that is substantially smaller than the diameter of the hemispherical form. Can you sketch a representative curve for such a traverse?

Actually, for rough approximation, we can think of the hemispherical form as two vertical contacts that do not extend far downward. As we cross each contact, we expect to see curves somewhat similar in form to vertical contact curves. However, the effect of the ρ_2 material is tempered by the presence of ρ_1 material

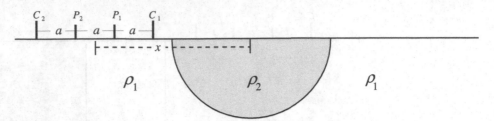

FIGURE 5.26 Hemispherical sink with diameter substantially greater than electrode spacing (*a*).

beneath it. Nevertheless, the curve shape is similar to that produced by two vertical contacts. If the traverse cuts the hemispherical form but crosses nearer to its edge, the bilaterally symmetric curve form still will be evident, but the cusps will be more rounded and less pronounced.

5.7.3 Dipping Interfaces

Because of the number of permutations involved with traverses across dipping interfaces, we will confine this brief discussion to one specific comparison. Our standard will be the vertical contact (or vertical interface) curve in Figure 5.23(c), which is reproduced in Figure 5.27(b). How will this curve change if we lower the dip to 15° as diagrammed in Figure 5.27(a)?

If the traverse proceeds from left to right, apparent resistivity values approaching ρ_1 (100 ohm·m) will be encountered first. As electrodes cross the contact separating ρ_1 material from ρ_2 material, we should expect similar behavior as observed for the vertical contact. However, due to the 15° dip, only a thin wedge of ρ_2 material is present near the contact. This will reduce the effect of this lower-resistivity material so that apparent resistivity values will be higher relative to those for the vertical contact. Also, as the electrode spread moves farther to the right away from the contact, the decrease in apparent resistivity values will not occur as quickly because ρ_1 material still is not far from the surface. If we sketch a qualitative representation of this behavior, our curve should appear somewhat similar to that in Figure 5.27(b). Cusps still are present due to electrodes crossing a boundary between materials of different resistivities, but the total variation in apparent resistivity from cusp to cusp is less than for the vertical contact. This may cause greater difficulty in defining the cusp position or its existence and, therefore, in locating the position of the contact and the amount of its dip.

5.8 FIELD PROCEDURES

Until now we have concentrated on developing a sufficient theoretical background for understanding the shape of qualitative and quantitative apparent resistivity curves connected to a variety of simple subsurface geometries. We have not as yet

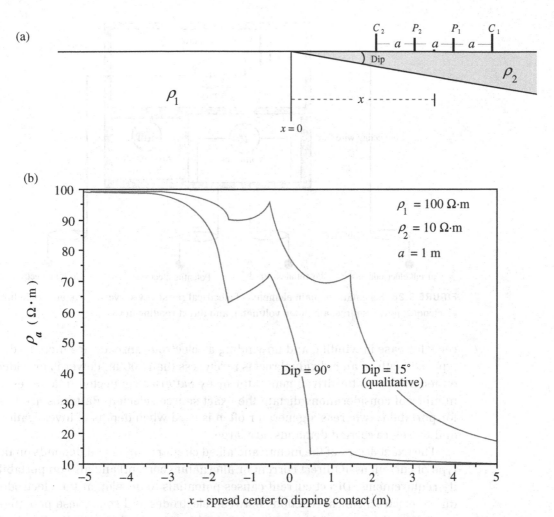

FIGURE 5.27 Comparison of apparent resistivity curves for a constant-spread traverse across an interface dipping 90° and one dipping 15°.

addressed the question of proceeding from actual field data to an interpretation of the subsurface. Before doing so we need to briefly address resistivity equipment, possible electrode geometries to use in data collection, and some strategies for acquiring useful resistivity data.

5.8.1   Equipment

Equipment for a typical resistivity survey consists of an ammeter, a voltmeter, a power source, electrodes, and connecting wire (Figure 5.28). Electrodes are metal stakes (copper, steel, or aluminum) with a cap to facilitate driving them into the ground. Wire to connect the electrodes to the power source and meters must be durable, light, of low resistance, and well insulated. The wire is placed on

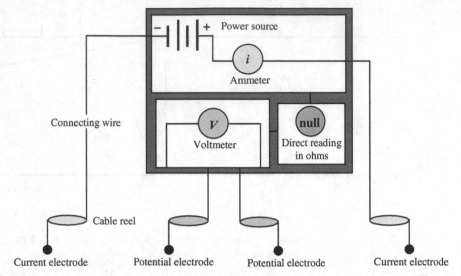

FIGURE 5.28 Schematic of main elements of electrical resistivity surveying system, including electrodes, power source, ammeter, voltmeter, and direct reading device.

reels for ease in winding and unwinding as electrode spreads are increased; a typical length on an individual reel is rarely less than 100 m. Power is provided either from a motor-driven generator or by batteries connected in a series. A number of considerations dictate the exact source selected. Batteries are best for portability, whereas a generator often is used when depths of investigation, and therefore current demands, are large.

The exact design of the meters and allied circuitry varies and depends on the type of current used (direct current or alternating current) and also on portability requirements. Direct current causes potentials to develop at the electrodes due to electrochemical reactions between electrodes and soil. These potentials are included in potential measurements unless special electrodes (*porous pot*) are used. Commutators, which frequently reverse the polarity of the direct current, alleviate this problem; but a common trend in modern instruments is to use low-frequency alternating current.

The circuits to measure potential and current sometimes are in separate units, which increases flexibility but reduces portability. A typical unit for standard shallow exploration exercises includes not only the power source but both ammeter and voltmeter. Obtaining a reading is straightforward with such instruments. Electrodes are placed in a selected pattern and connected to the instrument. A few preliminary adjustments are made as directed by the manufacturer, and current is then applied. Typically a meter needle is deflected from a zero or null position. Rotating a dial (that provides a direct readout in ohms) brings the needle back to the zero position. The ohmmeter reading is recorded, and electrodes are moved to the next position. Figure 5.29 pictures an instrument of this type. Readings

FIGURE 5.29 Photograph of a typical resistivity meter.

can be taken quickly, but moving electrodes and wire for the 10 or more readings required for any individual traverse is time-consuming.

5.8.2 Electrode Configurations

The electrode patterns used in resistivity surveying almost always are the Wenner, Schlumberger, or dipole–dipole. Many other arrangements have been tried, but these three are the survivors. The Wenner electrode geometry is illustrated in Figure 5.30(a). Spacing between all electrodes is equal and conventionally is denoted by the letter a. This configuration should appear familiar because it is the pattern we have used in all previous examples in this chapter. In conducting an expanding-spread Wenner survey, we move all electrodes along a straight line after every reading so the spacing between electrodes remains equal and takes on certain preselected values. Typically we try to achieve values of a that are equally spaced on a logarithmic scale and ideally number six points per decade (for example, 10.0, 14.7, 21.5, 31.6, 46.4, and 68.1). If you used RESIST, you may have noted that these are the intervals for which apparent resistivity values are calculated. RESIST, as is discussed later in this chapter, assumes field readings are acquired at these same intervals. The main reason for such spacings is that shallow materials always are included in any resistivity reading. Larger electrode spacings are therefore necessary to acquire a reasonable electrical sample of material at depth. Using the recommended electrode intervals produces equally spaced points on a graph of the field data.

A symmetric distribution of current and potential electrodes about a central point also is employed for the Schlumberger array (Figure 5.30(c)), but the

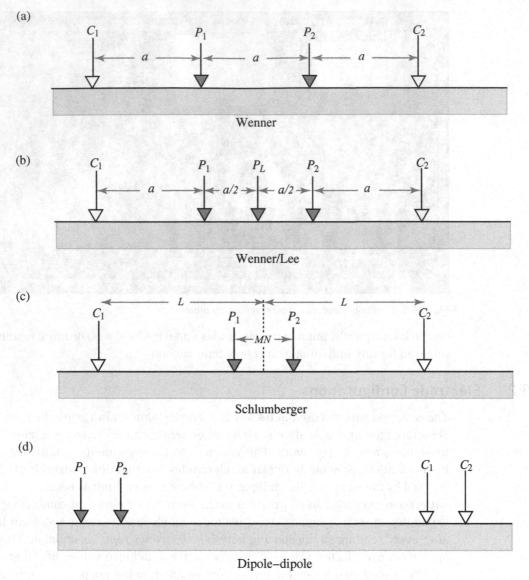

FIGURE 5.30 The most common electrode geometries used in resistivity surveying. (a) Wenner. (b) Wenner/Lee (Lee partitioning). (c) Schlumberger. (d) Dipole–dipole.

potential electrodes are spaced much more closely than the current electrodes. Spacings are selected to maintain the relationship $2L > 5MN$ and also follow the same numbering scheme as the Wenner array. Because of this particular geometry, meter sensitivity is exceeded after the current electrode spacing is increased several times. At this point the potential electrode spacing MN is increased to the next larger value in the numbering scheme (for example, 14.7 m if the previous value was 10.0 m) and the current electrode spacing L is reduced by two intervals (such as to 100 m if the previous reading was 215 m). This recording procedure produces overlapping curve segments on a plot of apparent resistivity versus electrode spacing (Figure 5.31).

The final common geometry is the dipole–dipole (Figure 5.30(d)). In this arrangement the potential electrodes and current electrodes function independently. Both sets tend to be fairly closely spaced with a significant distance between the sets. The potential electrodes are placed relative to the current electrodes in one of several basic patterns (not all along the same line). Because cable lengths between the electrodes are short, it is much easier to place the potential electrodes at large distances from the current electrodes, thereby facilitating deep investigations. At the same time larger currents are necessary to reach these depths. This geometry is used much less in North America than the Wenner and Schlumberger spreads, although use in the Soviet Union, especially for petroleum exploration, has been significant. Because of this relative lack of use in North America, we will concentrate on the Wenner and Schlumberger arrays.

The Wenner electrode array primarily has been used in North American surveys because Frank Wenner employed that geometry early in the development of the resistivity method. Most European workers use the Schlumberger spread, as do many American investigators. Each geometry has advantages and disadvantages. One advantage of the Wenner array is that the larger potential electrode spacing places less demand on instrument sensitivity. A second advantage is the simplicity of the apparent resistivity equation (Equation 5.20) when electrodes are equally spaced (Equation 5.35). Prior to the present accessibility and capability of computing facilities, this simplicity reduced the complexity of many equations and therefore demanded much less effort in quantitative approaches to resistivity interpretation (which is why we used it in previous sections). Frequent use led to many examples and techniques in the literature, which furthered more use.

The Wenner geometry has several shortcomings. Because all electrodes must be moved for each reading, which is not the case with the Schlumberger method, the Wenner array requires more field time. Perhaps more serious is that it is more sensitive to local, near-surface lateral variations than the Schlumberger geometry. Of course, Wenner data can be analyzed directly; but Schlumberger requires some pre-interpretation processing because the overlapping curve segments must be smoothed and new values obtained from the smoother curve. On the other hand, the curve segments follow a specific pattern of overlaps, which is clearly illustrated in Figure 5.31. Increasing MN spacing when apparent resistivity values are rising results in offset of the new values downward (positions 1 and 2 in Figure 5.31). When apparent resistivity values are decreasing, an increase in MN results in offset of the new values upward (positions 3 and 4 in Figure 5.31). If the overlaps depart from this pattern, lateral inhomogeneity is indicated and, if not too severe, can be corrected for in the smoothing process.

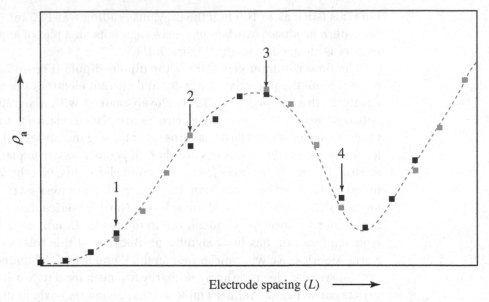

FIGURE 5.31 Apparent resistivity values illustrating effect of changing *MN* spacing during Schlumberger expanding-spread traverse. *MN* spacing changed when *L* values are equivalent to positions 1, 2, 3, and 4.

The susceptibility of the Wenner electrode geometry to near-surface lateral variations increases the likelihood that we may interpret these as variations at depth. One approach to reducing the possibility of such an erroneous conclusion is to place a third potential electrode at the center point of the electrode array. We take readings as usual; but in addition to the normal potential difference reading at P_1P_2, we also take readings for P_1P_L and P_LP_2. This approach usually is referred to as the *Lee modification*. All apparent resistivities are calculated using the Wenner equation. If no lateral variations are present, the Lee apparent resistivity curves should exhibit the same form and should mimic the Wenner curve. The Lee values will plot at half-values of ρ_a, which makes visual interpretation easier. [Note: When using the Lee partition array independently, change 2π to 4π in the Wenner equation (Equation 5.35).] Departures from lateral homogeneity should be evident from departures in the similarity of the Lee values. Once such variations are identified, we can make appropriate adjustments to the exploration plan. Figure 5.32 illustrates a case where the Lee modification values are very similar and almost perfectly mimic the Wenner apparent resistivity values. Hence lateral variations are not significant.

Because quantitative interpretations of Wenner and Schlumberger curves are very similar, we generally will continue to provide examples based on the Wenner array for reasons cited previously. RESIST provides options to use either array, so you may select either depending on the category of field data you have available. If you can gather data on your own, we suggest you take the time to use both arrays to better judge their strengths and weaknesses.

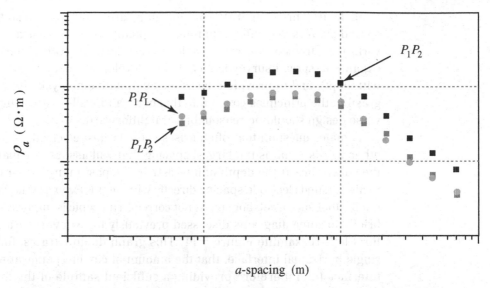

FIGURE 5.32 Apparent resistivity values obtained with a Wenner array near Hatfield, Massachusetts. Note that the Lee modification values are very similar and mimic the Wenner values closely.

5.8.3 Surveying Strategies

As in the case of seismic exploration, resistivity survey goals should be specifically defined before fieldwork begins. Equipment must be selected that is adequate for the task at hand. And, if at all possible, independent geologic controls should be available. It is difficult to derive detailed geologic information from resistivity surveys unless such control exists. The next step is to locate sampling sites and to decide whether to use constant-spread or expanding-spread traverses.

If the goal of the effort is primarily to seek information about apparent resistivity variations with depth to define the location of a high-resistivity-contrast interface, then expanding-spread traverses are the obvious choice. An example is a survey interested in mapping depth to a bedrock surface overlain by saturated silty sands. Of course, either Wenner or Schlumberger can be used. In either case, you always should run another traverse at right angles to the first to establish whether lateral variations are present. Remember that an expanding-spread traverse oriented parallel to a vertical contact can produce a curve almost identical to that for a two-layer case, whereas a traverse perpendicular to the contact reveals cusplike discontinuities that establish the presence of the contact without doubt.

If survey goals are to map lateral variations in resistivity, perhaps due to the presence of a gravel bar, then constant-spread traversing is required. Normal practice dictates that two or three electrode spacings are used at each station to develop three-dimensional information. Apparent resistivities for each electrode spacing are placed at each array center and contoured. If three spacings are used at each site, three separate contour maps are produced. Results of constant-spread surveys are difficult to interpret quantitatively, especially without some vertical control. For this reason at least one (and ideally more) fully implemented, expanding-spread survey is included to develop a sense of resistivity variation with depth and to provide a basis for quantitative analysis. If you have a good grasp of the principles covered to this point and well-defined survey goals, traverse design should be reasonably straightforward.

A major question that often arises is the largest electrode spacing to use. In other words, what is the largest spacing that will assure adequate sampling of the subsurface at the depth of interest? In the past many explorationists incorrectly equated electrode spacing directly with depth. Even at this stage you should realize that such a relationship is not correct. This topic is important, so we return briefly to some diagrams discussed previously to review several relationships. First let's reexamine Figure 5.11. This graph demonstrates, for the case of a single horizontal interface, that the amount of current penetrating beneath the interface (and therefore providing a sufficient sample of the lower material) depends not only on electrode spacing but also on the relative values of ρ_1 and ρ_2. In cases where the depth of the interface is equal to electrode spacing ($z = a$) and ignoring extreme resistivity contrasts, the percentage of current penetrating beneath the interface can vary from 20 to 80 percent.

Similar lessons are available from Figures 5.19 and 5.20. Figure 5.19 demonstrates that increased electrode spacings are required to completely define curves for interfaces at increasing depths. Resistivity contrast is held constant for each curve. Graphs such as this suggest that three decades of readings (from 1 m to 1000 m) are sufficient to define curves for most shallow interfaces. Now study Figure 5.20. Interface depth is constant for each curve. At an a-spacing of 100 m the curve for $\rho_2 = 20$ ohm·m is completely defined. At this same spacing, however, the curves for $\rho_2 = 200$ ohm·m and 500 ohm·m are barely distinguishable and still are rising at such a rate that they offer no insight into the resistivity value of the lower layer except that it is high. However, many experienced explorationists find the rule that a-spacing is equal to depth to be useful in a variety of geologic settings. It is perfectly acceptable to use this rule to gain a feel for the meaning of curve form during fieldwork as long as you remember this approximation is imperfect and eventually subject your field data to detailed computer analysis.

The best advice we can offer with regard to electrode spacing decisions is to plot apparent resistivity values as readings are taken. Unless you are positive that the form of the curve is sufficient to meet the goals of your survey, keep acquiring data until you reach the limits of your instrument, cable length, or available room. Of course, if many traverses are required, then time and expense may constrain the number of readings you can accumulate. More often than not, we find that we run out of room or cable before obtaining as many readings as we would like.

5.8.4 Other Considerations

A number of factors can affect resistivity measurements in addition to the resistivities of the rocks and sediments beneath the surface, and you should be aware of the most significant of these to aid in field survey design. Some factors are obvious. Buried pipelines or cables may be good conductors and, if so, will affect current flow. Electrode contact must be good, and resistances should be low and uniformly maintained. If dry surface materials are present, wetting the ground around the electrode often is all that is required.

Topography affects apparent resistivity measurements because it affects current flow. Perhaps the most obvious example of this is the vertical contact. Check Figure 5.23 and replace the ρ_1 material with air, which we assume to have infinite resistivity. This provides the example of the vertical cliff, which admittedly is an extreme case, but which illustrates that current flow would be affected and that apparent resistivities measured near the cliff would also be affected. Slight undulations and elevation changes are not a problem, but rugged topography should be avoided if at all possible. No "topographic corrections" are available as there were for seismic work.

Finally, many rocks and some sediments often possess different resistivities depending on the measurement direction. An obvious example is a rock unit containing a single, very well oriented fracture set filled with water. Traverses parallel to the strike of the fractures will measure lower resistivities than traverses normal to the fractures. Although this electrical anisotropy is not as serious an interpretation problem as buried objects and rugged topography, you should be aware that it can affect apparent resistivities. You generally attempt to collect data along two perpendicular traverses. Variations between curves for the two traverses normally reflect lateral variations in materials; but keep in mind that anisotropy in a given unit may be responsible in some cases.

5.9 QUANTITATIVE INTERPRETATION OF APPARENT RESISTIVITY CURVES

An exclusively quantitative interpretation of apparent resistivity curves often is difficult. This arises from the wide variations in resistivity possessed by geologic materials and the difficulty in developing theoretical expressions for apparent resistivities of all but the simplest geometries. Quantitative interpretation is best developed for cases in which layering is horizontal, so it is to these situations that we will confine our analysis. First, however, we consider the resistivities of sediments and rocks.

5.9.1 Electrical Resistivities of Geologic Materials

In the shallow subsurface electricity is conducted almost entirely by the fluid present. Thus the resistivities of sediments and rocks in this environment are controlled by the amount of water present and its salinity. Because this conduction is electrolytic and is related primarily to ions moving through the fluid, the cation exchange capacity of clay minerals increases conductivity. Although clay minerals are the most important in this regard, all fine-grained minerals possess an exchange capacity to some degree. This means that increasing silt or clay content in poorly sorted rocks or sediments will reduce resistivities. In saturated materials increasing porosity will reduce resistivities. Well-sorted materials have greater porosities than poorly sorted ones, but in well-sorted rocks or sediments with well-rounded grains, grain size alone does not influence porosity (Fetter 1980, 60–62). In addition, irregularly shaped grains will pack less tightly than sphere-shaped grains and consequently will have higher porosities. In general, finer-grained sediments will have higher porosities (and lower resistivities) than coarse-grained sediments. Based on these comments, however, it seems prudent to apply this relationship only in the most general way.

All the following *reduce* resistivities: increasing water content, increasing salinity of water, increasing clay content, and decreasing grain size. Assuming that water is available to fill voids, resistivity is lowered by increasing porosity, increasing number of fractures, and increasing weathering. Conversely, resistivities are raised by increasing compaction and lithification.

Because these factors vary so much in the natural environment, it is not surprising that resistivities vary greatly and that it will be difficult to correlate resistivities with source materials in the absence of other geologic information. Indeed, Zohdy, Eaton, and Mabey (1974, 9) note, "No other physical property of naturally occurring rocks or soils displays such a wide range of values." Although direct correlation will not be as straightforward as it was with seismic velocities, some generalized relationships often may be established. Bedrock almost always has higher resistivities than the saturated sediments lying above. Generally, the unsaturated sediments above the water table will have higher resistivities than the saturated sediments below the water table. However, the only reliable way to correlate resistivities with local geology is by using independent geologic information (such as an expanding-spread traverse located near a carefully logged drill hole). In this way you can establish correlations and carry them forward into unknown terrain.

The list of resistivities shown here is based on personal experience and is adjusted to include values provided by numerous sources. These values are supplied primarily to provide a starting point for interpretation, but beware. For every observation that falls within the values on this list, there probably is an observation that provides an exception.

Material	Resistivity ($\Omega \cdot$m)
Wet to moist clayey soil and wet clay	1s to 10s
Wet to moist silty soil and silty clay	Low 10s
Wet to moist silty and sandy soils	10s to 100s
Sand and gravel with layers of silt	Low 1000s
Coarse dry sand and gravel deposits	High 1000s
Well-fractured to slightly fractured rock with moist, soil-filled cracks	100s
Slightly fractured rock with dry, soil-filled cracks	Low 1000s
Massively bedded rock	High 1000s

A thorough discussion of factors controlling resistivities and a good presentation of resistivity ranges for common geologic materials is presented in Ward (1990, 148–51).

5.9.2 Empirical Methods

At least two empirical methods are cited frequently in texts dealing with interpretations of expanding-spread apparent resistivity curves. These are the *Moore cumulative resistivity method* (Moore 1945) and the *Barnes layer method* (Barnes 1952). Neither method should be used except for the most preliminary analysis or when the analytical methods described in the following sections cannot be utilized because the required materials are not available.

When using the Moore method, we collect apparent resistivity data at electrode spacings that are evenly spaced on a *linear* scale using a Wenner spread. For each *a*-spacing a cumulative resistivity value is calculated that is the sum of all apparent resistivities collected at all previous electrode spacings. These cumulative resistivities are plotted on the vertical axis; *a*-spacings are plotted on the horizontal axis. Straight lines are fit to the data points. The horizontal coordinates of the points of intersection of these lines are assumed to give the depths to horizontal interfaces.

The Moore method has succeeded in some locales and failed in others. It is possible to check this approach by using a program such as RESIST to generate apparent resistivity values for known models and then to interpret the data using the Moore method. If curiosity leads you to such an investigation, note that RESIST plots data that are equally spaced on a *logarithmic* scale, whereas the Moore method requires data equally spaced on a *linear* scale. Zohdy, Eaton, and Mabey (1974, 46) report that the method appears to work well for two-layer cases where the resistivity contrast is not too great. If this condition does not hold, the depth to the interface can be in error by as much as 50 percent. These investigators regard the application of the Moore method to data from three or more layers as highly questionable. In addition, the method does not provide resistivities for the derived layering.

You realize that each time electrode spacing is increased, the volume of earth is increased that affects an apparent resistivity determination. A basic premise of Barnes's method is that each measured apparent resistivity represents the average resistivity in a layer that extends from the surface to a depth equal to the electrode spacing. This premise leads to the following: When the electrode spacing is increased in equal intervals, the depth representing the average resistivity also increases in equal intervals. This assumption is not accurate, as you should realize at this point. Once again, you could test this method by generating data using RESIST, plotting and interpreting them using the Barnes procedure, and comparing the results with your initial model. Like the Moore method, the Barnes method works for certain subsurface configurations but not for others.

Both methods function best when data are collected at equal electrode-spacing intervals on a linear scale. The analytical methods to be discussed next assume that data are collected at electrode-spacing intervals that are equally spaced on a logarithmic scale. Because these analytical methods are preferred for quantitative interpretation, using the Moore or Barnes method will require additional field time to collect data or to interpolate existing data.

5.9.3   Analytical Methods—Curve Matching

In discussing curve-matching techniques, we restrict examples to the case of the single horizontal interface and an expanding-spread traverse. However, these techniques can be extended to any situation for which the appropriate equations exist (such as Equation 5.36). Curve-matching techniques as described here essentially have been supplanted by the computer techniques that are described in the following section. Nevertheless, we believe that understanding this approach and working with and observing many different types of curves provide useful exposure and practice. Some authors recommend curve matching as a preliminary step before using the computer.

Consider the following. If we have a field curve for which apparent resistivity is plotted against electrode spacing and can match this curve with a curve calculated by RESIST for a given model data set, we then have a solution to our field data. The actual subsurface geology must be very close to the model on which the theoretical curve is based. Therefore, given a field curve, all we have to do is find a similar theoretical curve and a solution is at hand. Unfortunately, because we are working with three variables (ρ_1, ρ_2, and the depth to the interface z_1), there exist an infinite number of combinations that create an infinite number of curves through which to search.

However, there is an elegant but simple solution to this problem. If we choose a model and use RESIST to calculate apparent resistivity data and then divide ρ_a values by ρ_1 and electrode-spacing values by z_1, we create a curve such as that labeled as $k = -0.4$ in Figure 5.33(a). If we now choose a new model with different values of ρ_1, ρ_2, and z_1, but constrain the resistivity values so we maintain

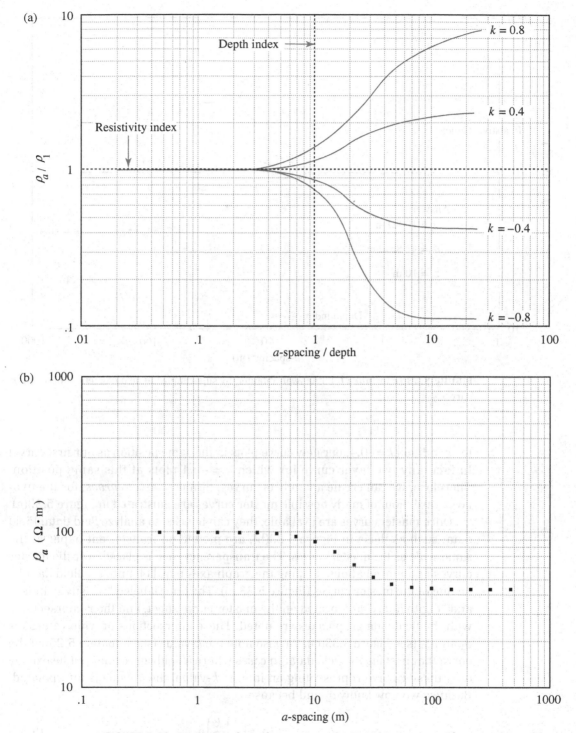

FIGURE 5.33 Basic procedures for curve matching. (a) Master curves for several possible k-values. (b) Field data plotted on the same scale as (a).

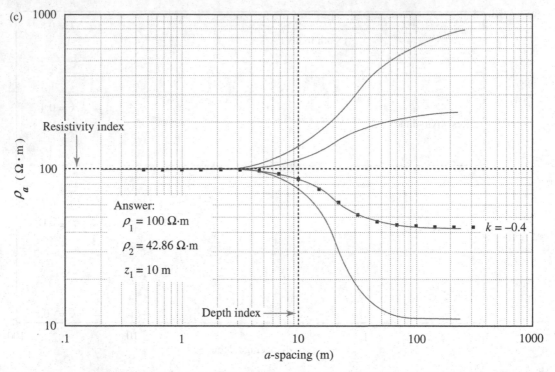

FIGURE 5.33 (continued) (c) Answer obtained by superimposing field data curve on master curves.

the condition $k = -0.4$, our new curve plots in the same position as our first curve! In fact, any two-layer curve for which $k = -0.4$ plots at this same position. Therefore, we can create a family of curves, termed *master curves*, for the two-layer case. Four of many possible master curves are illustrated in Figure 5.33(a).

Once master curves are available, they can be used to analyze field data. Field data, such as those in Figure 5.33(b), are plotted on transparent paper at the same scale as the master curves. The transparency then is placed over the master curves and moved about, keeping the graph axes parallel, until the field data lie on one of the master curves (Figure 5.33(c)). The lines labeled "resistivity index" and "depth index" are located on the master curve sheet, and their intersections with the field data graph axes are noted. These intersections provide values for ρ_1 and z_1 (see Figure 5.33(c)). We also note the k-value (see Equation 5.25) of the curve that best fits the field data. In cases where a field curve does not lie exactly on a master curve representing an integer k-value, the k-value is interpolated. Because we now know ρ_1 and because

$$\rho_2 = \left(\frac{1+k}{1-k}\right)\rho_1 \tag{5.37}$$

it is straightforward to determine ρ_2. Figure 5.33(c) provides values of $\rho_1 = 100$ ohm·m, $k = -0.4$, and $z_1 = 10$ m. Equation 5.37 then provides a value of 42.86 ohm·m for ρ_2.

Generating master curves for three, four, or five layers basically follows the same procedure but is more complicated. Many sets of curves are produced, and more searching is required to find proper fits between master curves and field data. Several sets of master curves have been published (Mooney and Wetzel 1956; Orellana and Mooney 1966; Fläthe 1963) and are available at libraries for those interested in working with this approach.

Two specific problems occur when we try to interpret field data based on comparisons with theoretical curves representing three or more layers. These problems are known as *equivalence* and *suppression*. Equivalence essentially means that various subsurface models can generate curves that are very similar (equivalent). In other words, a given interpretation is not necessarily unique. You already observed one case of equivalence in Figure 5.22, where a three-layer curve is similar to a two-layer curve. Another case is documented in Figure 5.34(a), in which two three-layer models generate similar curves. Given the noise inherent in field data, one of these likely could not be distinguished from the other.

Suppression occurs when the presence of a thin layer in a multilayer sequence cannot be recognized on an apparent resistivity curve. Suppression is controlled by the relative thickness of a layer (thickness of the layer ÷ depth of the layer) and the resistivity contrast between the thin layer and adjacent layers. Figure 5.34(b) illustrates curves produced by a three-layer model and a five-layer model that are very similar. Your major observation should center on the form of the Model 1 curve. Even though the model consists of five layers, the curve essentially is identical in form to a common type of three-layer curve.

Note that the problems of equivalence and suppression also are present when we fit model curves to field data by computer. We now turn to this approach.

5.9.4 Analytical Methods—Automated Curve Matching

As we noted at the beginning of our discussion of curve matching, computer analysis of field curves largely has supplanted the curve-matching technique. In performing a computer-based analysis we enter field observations and an initial model that we believe is responsible for the observed values. The initial model should produce a curve of the same type and as close as possible to the observed curve. Hence, if we are inexperienced in interpreting resistivity curves, a preliminary step using curve matching often saves computer time and results in more realistic solutions.

After we enter field data and an initial model, the computer program calculates an apparent resistivity curve based on the model and compares this curve to the observed curve. Depending on the variation between the two, adjustments are made to the model according to the mathematical procedure used in the

(a)

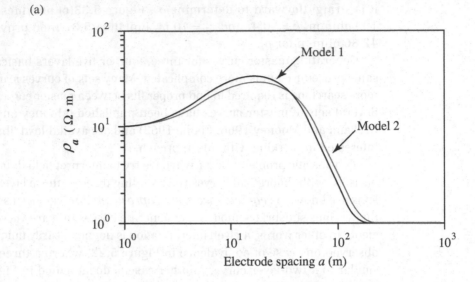

(b)

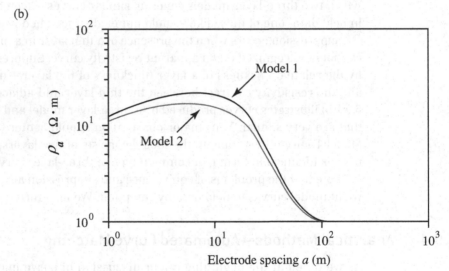

FIGURE 5.34 RESIST curves illustrating equivalence and suppression. (a) An example of equivalence. Model 1: layer 1 (3 m, 10 Ω·m); layer 2 (10 m, 100 Ω·m); layer 3 (1 Ω·m). Model 2: layer 1 (3 m, 10 Ω·m); layer 2 (20 m, 50 Ω·m); layer 3 (1 Ω·m). (b) An example of suppression. Model 1: layer 1 (1 m, 10 Ω·m); layer 2 (1 m, 40 Ω·m); layer 3 (1 m, 10 Ω·m); layer 4 (7 m, 40 Ω·m); layer 5 (1 Ω·m). Model 2: layer 1 (1 m, 10 Ω·m); layer 2 (13 m, 20 Ω·m); layer 3 (1 Ω·m).

computer algorithm, and a new model curve is calculated. This process continues until both curves are as similar in shape as specified by the person guiding the analysis or until the computer program is unable to make the curves any more similar. Based on the results of the automated interpretation, we may accept the final model or initiate another analysis by specifying a new model.

The program RESIST that is included with this text follows this general procedure. It performs analysis for either Wenner or Schlumberger data. It is possible to fix certain model parameters so they cannot be changed by the computer program during the analytical procedure. Full details about this program are in Appendix C.

A slightly different computer-based approach is suggested by Zohdy (1989). Zohdy notes that apparent resistivity curves tend to be out of phase with true resistivity–depth curves and that the values of observed apparent resistivities are always less than or equal to the actual resistivities. These and other properties of apparent resistivity curves for horizontally stratified materials lead to the following iterative approach. The observed apparent resistivities and electrode spacings are used as the beginning model (that is, the apparent resistivities are treated as true resistivities and the electrode spacings are used as depths). Of course we realize this is not the case, but we make this assumption to generate an initial curve. A theoretical curve is computed for this model. The theoretical curve is compared to the observed curve. If the agreement is not good (say within 2 percent), model depths are uniformly decreased by 10 percent, and a new theoretical curve is calculated. This continues until good agreement is reached. Resistivities used in the model then are adjusted by a similar interactive procedure until observed and computed curves agree to within set conditions. The final model represents a solution to the observed data.

Doubtless other computer-based procedures will appear. Because of equivalence and suppression and because of the many factors that affect curve details, a thorough knowledge of resistivity principles and as much experience as possible must go hand-in-hand with computer-derived solutions for field measurements.

5.10 APPLICATIONS OF ELECTRICAL RESISTIVITY SURVEYING

In terms of the material discussed until this point, you may wonder if electrical resistivity surveying is a viable tool for exploration. Although resistivity interpretation is not particularly straightforward, especially when compared to the techniques presented for seismic work, resistivity surveying can be extremely valuable. In some cases it is the method of choice simply because of cost. For example, in many areas it is possible to determine reasonably accurate depths to bedrock using resistivity. This may be a more cost-efficient procedure than investing in relatively expensive seismic equipment. However, resistivity applications most often are used when other geophysical methods simply cannot supply the desired information. We emphasize these cases in the following descriptions.

5.10.1 Applications Related to Aquifers

Many communities that depend on confined aquifers for their water supplies obviously want to prevent contamination of this supply or to have some idea of the extent of the aquifer for planning purposes. In many cases a buried aquifer (sand and gravel) is overlain by a clay layer that prevents most infiltration of solutions from above. Therefore any information about the lateral distribution of the clay–gravel contact, its depth, and the clay thickness is valuable. Refraction seismic methods often cannot map the contact because of a velocity inversion, and reflection methods, if they can detect the contact, may be too expensive or elaborate for use.

In Chapter 2 we introduced the Whately study and noted that resistivity methods also were used. In this case resistivity was effective in mapping bedrock depths, which tended to correlate with the location of a buried aquifer. Figure 5.35(a) illustrates a typical apparent resistivity curve from the Whately area. The steep downward segment of the curve is due to a thick clay layer (approximately 50 m), and the last upward segment reveals the presence of bedrock. Analysis of the curve yields a bedrock depth of 71 m, which is abnormally deep for the local area; but a well less than 1 km away along the trend of the buried aquifer gives a bedrock depth of 75 m. Although the gravel aquifer cannot be detected on the curve, the bedrock depth suggests it should be present.

Easthampton, Massachusetts, also depends for most of its water supply on a confined aquifer. The aquifer is a glacial sand and gravel deposit that rests for the most part on Triassic sedimentary rocks. The sand and gravel deposit is overlain by glacial lake clays that vary in thickness, thinning to zero in the recharge area where the aquifer is exposed at the surface. In 1981 a group of Smith College students and faculty conducted a study, supported by funds from the Shell Oil Company Foundation, to outline the thickness and extent of the aquifer. The study used both seismic refraction and resistivity methods and accomplished its objectives. Figure 5.35(b) is an expanding-spread profile from this study. This particular profile yields a clay thickness of 31 m, which, when compiled with many other such determinations, produced a contour map for clay thickness and distribution.

In some situations it might be of interest to map the depth of the water table. If the surface deposits are homogeneous, then the lower resistivity of the water-saturated material will impart a sufficiently great resistivity contrast for analysis. However, if the near-surface materials are not homogeneous, it will then be difficult or impossible to map the water table with any precision. Moreover, the extent to which the capillary fringe is developed also will affect how closely the actual depth of the water table can be determined.

Resistivity surveys have been used successfully at a number of sites to map buried stream channels. Most of these surveys utilized both constant-spread and expanding-spread traversing to gather sufficient information to locate a buried stream channel. Shallow reflection work might be able to produce superior results

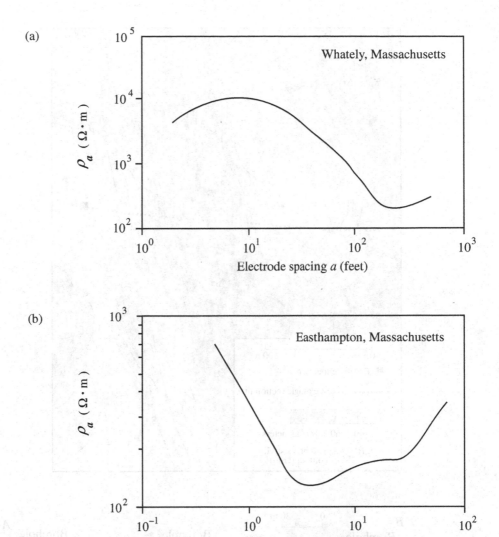

(a)

(b)

FIGURE 5.35 Expanding-spread, Wenner configuration field curves. In both cases the presence of a thick, low-resistivity clay layer and the presence of high-resistivity bedrock are evident. (a) Whately, Massachusetts. (b) Easthampton, Massachusetts.

but would entail at least equivalent field time as well as more expensive equipment. Figure 5.36(a) is an apparent resistivity map from a study in the Penitencia, California, area, which is northeast of San Jose (Zohdy, Eaton, and Mabey 1974, 47–50). This map was produced from many constant-spread traverses using an a-spacing of 6.1 m. Contours of equal apparent resistivity values define a zone of high resistivity oriented approximately east–west. A geologic section (Figure 5.36(b)) based on four expanding-spread traverses, the apparent resistivity profile along

(a)

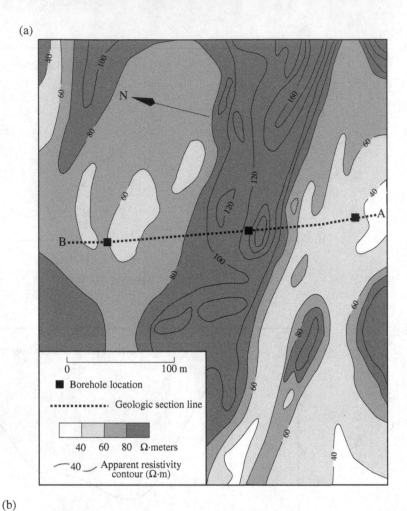

0   100 m

■ Borehole location

┄┄┄┄┄ Geologic section line

     40   60   80 Ω·meters

—40— Apparent resistivity
contour (Ω·m)

(b)

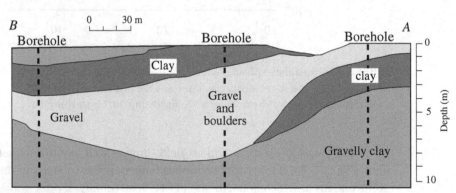

FIGURE 5.36 Mapping a buried stream channel using resistivity. (a) Apparent resistivity map, Penitencia, California, prepared using constant-spread traverses with electrode spacing = 6.1 m. (b) Geologic section along line *BA* based on resistivity and borehole information. (Modified from A.A.R. Zohdy, G.P. Eaton, and D.R. Mabey. 1974. Application of surface geophysics to groundwater investigations. *Techniques of water-resources investigation of the United States Geological Survey, Book 2*, chap. D1, 49–50.)

the section line *BA*, and three boreholes reveals that the high-resistivity trend on the apparent resistivity map is a zone of gravel and boulders that defines the location of a buried stream channel.

5.10.2 Applications Related to Contamination

Some of the most important successes of the resistivity method are based on contamination of normal groundwater. Usually contaminated groundwater, whether affected by leachate from a landfill or by saltwater intrusion, has greater conductivity than ordinary groundwater. Therefore, materials containing the contaminated water will possess lower resistivities than materials containing unaffected groundwater. If the water table is fairly shallow and the subsurface is relatively homogeneous, it is possible to map the extent of the contamination.

Ideally, traverses are implemented *before* contamination occurs. Such data document resistivity variations due to lateral and vertical variations in materials. Deviations from these values, based on traverses taken after contamination, provide a good basis for delineating the horizontal and vertical extent of the problem. Figure 5.37 summarizes how resistivity can define the leachate plume associated with contaminated waters leaking from landfills. It is based on patterns observed in studies of several landfills located at various sites throughout the United States. The resistivity contours are determined from variations in apparent resistivities compared to normal values associated with the landfill site.

5.10.3 Applications in Mapping Karst and Geologic Structures

Hubbert (1932, 1944) published two classic studies demonstrating the application of constant-spread traversing in defining karst relationships and mapping fault locations. Figure 5.38 illustrates a constant-spread profile over karst topography in Hardin County, Illinois (Hubbert 1944). Low apparent resistivity values are associated with clay at the surface, whereas somewhat higher apparent resistivity values are associated with masses of limestone at shallow depths that are overlain by clay. The highest apparent resistivity values correlate with limestone at the surface or large voids (caverns), which have infinite resistivity.

Hubbert (1932) also located faults between blocks of sandstone and limestone and a shear zone in sandstone by a constant-spread traverse (Figure 5.39). Note the similarity of the apparent resistivity curves in the vicinity of faults to the theoretical curves for vertical contacts (Figures 5.23 and 5.25). If the 30 m station interval was reduced, the increased station density likely would result in curves with even more similarities to theoretical curves. Both of Hubbert's studies confirm that, given sufficient resistivity contrast and relatively straightforward geology, constant-spread traversing is capable of locating vertical geologic contacts and similar features of interest.

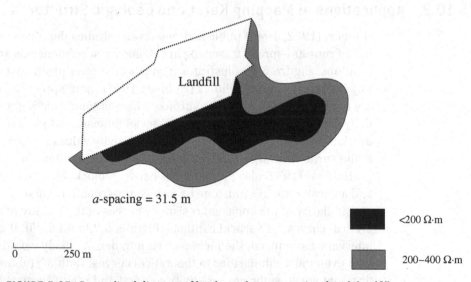

Landfill

a-spacing = 4.7 m

Landfill

a-spacing = 21.5 m

Landfill

a-spacing = 31.5 m

< 200 Ω·m

200–400 Ω·m

0   250 m

FIGURE 5.37 Generalized diagram of leachate plumes associated with landfills. Apparent resistivity contours determined by constant-spread traversing. Note the expanding form of the leachate plume with depth.

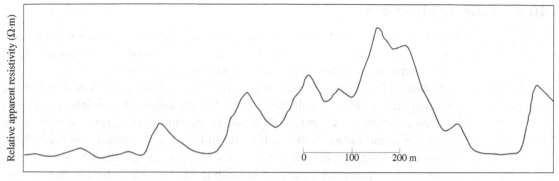

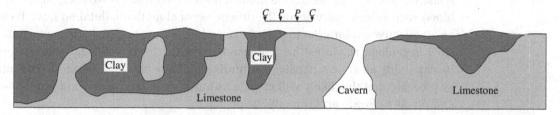

FIGURE 5.38 Apparent resistivity curve for a constant-spread traverse over karst topography, Hardin County, Illinois. Wenner array, a-spacing = 30.48 m, spread-center spacing = 30.48 m. Apparent resistivity highs correspond to limestone and voids; apparent resistivity lows correlate with clays. (Modified from M.K. Hubbert. 1944. An exploratory study of faults in the cave in Rock and Rosiclare districts by the earth resistivity method. In *Geological and geophysical survey of fluorspar areas in Hardin County, Illinois*. United States Geological Survey Bulletin 942, Part 2, 73–147.)

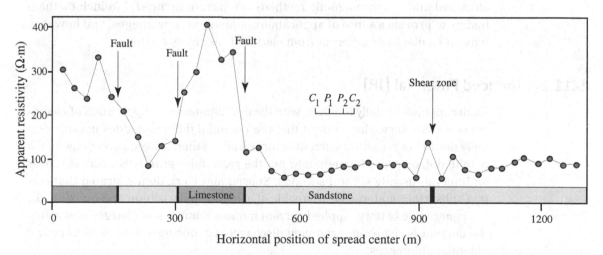

FIGURE 5.39 Apparent resistivity curve for a constant-spread traverse over faults in Illinois. Wenner array, a-spacing = 30.48 m, spread-center spacing = 30.48 m. Curve form is similar to that predicted over vertical contacts. (Based on data in M.K. Hubbert. 1932. *Results of earth-resistivity survey on various geologic structures in Illinois*. American Institute of Mining and Metallurgical Engineers Technical Publication 463, 16.)

5.10.4   Other Applications

Resistivity surveys have been used to outline new geothermal fields in areas of known geothermal potential. Geothermal waters are better conductors than normal waters due to their greater content of dissolved salts.

Because of the typically high resistivity contrast between rock and sediment, resistivity surveys often can delineate buried stone walls and foundations. Surveys also are sensitive to differences in moisture content of materials, so they often can differentiate between excavated and unexcavated ground. For both these reasons resistivity has been a frequent tool in archaeological studies.

The complexities inherent in resistivity interpretations and the nonuniqueness of potential solutions to observed data have tended to obscure the value of resistivity surveys. Although the method does have these drawbacks, when combined with seismic surveys or the other electrical methods detailed next, it can be an effective exploration tool. However, when we use resistivity to explore subsurface geologic relationships, independent control from well logs is mandatory. Its capability as an inexpensive monitoring tool has not been developed as fully as possible, but this likely will change as our society faces more contamination and needs to locate and remedy the problem.

5.11   OTHER ELECTRICAL METHODS

As noted at the beginning of this chapter, the following descriptions of additional electrical and electromagnetic methods are purposely brief. We include them mainly to provide a sense of applications, limitations, advantages, and how they may be similar to or different from electrical resistivity methods.

5.11.1   Induced Potential (IP)

Explorationists initially involved with the development and application of electrical resistivity surveying noticed that the potential difference does not immediately decline to zero after current is turned off. Rather, a large percentage of the potential disappears immediately, but the remainder gradually decreases over an interval, usually several seconds. Subsequent work demonstrated that certain subsurface materials are capable of electrical polarization and function like a rechargeable battery. Applied current induces a buildup of charges that, when the current is removed, gradually discharge, producing a flow of current and potential differences.

This decay of potential with time is referred to as *induced potential in the time domain* and is the phenomenon we discuss here. We also can investigate the effect of different frequencies of alternating current on apparent resistivity values. This approach is termed *IP in the frequency domain*. A common measure of the effect of IP in the time domain is *chargeability*, which is defined as the ratio of the area under the decay curve to the potential difference measured before the current was turned off. It is possible to demonstrate a theoretical relationship between apparent resistivity and chargeability. This is extremely valuable: It permits us to derive quantitative relationships between subsurface geometries and apparent chargeabilities if we know the appropriate equations for apparent resistivity, such as those that were presented for the horizontal interface.

Because of this specific relation between apparent resistivity and apparent chargeability, field procedures and data interpretation techniques are similar. Electrode arrays typically are the Schlumberger or dipole–dipole, and apparent chargeability is plotted against electrode spacing. Curve matching is commonly employed. Historically, IP has found much more application in exploration for base metals rather than in typical electrical resistivity applications such as ground-water studies. One reason is that equipment tends to be bulkier and data acquisition slower. Fieldwork is more expensive, and considering the relative lack of interpretive techniques, IP tends not to be cost effective unless the economic potential is substantial.

Equipment for conducting induced potential studies is similar to electrical resistivity equipment with two major exceptions. A timing circuit is required to switch the current on and off, and a recording unit is necessary to preserve the decrease of potential with time. Standard electrical resistivity equipment cannot, therefore, be used for IP studies; but equipment designed for IP work can be used to measure apparent resistivities.

Induced potential studies can be valuable assets to electrical resistivity studies. Based on our discussion of interpretive procedures for resistivity, it is apparent that curves that look like a single horizontal interface actually may contain the signature of several layers but cannot be differentiated. One or more of these hidden layers may possess a different chargeability. If apparent resistivity and apparent chargeability curves are plotted on the same graph, such distinctions may permit more informed, and more correct, interpretations of the subsurface. A good recent elaboration of this method is presented in Ward (1990).

5.11.2   Spontaneous Potential (SP)

Natural potential differences at the surface may arise from electrochemical reactions at depth that cause currents to flow. The most common example of this type of phenomenon is associated with ore bodies, part of which are above the water table. Oxidation takes place above the water table, whereas reduction takes place below it. These processes result in different charge concentrations that cause current flow. Sulfide and graphite bodies are by far the most common source of significant potentials of this sort, so it is not surprising that most SP surveys are conducted to locate ore bodies, especially sulfide ones.

Equipment demands are fairly straightforward. Two electrodes, wire, and a millivoltmeter constitute the normal setup. In this case porous-pot electrodes must be used because metal electrodes themselves produce a self-potential. A porous-pot electrode consists of a ceramic container with holes. A copper tube is surrounded by a solution of copper sulfate that slowly leaks through the holes into the ground. This solution disseminates the charges that otherwise would build up if the ground was in direct contact with a copper electrode.

Several characteristics conspire to negate the possibility of wide application of the self-potential method. Although surveys are quick and efficient and equipment demands are minimal, quantitative interpretation is difficult and depths of sources are difficult to model. Self-potentials are easily masked by other sources such as telluric currents. Because the water table is an important component of most sources, its typically shallow depth limits surveys to the same shallow terrain. And finally, of course, sources are limited to the specific types mentioned. SP exploration, therefore, is not a common shallow method except where we desire a relatively fast, inexpensive tool in searching for shallow sulfide bodies. A detailed discussion of SP exploration can be found in Corwin (1990).

5.11.3   Telluric and Magnetotelluric Methods

Natural electric currents flow in the Earth that are induced by the flow of charged particles in the ionosphere. At any given moment these currents follow uniform patterns over large areas. However, currents are flowing at depth, and based on what you know from this chapter, it seems reasonable to expect these currents to describe patterns based on the conductivity and geometry of the materials through which they flow. Naturally, we measure the potential differences produced by this current flow.

Because telluric currents vary with time, field operations include a base station at which two pairs of electrodes are placed at right angles. Voltage is continuously monitored, and we move two additional pairs of electrodes along a grid. Comparison between the base station readings and readings taken at grid locations yields a potential difference that is due to the subsurface effect on current density.

Magnetotelluric methods combine the measurement of magnetic field intensity variations with voltage measurements. Both sets of data are used to compute apparent resistivity functions, which then are compared with theoretical curves to determine thicknesses and true resistivities. A major requirement of this approach is a magnetometer capable of measuring weak magnetic field variations that fluctuate frequently.

Most surveys of this type have been in Europe and the Soviet Union. Survey objectives typically are directed at more regional objectives than we have mentioned to date, such as determining the geometry of sedimentary rock sequences in major depositional basins.

PROBLEMS

5.1 Determine the potential difference between the two potential electrodes for cases (a) and (b) below. Assume a current of 0.6 ampere.

(a)

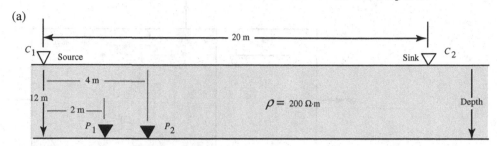

(b)

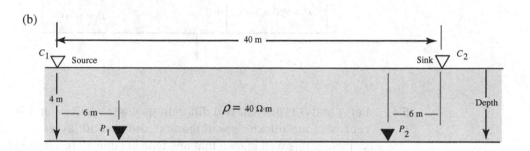

5.2 Construct the current flow lines beneath the interface in (a) and (b) below.

(a)

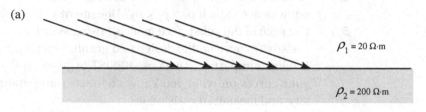

(b)

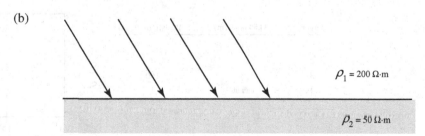

5.3 Calculate the potential at P_1 due to a current at C_1 of 0.6 ampere. The material in this section view extends to infinity in all directions. The bold line represents an interface between ρ_1 and ρ_2 material.

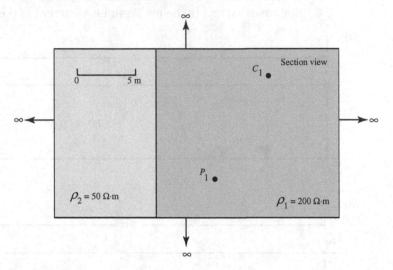

5.4 Let A and B represent two different geologic sections and let H_1 and H_2 represent the thicknesses of the first and second layers in a three-layer sequence. It is well known that one type of equivalence occurs for a three-layer case when $\rho_1 < \rho_2 > \rho_3$, $\rho_{1A} = \rho_{1B}$, $\rho_{3A} = \rho_{3B}$, $H_{1A} = H_{1B}$, and $\rho_{2A}H_{2A} = \rho_{2B}H_{2B}$. Demonstrate that this is true.

5.5 Assume the same general situation as explained in Problem 5.4. Does equivalence exist if $\rho_1 > \rho_2 < \rho_3$? Document your position.

5.6 For each of the subsurface models (a)–(c), sketch an appropriate apparent resistivity curve on the designated graph. The general shape of the curve is what is important. Don't use RESIST to generate the curves; rather, base your curves on what you know about current penetration, current density, and measured resistivities.

(a)

Layer	Thickness (m)	Resistivity ($\Omega \cdot m$)
1	1	100
2	5	10
3	5	100
4	Inflinite	1

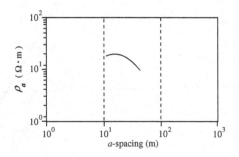

(b)

Layer	Thickness (m)	Resistivity (Ω·m)
1	10	100
2	10	10
3	Infinite	100

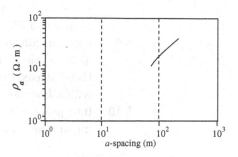

(c)

Layer	Thickness (m)	Resistivity (Ω·m)
1	5	1
2	10	100
3	5	1
4	Infinite	100

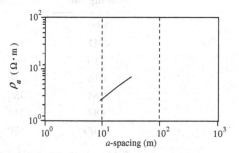

5.7 The data listed here were acquired using a constant-spread, Wenner traverse (a-spacing = 3 m). Interpret the data as completely as possible.

Horizontal Position (m)	ρ_a (Ω·m)	Horizontal Position (m)	ρ_a (Ω·m)	Horizontal Position (m)	ρ_a (Ω·m)
5.00	20.05	19.00	22.55	33.00	858.94
6.00	20.06	20.00	25.32	34.00	872.33
7.00	20.07	21.00	28.45	35.00	880.40
8.00	20.08	22.00	27.97	36.00	885.58
9.00	20.10	23.00	27.11	37.00	889.07
10.00	20.12	24.00	242.39	38.00	891.51
11.00	20.15	25.00	460.00	39.00	893.26
12.00	20.19	26.00	572.39	40.00	894.57
13.00	20.24	27.00	580.25	41.00	895.55
14.00	20.32	28.00	541.30	42.00	896.31
15.00	20.44	29.00	519.88	43.00	896.90
16.00	20.61	30.00	660.54	44.00	897.38
17.00	20.91	31.00	785.22	45.00	897.76
18.00	21.45	32.00	834.74		

5.8 Refer to Figures 5.9 and 5.30(c) and Equation 5.20. Derive an apparent resistivity equation for the Schlumberger array (see the derivation of Equation 5.35 for the Wenner array).

5.9 Using RESIST or Table 5.4, create data for a k-value of -0.6. Using these data, develop a master curve similar to those illustrated in Figure 5.33(a). Using at least three different models, demonstrate that all two-layer models with a k-value of -0.6 plot on the master curve.

5.10 Interpret the following data, which were obtained with a Wenner, expanding-spread traverse.

Electrode Spacing (m)	ρ_a ($\Omega \cdot$m)	Electrode Spacing (m)	ρ_a ($\Omega \cdot$m)	Electrode Spacing (m)	ρ_a ($\Omega \cdot$m)
1.00	984	14.68	72	215.44	327
1.47	955	21.54	53	316.23	432
2.15	883	31.62	63	464.16	554
3.16	742	46.42	87	681.29	686
4.64	533	68.13	124	1000.00	818
6.81	311	100.00	174		
10.00	150	146.78	241		

5.11 Interpret the following data, which were obtained with a Schlumberger, expanding-spread traverse.

Electrode Spacing (m)	ρ_a ($\Omega \cdot$m)	Electrode Spacing (m)	ρ_a ($\Omega \cdot$m)	Electrode Spacing (m)	ρ_a ($\Omega \cdot$m)
1.00	108	14.68	307	215.44	293
1.47	121	21.54	245	316.23	381
2.15	148	31.62	168	464.16	479
3.16	191	46.42	122	681.29	580
4.64	244	68.13	125	1000.00	675
6.81	295	100.00	162		
10.00	323	146.78	220		

5.12 The following data were gathered with a Wenner, expanding-spread traverse in an area with thick alluvial deposits at the surface. What is the likely depth to the water table?

Electrode Spacing (m)	ρ_a ($\Omega \cdot$m)	Electrode Spacing (m)	ρ_a ($\Omega \cdot$m)
0.47	198	6.81	84
0.69	160	10.00	82
1.00	140	14.68	92
1.47	112	21.54	101
2.15	95	31.62	100
3.16	84	46.42	102
4.64	79		

5.13 The following data were gathered with a Wenner, expanding-spread traverse in an area of thick deltaic sands. Bedrock depths are greater than 30 m. What is your best estimate of the depth to the water table in this area?

Electrode Spacing (m)	ρ_a ($\Omega \cdot$m)	Electrode Spacing (m)	ρ_a ($\Omega \cdot$m)
0.47	2590	6.81	8753
0.69	3288	10.00	7630
1.00	4421	14.68	4805
1.47	5198	21.54	2160
2.15	6055	31.62	995
3.16	6686	46.42	584
4.64	7782		

5.14 The following data were gathered with a Wenner, expanding-spread traverse in an area of dune sands underlain by lake clays that in turn are underlain by Triassic sedimentary rocks. Estimate a value for clay thickness.

Electrode Spacing (m)	ρ_a ($\Omega \cdot$m)	Electrode Spacing (m)	ρ_a ($\Omega \cdot$m)
0.69	1298.90	10.00	360.67
1.00	1398.06	14.68	240.11
1.47	1306.98	21.54	191.51
2.15	1153.02	31.62	153.81
3.16	925.27	46.42	116.38
4.64	762.40	68.13	98.03
6.81	554.13	100.00	86.08

REFERENCES CITED

Barnes, H.E. 1952. Soil investigations employing a new method of layer value determination for earth resistivity investigation. *Highway Research Board Bulletin 65*, 26–36.

Corwin, Robert F. 1990. The self-potential method for environmental and engineering applications. In *Geotechnical and environmental geophysics, volume 1: Review and tutorial,* ed. Stanley H. Ward, 127–45. Society of Exploration Geophysicists Investigations in Geophysics, no. 5.

Fetter, C.W. Jr. 1980. *Applied hydrogeology.* Columbus, OH: Merrill Publishing Co.

Fläthe, H. 1963. Five-layer master curves for the hydrogeological interpretation of geoelectric resistivity measurements above a two-story aquifer. *Geophysical Prospecting,* v. 11, 471–508.

Frohlich, Bruno, and Warwick J. Lancaster. 1986. Electromagnetic surveying in current Middle Eastern archaeology: Application and evaluation. *Geophysics,* v. 51, 1414–25.

Hubbert, M.K. 1932. Results of earth-resistivity survey on various geologic structures in Illinois. *American Institute of Mining and Metallurgical Engineers Technical Publication 463,* 9–39.

Hubbert, M. King. 1940. The theory of ground-water motion. *Journal of Geology,* v. 48, 785–944.

Hubbert, M.K. 1944. An exploratory study of faults in the cave in Rock and Rosiclare districts by the earth resistivity method. In *Geological and geophysical survey of fluorspar areas in Hardin County, Illinois.* United States Geological Survey Bulletin 942, Part 2, 73–147.

Keller, George V., and Frank C. Frischknecht. 1966. *Electrical methods in geophysical prospecting.* Oxford, England: Pergamon Press.

McNeill, J.D. 1990. Use of electromagnetic methods for groundwater studies. In *Geotechnical and environmental geophysics, volume 1: Review and tutorial,* ed. Stanley H. Ward, 191–218. Society of Exploration Geophysicists Investigations in Geophysics, no. 5.

Mooney, Harold M. 1958. A qualitative approach to electrical resistivity interpretation. *Pure and Applied Geophysics,* v. 40, 164–71.

Mooney, H.M., and W.W. Wetzel. 1956. *The potential about a point electrode and apparent resistivity curves for a two-, three-, and four-layer earth.* Minneapolis: University of Minnesota Press.

Moore, W. 1945. An empirical method of interpretation of earth resistivity measurements. *Transactions of the American Institute of Mining and Metallurgical Engineers,* v. 164, 197–223.

Orellana, Ernesto, and H.M. Mooney. 1966. *Master tables and curves for vertical electrical sounding over layered structures.* Madrid: Interciecia.

Telford, W.M., L.P. Geldart, and R.E. Sheriff. 1990. *Applied geophysics,* 2nd ed. Cambridge, England: Cambridge University Press.

Van Nostrand, Robert G., and Kenneth L. Cook. 1966. *Interpretation of resistivity data.* United States Geological Survey Professional Paper 499.

Ward, Stanley H. 1990. Resistivity and induced polarization methods. In *Geotechnical and environmental geophysics, volume 1: Review and tutorial,* ed. Stanley H. Ward, 147–89. Society of Exploration Geophysicists Investigations in Geophysics, no. 5.

Zohdy, A.A.R. 1989. A new method for the automatic interpretation of Schlumberger and Wenner sounding curves. *Geophysics,* v. 54, 245–53.

Zohdy, A.A.R., G.P. Eaton, and D.R. Mabey. 1974. Application of surface geophysics to groundwater investigations. *Techniques of Water Resources Investigation of the United States Geological Survey,* Book 2, chap. D1.

SUGGESTED READING

Bentley, L.R., and M. Gharibi. 2004. Two- and three-dimensional electrical resistivity imaging at a heterogeneous remediation site. *Geophysics*, v. 69, no. 3, 674–80.

Lane, J.W., F.P. Haeni, and W.M. Watson. 1995. Use of a square-array direct-current resistivity method to detect fractures in crystalline bedrock in New Hampshire. *Ground Water*, v. 33, no. 3, 476–85.

Mooney, Harold M. 1958. A qualitative approach to electrical resistivity interpretation. *Pure and Applied Geophysics,* v. 40, 164–71.

Morris, M., J.S. Ronning, and O.B. Lile. 1997. Detecting lateral resistivity inhomogeneities with the Schlumberger array. *Geophysical Prospecting*, v. 45, no. 3, 435–48.

Smith, D.L., and A.F. Randazzo. 2003. Application of electrical resistivity measurements to an evaluation of a potential landfill site in a karstic terrain. *Environmental Geology,* v. 43, no. 7, 743–51.

Stummer, P., H. Maurer, and A.G. Green. 2004. Experimental design: Electrical resistivity data sets that provide optimum subsurface information. *Geophysics*, v. 69, no. 1, 120–39.

Zohdy, A.A.R., G.P. Eaton, and D.R. Mabey. 1974. Application of surface geophysics to groundwater investigations. *Techniques of Water Resources Investigation of the United States Geological Survey,* Book 2, chap. D1.

CHAPTER 6

Exploration Using Gravity

Because an object on the Earth's surface is attracted by the mass of the Earth, the gravity exploration method detects variations in the densities of subsurface materials by measuring gravity at the surface and analyzing the differences in the recorded values. Using gravity to explore the subsurface appears to be straightforward compared with some of the difficulties we encountered when deriving relationships for electrical resistivity surveying. However, if gravity exploration was a panacea, the gravity method would be at the beginning of this text, and the text itself would be much thinner.

One difficulty becomes apparent if we consider the magnitude of the variations we must measure. An average value for gravity on the Earth's surface is 980 cm/s^2, but in routine surveys we need to detect variations as small as 0.00001 percent of average Earth gravity. This demands instruments with great sensitivity. In addition, factors other than subsurface density changes cause gravity to vary. All these causes must be identified and corrections for their effects applied to our data. For example, changes in elevation of the land surface lead to variations in gravity because observation points will be closer to or farther from the center of the Earth, and more or less material will be between the observation point and the Earth's center. Elevations of observation points must be known to reasonable accuracy—often to at least 0.25 m. This requirement adds time and cost to any survey but cannot be ignored if the data are to be useful.

Another factor affecting exploration using gravity is the complexity of the subsurface. All the material beneath an observation site affects a gravity value. Therefore, we constantly are attempting to remove the effects due to bodies in which we have no interest to isolate the effects of items of interest. This is a task for which we typically achieve only incomplete success. Computing the effect of subsurface masses of varying shapes and densities is not difficult and surely is easier than determining apparent resistivities for complex shapes; but we face a similar problem in that a given curve of observed gravity values can be produced by numerous subsurface configurations.

Although the approach of this chapter is similar to that of earlier ones, and we consider the same general topics, you should concentrate on the tasks and requirements that provide the most accurate values of gravity possible and then process these values so the information that remains is related to bodies of interest in the subsurface.

6.1   FUNDAMENTAL RELATIONSHIPS

6.1.1   Gravitational Acceleration

Newton's law of universal gravitation states that there is a force of attraction F between two particles with masses m_1 and m_2 separated by a distance r that is represented by the relationship

$$F = G\frac{m_1 m_2}{r^2} \tag{6.1}$$

where G is the universal gravitation constant. The value of G first was determined in 1798 by Lord Cavendish; the present value, which was determined in 1942, is equal to

$$6.6732 \times 10^{-11} \text{ nt} \cdot \text{m}^2/\text{kg}^2 \quad \text{in SI units}$$

or

$$6.6732 \times 10^{-8} \text{ dyne} \cdot \text{cm}^2/\text{g}^2 \quad \text{in cgs units}$$

If we assume that the Earth is spherical, the force exerted by the Earth on a spherical body with mass m resting on the Earth's surface is

$$F = G\frac{mM}{R^2} \tag{6.2}$$

where M is the Earth's mass and R is its radius. This also assumes that density varies only with distance from the center of the Earth (that the Earth is spherically symmetric) and that R is large compared to the size of the object. Values that we will use in future calculations include

Earth radius (at equator): 6.378×10^8 cm,

Earth radius (at 45° latitude): 6.367×10^8 cm, and

Earth mass: 5.976×10^{27} g.

Force also is given by Newton's second law of motion, which states that

$$F = ma \tag{6.3}$$

where a is acceleration. If we define g as gravitational acceleration, then, when the acceleration is caused only by the gravitational attraction of the Earth, we can write

$$F = mg = G\frac{mM}{R^2}$$

and

$$g = \frac{GM}{R^2} .$$ (6.4)

The dimensions of g are L/M², which are expressed as m/s² (SI) or cm/s² (cgs). In geophysics the normal unit of gravitational acceleration is the Gal (in honor of Galileo), which is 1 cm/s². Because the variations in gravitational acceleration g (or *gravity* for short) in which we are interested are so small, we often use the milliGal (1 mGal = 0.001 Gal) for exploration purposes. The gravity unit *gu* often is used instead of the mGal. It is equal to 0.000001 m/s², which is 0.1 mGal.

6.2 MEASURING GRAVITY

Ultimately we are interested in determining the acceleration due to gravity at various sites around the world. Instruments designed to measure gravity directly perform *absolute* measurements. Such devices are elaborate, are difficult to transport, and require considerable time to set up and use. Somewhat easier to design and use are instruments that determine the difference in gravity between two observing stations. These obtain *relative* differences in gravity and are the instruments we use in exploration surveys. Because this type of instrument is the one you are most likely to encounter, we discuss relative measurements first.

6.2.1 Relative Measurements Using a Pendulum

The classic instrument for measuring both absolute and relative gravity is the pendulum. An ideal, simple pendulum suspends a material point that is dimensionless from a massless string that does not stretch and is perfectly flexible. For such a device the period T is related to the length of the string l and gravitational acceleration by the relationship

$$T = 2\pi\sqrt{\frac{l}{g}} .$$ (6.5)

Such an ideal pendulum cannot be constructed, as should be obvious from its description. For a real pendulum Equation 6.5 becomes

$$T = 2\pi\sqrt{\frac{K}{g}} \tag{6.6}$$

where K is a constant that represents the characteristics of a particular pendulum system. If we are able to determine T and K accurately, then we have a measured or observed value for gravity, which we refer to as g_{obs}. Unfortunately, K cannot be determined accurately, and absolute values measured in this fashion have an accuracy of no better than 1 mGal and often are not that good. In routine surveys we need to know g_{obs} to at least 0.1 mGal.

However, all is not lost. If we use the same pendulum at two observation sites, say x and y, then we have

$$g_{obs_x} = \frac{4\pi^2 K}{T_x^2} \quad \text{and} \quad g_{obs_y} = \frac{4\pi^2 K}{T_y^2}. \tag{6.7}$$

Both values will be inaccurate due to the imprecision of K; but because the same value of K enters both measurements, the gravity values will vary from the true value of gravity by the same factor. Therefore, we can use them to determine the relative difference in gravity (Δg) by accurately measuring T at each site as

$$\Delta g = g_{obs_x} - g_{obs_y}. \tag{6.8}$$

Precision is on the order of 0.1 mGal; however, pendulum-based instruments not only are bulky but require substantial amounts of time to acquire a measurement. They have been superseded by the gravimeter, which is very portable. Reading times are short, and precision is on the order of 0.01 mGal.

6.2.2 Relative Measurements Using a Gravimeter

As in the case of the pendulum, the principle on which gravimeter design is based is quite simple. If a mass is placed on a spring and this assembly is moved from one position to another, the spring will lengthen or shorten by a small amount due to variations in gravity. However, the small changes in gravity from site to site result in only very small displacements of the spring. If we employ a very weak spring, it will extend only 1 micron for a gravity increase of 1 mGal. To accurately measure changes for such small gravity variations, a spring several tens of meters in length is required. The length of such a spring clearly is impractical, but it helps to illustrate another factor in gravimeter design. A mass suspended from such a long spring will oscillate up and down for quite a while after being disturbed. Such behavior does not facilitate rapid readings, so a high degree of stability is crucial if fieldwork is to progress at an acceptable rate.

A great deal of ingenuity, careful design, and much trial and error have produced various combinations of special springs attached to a mass at the end of a beam. These assemblies typically are held together by various hinges and levers. Not only do the springs have to be short and sensitive, but the design must ensure oscillation periods of only a few seconds. A critical component of most modern gravimeters is the *zero-length* spring. Tension is placed on such a spring during manufacture so that the spring is much more sensitive than normal and amplifies the displacements caused by small variations in gravity. A schematic example of the interior of one gravimeter (the Worden) with an arrangement of this type is illustrated in Figure 6.1. Figure 6.2 is a photograph of the actual gravimeter.

We null the instrument by turning a calibrated dial until a beam produced by a pointer (Figure 6.1) is brought to the center of the magnifying eyepiece. We note the setting of the dial, and then we move the instrument to a new location and level it. Because the change in gravity moves the mass attached to the zero-length spring and the pointer, the beam no longer is in the center of the eyepiece. We again null the instrument by turning the calibrated dial. The amount of restoring force necessary to move the beam back to the center of the eyepiece

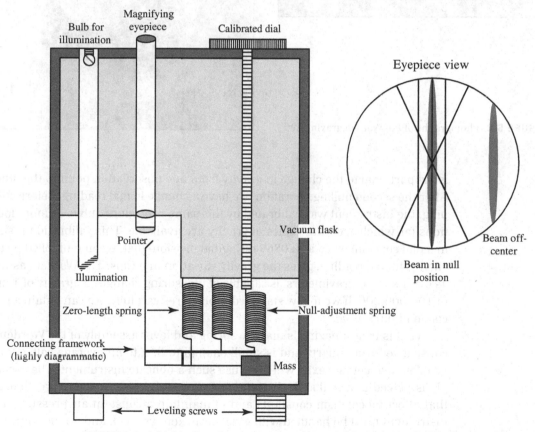

FIGURE 6.1 A highly diagrammatic representation of the interior of a Worden gravimeter.

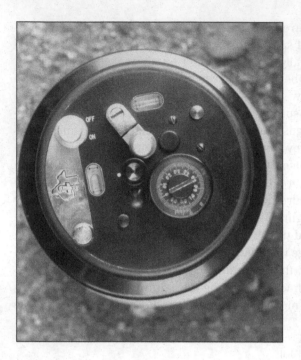

FIGURE 6.2 Photographs of the Worden gravimeter.

is proportional to the change in gravity from one observation point to the other. After the second nulling operation, we have a change in dial reading. Before shipping, the instrument was calibrated by this same procedure at two or more locations for which accurate values of gravity are available. This calibration results in a dial constant (such as 0.0869 mGal/dial division) that, when multiplied by the change in dial reading, gives the gravity variation in mGals. The Worden, as well as other similar gravimeters, is capable of measuring changes in gravity of 1 unit in 100,000,000. If we follow standard procedures carefully, we can obtain a precision of 0.01 mGal.

What is truly amazing is that the spring and lever assembly of the Worden is constructed from quartz and is small enough to fit into an average coffee mug. Problems, of course, exist when we use such a delicate instrument. The assembly is placed in a sealed flask at the factory to minimize temperature changes that affect mechanism constants and to maintain a constant air pressure. The instrument must be handled with great care, kept vertical, and not be subjected to sudden accelerations. The major problem during routine use is what is termed meter *drift*. Over time the physical properties of the springs change very slightly. Because the springs are so sensitive, these changes are large enough to affect readings, so procedures must be devised to check for and correct for these variations. Corrections for instrument drift are described in a following section that discusses field procedures.

6.2.3 Absolute Measurements

As noted previously, pendulums can measure gravity directly but not with enough precision. Today most absolute measurements use what is called the *falling-body* method. The first accurate determinations using this approach were made in 1952, and by the 1960s the availability and sophistication of lasers and electronic measuring and timing devices gave a precision of 0.01 mGal.

All methods essentially measure the amount of time t it takes an object to fall a distance z and use the relationship

$$z = \frac{gt^2}{2}.$$

(6.9)

One such approach throws a body upward. The object passes two established levels on its way up and again on its way down. If the positions of the levels are z_1 and z_2, and if t_1 and t_2 are the times at which the object passes these levels going up and t_3 and t_4 are the times on the way down, then it is straightforward to demonstrate that

$$g = \frac{8(z_2 - z_1)}{(t_4 - t_1)^2 - (t_3 - t_2)^2}.$$

(6.10)

We need only to accurately measure the distance $z_2 - z_1$ and the four times. Such instruments clearly are not appropriate as field instruments; but because gravimeters can achieve the same level of precision in determining relative variations in gravity, we require only a few sites where gravity is known accurately.

Excellent detailed descriptions of instruments for determining absolute and relative gravity are presented in Robinson and Coruh (1988) and Tsuboi (1979). If you are interested in more detail about design and operating principles, you should consult these sources.

6.2.4 International Gravity Standardization Net 1971 (IGSN71)

If a gravity survey is searching for shallow anomalous masses in a restricted geographic area, it does not need access to an accurate gravity value. Exploration methods use variation from a norm, so such a survey could assume a value for g, make accurate relative determinations, calculate absolute values based on the assumed *g*, and then map the departures from this value. However, once we decide to compare gravity values from various areas, especially worldwide, we require a network of observation sites at which gravity is known accurately.

Work to establish such a worldwide network began in the 1950s under the leadership of Professor George P. Woollard of the University of Wisconsin. This survey, based on very painstaking measurements using pendulums and gravimeters, was completed in 1963. Further refinements, primarily due to more accurate values from instruments utilizing the free-fall approach, necessitated adjustments to the 1963 determinations. Values from more than 1800 sites around the world were published in 1971 as a unified, internally consistent network that is referred to as the *International Gravity Standardization Net 1971 (IGSN71)*. Every gravity survey should be tied to one of these sites.

6.3   ADJUSTING OBSERVED GRAVITY

If the Earth had perfect spherical symmetry and did not rotate, then gravity should be the same everywhere on the surface. Of course, we know this is not so, and therefore it is essential to identify the reasons that gravity varies so we can correct for them in our quest to use gravity to explore the subsurface. One factor clearly is position on the Earth's surface: The absolute value of gravity at the equator is 978.0 Gals, and at the poles gravity equals 983.2 Gals.

6.3.1   Variation in *g* as a Function of Latitude

The rotation of the Earth produces an outward-directed centrifugal force that acts in a direction opposite to gravity and therefore diminishes the measured value of *g*. The effect of this centrifugal force is greatest at the equator and diminishes to zero at the poles of the Earth's rotational axis. As a direct consequence of this force, *g* is greater at the poles than at the equator by 3.4 Gals.

The long-term behavior of the Earth is that of a fluid, so the Earth's rotation produces centrifugal effects that cause its shape to be an *ellipsoid of revolution*. This means that the Earth is not spherical but is flattened at the poles. In other words, the length of the Earth's radius is greater at the equator than at the poles. This distance factor causes *g* to increase from equator to pole by 6.6 Gals because the surface is closer to the center of mass (the center of the Earth) at the poles. However, because radius length is greater at the equator, more mass is positioned between the surface and the Earth's center at the equator than at the poles. This mass factor causes *g* to decrease by 4.8 Gals from equator to pole.

The effect of rotation, the distance factor, and the mass factor result in a net increase in gravity of 5.2 Gals as one travels from the equator to the poles. This explains the variation of gravity from 978.0 Gals at the equator to 983.2 Gals at the poles. Recall that our ultimate goal in gravity exploration is to obtain values of gravity for which variations are entirely due to subsurface density distributions. Therefore, we must derive an equation that removes position on the Earth's surface as a cause of gravity difference. Differences that remain in our measured gravity values then will be due to circumstances other than position.

6.3.2 Correcting for the Latitude Effect

Consider an imaginary, regular surface that corresponds to the true, gross shape of the Earth. This is the surface we referred to previously as an ellipsoid of revolution, and it coincides closely with sea level over the open oceans. This surface also often is referred to as the *reference ellipsoid,* which is the term we adopt for this discussion. In detail, however, the reference ellipsoid does not coincide exactly with the sea-level surface. If we could crisscross the continents with canals open to the oceans, the water level in these canals also would be at sea level. The surface passing through the sea-level surface around the world is known as the *geoid* and contains broad undulations that are minor compared to its overall dimensions. Although the reference ellipsoid does not fit the geoid exactly, it approximates it to a degree that is acceptable for most gravity exploration.

We now need an equation that corrects for position on the reference ellipsoid. Equation 6.11 specifies such a relationship.

$$g_n = g_e(1 + A\sin^2\phi - B\sin^2 2\phi) \text{ cm/s}^2 \tag{6.11}$$

The derivation of this equation is quite lengthy and is not relevant to what follows, so we do not include it here. A good development is presented in Tsuboi (1979, 68–78). A and B are constants that take into account the Earth's angular velocity of rotation, its size, and its ellipticity. Values for A and B are determined from careful analysis of satellite orbits and from thousands of measurements of g at various latitudes. Note that longitude does not enter the computations. Latitude is represented by ϕ, g_e stands for gravity on the equator at sea level, and g_n represents normal or standard gravity. The value for g_e is determined from actual measurements at coastal sites on the equator and from carefully adjusted values from other sites at various latitudes. Current values for A, B, and g_e were adopted by the International Association of Geodesy in 1967. When we place these values into Equation 6.11, the result is the *geodetic reference system formula of 1967* or *GRS67,* which is presented as Equation 6.12:

$$g_n = 978.03185(1 + 0.005278895\sin^2\phi + 0.000023462\sin^4\phi) \text{ cm/s}^2 \tag{6.12}$$

Normal gravity g_n represents the value of gravitational acceleration that we should observe when measuring gravity at various positions on the Earth's surface as Equation 6.12 corrects for position. Typically, the value of gravity that is measured, g_{obs} (observed gravity), does not match the value predicted by the geodetic

TABLE 6.1	LATITUDE, NORMAL GRAVITY, AND POSITION REQUIREMENTS		
Latitude:	**50**	Normal gravity (mGal):	981068.6407
Latitude:	51	Normal gravity (mGal):	981157.5477
Length of 1 degree of latitude (km):			111
Variation in g per degree (mGal):			88.91
Meters per milliGal:			1248.50
MilliGal per meter:			0.000800964
Position (m) for 0.1 mGal accuracy:			125
Position (m) for 0.05 mGal accuracy:			62

reference system formula. The difference, $g_{obs} - g_n$, is thus anomalous and is called the *gravity anomaly*.

How accurately do we need to know our position on the Earth's surface when making a gravity measurement? Table 6.1 provides an answer. At a latitude of 50° the variation in normal gravity is 88.91 mGal for a degree of latitude. If we take the length of a degree of latitude to be 111 km (it actually varies from about 110.5 km at the equator to 111.5 at the poles), this variation translates to 1 mGal in 1248.5 m. If we want to maintain an accuracy of 0.1 mGal in our work, we must be able to locate our position to within 125 m. On a map with a scale of 1:25,000 this translates to a location within 0.5 cm.

> Are positioning requirements more severe at low latitudes than at middle latitudes? At high latitudes than at middle latitudes?

Of course, most of our measurements will not be obtained at sea level, and, as we noted in the introduction to this chapter, variations in elevation substantially affect observed gravity. Once again we must correct observed gravity before such data are useful.

6.3.3 Elevation Correction 1: The Free-Air Correction

Let's again return to the ideal Earth, which is spherical and nonrotating. We've already established that the gravitational acceleration imparted by the Earth is

$$g = \frac{GM}{R^2}. \tag{6.4}$$

If we are interested in how gravity varies with elevation, we can determine the vertical gradient by taking the first derivative of Equation 6.4. Thus

$$\frac{dg}{dz} = \frac{dg}{dR} = -2\frac{GM}{R^3} = -g\frac{2}{R}. \tag{6.13}$$

If we substitute representative values at 45° latitude for g at sea level and R, the result is –0.3086 mGal/m. This value tells us what we already know—that gravity decreases as we increase our distance from the Earth's center—but now we have an idea of the magnitude of the decrease. Equation 6.13 gives us a good approximation of the value we seek, but remember the simplicity of our initial assumption. As in the case of the latitude correction, we must use a more adequate representation of the true character of the Earth and take a derivative of an equation that includes factors for rotation and ellipsoidal shape. Because the equation and its derivative are complex, they are not presented here. The final relationship (Grant and West 1965, 237–38) is

$$\frac{dg}{dR} = -0.3086 - 0.00023\cos 2\phi + 0.00000002z \tag{6.14}$$

where ϕ is latitude, z is elevation, and the units are mGal/m. In any given survey of modest extent the latitude adjustment (second term) in Equation 6.14 is very small. The correction for the vertical gradient due to increasing distance from the ellipsoid surface (third term) becomes significant only in mountainous regions where there is extreme variation in elevation. In practice both terms usually are ignored, and the value of 0.3086 mGal/m is the only value used. Note that this correction considers only elevation differences relative to a datum and does not take into account the mass between the observation point and the datum. For this reason the correction is termed the *free-air correction*.

Normally the datum used for gravity surveys is sea level. Because gravity decreases 0.3086 mGal for every meter above sea level, the free-air correction is added to observed gravity. If the elevation of the observation point is below sea level, the free-air correction is subtracted. If the datum chosen is not sea level, the same procedure holds relative to the selected datum. Once again it is important to note that if we want to maintain an accuracy of 0.1 mGal in our final gravity values, we must know the elevation of an observation point to at least 33 cm.

The value remaining after we add the free-air correction FA_{corr} and subtract normal gravity from observed gravity is termed the *free-air anomaly*. We can express the free-air anomaly Δg_{FA} as

$$\Delta g_{FA} = g_{obs} - g_n + FA_{corr}. \tag{6.15}$$

Let's turn now to an analysis of how to handle the mass lying between an observation station and the selected datum.

6.3.4   Elevation Correction 2: The Bouguer Correction

Consult Figure 6.3. If a gravity observation point is at A, which is at sea level, no free-air correction is necessary. If a station is at B, then a correction is applied for the difference z between sea level and the elevation of station B. At this point we have not corrected for the mass lying between B and sea level, assuming sea level is our datum. Even if there are no subsurface density variations between stations A and B, the gravity at B will be greater due to the presence of the material between B and sea level. Recall once again that our purpose is to correct for all factors that cause gravity to vary from one station to another, so that in the end the only reason for gravity differences must be subsurface density variations.

The usual approach to correcting for the excess material beneath B is to compute the attraction of an infinite slab of density ρ and thickness z. This correction is called the *Bouguer correction* in honor of Pierre Bouguer (1698–1758), who was the first to attempt to measure the horizontal gravitational attraction of mountains. A cursory examination of Figure 6.3 clearly indicates that this is an approximation, but an additional correction that we present in the next section addresses this concern.

We now develop the Bouguer correction. Although this is a somewhat lengthy derivation, we include it here because we will use one of the equations again when computing gravity anomalies over buried objects of tabular shape. Figure 6.4 illustrates the critical relationships. Our goal is to derive a relationship for g due to a very thin rod. We do this by summing the individual contributions of very small-volume elements. We then use the relationship derived for the rod to compute g for a thin sheet. This is achieved by summing the individual contributions of an infinite number of thin rods. Because an infinite slab is simply an assembly of a number of infinite sheets, this provides the desired result. To use this approach, we employ integral calculus. Even if you are not familiar with this approach, work through the derivation; you will be able to follow most steps except for the evaluation of the integral.

We first want to determine the small value of gravity Δg at point P due to a very small-volume element of volume $dxdydz$ and density ρ. We can use the general form of Equation 6.4 to state the relationship for g along the direction r, which gives us Δg_r.

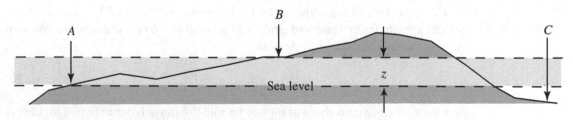

FIGURE 6.3 The infinite slab used to correct for excess mass beneath point B.

$$\Delta g_r = \frac{G(\rho dx dy dz)}{r^2} \qquad (6.16)$$

Because we always are interested in the vertical downward attraction of gravity Δg_z, we multiply Equation 6.16 by $\cos\theta$ to produce

$$\Delta g_z = \frac{G(\rho dx dy dz)}{r^2}\cos\theta. \qquad (6.17)$$

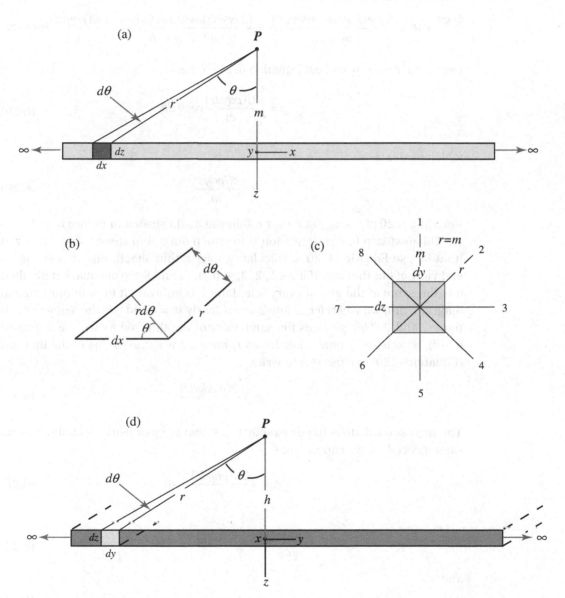

FIGURE 6.4 Relationships and notation used to derive the Bouguer correction.

The attraction of a thin rod is the sum of all these components, so we write

$$g = \int_{x=-\infty}^{x=\infty} \frac{G(\rho dx dy dz)}{r^2} \cos\theta .$$ (6.18)

Before evaluating the integral, we note that $r = m/\cos\theta$. Also (see Figure 6.4(b)), because the length of arc for a small angle $d\theta$ equals the radius r times the change in angle $d\theta$, it is straightforward to demonstrate that $dx = r\, d\theta/\cos\theta$. Using these relationships, we show that

$$\frac{G(\rho dx dy dz)}{r^2} \cos\theta = \frac{G(\rho dy dz) r d\theta \cos\theta (\cos\theta)^2}{m^2 \cos\theta} = \frac{G(\rho dy dz) m d\theta \cos\theta (\cos\theta)^2}{m^2 \cos\theta \cos\theta} = \frac{G(\rho dy dz)}{m} \cos\theta\, d\theta .$$

Using this result, we adjust Equation 6.18 so that

$$g = \int_{-\frac{\pi}{2}}^{\frac{\pi}{2}} \frac{G(\rho dy dz)}{m} \cos\theta\, d\theta$$ (6.19)

which evaluates to

$$g = \frac{2G\rho dy dz}{m} .$$ (6.20)

Equation 6.20 gives us g at P over a thin rod as illustrated in Figure 6.4(a).

The next step in our derivation is to sweep out a thin sheet using a thin rod (that is, use Equation 6.20 to calculate g over a thin sheet). Figure 6.4(c) is an end view of our thin rod. If lines 1, 2, 3, 4, 5, 6, 7, and 8 are of equal length, then g is the same at the end of every line. Line 1 is equivalent to m in our original diagram, and we will refer to line 2 as r. Clearly $m = r$ and g is the same at positions 1 and 2. This gives us the same view of the thin rod as we see in Figure 6.4(d), where line 2 now is labeled as r. Because we know g due to the thin rod (Equation 6.20), we use this to write

$$\Delta g_r = \frac{2G(\rho dy dz)}{r} .$$ (6.21)

The next several steps mimic exactly those that we just listed, so following the same procedure we can say that

$$\Delta g_z = \frac{2G(\rho dy dz)}{r} \cos\theta$$ (6.22)

$$g = \int_{-\frac{\pi}{2}}^{\frac{\pi}{2}} 2G(\rho dz) d\theta$$ (6.23)

and

$$g = 2\pi G(\rho dz).$$ (6.24)

Thus the Bouguer effect, or g due to a slab of infinite extent and thickness z, is $2\pi G(\rho z)$, and the value of the correction is $0.04193(\rho z)$ mGal/m.

Because the Bouguer correction is supposed to remove the effect of the additional mass above the datum that adds to observed gravity, the correction is subtracted from g_{obs}. If an observation station is at a point below the datum, such as C in Figure 6.3, the Bouguer correction must be added to observed gravity. Following the same line of reasoning we used in the case of the free-air correction, residual gravity remaining after the Bouguer correction B_{corr} is referred to as the *Bouguer anomaly* (Δg_B) and is equal to

$$\Delta g_B = g_{obs} - g_n + FA_{corr} - B_{corr}. \tag{6.25}$$

It can be confusing to remember whether the Bouguer and free-air corrections should be added or subtracted, so it is best to record elevation differences from a datum as positive when above the datum and negative when below the datum. If this procedure is followed, the differences can be substituted directly into Equation 6.26, which is an expanded version of Equation 6.25, and the correction always will be in the proper sense.

$$\Delta g_B = (g_{obs} - g_n + 0.3086z - 0.04193\rho z) \text{ mGal} \tag{6.26}$$

Note that the units in Equation 6.26 are in milliGals, so g_{obs} and g_n also must be in milliGals.

The majority of topographic maps in the United States still present elevations in feet, so many explorationists find it more convenient to retain feet as a unit when correcting gravity readings. In this case the proper form of the Bouguer anomaly equation becomes

$$\Delta g_B = (g_{obs} - g_n + 0.09406z - 0.01278\rho z) \text{ mGal}. \tag{6.27}$$

Because Bouguer anomaly values are the ones most often used for interpretation in gravity surveys, Equation 6.26 is presented in its complete expanded form as Equation 6.28. This incorporates the GRS67 formula and the free-air and Bouguer corrections.

$$\Delta g_B = \begin{cases} g_{obs} - [978031.85(1 + 0.005278895\sin^2\phi + 0.000023462\sin^4\phi)] + \\ [(0.3086 - 0.04193\rho)z] \end{cases} \text{mGal} \tag{6.28}$$

Table 6.2 provides an opportunity to study the process of adjusting observed gravity values (gravity reduction). By selectively changing values for observed gravity, latitude, elevation, and Bouguer density, you can study the effect of each of these on the final anomaly values. It's also instructive to change the amount of error for elevation or latitude (or both) to see how much the Bouguer anomaly would change if such errors were present in your data. Remember that we are striving to keep the accuracy of our final values to within 0.1 mGal and would like to do better if possible.

TABLE 6.2   EXAMPLE OF GRAVITY REDUCTION

Observed gravity	980658.67	Observed gravity	980658.67
Normal gravity	980674.39	Latitude (ϕ)	45.62
Free-air correction	30.93	Elevation (m)	100.24
Bouguer correction	11.22	Bouguer density (g/cm^3)	2.67
Free-air anomaly	15.22		
Bouguer anomaly	4.00		
Elevation error (m)	**0.33**	Latitude error (ϕ)	**0.01**
Bouguer anomaly error	0.06	Bouguer anomaly error	0.90

(All gravity values are in milliGals.)

The most vexing problem in the gravity reduction process is selecting an appropriate value for density in the Bouguer correction. Many surveys simply select the value of 2.67 g/cm^3, which is the average density for crustal rocks. In many areas such a value is appropriate and is useful if we want to compare gravity values from numerous surveys over an extended area. If a survey is of limited extent and has a specific target, 2.67 g/cm^3 may not be appropriate. Selecting the best value is possible only by extensive sampling and determining densities for many specimens or by employing special field procedures during the gravity survey. An example of this latter method is presented in the section addressing field procedures.

6.3.5   Elevation Correction 3: The Terrain Correction

As you already may have sensed, the Bouguer correction is somewhat crude in its assumption of an infinite slab. Figure 6.5 illustrates the Bouguer slab for a gravity observation station at B. To reduce observed gravity at B to sea level, we subtract the effect of the mass in this slab. Note that at x no rock or sediment is present because the shaded area is above the land surface. This means we have overcorrected with respect to region x by subtracting a mass effect where no mass existed in the first place. Therefore, we must *add* a small amount to observed gravity to adjust for this overcorrection.

If we consider the shaded region at y in Figure 6.5, we see that this presents a different problem. The mass at y exerts an attraction on the gravimeter mass at B. The vertical component of this attraction reduces observed gravity at B. Because observed gravity is reduced, we must *add* a small amount to observed gravity to negate the effect of y. In both cases the correction factor is added to

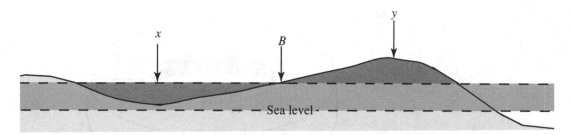

FIGURE 6.5 An illustration of terrain corrections required at regions x and y due to the oversimplified nature of the Bouguer correction applied at point B.

observed gravity if the observation station is above sea level (assuming sea level is the datum being used). These corrections that account for the undulations of topography above and below the elevation level of an observation point are referred to as the *terrain correction*. If TC represents the terrain correction, the Bouguer anomaly formula becomes

$$\Delta g_B = g_{obs} - g_n + FA_{corr} - B_{corr} + TC. \tag{6.29}$$

The question now becomes one of deciding whether a terrain correction is necessary and how to go about the process in an area of irregular topography. Certainly the most widely used method is one proposed by Hammer (1939), which many geophysicists still employ. Hammer's approach considers the gravity effect of a ring (Figure 6.6(a)). If the ring has a thickness z, an inner radius R_i, and an outer radius R_o, the equation expressing the gravitational attraction of the ring at a point at its center and on the same level as the top (or bottom) is

$$g_{ring} = 2\pi G\rho\left[R_o - R_i + (R_i^2 + z^2)^{\frac{1}{2}} - (R_o^2 + z^2)^{\frac{1}{2}}\right]. \tag{6.30}$$

By dividing the ring into an equal number n of compartments or sectors, the attraction of each sector becomes g_{ring}/n. Because we are interested in the effect of a portion of land above (or below) the level of our observation point (Figure 6.6(b)), we treat that segment of land as a ring sector, and z becomes the difference in elevation between the level of the observation point and the top (or bottom) of the ring sector.

Hammer calculated the sizes of ring radii and sectors that give the most accurate results at increasing distances from the gravity observation point in question. In practice a template consisting of concentric circles is drawn on a clear overlay. Each pair of adjacent circles outlines a ring and is referred to as a *zone*. Each zone is divided into varying numbers of sectors. The rings are drawn to the scale of the map using Hammer's original ring radii. The overlay is placed on a topographic map so that the center of the circles is located on the observation point for which observed gravity is being reduced. Figure 6.7 illustrates a simplified

(a)

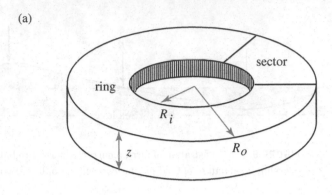

(b)

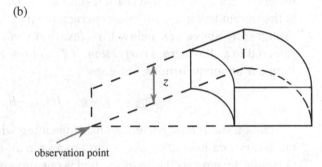

observation point

FIGURE 6.6 The basic approach to representing masses for terrain corrections. (a) A cylindrical ring with inner radius R_i, outer radius R_o, and thickness z. A section of the ring is referred to as a *sector*. (b) A sector represents a topographic feature above the level of a gravity observation point.

template placed on a topographic map with only a few contours shown for clarity. The average elevation in each sector then is noted. For example, in sector H9 the average elevation is about 500 m. The absolute value of the difference between the average elevation and the elevation of the gravity station is computed. Note that sign is not important because the terrain correction always is added to observed gravity. The next step typically is to consult a table to determine the gravitational effect of the sector. We repeat this process for each sector in a ring. Then we add together all sector effects to determine the total terrain correction for the gravity station.

It's obvious that this is a very labor-intensive (and boring) chore. In fact, it is the most time-consuming aspect of obtaining and reducing gravity values. There are various ways to partially automate the process. One method that is useful for our purposes is to employ a spreadsheet instead of a lookup table. Table 6.3 is an example of how this might be done but is oversimplified and would need to be expanded to serve such a purpose. It is useful, however, to work with Table 6.3 to get a sense of the magnitude of terrain corrections for various degrees of topographic relief. Eight of the 12 zones defined by Hammer are included in the table.

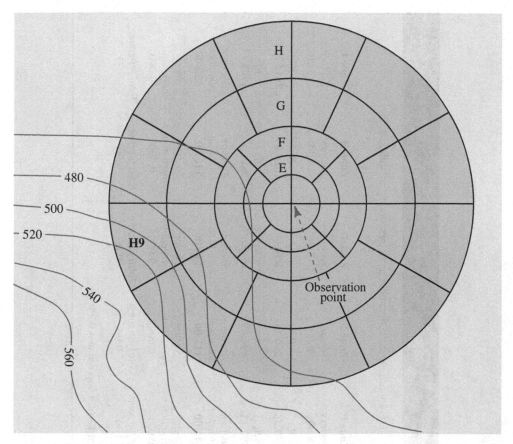

FIGURE 6.7 A terrain correction template placed on a topographic map with the center at a gravity observation point. Most contours are not shown for clarity. Only zones E, F, G, and H are illustrated.

The term *equation constant* refers to $2\pi\rho G$ in Equation 6.30 and is adjusted so the results of calculations are in milliGals. In Table 6.3 the number of sectors and the ring radii are designated for each zone. The elevation of the gravity station is entered in the appropriate cell. Next the average elevation is entered into the designated cell for a sector in the zone in question. The value of the sector correction in milliGals then can be observed.

Usually corrections for Zones B and C are determined in the field. The remainder are determined from topographic maps. For illustration, the identical value for the average elevation in each sector was placed in the spreadsheet to show how a 5 m difference between station elevation and sector elevation relates to the size of the sector correction. As you should realize, the significance of such an elevation difference diminishes rapidly with increasing distance from the gravity station. In Table 6.3 each sector correction is multiplied by the number of sectors in the zone to compute a total terrain correction. This is done just to give

TABLE 6.3 TERRAIN CORRECTIONS

Elevations in meters: distances in meters; gravity in milliGals

Elevation of gravity station 120.00
Equation constant 0.11

	Zone B	Zone C	Zone D	Zone E	Zone F	Zone G	Zone H	Zone I
Sectors	4.00	6.00	6.00	8.00	8.00	12.00	12.00	12.00
R_i	2.00	16.64	53.34	170.08	390.14	894.89	1529.49	2614.57
R_o	16.64	53.34	170.08	390.14	894.89	1529.49	2614.57	4453.74
Average elevation	115.00	115.00	115.00	115.00	115.00	115.00	115.00	115.00
Sector correction	0.074	0.009	0.003	0.001	0.000	0.000	0.000	0.000
Total for zone	0.30	0.06	0.02	0.00	0.00	0.00	0.00	0.00

Total Terrain Correction = 0.38

you a sense of the size of the terrain correction for various topographic relief factors. Of course, in an actual case the average elevation in each sector of a zone would vary, sometimes by quite a lot.

> Approximately what average elevation must be present in each sector of Zone H to result in a total zone correction of 0.25 mGal?

If digitized versions of topographic maps are available for a region in which a gravity survey is taking place, the terrain correction process can be almost wholly automated. One must be sure, however, that the digital elevation product is at a suitable scale for the precision required. In some cases we ignore terrain corrections because of the time involved. Is this justified? It depends. The best approach is to undertake a few corrections in an area to appreciate their magnitudes. Once these are known, they should be compared with the objectives of the survey to determine the effect of neglecting the corrections.

6.3.6   The Isostatic Anomaly

The general flotational equilibrium of lithospheric segments is known as *isostasy*. Most geologists and geophysicists are aware that high mountains have extensive "roots" of lower-density material that extend into the higher-density mantle. The mountains appear to be floating in a manner that is similar to icebergs. This flotational equilibrium is revealed in Bouguer anomalies that are negative over high areas and positive over the oceans. Near sea level the average is close to zero. The negative Bouguer anomalies arise due to the excess of low-density material at depth in mountainous regions when compared to rocks at depth in areas of low relief near sea level.

It is possible to correct for the isostatic effect by assuming a certain compensating mechanism and computing a correction. Residual gravity left after applying an isostatic correction I_{corr} is referred to as the *isostatic anomaly* Δg_I and is expressed as

$$\Delta g_I = g_{obs} - g_n + FA_{corr} - B_{corr} + TC \pm I_{corr}. \qquad (6.31)$$

For exploration purposes the Bouguer anomaly generally is the value used, so the isostatic correction is not applied. If variations in Bouguer anomalies are due to isostatic effects in an exploration area, they will be present as gradual, regional trends due to their deep-seated causes. As such, they may be removed during the process of anomaly separation (discussed in a following section). If so, the isostatic effect is deleted directly as part of a regional trend and not by considering isostasy as a cause and computing a compensating mechanism.

6.4 BASIC FIELD PROCEDURES

Most of the topics in this section were already mentioned during the preceding discussion. The following material provides additional detail and gives insight into how certain field requirements and procedures typically are accomplished. This is not intended to be an exhaustive survey, and you should consult more detailed discussions before commencing a full-scale effort involving many gravity readings. What is presented here is sufficient for beginning efforts involving a small area and a limited number of measurements.

6.4.1 Drift and Tidal Effects

If we placed a gravimeter in one position and took readings every hour or so, the values we obtained would vary. This variation is due to two causes. One is instrument *drift*, which is caused by small changes in the physical constants of gravimeter components. The other is due to tidal effects that are governed by the positions of the sun and moon relative to the Earth.

Due to the extreme sensitivity of gravimeters, instrument drift affects readings to such an extent that it cannot be ignored. Neither can we predict the changes, so we must return a meter to a reference point every so often to establish whether drift occurred and, if so, its magnitude. The simplest technique is to obtain a reading at a selected site, visit several other sites, and then to return to the original site. This reading sequence is referred to as *looping*. Figure 6.8 illustrates the results of such a loop.

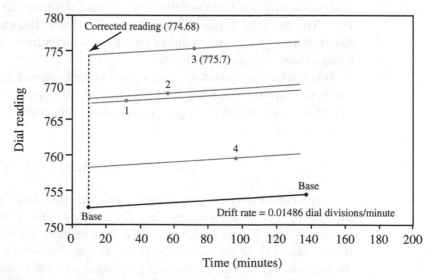

FIGURE 6.8 A typical drift curve for gravity observations at base and stations 1, 2, 3, and 4.

The sequence in the loop is base—1—2—3—4—base. If we return to base every one or two hours, we can assume linear drift. We then establish the drift rate simply by dividing the difference in the initial and final base values by the elapsed time between base station readings. The remaining stations in the loop are adjusted using this drift rate. For example, in Figure 6.8 all stations are referred to the original base reading. The drift rate is illustrated by the solid line connecting the two base readings. Dotted lines parallel to the solid line are drawn through stations 1, 2, 3, and 4 to illustrate how these values should be corrected for the observed drift. Note that at this stage in the process we have not yet changed gravimeter dial readings into gravity values. Because we are dealing only with relative changes in g among stations, we can continue to work with dial divisions if we wish.

Tidal variations produce an effect on gravimeter mass that varies by ± 0.15 milliGal from a mean value and can have a rate of change as high as 0.05 mGal/hour. Because these are substantial values relative to the 0.01 precision of most gravimeters, a correction clearly is called for. Tidal effects can be predicted accurately, so it is relatively straightforward to devise a computer program to produce values for any location at any time. Figure 6.9 is a representative curve for tidal variations produced by such a program (Rudman, Ziegler, and Blakely 1977).

Observe the scale lengths for one- and two-hour durations in Figure 6.9. The tidal variation curve is essentially linear within one-hour segments. Thus if we construct a drift curve such as in Figure 6.8 and return to base every hour, such a curve will remove both the tidal effect and instrument drift with good precision. However, returning to a base station every hour is not only inconvenient

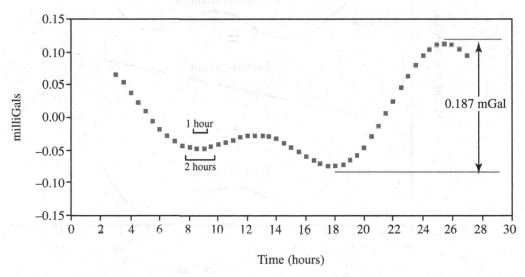

FIGURE 6.9 Variation of gravity values due to tidal effects. This is a representative curve for middle latitudes.

but reduces significantly the number of gravity stations that can be visited. It simply is not very efficient. We can increase the return interval to two hours, but with the help of the two-hour scale in Figure 6.9, it becomes evident that at some times during the tidal cycle our assumption of linear change does not hold. However, due to the relatively small gravity change from tidal effects in a two-hour period, we could employ the standard drift curve technique to correct for both tidal effects and instrument drift with only a slight loss in precision if we remain within the two-hour maximum. Many gravity surveys have used this approach.

If it is impossible or extremely inconvenient to return to a base location every two hours, then it is imperative to employ tidal correction programs. Actually, due to the ready availability of such programs and the present power of personal computers, it seems reasonable to use such programs for all gravity surveys. Figure 6.10 illustrates how gravimeter readings would vary at one position due to both the tidal effect and instrument drift. Given a standard loop consisting of base stations and gravity readings, one first removes the tidal variation effect by the appropriate program. What remains is therefore instrument drift and is removed by the aforementioned procedure.

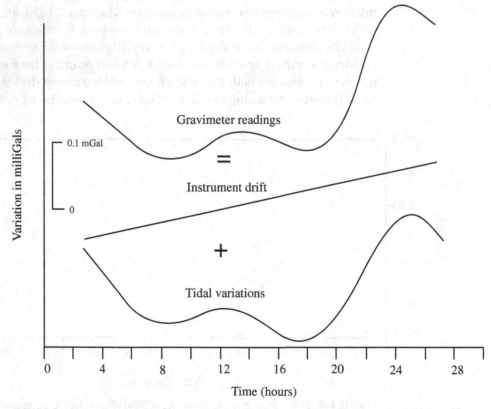

FIGURE 6.10 An illustration of how gravity values measured at a stationary position would vary due to a combination of tidal effects and linear instrument drift.

The question now becomes how often we must return to base to correct for instrument drift because we can predict the tidal effect as far into the future as we wish. The answer depends on instrument characteristics. Some instruments have very stable drift rates, whereas others are less predictable. The best course is to establish the drift behavior of your instrument and then to plan a looping procedure accordingly. We tend to favor applying tidal corrections and returning to base every two to three hours. Although this slows down fieldwork somewhat, we are confident that we are maintaining a high-precision survey. To be possible, this approach demands a number of base stations throughout a survey area. We now turn to the question of how these base stations are established.

6.4.2 Establishing Base Stations

It always is preferable to tie gravity measurements to a location where absolute gravity is known accurately. Therefore, we need to find the closest IGSN71 locality and then relate measurements to this IGSN71 value. Although it is not necessary to do this for limited, local surveys because such surveys seek only variations in gravity, you always should make such a link to add to the research base by making your measurements part of the system that everyone else uses.

Figures 6.11 and 6.12 illustrate the standard procedure. For simplicity Figure 6.12 assumes linear drift rates and time intervals short enough so that tidal variations are included in the drift curves as linear changes. The looping procedure begins at the IGSN71 station. After obtaining a reading there, we occupy the Base 1 position and return to the IGSN71 station. This sequence provides the relative gravity difference between the two stations after we correct for drift. Because we know the absolute gravity value for the IGSN71 station, we now have an absolute gravity value for Base 1. We return to Base 1 and use a similar loop to determine absolute gravity at Base 2. Following this loop, we determine absolute gravity at Base 3. The final operation uses Base 3 to acquire a value for the IGSN71 station. Rarely will this value be the same as the established value, and the difference shows the amount of error in our measurements that provided values for Bases 1 through 3. We hope that the accumulated error is small, so that each base value can be adjusted slightly. One adjustment technique is simply to allocate the error equally among all the bases. For surveys with many such base loops, the best procedure is to tie bases together through many interconnected loops and to allocate error through more sophisticated statistical techniques (Nettleton 1976, 84–87).

6.4.3 Determining Elevations

We've already seen that elevations must be known to within 25–30 cm to maintain Bouguer anomaly values that are accurate to better than 0.1 mGal. Normally, such control can be achieved only by surveying procedures. These typically include using a rod and transit and consume considerable time. Because this increases

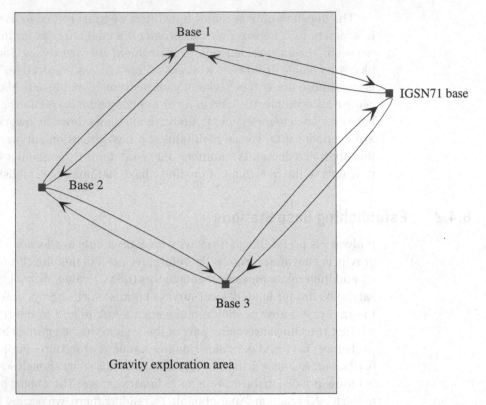

FIGURE 6.11 Establishing absolute gravity stations in an exploration area by referring them to an IGSN71 station and following a looping procedure.

expense, elevation requirements may preclude many gravity surveys that otherwise make sense in terms of acquiring useful information.

Topographic maps show the position of *benchmarks*, which are points where elevations have been measured to a high order of precision. The benchmarks are brass cylinders encased in concrete and are highly reliable as starting points for transit surveys. Other points (referred to as *spot elevations*) indicated on most topographic maps are points where elevations have been determined to the nearest foot (assuming the typical nonmetric map). Even though these points are found at easily identified locations such as road intersections, the exact point is not marked in the field, so considerable error in elevation is possible. For some low-precision surveys it is possible to interpolate between contour lines, but the accuracy of elevations will be no better than half the contour interval. In rugged areas or for regions where topographic control is not available, some surveys use aneroid barometers. Because the barometers are affected by pressure variations, we can use a looping procedure similar to that for measuring relative gravity differences. Such a procedure would be very time-consuming. A better

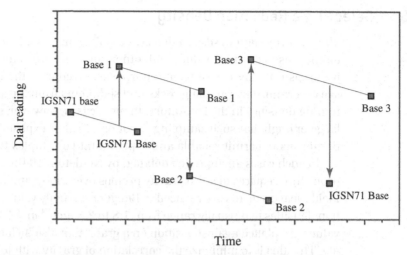

FIGURE 6.12 Drift curves resulting from the process illustrated in Figure 6.11.

solution is to employ a stationary recording altimeter to record pressure variations that occur during a day's work. However, elevations determined in this manner normally are subject to errors of a meter or more.

If you are involved in a gravity survey taking place in a relatively urban environment, be sure to consult with city and town engineering departments. Such departments often have large-scale topographic maps used for civil construction that contain many surveyed elevations possessing the required accuracy. Although tracking these maps down may take several days, the time savings can be enormous.

The best alternative to time-consuming rod-and-transit surveys is the Global Positioning System (GPS). GPS surveying uses satellites to achieve horizontal and vertical controls on the order of 1 centimeter. Although the equipment still is expensive, its speed and ease make this an attractive alternative, and GPS is becoming the dominant approach.

6.4.4 Determining Horizontal Position

Horizontal position usually can be determined accurately enough by employing topographic maps with a scale of 1:25,000 or larger and by careful location of gravity stations. In the absence of suitable computer digitizing capabilities, latitudes are best measured by using a clear plastic overlay on which is inscribed a grid with cells at 0.1-minute intervals. This makes quick work of reading latitudes to 0.01 minute, which results in a Bouguer accuracy on the order of 0.02 mGal. If more accuracy is needed, larger-scale maps must be used, positions must be surveyed from benchmarks, or GPS must be employed.

6.4.5 Selecting a Reduction Density

Many gravity surveys select a density of 2.67 g/cm^3 for the Bouguer reduction. This enables us to combine values with other surveys for production of regional gravity maps but may not suit some purposes. Of course, the best reduction values are determined by sampling rocks and sediments from the survey area and determining densities in the laboratory. In most cases, however, surveys cover an area large enough that such sampling is not practical or exposures are not sufficiently numerous to permit reliable and representative samples to be gathered.

In such cases an approach detailed by Nettleton (1942) can be attempted. This procedure requires detailed gravity profiles over topographic features such as hills. Field measurements are reduced to Bouguer anomaly values using different reduction densities that usually range from 1.8 to 2.8 g/cm^3 in 0.1 intervals. The reduced values are plotted against position on a graph that also includes a topographic profile. The idea is to minimize the correlation of gravity with topography because we are searching for gravity anomalies arising from subsurface density variations. Therefore, the density chosen for the final Bouguer reduction is the one that produces the gravity profile illustrating the least correlation with topography.

The main problem with this approach occurs when topography is correlated with subsurface structure. In such cases we end up removing the anomaly that the survey is intended to locate. With inexpensive computer access and good contouring software and graphics output, the best approach probably is to use several Bouguer reduction densities, including the standard of 2.67. Several contour maps can be prepared, and if we are confident that there is no subsurface control of topography, then the map illustrating the least correlation of topography with gravity values is the one chosen for interpretation.

6.4.6 Survey Procedure

Table 6.4 contains gravity data obtained by an undergraduate geophysics class at Smith College. The purpose of the traverse was an attempt to map bedrock topography under the floodplain of the Connecticut River near Northampton, Massachusetts. The first step in the project was to determine elevations by rod and transit at preselected locations on a 1:24,000 topographic map to obtain relatively equidistant gravity stations. Next a base station position was selected and linked to an IGSN71 location. Gravity stations then were occupied.

At each station the date, time, and dial reading of the gravimeter were recorded. This was repeated until three consecutive readings were within 0.1 dial division to make sure the meter was stable. The base station was occupied every 60 to 90 minutes to allow for drift corrections without separately calculating tidal corrections.

After all stations had been occupied, the data were reduced. Sea level was selected as a datum both because elevations in the study area averaged only about 40 meters above sea level and because the gravity stations would be added

TABLE 6.4 DATA FOR A GRAVITY TRAVERSE ON THE CONNECTICUT RIVER FLOODPLAIN

Station	Latitude (deg. min.)	Elevation (feet)	Dial Reading	Terrain Correction (mGal)	Normal Gravity (mGal)	Observed Gravity (mGal)	Free-Air Anomaly (mGal)	Bouguer Anomaly (mGal)
SC base	42. 19.05	173.65	1603.10	0.23	980,387.63	980,370.56	-0.73	-9.20
SC1	42. 24.77	177.00	1750.32	0.30	980,396.25	980,383.34	3.74	-4.80
SC2	42. 22.85	310.00	1666.90	0.53	980,393.31	980,376.10	11.95	-0.94
SC3	42. 24.16	468.00	1545.17	0.37	980,395.31	980,365.53	14.24	-4.26
SC4	42. 24.70	169.89	1752.25	0.22	980,396.13	980,382.63	2.48	-5.90
SC5	42. 24.54	131.00	1767.15	0.17	980,395.88	980,383.92	0.37	-6.72
SC6	42. 24.62	130.00	1767.78	0.16	980,396.00	980,383.97	0.20	-6.86
SC7	42. 24.62	130.00	1774.10	0.29	980,396.00	980,384.52	0.75	-6.18
SC8	42. 24.50	130.00	1779.99	0.19	980,395.81	980,384.85	1.27	-5.76
SC9	42. 24.41	129.00	1779.25	0.17	980,395.69	980,384.79	1.24	-5.78
SC10	42. 24.65	167.00	1759.14	0.23	980,396.00	980,383.05	2.75	-5.51
SC11	42. 24.45	154.00	1724.35	0.18	980,395.81	980,380.94	-0.38	-8.25
SC12	42. 24.59	158.00	1706.48	0.17	980,395.88	980,379.39	-1.62	-9.64
SC13	42. 24.59	158.00	1702.38	0.24	980,395.88	980,379.04	-1.98	-9.92
SC14	42. 24.58	161.00	1711.49	0.19	980,395.88	980,379.83	-0.90	-9.00
SC15	42. 24.61	314.00	1679.86	0.25	980,396.00	980,376.37	9.91	-3.41
SC16	42. 24.57	310.00	1683.79	0.32	980,395.88	980,376.72	10.00	-3.11

to a regional database. Because of the short time loops, tidal variations were assumed to be linear. Therefore, standard drift curves were used to correct for both instrument drift and tidal effects. Terrain corrections were calculated by overlaying templates on topographic maps of the region. The Bouguer reduction density was 2.67 g/cm^3. Because the Bouguer anomaly values in Table 6.4 show some substantial variations, we will attempt to interpret them in a later section.

6.5 GRAVITY EFFECTS OF SIMPLE GEOMETRIC SHAPES

We can now measure gravity and apply corrections to remove many factors that cause gravity to vary. Ideally, we are left with values (Bouguer anomalies) that are due to density variations within the subsurface. Our ultimate goal is to explain these anomalous values by invoking valid models of the subsurface. The first step toward this goal is to document the gravity effects of simple shapes and to thoroughly understand these effects before proceeding to a more advanced analysis.

6.5.1   Rock Densities

Before we address these gravity effects, it makes sense to spend a moment considering rock densities because these will figure prominently in our analysis. The total variation in rock densities is quite small relative to the other physical properties we have discussed, and, as in the case of electrical resistivities, considerable overlap exists. The values in Table 6.5 consist of a compilation of density ranges from several sources. These should be used only in a very general way, such as answering some of the questions we ask when considering the gravity effects of various models. Clearly you will need to establish much more precise values for any area in which you work.

The bulk density of rocks and sediments is controlled by the densities of the minerals present, the amount of open space in the rock or sediment, and the degree to which fluids fill these spaces. If you examine the densities for some of the common minerals in Table 6.5, the densities of the rocks listed make sense. Use Table 6.5 to experiment with various combinations of density and porosity. For example, if we select a clay mineral with a density of 2.7 g/cm^3 and gradually bury it to greater and greater depths so that the porosity is reduced from 80 percent to 40 percent to 0 percent, we in effect observe a density change from mud to shale to metamorphic rock. The equation utilized in Table 6.5 to calculate the density of saturated material is

$$\rho_{sm} = \rho_{min}(1 - p_\% / 100) + p_\% / 100 \tag{6.32}$$

where ρ_{sm} is the density of saturated material, ρ_{min} is the density of the mineral of which the material is composed, and $p_\%$ is the percentage of porosity.

TABLE 6.5	DENSITIES OF COMMON GEOLOGIC MATERIALS

Material	Density (g/cm³)
Quartz	2.65
Feldspars	2.5–2.7
Micas	2.8
Calcite	2.72
Clay minerals	2.5–2.8
Unconsolidated sediments	1.7–2.3
Sandstones	2.0–2.6
Shales	2.0–2.7
Limestones	2.5–2.8
Dolomites	2.3–2.9
Granitic rocks	2.5–2.8
Basaltic rocks	2.7–3.1
Metamorphic rocks	2.6–3.0

Density of mineral:	**2.7**
Percentage porosity:	**50**
Density of saturated material:	1.85

6.5.2 Gravity Effect of a Sphere

Due to its simplicity we begin our analysis with the sphere. Although we admit that few subsurface forms are spherical, the effect of a roughly equidimensional ore body is quite similar to that of a sphere, so our initial model is not entirely inappropriate. Because we seek to explain Bouguer anomalies in terms of density variations, we speak in terms of the *density contrast* ρ_c when considering gravity effects of models. The density contrast is the density of the model minus the density of the remaining material, which we assume to be homogeneous. Thus a sphere with density of 2.0 g/cm³ in material with a density of 2.6 g/cm³ has a density contrast of –0.6 g/cm³.

Derivation of an equation for the gravity effect of a sphere is relatively straightforward because the effect is the same as if all the mass is concentrated at the sphere's center. Figure 6.13 illustrates the pertinent notation. Depth to the center of the sphere is z, sphere radius is R, distance from the observation point to the sphere's center is r, and g_z represents the vertical component of gravity due to the attraction of the sphere. The point $x = 0$ on the surface is directly over the center of the sphere. Given a density contrast ρ_c, the excess mass due to the sphere is equal to $4\pi R^3 \rho_c/3$. Using Equation 6.4 we see that the gravitational attraction of the sphere in the direction of r is

$$g_{sphere(r)} = \frac{Gm}{r^2} = \frac{G4\pi R^3 \rho_c}{3r^2} = \frac{G4\pi R^3 \rho_c}{3(x^2 + z^2)}. \qquad (6.33)$$

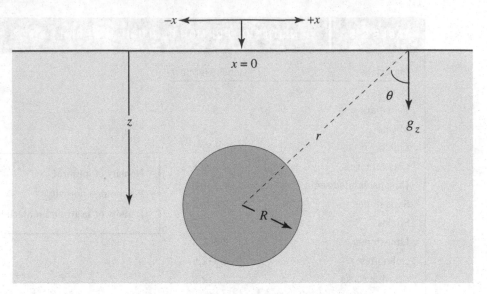

FIGURE 6.13 Notation used in deriving the gravity effect of a buried sphere. The same notation is used for a traverse at right angle to the strike of a horizontal, infinitely long cylinder.

We seek the vertical component of $g_{sphere(r)}$ because gravimeters respond only to gravity in the vertical direction. In general, unless noted otherwise, the component in the vertical direction g_z is implied. Therefore,

$$g_{sphere} = \frac{G4\pi R^3 \rho_c}{3(x^2+z^2)}\cos\theta = \frac{G4\pi R^3 \rho_c}{3}\frac{z}{(x^2+z^2)^{\frac{3}{2}}}. \tag{6.34}$$

Now that we have the required equation, let's study some attributes of gravity over a buried sphere. Table 6.6 uses Equation 6.34 and allows us to directly compare gravity over two spheres so we can vary parameters in each and observe the differences. In case you are interested in making similar computations for other equations we present, Equation 6.35 illustrates units conversion so g is in milliGals when density is in g/cm^3, distances are in meters, and G is in dyne·cm²/g².

$$g(\text{milliGals}) = 6.67\times10^{-8}\frac{\text{dyne}\cdot\text{cm}^2}{g^2}\cdot\frac{g}{cm^3}\cdot\frac{m}{1}\cdot\left(\frac{g\cdot cm\Big/s^2}{\text{dyne}}\cdot\frac{100\ cm}{m}\cdot\frac{\text{Gal}}{cm\Big/s^2}\cdot\frac{1000\ \text{mGal}}{\text{Gal}}\right) \tag{6.35}$$

As you probably would predict after studying Figure 6.13 and Equation 6.34, maximum gravity is directly above the sphere's center, and the Bouguer anomaly curve is symmetric about this position. Especially note the form of the two curves in Figure 6.14(a) that represent the values in Table 6.6. The curve for the shallower sphere not only possesses higher gravity values, as one would expect, but is much sharper in form. Gravity values decrease at a much greater rate

TABLE 6.6	GRAVITY OVER A SPHERE		

	Sphere A		Sphere B
Sphere radius (m)	200	Sphere radius (m)	200
Depth to sphere center	500	Depth to sphere center	1000
Density contrast (g/cm^3)	0.4	Density contrast (g/cm^3)	0.4
Horizontal increment	100	Horizontal increment	100

Horizontal Position (m)	Gravity (mGal)	Horizontal Position (m)	Gravity (mGal)
−1200	0.0203	−1200	0.0235
−1100	0.0253	−1100	0.0272
−1000	0.0320	−1000	0.0316
−900	0.0410	−900	0.0367
−800	0.0532	−800	0.0426
−700	0.0702	−700	0.0492
−600	0.0938	−600	0.0564
−500	0.1264	−500	0.0640
−400	0.1703	−400	0.0716
−300	0.2255	−300	0.0786
−200	0.2862	−200	0.0843
−100	0.3372	−100	0.0881
0	0.3576	0	0.0894
100	0.3372	100	0.0881
200	0.2862	200	0.0843
300	0.2255	300	0.0786
400	0.1703	400	0.0716
500	0.1264	500	0.0640
600	0.0938	600	0.0564
700	0.0702	700	0.0492
800	0.0532	800	0.0426
900	0.0410	900	0.0367
1000	0.0320	1000	0.0316
1100	0.0253	1100	0.0272
1200	0.0203	1200	0.0235

(a)

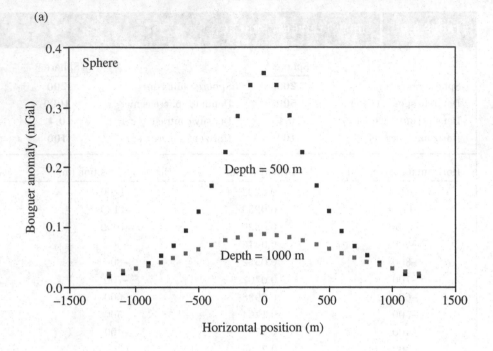

(b)

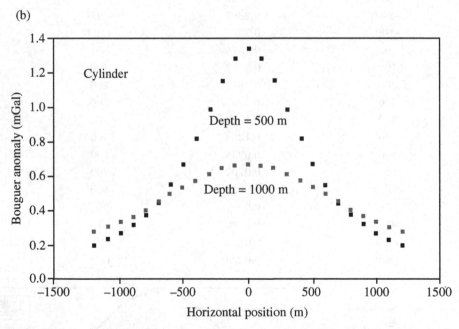

FIGURE 6.14 Bouguer gravity anomalies over a sphere and a horizontal cylinder. (a) Anomaly curves for spheres with centers at depths of 500 m and 1000 m. (b) Anomaly curves for horizontal cylinders with centers at depths of 500 m and 1000 m.

along a traverse extending away from a sphere's center for the shallower sphere than for the deeper sphere. Later we will use this characteristic to estimate depths to sources of certain anomalies.

Also note that we could change the radius in one sphere in Table 6.6, and if we change the density contrast in the other sphere by an appropriate amount, gravity values would be identical. This is obvious from the equation for the sphere because the gravity effect is due to the excess mass of the sphere. Even if we know the depth of the sphere, we cannot tell its size unless we also know the density contrast very accurately (usually we don't). For the examples in Table 6.6 we use a positive density contrast. If we substitute negative values and keep all other parameters constant, the Bouguer anomaly curves in Figure 6.14(a) will have exactly the same shape but will be inverted.

If you use Table 6.6 as a dynamic table, it is instructive to vary parameters, predict the results, and then graph the table values to study curve shapes and the magnitude of Bouguer values.

6.5.3 Gravity Effect of a Horizontal Cylinder

It is not too difficult to envision geologic structures that are relatively equidimensional in cross section and are many times longer than wide. In such cases we can model the gravity effect by employing a horizontal cylinder with infinite length. Even though geologic forms do not possess infinite length, the gravity effect diminishes so quickly with distance from an observation point that the infinite length of our model provides a sufficient representation for many actual subsurface configurations.

Fortunately, we already have done most of the work in developing an equation for the horizontal cylinder. Equation 6.20 is the equation for a thin rod that was an intermediate result during our derivation of an equation for the Bouguer correction, and we adopt that form for our not-so-thin cylinder. We can use the same notation as for the sphere, so we only need to pretend that Figure 6.13 is a cross section through a horizontal cylinder. The section is oriented perpendicular to the axis of the cylinder. Because the area of the circle in Figure 6.13 equals πR^2, substituting into Equation 6.20 gives us, for gravity in the direction of r,

$$g_{cylinder(r)} = \frac{G 2\pi R^2 \rho_c}{r} .$$ (6.36)

Thus the equation for the vertical component becomes

$$g_{cylinder} = \frac{G 2\pi R^2 \rho_c}{r} \cos\theta = \frac{G 2\pi R^2 \rho_c z}{r^2}$$

and

$$g_{cylinder} = G 2\pi R^2 \rho_c \frac{z}{(x^2 + z^2)} .$$ (6.37)

TABLE 6.7   GRAVITY OVER A HORIZONTAL CYLINDER

	Cylinder A		Cylinder B
Cylinder radius (m)	**200**	Cylinder radius (m)	**200**
Depth to cylinder center (m)	**500**	Depth to cylinder center (m)	**1000**
Density contrast (g/cm³)	**0.4**	Density contrast (g/cm³)	**0.4**
Horizontal increment	**100**	Horizontal increment	**100**

Horizontal Position (m)	Gravity (mGal)	Horizontal Position (m)	Gravity (mGal)
−1200	0.1984	−1200	0.2748
−1100	0.2296	−1100	0.3034
−1000	0.2682	−1000	0.3353
−900	0.3163	−900	0.3705
−800	0.3767	−800	0.4089
−700	0.4531	−700	0.4500
−600	0.5496	−600	0.4930
−500	0.6705	−500	0.5364
−400	0.8177	−400	0.5781
−300	0.9861	−300	0.6152
−200	1.1561	−200	0.6448
−100	1.2895	−100	0.6639
0	1.3411	0	0.6705
100	1.2895	100	0.6639
200	1.1561	200	0.6448
300	0.9861	300	0.6152
400	0.8177	400	0.5781
500	0.6705	500	0.5364
600	0.5496	600	0.4930
700	0.4531	700	0.4500
800	0.3767	800	0.4089
900	0.3163	900	0.3705
1000	0.2682	1000	0.3353
1100	0.2296	1100	0.3034
1200	0.1984	1200	0.2748

Table 6.7 uses Equation 6.37 to calculate the gravity effect of a horizontal cylinder. Because the equations for the sphere and cylinder are similar in many respects, you might predict that the anomaly curves would be similar also. A quick glance at Figure 6.14(b), which is a graph of the values in Table 6.7, would prove you correct for the curve form. The major difference is the greater magnitude of the gravity values for the cylinder because much more mass is present in this case than if we select the same radius for the sphere and the cylinder.

Most other relationships remain the same, such as broader and more attenuated curves as the cylinder is buried to greater depths. This might be a good time to explore another practical use of these equations and tables. Let's assume we decide to explore for a buried ore deposit and that the lower half of a horizontal cylinder is an appropriate preliminary model. Bedrock density is 2.3 g/cm³, and the ore has a density of 3.0 g/cm³. The precision of the Bouguer anomaly values measured by the survey will be 0.1 mGal but no better. Seismic information suggests that ore depths are on the order of 100 m. How large will the deposit have to be before we can reliably detect it? Figure 6.15 is a plot of results for cylinders of various radii. Gravity values for each curve represent one-half the values for the respective cylinder. Real values would be lower still because our crude assumptions essentially are representing the deposit by the lower half of a cylinder. Therefore, a deposit would have to possess a "radius" of at least 40 m to be detected. Although admittedly somewhat crude, the real point of this exercise is to demonstrate how even simple models can give us some idea of the size of mass excesses we are likely to detect under stated conditions.

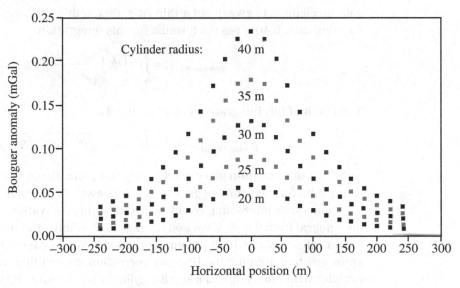

FIGURE 6.15 Bouguer anomaly curves for horizontal cylinders with various radii.

6.5.4 Gravity Effect of a Vertical Cylinder

Another useful model is the vertical cylinder. Assume we want to determine gravity directly above the axis of the cylinder (at point P in Figure 6.16). We can treat this derivation in a similar fashion as our Bouguer derivation. We select a small-volume element on the upper margin of the cylinder with volume equal to $drdzrd\phi$. The gravitational attraction of this volume element at point P (first in the direction s and then the vertical component) is

$$\Delta g_{vertical\ cylinder(s)} = \frac{G\rho_c dz(rd\phi)dr}{s^2},$$ (6.38)

$$\Delta g_{vertical\ cylinder} = \frac{G\rho_c dz(rd\phi)dr}{s^2}\cos\theta = \frac{G\rho_c dz(rd\phi)dr}{s^3}z,$$

and

$$\Delta g_{vertical\ cylinder} = \frac{G\rho_c(zdz)(rdr)d\phi}{(r^2+z^2)^{\frac{3}{2}}}.$$ (6.39)

Next we determine the gravitational attraction of the cylinder by using the volume element to sweep out a thin ring, then a thin disk, and finally the cylinder. Equation 6.40 gives the formula for this integration.

$$g_{vertical\ cylinder} = \int_{h_1}^{h_2}dz\int_0^R dr\int_0^{2\pi}\frac{G\rho_c(z)(r)}{(r^2+z^2)^{\frac{3}{2}}}d\phi.$$ (6.40)

The result of this integration is Equation 6.41.

$$g_{vertical\ cylinder} = G2\pi\rho_c\left(h_2 - h_1 + \sqrt{R^2+h_1^2} - \sqrt{R^2+h_2^2}\right).$$ (6.41)

Note that this equation gives gravity only at a point above the cylinder axis and is not valid for determining gravity at points away from the axis.

Some other interesting results arise from this derivation. When evaluating the integral from 0 to R, if we assume R goes to infinity, the result is an equation for an infinite slab (the Bouguer correction). Hence we could have used this approach to determine the Bouguer correction. In addition, if we have a large cylinder with radius R_o and a smaller cylinder (with radius R_i) with an axis coincident with the larger cylinder and with coincident top and bottom, we can deter-

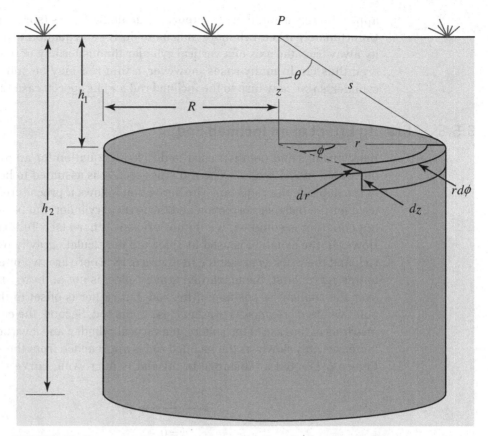

FIGURE 6.16 Notation used to derive the gravity effect of a vertical cylinder at a point P over its axis.

mine the effect of a ring by subtracting the effect of the smaller cylinder from the larger. Let P be on the top of the cylinder so that $h_1 = 0$, and let the bottom of the cylinder be a distance z from the top. Then

$$g_{ring} = G2\pi\rho_c\left[\left(z + \sqrt{R_o^2} - \sqrt{R_o^2 + z^2}\right) - \left(z + \sqrt{R_i^2} - \sqrt{R_i^2 + z^2}\right)\right]$$

and

$$g_{ring} = G2\pi\rho_c\left(R_o - R_i - \sqrt{R_o^2 + z^2} + \sqrt{R_i^2 + z^2}\right). \tag{6.42}$$

Equation 6.42 is the *terrain correction*

Although the vertical cylinder equation is useful in finding the maximum gravity associated with such a form (which is over its axis) and for demonstrating how the terrain correction equation can be developed, we often will want to know the gravity curves for traverses across vertical cylinders because this geometry approximates a number of important geologic forms (such as salt domes). Unfortunately, it is much more difficult to derive an equation to determine gravity away from the axis of a vertical cylinder than is for any of the forms considered thus far. In many cases, however, a thin rod may be substituted for the cylinder, so we now turn to the inclined rod and its special case, the vertical rod.

6.5.5 Gravity Effect of an Inclined Rod

Relationships and notation used to derive an equation for an inclined rod are illustrated in Figure 6.17. The rod cross section is assumed to be small relative to the depth to the rod's top. The derivation follows a procedure similar to that used for the Bouguer correction and the vertical cylinder; but because the resulting equation is complex, we do not present it here (see Telford et al. 1976). However, the equation is used in Table 6.8 to calculate gravity over an inclined rod, and the result is presented in Figure 6.18. There are two important aspects to this curve. First, the maximum gravity effect is not at the $x = 0$ point directly over the shallowest position of the rod, but rather is offset to the right of this point where more mass is present close to the rod. Second, the curve is not symmetric as in the case of the sphere, the vertical cylinder, and a vertical rod. Gravity increases only slowly as the inclined rod is approached from the $-x$ direction (in Figure 6.17); but as distance to the rod is narrowed, curve slope increases

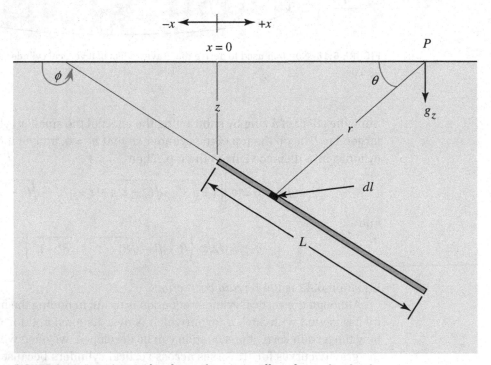

FIGURE 6.17 Notation used to derive the gravity effect of an inclined rod.

TABLE 6.8	GRAVITY OVER INCLINED AND VERTICAL RODS

	Inclined Rod		Vertical Rod
Rod radius (m)	5	Rod radius (m)	5
Depth to rod top (m)	10	Depth to rod top (m)	10
Rod length (m)	100	Rod length (m)	100
Rod inclination (degree)	135	Rod inclination (degree)	**********
Density contrast (g/cm³)	0.4	Density contrast (g/cm³)	0.4
Horizontal increment	2	Horizontal increment	2

Horizontal Position (m)	Gravity (mGal)	Horizontal Position (m)	Gravity (mGal)
−24	0.002	−24	0.006
−22	0.002	−22	0.007
−20	0.002	−20	0.007
−18	0.003	−18	0.008
−16	0.003	−16	0.009
−14	0.003	−14	0.010
−12	0.004	−12	0.012
−10	0.004	−10	0.013
−8	0.005	−8	0.014
−6	0.006	−6	0.016
−4	0.007	−4	0.018
−2	0.008	−2	0.019
0	0.009	0	0.019
2	0.011	2	0.019
4	0.013	4	0.018
6	0.015	6	0.016
8	0.017	8	0.014
10	0.019	10	0.013
12	0.021	12	0.012
14	0.021	14	0.010
16	0.021	16	0.009
18	0.020	18	0.008
20	0.019	20	0.007
22	0.018	22	0.007
24	0.017	24	0.006

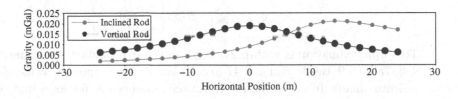

markedly. After the curve maximum is passed, curve slope is less than that immediately before the maximum.

In the case of a vertical rod (inclination = 90°), the equation becomes less complex and is

$$g_{vertical\ rod} = G\rho_c \pi R^2 \left\{ \frac{1}{(z^2+x^2)^{\frac{1}{2}}} - \frac{1}{[(z+L)^2+x^2]^{\frac{1}{2}}} \right\} \tag{6.43}$$

where πR^2 is the cross-sectional area of the rod. Example values for the same parameters as an inclined rod (except for inclination!) are presented in Table 6.8. Telford et al. (1976, 59) note that the vertical rod gives off-axis values within 6% of a vertical cylinder when the diameter is less than the depth to the top.

So far we have a few basic shapes we can use as models to provide information about the gravitational effects of simple geologic structures. However, one shape that is very common in the subsurface and for which we do not yet have a model is a sheetlike form.

6.5.6   Gravity Effect of a Horizontal Sheet

When deriving the Bouguer effect, we created a thin slab by moving a thin rod from $+\infty$ to $-\infty$ (integrating from $\pi/2$ to $-\pi/2$; see Equation 6.23). If instead we integrate from θ_1 to θ_2 (see Figure 6.19(a)), we have

$$g = \int_{x=\theta_1}^{x=\theta_2} 2G(\rho dz)d\theta \tag{6.44}$$

and

$$g = 2G\rho_c(dz\bar{\theta}) \tag{6.45}$$

where $\bar{\theta}$ represents the included angle in radians. If we keep the slab thin with respect to its depth, we can write

$$g_{truncated\ slab} = 2G\rho_c t(\bar{\theta}) \tag{6.46}$$

where t is the thickness of the truncated slab (Figure 6.19(a)). Our goal in this section is to derive a relationship for a semi-infinite slab (Figure 6.19(b)). For this case the included angle $\bar{\theta}$ is equal to $\pi/2 + \tan^{-1}(x/z)$, and therefore we have

$$g_{semi-\infty\ sheet} = 2G\rho_c t \left(\frac{\pi}{2} + \tan^{-1} \frac{x}{z} \right). \tag{6.47}$$

This approximation is within 2% if $z \geq t$ (Telford, Geldart, and Sheriff 1990, 40). Table 6.9 uses Equation 6.47 to calculate Bouguer anomaly values for semi-infinite sheets. In this table the input parameters are for extremely shallow and thin sheets. Note that gravity values for such a model are quite small and

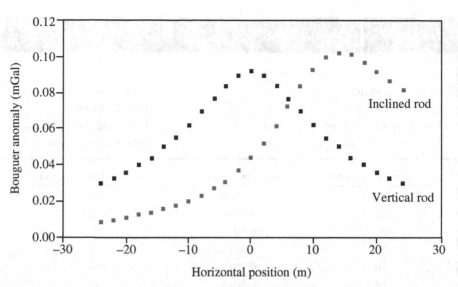

FIGURE 6.18 Bouguer anomalies over a thin vertical rod and a thin rod inclined at 135° (dipping at 45°). See Table 6.8 for the rods' parameters.

(a)

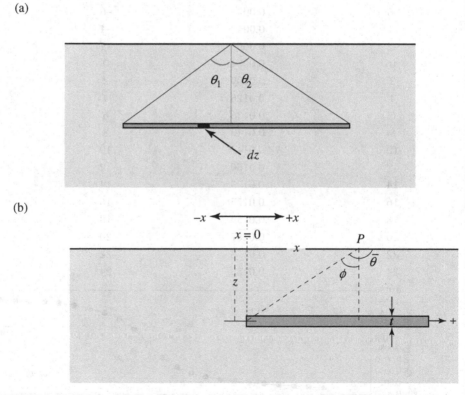

(b)

FIGURE 6.19 Relationships and notation for derivation of semi-infinite sheet relationships. (a) Diagram for a thin, truncated slab. (b) Diagram for a semi-infinite sheet.

TABLE 6.9	GRAVITY OVER SEMI-INFINITE SHEETS

	Sheet A		Sheet B
Depth to sheet (m)	4	Depth to sheet (m)	8
Sheet thickness (m)	1	Sheet thickness (m)	1
Density contrast (g/cm^3)	0.4	Density contrast (g/cm^3)	0.4
Horizontal increment	2	Horizontal increment	2

Horizontal Position (m)	Gravity (mGal)	Horizontal Position (m)	Gravity (mGal)
−24	0.0009	−24	0.0150
−22	0.0010	−22	0.0149
−20	0.0011	−20	0.0147
−18	0.0012	−18	0.0145
−16	0.0013	−16	0.0143
−14	0.0015	−14	0.0140
−12	0.0017	−12	0.0136
−10	0.0020	−10	0.0132
−8	0.0025	−8	0.0126
−6	0.0031	−6	0.0118
−4	0.0042	−4	0.0109
−2	0.0059	−2	0.0097
0	0.0084	0	0.0084
2	0.0109	2	0.0071
4	0.0126	4	0.0059
6	0.0136	6	0.0049
8	0.0143	8	0.0042
10	0.0147	10	0.0036
12	0.0150	12	0.0031
14	0.0153	14	0.0028
16	0.0155	16	0.0025
18	0.0156	18	0.0022
20	0.0157	20	0.0020
22	0.0158	22	0.0019
24	0.0159	24	0.0017

that these sheets could not be detected with the normal accuracy of gravity surveys. Nevertheless, the curve shapes still are valid, and we are interested primarily in curve shapes and relationships at this point.

Study the thin, semi-infinite sheet pictured in Figure 6.19(b) and try to predict the shape of a Bouguer anomaly curve for such a body. As you probably realize, such a curve has a maximum value above the sheet and far removed from the edge location ($+\infty$) and a minimum value of zero far removed from the edge in the opposite direction ($-\infty$) (see the curve for sheet A in Figure 6.20(b)). The value of gravity directly above the edge of the sheet is half the maximum value. These relations become quite clear if we allow x to assume very large positive values and very large negative values in the quantity ($\pi/2 + \tan^{-1} x/z$) from Equation 6.47. The semi-infinite sheet model also can be used to simulate a horizontal bed truncated by a high-angle fault where the displacement effectively removed the other half of the bed from consideration. Other models, such as a sill termination, are equally plausible.

By a minor bit of manipulation we also can use Equation 6.47 to compute the effect of a vertical fault cutting a horizontal sheet where both displaced sheet segments are present. We simply compute the effect of each sheet as illustrated in Figure 6.20(a) using Equation 6.47. Then, because we have to orient the sheets in opposite directions to simulate a faulted bed, we reverse the sense of the x axes for one sheet and add the effects of both sheets. This is the scheme used in Table 6.9. Computations for each sheet are independent, but the signs of the *horizontal position* values are reversed. Therefore we can add the values in the *gravity* columns directly if we want to use the table to simulate a faulted sheet. The results in Table 6.9 are graphed in Figure 6.20(b). Here we see the curve for sheet A (assuming sheet B is not present), sheet B (assuming sheet A is not present), and sheets A and B (the sum of curves A and B for the geometry in Figure 6.20(a)). The purpose of this exercise is to demonstrate again how we can devise models using the simple shapes already at our disposal.

Although interesting, the vertical fault model is somewhat limited. It is possible to derive an equation for a similar geometry to the semi-infinite sheet but with the added flexibility of varying fault angle. We do not derive the equation here but will use it to investigate some important relationships. A derivation can be found in Telford, Geldart, and Sheriff (1990, 40–43). The equation for the faulted, horizontal sheet is

$$g_{faulted\ sheet} = 2G\rho_c t\left[\pi + \tan^{-1}\left(\frac{x}{z_1} + \cot\phi\right) - \tan^{-1}\left(\frac{x}{z_2} + \cot\phi\right)\right] \qquad (6.48)$$

where x is the horizontal position of the measurement location, t is sheet thickness, ϕ is fault inclination, z_1 is depth to upthrown block, and z_2 is depth to downthrown block. Note that the relative positions of the faulted sheet are constant (that is, the upthrown block is always toward the $+x$ coordinate axis). Table 6.10 is based on Equation 6.48. Figure 6.21(b) illustrates Bouguer gravity curves for

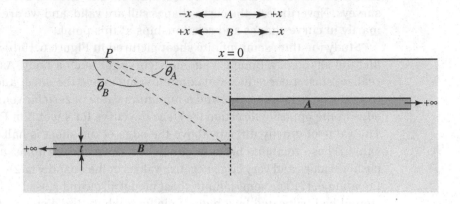

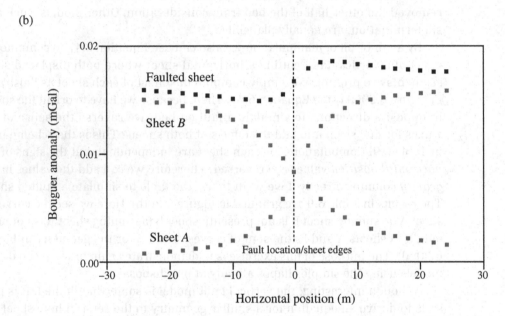

FIGURE 6.20 Using semi-infinite sheets to simulate a faulted horizontal bed. (a) Notation used in Table 6.9 to derive gravity values. (b) Graph illustrating effects of sheet A, sheet B, and sum of A and B (the faulted relationship shown in (a)).

fault inclinations of 60°, 90°, and 120° as defined in Figure 6.21(a) and calculated by Table 6.10.

The input parameters in Table 6.10 are the same as in Table 6.9 (fault inclination of 90°). Note that the sums of the *gravity* columns in Table 6.9 equal the values in the *gravity* column in Table 6.10, as should be the case because both are computing the same model. Let's use the gravity curve for the vertical fault to compare fault inclinations of 60° and 120°. A ϕ value of 120° models a high-angle

TABLE 6.10	GRAVITY OVER A FAULTED HORIZONTAL SHEET

Horizontal Position (m)	Gravity (mGal)
−24	0.0159
−22	0.0159
−20	0.0158
−18	0.0157
−16	0.0156
−14	0.0155
−12	0.0153
−10	0.0152
−8	0.0150
−6	0.0150
−4	0.0150
−2	0.0156
0	0.0168
2	0.0179
4	0.0185
6	0.0186
8	0.0185
10	0.0183
12	0.0182
14	0.0180
16	0.0179
18	0.0178
20	0.0177
22	0.0177
24	0.0176

Depth to downthrown sheet (m)	**8**
Depth to upthrown sheet (m)	**4**
Sheet thickness (m)	**1**
Fault inclination (degrees)	**90**
Density contrast (g/cm³)	**0.4**
Horizontal increment (m)	**2**

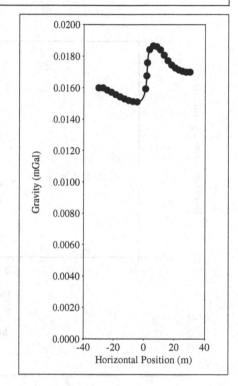

reverse fault (Figure 6.21(a)). This geometry produces an overlapping of the sheets so that double the excess mass is present in the overlap zone. This doubling accounts for the pronounced maximum on the 120° curve (Figure 6.21(b)) and the greater gravity values for this curve. In contrast, a ϕ value of 60° (Figure 6.21(c)) models a high-angle normal fault, which creates a gap where no sheet is present. This mass deficiency leads to the distinct minimum on the 60° curve and accounts for the lower gravity values of this curve (as compared to the 90° and 120° curves).

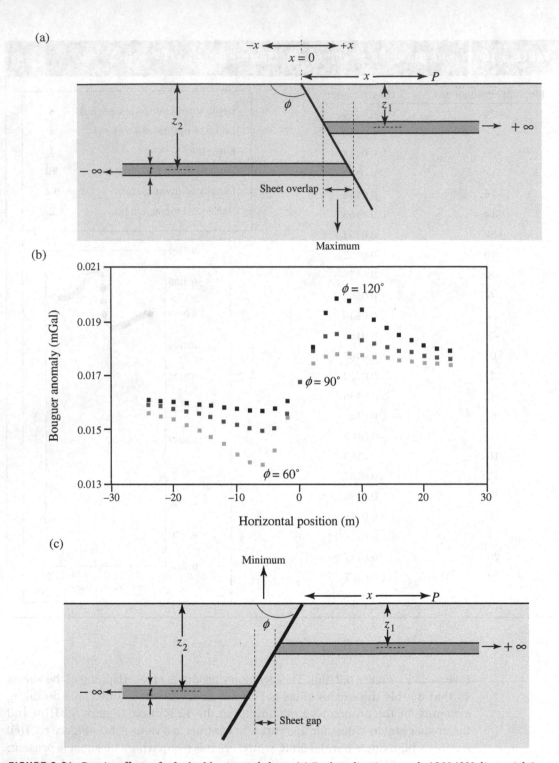

FIGURE 6.21 Gravity effects of a faulted horizontal sheet. (a) Fault inclination equals 120° (60° dip to right). (b) Bouguer anomalies for fault inclinations of 60°, 90°, and 120°. (c) Fault inclination equals 60° (60° dip to left).

This concludes our analysis of the gravity effect of simple shapes. We could continue with numerous other shapes, but this does not seem warranted. Our goal in this section is to acquaint you with the general effect for a number of different shapes, to demonstrate that more complex shapes can be constructed from combining simple shapes, and to start you thinking about issues such as detectability of some models. Before considering some quite different issues in gravity exploration, we introduce a computer program that permits you to compute the gravity effect of some reasonably complicated models.

6.5.7 GRAVMAG

GRAVMAG calculates the gravity effect for solids of infinite strike length and polygonal cross section. Because each polygon can have up to 20 vertices and several polygons can comprise a model, subsurface models of substantial complexity are possible. We do not derive here the equations used in the program, but the fundamental equation is derived following a procedure much like that we employed to determine the Bouguer correction, the semi-infinite sheet, and the vertical cylinder. If we have a small cell of infinite length in the y dimension with side lengths dz and $d\theta$, we can easily determine the gravity effect of such a cell by integration. Thus we have

$$g_{cell} = 2G\rho_c \int_{\theta_1}^{\theta_2} d\theta \int_{z_1}^{z_2} dz \qquad (6.49)$$

or

$$g_{cell} = 2G\rho_c (\theta_2 - \theta_1)(z_2 - z_1). \qquad (6.50)$$

The notation is defined in Figure 6.22(a). All we need now do is to represent a polygon such as that illustrated by Figure 6.22(b) as an accumulation of such cells. If a polygon has n such cells, the gravity effect of the polygon is

$$g_{polygon} = 2G\rho_c n\Delta\theta\Delta z. \qquad (6.51)$$

The main challenge is that of carrying out this summation given the coordinates of a polygon. A solution to this problem was published by Hubbert (1948) and was adapted for computer use by Talwani, Worzel, and Landsman (1959). Because the mathematics are fairly complex, the solution and computational steps are not given here.

Instructions for using GRAVMAG are in Appendix D. If you can work with GRAVMAG, attempt a number of polygons in various configurations. Always try to predict the rough form of the anomaly curve before computation begins. Be sure you understand why the curve has the shape it does. Remember that we are dealing with polygonal cross sections of bodies with infinite strike length. We can therefore simulate geologic forms that have roughly constant cross-sectional

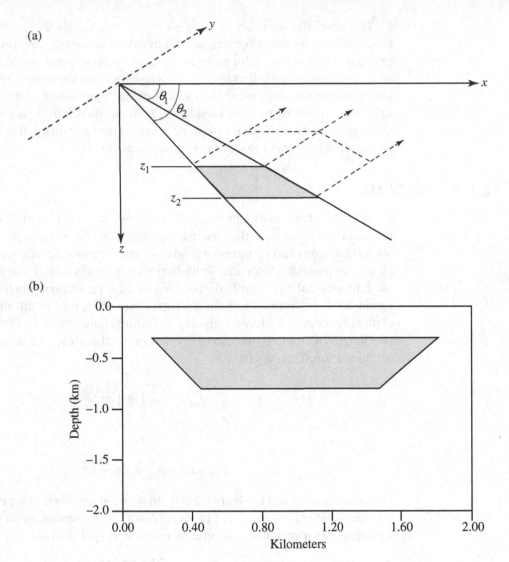

FIGURE 6.22 Calculating gravity anomalies due to irregular shapes using GRAVMAG.
(a) Notation used to develop the fundamental equation. (b) An example of a simple polygonal
shape drawn in GRAVMAG.

shape along their length and have lengths that are several times their cross-
sectional width.

Other solutions are available to model shapes that do not fit within the constraints of the Hubbert–Talwani approach. Three-dimensional modeling programs let us model essentially any three-dimensional shape and do not impose

constraints such as infinite strike lengths on the model. The computational load of complex three-dimensional modeling algorithms is easily handled by large computers and is not beyond personal computers. The theoretical approach is not all that different from the one used by GRAVMAG, but, of course, substantially more input data are required to define forms being modeled. If your mathematical background is good and you are interested in learning more about three-dimensional modeling, consult Plouff (1976).

6.6 ANALYZING ANOMALIES

Now that we are prepared to begin analysis of Bouguer anomaly curves, let's pause to emphasize that we face a somewhat different situation than in the case of seismic and resistivity data. When looking at seismic data, we knew that the horizontal distance traversed by recorded wave energy was restricted to the length of the geophone spread and that the depth sampled was controlled by the shot–receiver distances, the magnitude of the energy source, and the subsurface geometry. Similar controls existed in resistivity surveying.

On the other hand, when measuring gravity, we know that the value recorded contains contributions from sources of interest at shallow depths as well as contributions from much deeper sources. Even mass excesses at considerable horizontal distances from an observation point affect measured values. Thus deeper, larger sources may mask or obscure sources of interest. This section examines strategies for dealing with this problem.

6.6.1 Regionals and Residuals

Larger features produce Bouguer anomalies that are smooth over considerable distances. These smooth trends usually are referred to as *regional trends* or simply as *regionals*. Smaller, more local sources account for sharper anomaly shapes of more restricted areal extent. If regionals are removed from gravity data, the remaining or residual values are related to these local disturbances. These are referred to as *residual anomalies* or *residuals*. Of course, what is a regional and what is a residual depends on the exploration target. Once again we find that the goals of a survey must be firmly set before data are acquired. Station density must be sufficient to completely define residual anomalies and must be extensive enough to correctly describe regionals.

How can we determine regionals? The oldest method essentially involves constructing a series of profiles at regular intervals across a contoured Bouguer anomaly map. Each profile is a plot of Bouguer anomaly values against distance. A smoothly varying regional then is defined using as much geologic information as possible. Once defined, the regional trends along each profile are subtracted from the Bouguer values along the same profile. This produces a collection of residual values that can once again be contoured to define the form of the residual anomalies. The procedure necessarily is quite interpretative because regional definition depends on the skill and knowledge of the geophysicist. However, when adequate geologic information and interpreter skill are available, the method can be quite useful.

As a simple and somewhat contrived example of this procedure, let's confine our analysis to a single profile (this is usually inappropriate in the real world). For illustration we also will work with subsurface geology that we create. Figure 6.23 represents a model defined using GRAVMAG. Resulting Bouguer anomaly values are plotted as small crosses (Curve 1) in Figure 6.24(a). The model consists of a buried river channel filled with sands and gravels, a prism of sedimentary rock that thickens to the east, and a crystalline basement that is present beneath the sedimentary sequence and abuts against the sedimentary rocks on both east and west margins. Our target is the buried river channel (density contrast = –0.4), so the regional in this case will be due to the sedimentary rocks (density contrast = –0.2). We assume we know enough about the geology of the area represented by the model to be aware of the general thickening of the sedimentary rocks to the east. Therefore, the smooth decline of Bouguer values to the east reflects this thickening and is subtracted to obtain residual values.

Because we are using GRAVMAG to produce these curves, we simply subtract the effect of the sedimentary rocks, which is defined by the gray squares (Curve 2) in Figure 6.24(b). The residual (Curve 3) clearly illustrates the small decline in gravity values due to the effect of the buried channel. Of course, the geology in this example is quite simple (and we know what it is!). The general regional trend (the west–east decline in Bouguer values) is quite obvious, but the increase in values at 19 and 20 km (Figure 6.24(a)) might be difficult to recognize as part of the regional. This increase is due to the termination of the sedimentary sequence at 21 km in the model. Outcrops of crystalline basement at 21 km would explain the increase

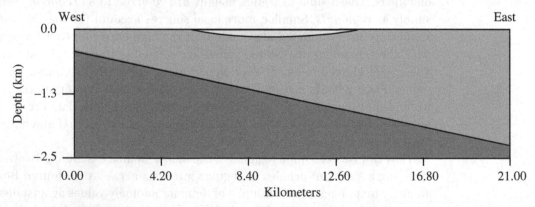

FIGURE 6.23 Model used for regional–residual discussion. A buried river channel is incised into a sedimentary rock sequence, which overlies a crystalline basement.

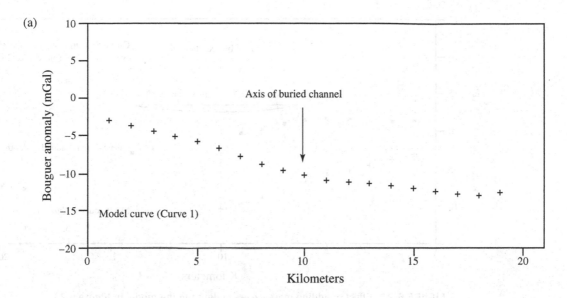

(a)

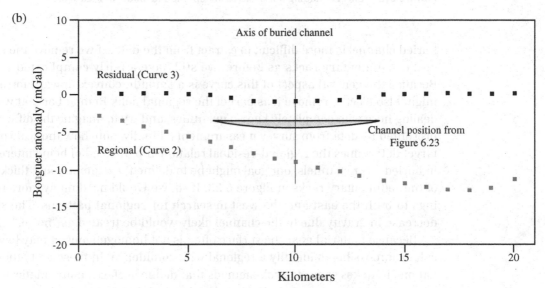

(b)

FIGURE 6.24 (a) Bouguer anomaly values calculated for the model in Figure 6.23.
(b) Regional and residual curves based on geology in Figure 6.23.

in values to a knowledgeable geophysicist, and these increases then could be includod in the regional. This simple example illustrates the importance of both geologic information and interpreter skill in appropriately identifying regionals.

Let's add another complication to our model. Assume that a domelike mass (density contrast = 0.4) is present at depth. This added mass excess would complicate our previous curve by warping it as illustrated by Curve 4 in Figure 6.25. Identifying an adequate regional now is more difficult, and the presence of the

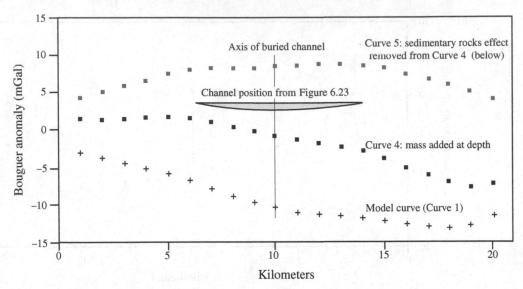

FIGURE 6.25 Effect of adding mass excess at depth to the model in Figure 6.23.

buried channel is more difficult to extract from the data. If we remove the effect of the sedimentary rocks as before, we still have a fairly complicated curve. Because the general aspect of this curve is a smooth, convex-upward form, we might also elect to remove it as part of the regional field. Remember that we are dealing here with completely known quantities, and try to imagine the difficulties presented by data from survey measurements. Finally, note how the scale of the target determines the regional–residual relationship. Instead of being interested in buried river channels, our goal might be to define the geometry and thickness of the sedimentary rocks in Figure 6.23. If so, we would need many more readings to both the east and the west to search for regional patterns. The small decrease in gravity due to the channel likely would be treated as "noise."

Because material near the surface often is not homogeneous, it may be difficult or impossible to identify a regional with confidence. In these and other situations it makes sense to seek methods that define surfaces mathematically.

6.6.2 Trend Surfaces

Recall that our goal is to determine regional trends in our data. One approach is to generate *trend surfaces*. These are mathematical surfaces that represent a surface of best fit to data points distributed in x–y–z space. Such surfaces are defined by polynomial functions, and the surface is fit to the data points using the method of least squares. Usually surfaces of various orders (a first-order surface is a plane, a second-order surface is a paraboloid, and so on) are computed so we can select one (or more) to represent the regional trend. Once we choose a trend surface to represent the regional, we subtract it from the Bouguer anomaly map to produce a residual map. The major deficiency of using trend surfaces to define regionals is that the former are mathematically defined and therefore take no geologic knowledge into account. This also could be regarded as a plus because no bias enters the determination. For good discussions of trend surfaces consult Robinson (1982, 72–81) and Maslyn (1984, 12–19).

The relation of trend surface to gravity data initially is somewhat difficult to visualize in three dimensions, so let's continue our string of two-dimensional examples. Figure 6.26 illustrates the Bouguer anomaly curve (Curve 1) that we've examined previously. Three curves are fit to this data: a first-order polynomial (straight line) and third- and fifth-order polynomials. The first-order polynomial curve provides a good generalized representation of the regional field. If these values were subtracted from Curve 1, the anomaly effect of the buried river channel would be well delineated. Substantial residuals also would be present near the beginning and end of Curve 1. A third-order polynomial gives a much better fit to the Bouguer data but would not define the residual anomaly due to the channel as well as the first-order curve. This residual anomaly still would be preserved, however, and the other residuals would be reduced in intensity. A fifth-order curve fits the data so well that the buried channel anomaly would all but disappear as a residual.

Typically surfaces of order greater than four are not used to define regionals. In our two-dimensional example a little knowledge of local geology probably would allow us to define the true regional more precisely than the trend-surface approach. However, Bouguer maps are considerably more complex and regionals are much more difficult to define, so trend surfaces may be the best approach if the Bouguer map is reasonably irregular. Figure 6.27 illustrates a Bouguer map of moderate complexity for two 7½-minute quadrangles in west-central Massachusetts. Figure 6.28 is a third-order, trend-surface map computed from the data points used to produce Figure 6.27. A strong north-to-south decline in Bouguer values is quite well defined by this surface and provides a good indication of the regional for this area.

6.6.3 Upward and Downward Continuation

In the next chapter, which addresses magnetic surveying, we note that both gravity and magnetic fields are potential fields and explore some consequences of this property. Because potential fields are continuous, it is possible to determine mathematically how a given potential field would appear if it was measured from a different level. Therefore, we can project Bouguer anomaly values to various levels to enhance both regional anomalies and local anomalies.

Upward continuation projects a Bouguer anomaly map to a higher level, which results in a map illustrating how the gravity field would appear if we measured it from a higher level. If we measured the field from a higher level (greater elevation),

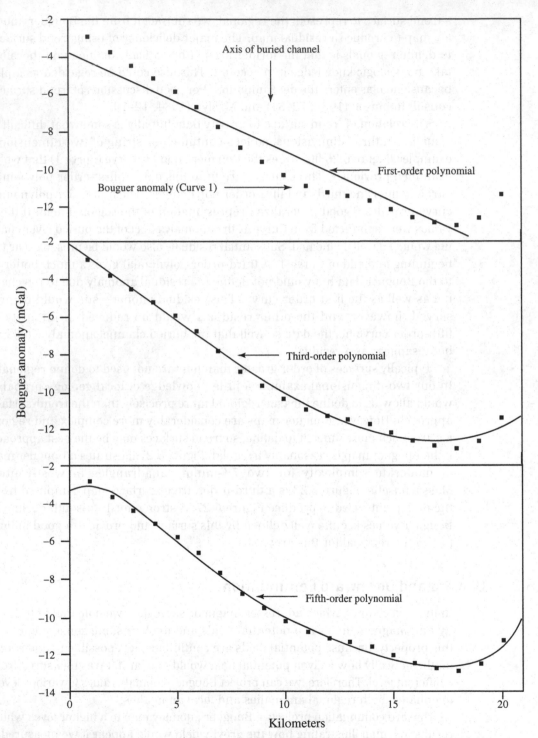

Axis of buried channel

First-order polynomial

Bouguer anomaly (Curve 1)

Third-order polynomial

Fifth-order polynomial

Bouguer anomaly (mGal)

Kilometers

FIGURE 6.26 First-, third-, and fifth-order polynomials fit to Bouguer anomaly data.

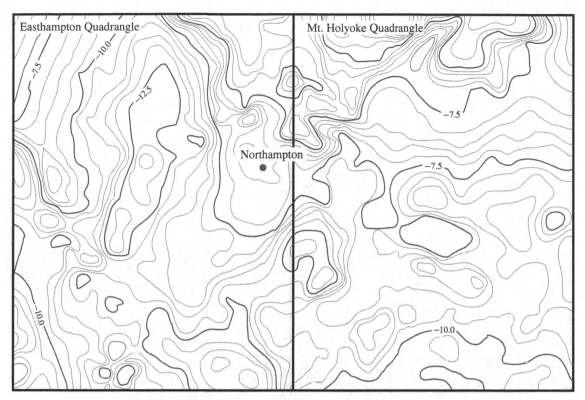

FIGURE 6.27 Bouguer anomaly map for the Easthampton and Mt. Holyoke Quadrangles, west-central Massachusetts. Contour interval = 0.5 mGal.

local, shallow sources would be suppressed and the regional field enhanced. Such a procedure provides a good idea of the regional anomaly for an area. We can simulate the effect of upward continuation by selecting the model in Figure 6.23 and moving it to greater depths. If we move the whole model one kilometer deeper, the effect is the same as if we moved our instruments to a plane one kilometer higher. In both cases we increase the distance from observational position to model coordinates.

Figure 6.29 portrays the results of increasing this distance. Once again the Bouguer anomaly curve for the model in Figure 6.23 is illustrated. Study the curve labeled "Up 2 km." Note that it is virtually identical in shape to the "true" regional as defined by Curve 2 in Figure 6.24 and that no trace of the anomaly due to the buried channel is present. If we used the "Up 2 km" curve to define the shape of the regional in Curve 1, a strong residual anomaly for the buried channel would result. Figure 6.30 is an upward continuation map for the Bouguer anomaly map (Figure 6.27) previously presented. Once again considerable smoothing reveals a pattern very similar to that produced by a third-order trend surface (Figure 6.28),

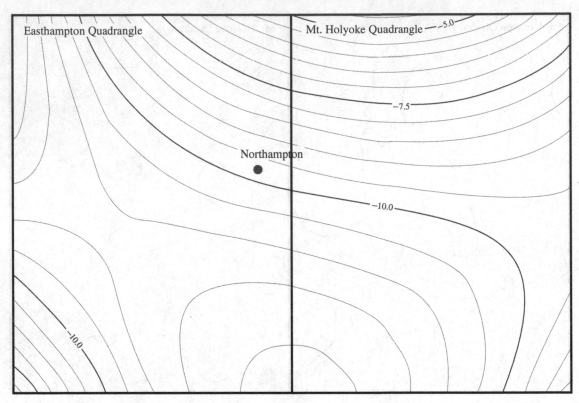

FIGURE 6.28 Third-order, trend-surface map derived from Figure 6.27. Contour interval = 0.5 mGal.

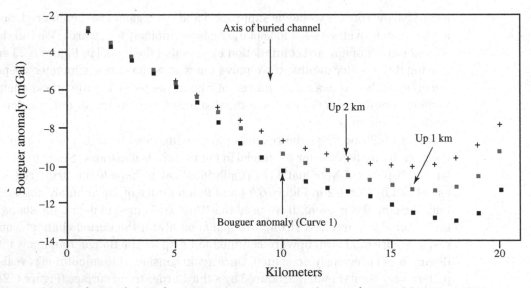

FIGURE 6.29 Simulation of upward continuation. Bouguer anomaly curve from model in Figure 6.23.

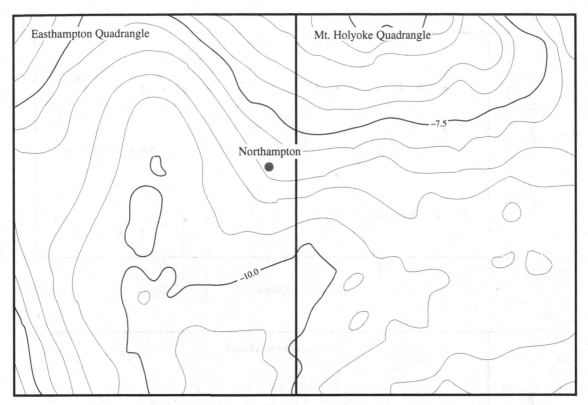

FIGURE 6.30 Upward continuation map (1.5 km) derived from Figure 6.27. Contour interval = 0.5 mGal. Compare the pattern with that in Figure 6.28.

which strongly reinforces the shape of this area's regional. In this case the regional is due to a pronounced southward, as well as eastward, thickening of Triassic and Jurassic sedimentary rocks in the Hartford Basin. The gross appearance of this rift basin is not too different from the sediment thickness pattern in our original model (Figure 6.23).

Downward continuation is employed less frequently. It effectively brings the observation position closer to shallow sources, thereby enhancing them and subduing the anomalies related to deeper sources. This procedure sometimes is effective in outlining the sources of shallow anomalies but can lead to anomalous and wildly fluctuating values if the observation level is placed deeper than the top of the source.

6.6.4 Second Derivatives

The first derivative is a measure of slope. The second derivative provides the rate of change of slope, which therefore measures the curvature of the gravity field. Areas where the curvature of the field is greatest will have the greatest second

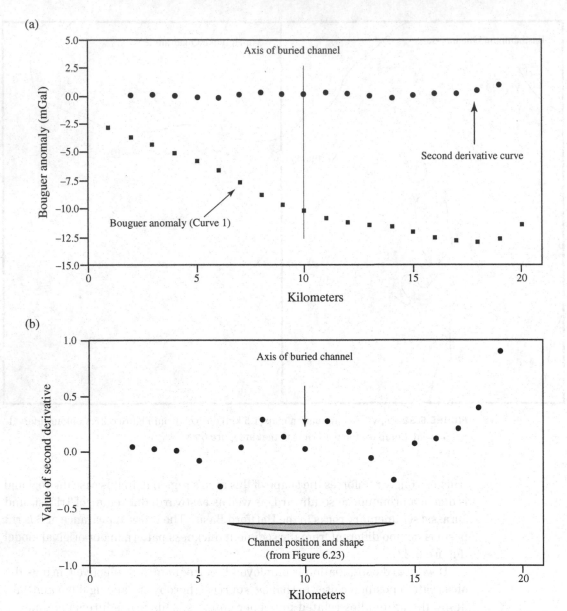

FIGURE 6.31 Example of using second derivatives to remove regional effects and emphasize local anomalies. (a) Bouguer anomaly curve from the model in Figure 6.23 with second derivative curve superimposed. (b) Second derivative values plotted on an expanded scale to emphasize value differences.

derivative values. This characteristic is useful for identifying the positions of small, shallow anomalies and suppressing regionals. When using second derivative maps, it is important to remember that this is a strictly mathematical procedure and that relatively small changes in the gravity field can produce significant second derivative values. Although second derivative maps therefore must be used with care, they can bring to attention small anomalies that normally are hidden by strong regionals.

To provide an example of second derivative use, we have calculated second derivative values for the Bouguer anomaly curve (Curve 1) that has functioned as our standard in the last few sections. Figure 6.31(a) illustrates this curve once again as well as containing an overlay of second derivative values. Note that these values are constant for areas of constant slope and show the most variation when slope changes quickly. Figure 6.31(b) is a more detailed presentation of these second derivative values. The position of the buried channel is superimposed on this graph in its correct spatial position to demonstrate how effectively the second derivative approach "extracts" this small anomaly from the strong regional. It's reasonably easy to visualize how well the position of this channel would be outlined on a second derivative map.

6.6.5 Filtering

A gravity anomaly can be represented by summing an assemblage of cyclic curves that have different wavelengths, amplitudes, and starting positions. Once such an assemblage is generated for a given anomaly, wavelengths of interest can be enhanced and others suppressed by subtracting from the assemblage those we wish to suppress. This process is referred to as *wavelength filtering*. Clearly filtering should be effective in enhancing anomalies due to local, shallow sources (short wavelengths) by removing regional effects (long wavelengths) or vice versa. Filtering is applicable to gravity maps as well as gravity profiles.

6.7 GRAVITY INTERPRETATION

Let's assume that we have conducted a gravity survey and are ready to interpret the data. How do we proceed? If we defined our objectives carefully, our data have good areal distribution and sufficient observational density to satisfy survey goals. The data are properly reduced and have the required accuracy. Regionals, if present, have been removed. Sufficient information about rock densities is available, and we have all relevant geologic information at our disposal. Normally we then use a modeling program such as GRAVMAG or one that accepts three-dimensional models. A model is constructed that fits the known geology as closely as possible. We compare the Bouguer anomaly values calculated by the model with measured values to evaluate the model. If agreement is not good, we alter the model, again following constraints imposed by geologic information or other geophysical information, to improve the agreement between the model and observed values.

When an acceptable fit is achieved between the model curve and the Bouguer anomaly curve, one possible geologic configuration is in hand. Note that, just as in the case of electrical resistivity, many different models can produce similar gravity values. There are no unique solutions. However, if sufficient geologic and geophysical controls are available, a good model fit likely provides a fairly close approximation to the actual subsurface geology. As geologic controls diminish, the number of possible geometries that will produce acceptable curves increases rapidly. A number of techniques are available to help narrow the range of possible choices for model parameters. In this section we describe two approaches to give you a sense of what is possible.

6.7.1   Half-Maximum Technique

In our previous discussion of gravity over a buried sphere, we discovered that maximum gravity occurs directly over the center of the sphere and that there seems to be a relationship between the depth of the sphere's center and the relationship between maximum gravity and anomaly width. Let's see if we can develop a quantitative expression that directly utilizes these relationships.

Because gravity is at a maximum over a sphere's center, we can write (using Equation 6.34)

$$g_{max_{sphere}} = \frac{G4\pi R^3 \rho_c}{3} \frac{z}{(z^2)^{\frac{3}{2}}} \tag{6.52}$$

because $x = 0$ over the sphere's center according to our coordinate system convention. At some horizontal position gravity declines to one-half its maximum value. If we use the notation $x_{1/2max}$ to refer to this special horizontal position, then the value of one-half the maximum gravity of the sphere is given by Equation 6.34 at the special position $x_{1/2max}$. One-half the maximum also is given by one-half the value of Equation 6.52. Therefore, we can equate these two statements, such that

$$g_{\frac{1}{2}max_{sphere}} = \frac{1}{2}\left(\frac{G4\pi R^3 \rho_c}{3} \frac{z}{(z^2)^{\frac{3}{2}}} \right) = \frac{G4\pi R^3 \rho_c}{3} \frac{z}{(z^2 + x^2_{\frac{1}{2}max})^{\frac{3}{2}}}. \tag{6.53}$$

Next we extract from Equation 6.53 the desired relationship between depth z and $x_{1/2max}$:

$$\frac{1}{2}\left(\frac{1}{(z^2)^{\frac{3}{2}}} \right) = \frac{1}{(z^2 + x^2_{\frac{1}{2}max})^{\frac{3}{2}}}$$

and

$$z = 1.305 x_{\frac{1}{2}max}. \tag{6.54}$$

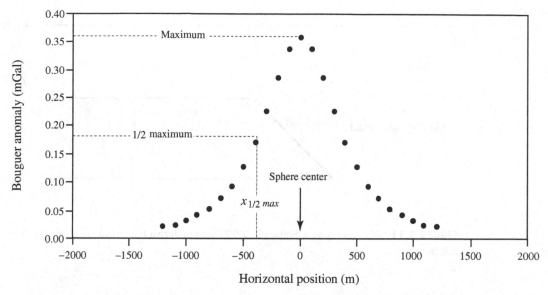

FIGURE 6.32 Diagram of gravity anomaly over a sphere illustrating locations of parameters used to determine depth to sphere's center.

Let's check our results on the data in Table 6.6 (Sphere A), which are diagrammed in Figure 6.32. Maximum gravity is 0.3576 mGal, so half that value is 0.1788 mGal. Locating this value as best we can given the unit divisions in Figure 6.32, we find that this value occurs at a horizontal position of 380 m ($x_{1/2max}$) from the sphere's center at $x = 0$. Substituting 380 m into Equation 6.54 produces a value of 496 m for the depth to the sphere's center. This compares well with the value of 500 m that was used in Table 6.6, considering that the horizontal and vertical scale coordinates in Figure 6.32 cannot be read very accurately.

If we have a gravity anomaly with a form suggesting the source is roughly equidimensional, then Equation 6.54 provides a reasonably good value with which to begin our modeling exercise. If an anomaly suggests a source that can be approximated by a horizontal cylinder, we can use

$$z = x_{\frac{1}{2}max} .$$ (6.55)

Similar expressions can be developed for other simple shapes (Telford, Geldart, and Sheriff 1990). Although most other half-maximum expressions are not as simple as Equations 6.54 and 6.55, they are valuable in providing estimates of source depths.

6.7.2   Second Derivative Techniques

Based on our application of the second derivative when discussing the regional–residual problem, it should not be surprising that we can use this technique to

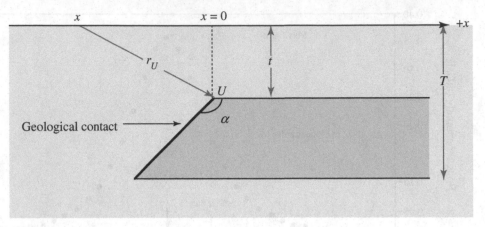

FIGURE 6.33 Notation used for Stanley's (1977) second derivative technique over a geologic contact.

derive information about the anomaly source itself. Although a number of second derivative techniques exist, we concentrate here on an approach developed by Stanley (1977) to present a specific example of this particular method. A truncated plate can represent a geologic contact if the depth to the top of the plate t is small relative to the depth to the bottom of the plate T (Figure 6.33). In such a case the equation for the second derivative is

$$\frac{d^2 g_{truncated\, plate}}{dx^2} = \frac{2Gp_c \sin\alpha (t\cos\alpha - x\sin\alpha)}{r^2{}_U}.$$ (6.56)

Stanley demonstrated that second derivative curves over such forms have several characteristic points that can be used to solve for contact parameters. These points are the maximum value g''_{max}, which is at the horizontal position x_{max}, the minimum value g''_{min}, which is at x_{min}, and $x = 0$, which is directly above the point U in Figure 6.33. This latter point is defined by the intersection of the second derivative curve and a line joining g''_{max} and g''_{min}. Note that x_{max} and x_{min} are defined relative to the point $x = 0$. The dip of the contact α and depth t (Figure 6.33) are related to these parameters by the following equations (Stanley 1977, 1233):

$$\frac{(g''_{max} + g''_{min})}{(g''_{max} - g''_{min})} = \cos\alpha$$ (6.57)

and

$$x_{max} - x_{min} = \frac{-2t}{\sin\alpha}.$$ (6.58)

An example of Stanley's technique is presented in Figure 6.34. The gravity data from which this graph is constructed were collected by Chandler (1978, 147) during a traverse across a fault on the eastern margin of the Deerfield

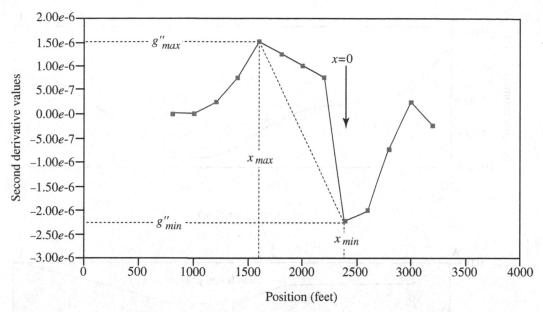

FIGURE 6.34 Second derivative curve for a gravity traverse across a fault block near Amherst, Massachusetts.

Basin (a Mesozoic rift valley) in west-central Massachusetts. The graph establishes that $g''_{max} = 1.50 \times 10^{-6}$, $g''_{min} = -2.25 \times 10^{-6}$, $x_{max} = -800$, and $x_{min} = 0$. Note that $x_{max} = -800$ feet is based on the location of $x = 0$ on the horizontal axis (see Figure 6.34). Placing these values in Equations 6.57 and 6.58 yields $\alpha = 101.5°$ (dip = 78°) and depth $t = 408$ feet (124 m). These results provide a good starting basis for using a program similar to GRAVMAG in an attempt to develop a model of subsurface geology for this region.

6.7.3 Revisiting Some Bouguer Anomaly Values

Earlier in this chapter we presented a sample of gravity data (Table 6.4) when discussing gravity survey procedures. These data are graphed in Figure 6.35(a). The purpose of the traverse was to determine bedrock depths beneath the Connecticut River floodplain. Seismic data are available at two points along the western portion of the traverse and limit bedrock depths to no greater than 100 m. Mesozoic sedimentary rocks constitute basement along the extent of the traverse but thin rapidly to the east and west. Paleozoic metamorphic and igneous rocks are exposed at the surface approximately 1 km from both ends of the traverse. Density contrasts are –0.5 g/cm³ for the sediments above bedrock and –0.2 g/cm³ for the Mesozoic sedimentary rocks.

Figure 6.35(b) illustrates one possible model that provides gravity values that are quite close to the observed Bouguer anomaly values. The model follows all

(a)

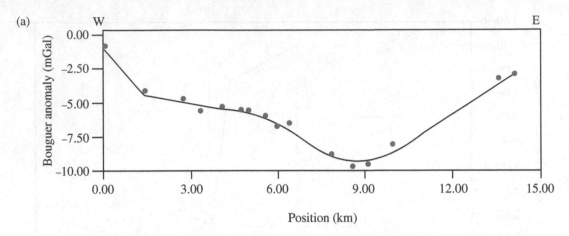

(b)

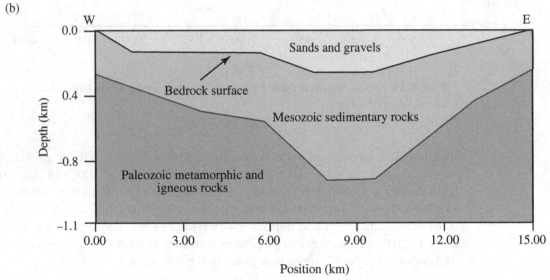

FIGURE 6.35 Gravity values (•) obtained across a buried river channel on the floodplain of the Connecticut River near North Hatfield, Massachusetts. (a) Measured values (•) compared with values (continuous curve) calculated for the model in (b).

known geologic and geophysical constraints and provides a reasonable solution. However, it is important to realize that it is not a unique solution. For instance, errors in the selected density contrasts would affect the size of the model polygons significantly. If the Mesozoic rocks beneath the central portion of the traverse are thicker than modeled, and assuming the density contrasts are correct, then the bedrock surface must be much shallower. Conversely, thinner Mesozoic rocks would necessitate thicker sands and gravels. For simplicity we assume that the Paleozoic rocks have an average density of 2.67 g/cm³ (the Bouguer reduction density). If this assumption is in error, this clearly will affect the present model. If anything, the Paleozoics are likely to have a positive density contrast, which would require increased thicknesses of sediments and/or sedimentary rocks to keep the same general model gravity values.

The purpose of this final presentation simply is to emphasize again the nonuniqueness of a model that provides gravity values that match observed values and to point out how crucial additional independent evidence is to create solutions that most likely are similar to actual subsurface geology.

6.8 APPLICATIONS OF THE GRAVITY METHOD

One of the most common applications of gravity surveys in shallow exploration is mapping bedrock depth. Knowledge of bedrock depth is required in many engineering endeavors as well as in efforts to define aquifer geometries for planning groundwater use or identifying new groundwater sources. Other applications involving detailed gravity work using modern gravimeters include detection of sinkholes and other subsurface voids and establishing landfill boundaries. Recent efforts have been successful in demonstrating gravity variations due to removal of groundwater during pumping tests. The examples we present here represent several of these typical applications.

6.8.1 Bedrock Depths

Kick (1985) describes a survey concerned with bedrock depths in an account of exploration work in the urban setting of Reading, Massachusetts. The objective of the study was to determine depths to bedrock along two proposed routes for a tunnel needed to alleviate sewage transport problems. Engineering design studies required detailed bedrock depths because the tunnel needed to be sited in bedrock rather than in the overlying soil, which would result in much higher costs.

Drilling was not feasible due to both the urban environment and the highly irregular bedrock surface. Normally seismic refraction (or electrical resistivity) would be the method of choice, but the local environment imposed widely paved areas, high noise levels, frequent obstructions such as buildings and fences, and underground pipe complexes. Therefore, gravity was selected because it would cause minimum disturbance to the surroundings and would be relatively unaffected by the urban factors. Readings could be obtained even inside buildings!

Detailed control was established by siting gravity stations at 40-foot intervals. In addition, stations were occupied well beyond the survey boundaries to adequately define regional variations. Careful surveying ensured vertical control to ± 0.1 foot and horizontal control to ± 4 feet. Corrections for instrument drift and terrain effects were calculated carefully. Such attention to survey methods ensured Bouguer values accurate to ± 0.02 mGal. After Bouguer gravity was in hand,

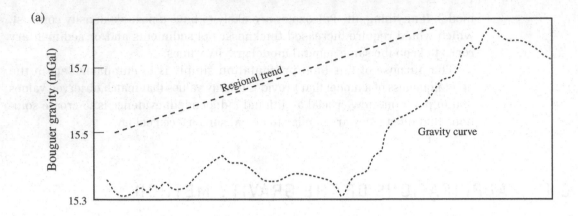

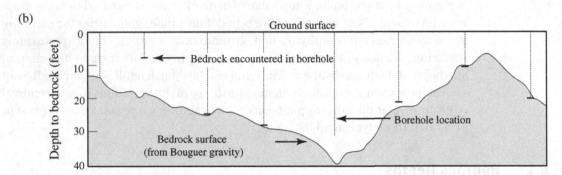

FIGURE 6.36 Bouguer gravity and depth to bedrock along a street in Reading, Massachusetts. (a) Bouguer anomaly curve. (b) Bedrock surface derived from Bouguer gravity. Depths at which wells encountered bedrock are shown by horizontal marks at well bottoms. Note that depths from Bouguer gravity are very close to those in the wells except at one location. (Modified from John F. Kick. 1985. Depth to bedrock using gravimetry. *Geophysics: The leading edge of exploration.* Society of Exploration Geophysicists, v. 4, 40.)

regionals were removed by manual smoothing. Also, tests indicated that bedrock density varied little within the study area, so residuals were believed to be solely a function of soil thickness. The results of this survey are illustrated in Figure 6.36. Although most traverses don't have the degree of borehole control illustrated, this traverse was chosen to demonstrate the remarkably good fit (with one exception) between bedrock depths calculated from Bouguer gravity and those confirmed by borehole information.

6.8.2 Subsurface Voids

A study by Butler (1984) at Medford Cave, Marion County, Florida, demonstrates how a carefully designed gravity survey can detect subsurface voids. This work utilized a LaCoste–Romberg gravimeter to obtain 420 readings spread over an approximately 80 m by 80 m area. Standard tidal and drift corrections were applied to the data as were Bouguer and terrain corrections. A planar regional field was determined by inspection and removed. Resulting residual gravity anomalies correlate well with features of the known cave system. Figure 6.37(a) illustrates a profile plotted from the residual map compared to a geologic section (Figure 6.37(b)) along a traverse containing numerous boreholes. Gravity highs correlate well with limestone pinnacles and lows correlate with depressions in limestone occupied by clay deposits. The most significant gravity low on the profile is due to a known void at depth. This study illustrates how careful fieldwork can use gravity to detect limestone topography (sinkholes and caves) in an area of karst. Figure 6.37 also emphasizes the importance of borehole information to establish correlations between gravity highs and lows and subsurface features. By comparing Figure 6.37 with Figure 5.38, we can visualize how a resistivity survey might be combined with a gravity survey to provide a powerful multitechnique approach to karst regions. Recall that in the area represented by Figure 5.38, high apparent resistivities correlated well with limestone pinnacles and voids and apparent resistivity lows correlated with bedrock depressions containing clay. These studies emphasize that independent geologic information is critical (especially in gravity and resistivity work) for reliable interpretation of geophysical data and that applying multiple geophysical methods in an exploration exercise often is extremely rewarding in terms of constraining interpretations.

6.8.3   Landfill Geometry

Roberts, Hinze, and Leap (1990) undertook a gravity survey of a multicomponent waste landfill to investigate the effectiveness of the method in establishing the lateral boundaries and the vertical extent of the landfill. The landfill is located on the Thomas Farm site in northwestern Indiana (Section 23; T23N, R5W). An evaluation of the effectiveness of the gravity study was possible due to the existence of prelandfill and postlandfill topographic maps. These maps supported construction of an isopach map (Figure 6.38) that illustrates landfill thickness variations. Boreholes surrounding the landfill site provide thicknesses of glacial deposits and bedrock depths.

The survey consisted of approximately 200 gravity stations spaced at 5 m to 10 m intervals. Measurements were made with a LaCoste–Romberg gravimeter, and standard field corrections were applied to the data. Bouguer and terrain corrections were made using a density determined from a gravity profile along a traverse of marked topographic relief. Finally, a regional trend was defined and removed by fitting a third-degree polynomial surface to gravity stations located outside the landfill boundaries. The residual gravity anomaly map is presented in Figure 6.39. A comparison of Figures 6.38 and 6.39 reveals the close correlation between the landfill boundary and the fill thickness and the residual gravity.

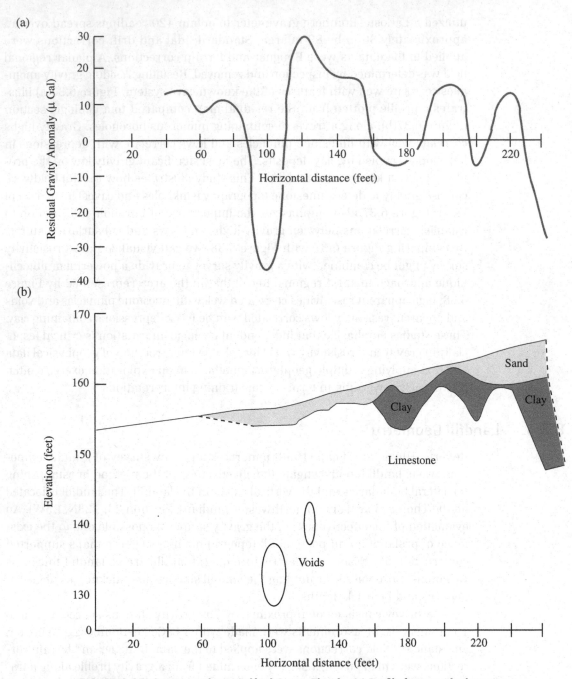

FIGURE 6.37 Gravity study at Medford Caves, Florida. (a) Profile from residual gravity map. (b) Geologic section determined from borehole information. Gravity highs correlate with limestone pinnacles and voids and lows correlate with depressions in limestone occupied by clay deposits. (Modified from Dwain K. Butler. 1984. Microgravimetric and gravity gradient techniques for detection of subsurface cavities. *Geophysics*, v. 49, 1084–96.)

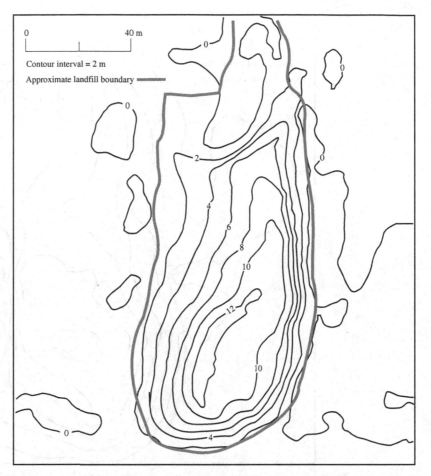

FIGURE 6.38 Isopach map of the Thomas Farm landfill, northwestern Indiana. (Modified from R.L. Roberts, W.J. Hinze, and D.I. Leap. 1990. Application of the gravity method to investigation of a landfill in the glaciated midcontinent U.S.A. In *Geotechnical and environmental geophysics, volume 2: Environmental and groundwater,* ed. Stanley H. Ward, 253–66. Society of Exploration Geophysicists Investigations in Geophysics, no. 5.)

Roberts, Hinze, and Leap (1990) demonstrate, in addition, that under suitable conditions gravity can be used to evaluate the gross nature of landfill material. We have mentioned previously, and we hope emphasized, that independent geologic information is very important when we evaluate gravity data. To focus further attention on this requirement, we conclude the discussion in this chapter with a quotation from Roberts, Hinze, and Leap (1990, 259): "However, the gravity method depends on precise gravity and surface elevation measurements, careful computations, and constraining information obtained from collateral geologic and geophysical studies for the interpretations to be valid."

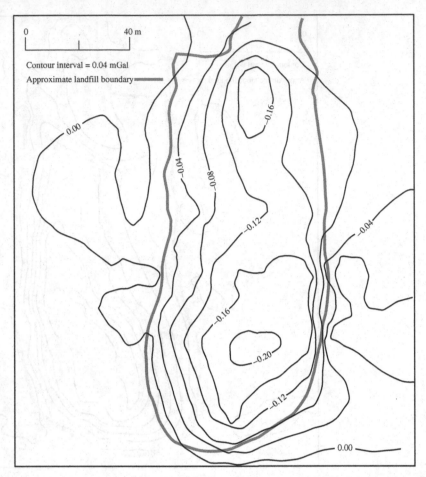

FIGURE 6.39 Residual gravity anomaly map of the Thomas Farm landfill, northwestern Indiana. (Modified from R.L. Roberts, W.J. Hinze, and D.I. Leap. 1990. Application of the gravity method to investigation of a landfill in the glaciated midcontinent U.S.A. In *Geotechnical and environmental geophysics, volume 2: Environmental and groundwater,* ed. Stanley H. Ward, 253–66. Society of Exploration Geophysicists Investigations in Geophysics, no. 5.)

PROBLEMS

6.1 If a gravity determination was made at an elevation of 152.7 m, what is the value of the free-air correction (assuming sea level as the datum)? The Bouguer correction (assuming a 2.50 g/cm³ reduction density)?

6.2 A gravity station at a location of 0 m is located in the center of an erosional basin. The floor of the basin has virtually no relief. An escarpment of a plateau is located at a distance of 450 m from the gravity station. The surface of the plateau has a relatively constant elevation of 400 m. Will terrain corrections be necessary?

6.3 Prepare a drift curve for the following data and make drift corrections. Convert your corrected data to milliGals. The data were collected by a gravimeter with a dial constant of 0.0869 mGal/dial division.

Station	Time	Reading in Dial Divisions
Base	11:20	762.71
GN1	11:42	774.16
GN2	12:14	759.72
GN3	12:37	768.95
GN4	12:59	771.02
Base	13:10	761.18

6.4 You are measuring values of g in a drill hole. Refer to the diagram below and calculate the free-air and Bouguer corrections in measured g at positions 1, 2, 3, and 4. Sea level is the reduction datum. The mass loss due to the drill hole is negligible.

6.5 What is the radius of the smallest equidimensional void (such as a cave) that can be detected by a gravity survey for which Bouguer gravity values have an accuracy of 0.05 mGal? Assume the voids are in limestone (density = 2.7 g/cm³) and that void centers never are closer to the surface than 100 m.

6.6 In a region of gentle, nonplunging folds ore deposits are localized in anticlinal crests of limestone beds. The center of the limestone bed in the crests is 1.5 km beneath the surface. The form of the deposit can be approximated by a horizontal cylinder with a radius of 0.25 km. Because of time constraints, a gravity survey can achieve an accuracy of only 0.1 mGal for Bouguer anomaly values. What is the smallest density contrast for which the ore deposits can be reliably located?

6.7 Determine the maximum value of the gravity anomaly due to the faulted basalt sill diagrammed below.

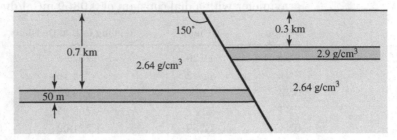

6.8 The curve in the diagram below represents a traverse across the center of a roughly equidimensional ore body. The anomaly due to the ore body is obscured by a strong regional anomaly. Remove the regional anomaly and then evaluate the anomaly due to the ore body, which has a density contrast of 0.75 g/cm³. (Graph values are supplied in the accompanying table.)

Horizontal Position (km)	Bouguer Anomaly (mGal)
0.1	−1.45
0.2	−1.36
0.3	−1.27
0.4	−1.17
0.5	−1.07
0.6	−0.96
0.7	−0.85
0.8	−0.74
0.9	−0.64
1.0	−0.56
1.1	−0.49
1.2	−0.44
1.3	−0.40
1.4	−0.36
1.5	−0.32
1.6	−0.27
1.7	−0.22
1.8	−0.16
1.9	−0.10
2.0	−0.03

6.9 Determine the depth to the center of each of the three spheres represented by anomaly curves 1, 2, and 3 shown below.

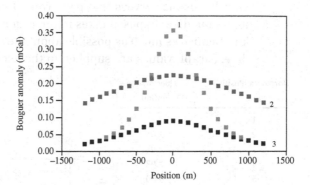

6.10 (a) Derive the half-maximum relationship for a horizontal cylinder (Equation 6.55) using Equation 6.37. (b) Determine the depth to the centers of the two horizontal cylinders represented by anomaly curves 1 and 2.

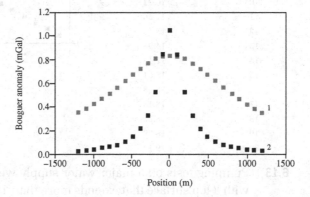

6.11 New boring information obtained along the traverse illustrated in Figure 6.35 fixes bedrock depth at 50 m at 3 km along the traverse and at 100 m at position 8 km. Assuming all other parameters are correct as discussed in the section "Revisiting Some Bouguer Anomaly Values," discuss how the maximum probable thickness of the Mesozoic sedimentary rocks is affected by this new information. The gravity data are in file Conn_Data (Chapter 6 folder, Dynamic Tables folder) found on the resources website accompanying this book.

6.12 The following Bouguer gravity curve was obtained on a traverse oriented perpendicular to a vertical dike. The dike has a density contrast of 0.9 g/cm³. Seismic surveys that penetrated to 0.5 km did not encounter the top of the dike. Exposed dikes in the region tend to be no wider than 0.4 km. Deduce as much as possible about the dimensions and position of the dike. (Graph values are supplied in the accompanying table.)

Horizontal Position (km)	Bouguer Anomaly (mGal)
1	1.11
2	2.04
3	2.98
4	3.93
5	4.90
6	5.88
7	6.89
8	7.93
9	9.04
10	10.23
11	11.56
12	13.10
13	14.96
14	17.12
15	18.61
16	18.48
17	18.18
18	18.25
19	18.59
20	19.12

6.13 Pumping tests on a major water supply well create a cone of depression with a top surface that extends more than 1000 m from the well. The area is underlain with totally saturated sand possessing a 40% porosity and a density of 2.60 g/cm³. The specific yield of the pumping is estimated at 30% (30% of the fluid occupying the 40% void space). Can the removal of water be detected by a gravity measurement with an accuracy of 0.01 mGal? For modeling purposes assume that the cone of depression can be represented sufficiently by a horizontal cylinder with a radius of 300 m and a center at a depth of 300 m.

6.14 The following data represent a traverse across a faulted contact between an uplifted crystalline basement complex and an overlying sedimentary sequence. Use a second derivative technique to determine the dip of the fault and the depth to the top of the uplifted basement.

Position (km)	Bouguer Anomaly (mGal)
19	0.530
20	0.597
21	0.677
22	0.775
23	0.897
24	1.048
25	1.239
26	1.478
27	1.777
28	2.143
29	2.569
30	3.018
31	3.415
32	3.719
33	3.944
34	4.110
35	4.235
36	4.331
37	4.406
38	4.464
39	4.508
40	4.543
41	4.568

REFERENCES CITED

Butler, Dwain K. 1984. Microgravimetric and gravity gradient techniques for detection of sub-surface cavities. *Geophysics*, v. 49, 1084–96.

Chandler, William E. Jr. 1978. *Graben mechanics at the junction of the Hartford and Deerfield Basins of the Connecticut Valley, Massachusetts*. University of Massachusetts, Department of Geology and Geography Contribution no. 33.

Grant, F.S., and G.F. West. 1965. *Interpretation theory in applied geophysics*. New York: McGraw-Hill.

Hammer, Sigmund. 1939. Terrain corrections for gravimeter stations. *Geophysics*, v. 4, 184–94.

Hubbert, M. King. 1948. Line-integral method of computing the gravimetric effects of two-dimensional masses. *Geophysics*, v. 13, 215–25.

Kick, John F. 1985. Depth to bedrock using gravimetry. *Geophysics: The Leading Edge of Exploration*, Society of Exploration Geophysicists, v. 4, 38–42.

Maslyn, R. Mark. 1984. Case histories of computers in oil exploration—polynomial trend surface analysis. *Computer Oriented Geological Society Computer Contributions*, v. 1, 12–19.

Nettleton, L.L. 1942. Gravity and magnetics calculations. *Geophysics*, v. 7, 293–310.

Nettleton, L.L. 1976. *Gravity and magnetics in oil prospecting*. New York: McGraw-Hill.

Plouff, Donald. 1976. Gravity and magnetic fields of polygonal prisms and applications to magnetic terrain corrections. *Geophysics*, v. 41, 727–39.

Roberts, R.L., W.J. Hinze, and D.I. Leap. 1990. Application of the gravity method to investigation of a landfill in the glaciated midcontinent U.S.A. In *Geotechnical and environmental geophysics, volume 2: Environmental and groundwater*, ed. Stanley H. Ward, 253–66. Society of Exploration Geophysicists Investigations in Geophysics, no. 5.

Robinson, Edwin S., and Cahit Coruh. 1988. *Basic exploration geophysics*. New York: John Wiley & Sons.

Robinson, Joseph E. 1982. *Computer applications in petroleum geology*. Stroudsburg: Hutchinson Ross Publishing Company, 72–81.

Rudman, Albert J., Robert Ziegler, and Robert F. Blakely. 1977. *FORTRAN program for generation of Earth tide gravity values*. Indiana Geological Survey Occasional Paper 22.

Stanley, John M. 1977. Simplified gravity interpretation by gradients—the geological contact. *Geophysics*, v. 42, 1230–35.

Talwani, M., J.W. Worzel, and M. Landsman. 1959. Rapid gravity computations for two-dimensional bodies with application to the Mendocino submarine fracture zone. *Journal of Geophysical Research*, v. 64, 49–59.

Telford, W.M., L.P. Geldart, R.E. Sheriff, and D.A. Keys. 1976. *Applied Geophysics*. London: Cambridge University Press.

Telford, W.M., L.P. Geldart, and R.E. Sheriff. 1990. *Applied Geophysics*, 2nd ed. London: Cambridge University Press.

Tsuboi, Chuji. 1979. *Gravity*. London: Allen & Unwin.

SUGGESTED READING

Allen, D.M., and F.A. Michel. 1996. The successful use of microgravity profiling to delineate faults in buried bedrock valleys. *Ground Water*, v. 34, no. 6, 1132–40.

Benson, A.K., and A.R. Floyd. 2000. Application of gravity and magnetic methods to assess geological hazards and natural resource potential in the Mosida Hills, Utah County, Utah. *Geophysics*, v. 65, no. 5, 1514–26.

Casten, U., and Z. Fajklewicz. 1993. Induced gravity anomalies and rock burst risk in coal mines—a case history. *Geophysical Prospecting*, v. 41, no. 1, 1–13.

Smith, R.A. 1959. Some depth formulae for local magnetic and gravity anomalies. *Geophysical Prospecting*, v. 7, 55–63.

Tsuboi, Chuji. 1979. *Gravity*. London: Allen & Unwin.

CHAPTER 7

Exploration Using the Magnetic Method

The magnetic method as used in geophysical exploration is possibly the most versatile of the methods discussed in this text: It can be applied to both deep and shallow structures, and relative to other methods, measurements can be obtained cheaply for both local and regional studies. On the other hand, quantitative interpretation of magnetic information is much more difficult than interpretation of gravity data even though these two methods share many similarities.

Much exploration of the shallow subsurface is directed toward characteristics of the sediments above bedrock or a definition of the bedrock surface, and in the majority of these cases magnetic information is not very useful. This arises in part from the difficulty in obtaining useful quantitative information from magnetic data but also is due to the magnetic characteristics of rocks and sediments. For the most part, magnetic signatures depend on magnetite content, which tends to be low in most sediments and in many rocks constituting basement. In general, therefore, magnetic fieldwork is not commonly employed for many shallow subsurface objectives. Nevertheless magnetic data can contribute to certain exploration efforts when targets are not deep. Buried ore deposits constitute an obvious example. Others include the mapping of rock units with sufficiently large magnetic signatures to be traceable beneath glacial or other cover and the discovery and delineation of a variety of features important to archaeological investigations.

In this chapter our goal is to provide a basic, practical grounding in magnetic surveys rather than to develop an in-depth theoretical basis. We cover fundamental relationships that control the nature of magnetic anomalies, govern field methods, and affect interpretation efforts. We dwell on a few relatively straightforward examples in detail so you can appreciate the applications of magnetic surveying and, we hope, understand what can and cannot be achieved. Although you are more likely to engage in seismic exploration than magnetic work, the magnetic method could be the most appropriate method to apply during a particular assignment, and we want to prepare you for that eventuality should it occur.

7.1 FUNDAMENTAL RELATIONSHIPS

Virtually everyone is familiar with the simple bar magnet and its behavior when placed near a second similar magnet. You likely also remember the great delight of sprinkling iron filings on a glass plate above a strong bar magnet. The iron filings (which became very small magnets) aligned themselves in a pattern about the magnet (Figure 7.1). This pattern outlines what we refer to as *magnetic lines of force,* which follow curved trajectories and are most closely spaced at the ends of the magnet. If you were lucky, you also were able to map these lines of force by placing a small compass at various positions above the magnet and recording the orientation of the compass needle. The pattern that results from such an investigation confirms the pattern of iron filings and demonstrates that the lines of force converge at the ends of the magnet. These points of convergence are referred to as *magnetic poles.* The lines of force record the orientation of the magnetic field produced by the magnet. At any point the orientation of the field is illustrated by a line drawn tangent to the lines of force. The spacing of the lines of force indicates the magnetic force at that position. The force is greatest where the lines of force are closest together.

Each bar magnet has two poles and is referred to as a *dipole.* In fact, a single pole cannot exist—poles always come in pairs. If a bar magnet is broken in half, each half still has two poles. Because the properties of magnets and their associated magnetic fields are closely related to magnetic fields produced by electric current, and by analogy with electric charges, the magnetic poles are referred to as positive ($+m$) and negative ($-m$). Poles of like sign repel and those of opposite sign attract. The positive (north-seeking) pole of a compass needle points toward the north magnetic pole of the Earth.

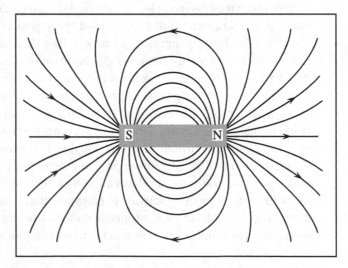

FIGURE 7.1 Magnetic lines of force produced by a simple bar magnet.

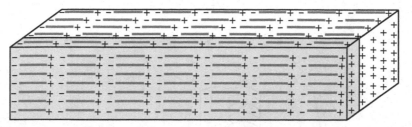

FIGURE 7.2 Representation of a magnet as an assemblage of small dipoles.

It is convenient in visualizing many magnetic effects to conceive of a magnet, or any magnetized body, as composed of a very large number of small dipoles that are similarly aligned and stacked closely together (Figure 7.2). Within the magnet effects of the positive and negative poles cancel; but one end has an accumulation of positive poles and the other end has an accumulation of negative poles, which creates the dipolar nature of the large magnet. If we cut the magnet across its length, we simply produce two new surfaces, and each has its own accumulation of poles of like sign. We assume, of course, that the dipoles within the magnet are too small to be cut, or, if cut, are themselves composed of yet smaller dipoles.

7.1.1 Magnetic Force

We've mentioned attraction and repulsion, but what is the actual force exerted between two magnetic poles? This force is stated in Coulomb's Law

$$F = \frac{1}{\mu} \frac{m_1 m_2}{r^2} \tag{7.1}$$

where r is the distance between poles of strength m_1 and m_2. The remaining factor μ, or *magnetic permeability*, is a property of the medium in which the poles are located. Because the magnetic permeability is nearly equal to unity in water and air, for simplicity we do not include the term in the remaining equations in this chapter. However, in some instances the permeability of the medium is important, so its effect cannot be neglected in all cases. Consider Figure 7.3, which illustrates an iron bar placed within a wire coil. When current passes through the wire, a magnetic field is generated that is greater than the field that would have been generated had the iron bar not been present. Thus the strength of the field generated depends on some property (magnetic permeability) of the medium (iron bar) in which the field is generated.

7.1.2 Magnetic Field Strength

Although poles cannot exist individually and always occur in pairs, it is convenient to utilize a unit magnetic pole m'. Such a pole repels a similar pole with a force

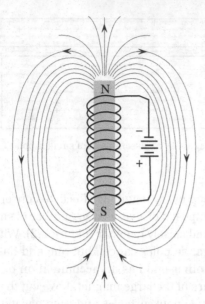

FIGURE 7.3 Effect of magnetic permeability on magnetic field strength.

of 1 dyne when separated by a distance of 1 cm. You can visualize this situation by considering two very long, slender bar magnets with their two positive (or negative) ends close together. In this arrangement we effectively are dealing with the interaction of two poles because the remaining poles are far removed from the region of interaction.

The *magnetic field strength H* is the force a unit magnetic pole would experience if placed at a point in a magnetic field that is the result of some pole strength m and where r is the distance of the point of measurement from m:

$$H = \frac{F}{m'} = \frac{m}{r^2}. \tag{7.2}$$

H is a vector quantity having a magnitude given by Equation 7.2 and with a direction determined by assuming our unit pole is *positive*. The SI unit of magnetic field strength is the *nanotesla* or *nT*, whereas the cgs unit of magnetic field strength is the *oersted*. The oersted is 1 dyne/unit pole strength and is equal to 10^5 *gammas*. The gamma is numerically equivalent to the nanotesla. Until recently the gamma was the common unit employed in geophysics, but current use favors the nanotesla.

7.1.3 Magnetic Moment

If a bar magnet is placed in a uniform magnetic field H (Figure 7.4) it will experience a pair of equal forces acting parallel to each other but in opposite directions (a *couple*). The magnitude of the couple is

$$C = 2(ml)H\sin\theta \tag{7.3}$$

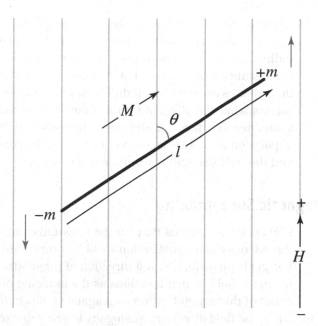

FIGURE 7.4 Diagram illustrating the role of magnetic moment in determining forces acting on a bar magnet placed in a magnetic field.

where θ specifies the original orientation of the magnet in the field. The motion produced by the couple depends on the magnitude of H as well as the value of θ (no motion is produced if $\theta = 0$). The other quantity (ml) that also affects the magnitude of the couple is termed the *magnetic moment* with magnitude typically represented by M so that

$$M = ml. \tag{7.4}$$

7.1.4 Intensity of Magnetization

A bar magnet possesses a fundamental property per unit volume known as the *intensity of magnetization I*. The magnitude of I is defined as the magnetic moment M per unit volume or

$$I = \frac{M}{\text{Volume}}, \tag{7.5}$$

and, therefore,

$$I = \frac{ml}{\text{Volume}} = \frac{m}{\text{Area}}. \tag{7.6}$$

To appreciate the relation of intensity of magnetization, consider once again the bar magnet in Figure 7.2. If we cut the magnet into unequal pieces perpendicular to its length, the force at a distance is proportional to the length of each piece.

We have not altered the fundamental magnetic properties of the material but simply have changed the volume of each. Similarly, if we cut the magnet longitudinally into two sections (as suggested by the shading in Figure 7.2), the force at a distance is proportional to the cross-sectional area of each section. Of course, the volumes are once again different; but because the lengths are the same, we can visualize the effect in terms of the cross-sectional areas, which are not the same. Recall Figure 7.2 and our visualization of the magnet as a concentration of poles on its end surfaces. As I increases, the concentration of poles increases, and the pole strength per unit area is greater.

7.1.5   Magnetic Susceptibility

If we place a material that can be magnetized in an external magnetic field H, the intensity of magnetization I will be proportional to the strength of the field. For geologic materials, the direction of magnetization is parallel to the Earth's magnetic field, which functions as the inducing field. We thus can consider the result of this magnetization as magnetic poles induced on surfaces perpendicular to the field direction, analogous to our representation of the bar magnet in Figure 7.2. The magnitude of I is

$$I = kH \qquad (7.7)$$

where k is the constant of proportionality, generally referred to as *magnetic susceptibility*. We'll examine representative values of magnetic susceptibility when we discuss the magnetic effects of various geologic bodies, but first let's take a moment to examine the process of magnetization in a bit more detail.

Based on their magnetic susceptibilities, minerals are classified as diamagnetic or paramagnetic. *Diamagnetic minerals* have negative susceptibilities, but the values are so low that these have no effect with regard to subsurface exploration. Quartz and feldspar are common diamagnetic minerals. *Paramagnetic minerals* have positive susceptibilities, but in general the values are quite low. Examples include the Fe–Mg silicates such as pyroxene, amphibole, and olivine. However, in a few paramagnetic materials the magnetic moments imparted to atoms by orbital motion and spin of electrons interact strongly. The result is the alignment of magnetic moments in areas with dimensions on the order of 10^{-4} cm. These small areas are referred to as *magnetic domains*. If these domains are parallel (as they are in iron, nickel, and cobalt), the material is termed *ferromagnetic* (Figure 7.5(a)). Ferromagnetic materials have very high susceptibilities, but they do not occur naturally on Earth (although they are present in meteorites) and so are not of interest within the realm of geophysical exploration. If the domains are parallel and antiparallel in equal numbers (Figure 7.5(b)), the net magnetic moment is zero, and the material is said to be *antiferromagnetic*. An example of an antiferromagnetic mineral is hematite, which certainly is common; but because susceptibilities of antiferromagnetic materials are low, such minerals need not be considered for exploration purposes.

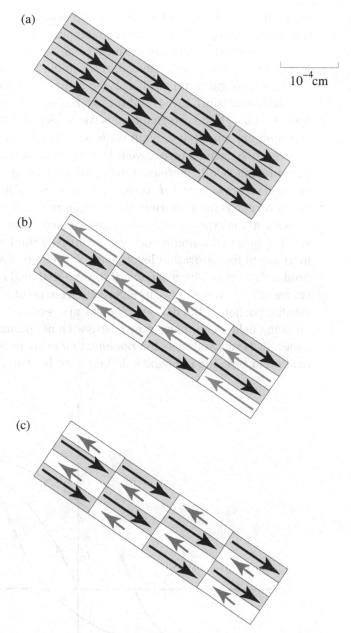

FIGURE 7.5 Schematic representation of magnetic domains in (a) ferromagnetic, (b) antiferromagnetic, and (c) ferrimagnetic materials.

Of primary importance to the exploration effort are *ferrimagnetic* minerals. In these few minerals (magnetite, titanomagnetite, ilmenite, and pyrrhotite are the common ones) the domains are oriented as in antiferromagnetic materials, but one direction of orientation is preferred (Figure 7.5(c)). This imparts a net magnetic moment to such materials and results in relatively high susceptibilities. These susceptibilities depend on temperature and the strength of the inducing field. It is these minerals, the most common of which by far is magnetite, that impart induced magnetizations to rocks in the Earth's crust.

If a ferrimagnetic mineral is placed in an external magnetic field, an induced magnetization is produced, and the external field causes some magnetic domain walls to move. If the external field is weak, as is the Earth's magnetic field, only a limited movement of domain walls occurs, and no permanent magnetization occurs when the field is removed. This behavior is represented in Figure 7.6. As the inducing field is increased to H_1, an induced magnetization I_i is generated and will return to zero if H_1 is removed. However, if the external field is raised to much larger values, more domain walls move, favorably oriented domains increase at the expense of unfavorably oriented domains, and reordering of unfavorably oriented domains occurs. Eventually the induced magnetization will increase to the saturation level (point S in Figure 7.6). Further increases in H produce no further increase in I. If the external field now is reduced to zero, the values of I follow the curve in Figure 7.6 from point S to point I_r. Because a substantial portion of the domain changes are retained during the decrease in the inducing field, the mineral now possesses a net permanent magnetization. This generally is referred to as the *remanent magnetization* I_r. To return the magnetization to zero, the inducing field first must be applied with a reversed polarity

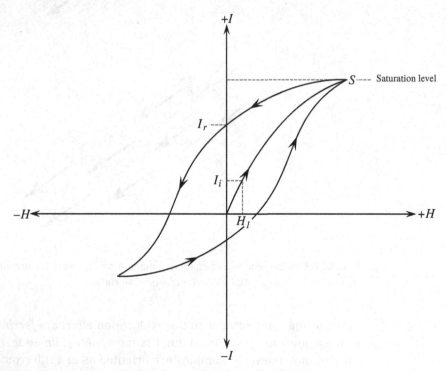

FIGURE 7.6 Hysteresis curve for a ferrimagnetic material in the presence of a magnetizing field.

and then returned to its normal polarity as indicated by the curve in Figure 7.6. This path defined by the values of induced magnetization and external field values is known as a *hysteresis curve*.

Because the Earth's field is weak, need we be concerned with remanent magnetizations? When we measure the magnetic effect of magnetizations induced in geologic bodies, any remanent magnetizations with an order of magnitude similar to the induced magnetization will either increase or decrease the total anomalous field depending on the respective orientations of the two magnetizations. Remanent magnetizations in rocks can be produced in a number of ways, all of which are important in paleomagnetic studies. But one especially strong remanent magnetization is *thermoremanent* magnetization, which typically is produced when ferrimagnetic minerals crystallize and cool during the solidification of igneous rocks. Because the ferrimagnetic minerals are initially at high temperatures, even a weak field such as the Earth's can facilitate preferential alignments of magnetic domains, thus creating a substantial I_r. In general, rock susceptibilities are controlled by magnetite content, and many rocks with sufficient magnetite to produce magnetic anomalies have an igneous origin. In many situations, therefore, thermoremanent magnetization may be present.

Field instruments used in magnetic surveying normally do not determine the magnitude and orientation of remanent magnetizations. Such information more commonly is determined in the laboratory. Typically, however, we interpret survey results by assuming that the recorded anomalous field is due entirely to induced magnetization, and possible effects of remanent magnetizations are ignored. Such neglect is not as serious as it might at first appear because of the inherent difficulty in obtaining quantitative results from magnetic surveys. In many situations the qualitative information sought by a survey is not sufficiently affected by the presence of remanent magnetizations to warrant the extra time and expense of laboratory analysis for determining remanent magnetizations and to then correct field data for their effect. When discussing magnetic anomalies due to geologic forms of various shapes and compositions, we assume all magnetization is induced only.

7.1.6   Magnetic Potential

Magnetic, gravitational, and electrical fields all are potential fields. A characteristic of such fields is that the work done by moving a pole (or mass or charge) from one point to another is path independent. If a complete circuit is made by returning the pole to its original position, then the work done is zero. It is convenient to define the potential V as the negative of the work done on a unit pole in a magnetic field as the pole is moved against the field from infinity to a point in the field. Using Equation 7.2 and remembering that we assume $\mu = 1$, we have

$$V = -\int_{\infty}^{r} \frac{m}{r^2} = \frac{m}{r}. \tag{7.8}$$

An especially useful feature of the potential is that we can find the magnetic field in a given direction by taking the negative of the derivative of the potential in that direction. We use this feature in the following section to derive equations for a dipole.

7.2 THE EARTH'S MAGNETIC FIELD

The Earth's magnetic field at any point on the Earth's surface is a vector quantity defined by its total intensity and direction. Intensity can be measured by any number of instruments, some of which are described in a following section. We can readily determine orientation by allowing a compass needle to rotate freely in all directions. Just as in the case of a bar magnet, the needle will rotate into parallelism with the Earth's field.

7.2.1 Field Elements

The total-field vector is defined by its intensity F_E, its *inclination i,* which is the angle the vector makes with a horizontal plane, and its *declination d,* which is the angle the vertical plane containing the total-field vector makes with geographic north. These relationships are illustrated in Figure 7.7. F_E can be resolved into a vertical component Z_E and a horizontal component H_E. The vertical plane containing F_E, Z_E, and H_E is a magnetic meridian. H_E also can be resolved into horizontal components directed toward geographic north (X_E) and geographic east (Y_E). These seven geomagnetic elements are interrelated in several ways that are readily apparent from Figure 7.7. Any three elements are sufficient to determine the remaining four. A thorough grasp of these relationships is essential for later derivations, so be sure you can reproduce each of the following:

$$F_E = \sqrt{H_E^2 + H_Z^2} = \sqrt{X_E^2 + Y_E^2 + Z_E^2}$$

$$Z_E = F_E \sin i, \quad H_E = F_E \cos i, \quad \text{and} \quad \tan i = \frac{Z_E}{H_E} \tag{7.9}$$

$$X_E = H_E \cos d \quad \text{and} \quad Y_E = H_E \sin d$$

The positions on the Earth's surface where $i = 90°$ are known as the *magnetic dip poles* (see Figure 7.11), and the *magnetic equator* is defined by positions of $i = 0°$. At the dip poles $Z_E = F_E$, and the intensity is approximately 70,000 nT (nanotesla). At the magnetic equator $H_E = F_E$, and the intensity is approximately 30,000 nT (nanotesla). Note that the Earth's magnetic field varies in intensity by more than 200 percent, whereas the gravity field varies only by approximately 0.5 percent.

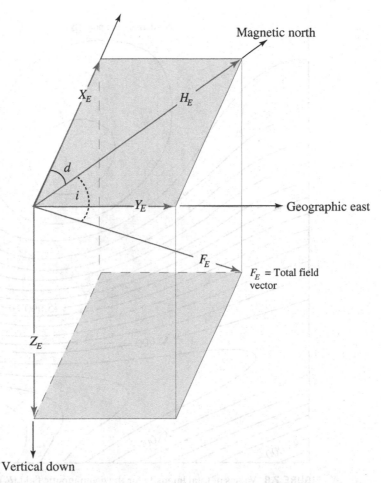

Geographic north

Magnetic north

Geographic east

F_E = Total field vector

Vertical down

FIGURE 7.7 The elements of the Earth's magnetic field: F_E = total-field vector, H_E = horizontal component, Z_E = vertical component, d = declination, and i = inclination.

7.2.2 Dipolar Nature of the Earth's Field

Magnetic data collected at many points on the Earth's surface can be displayed on a map of the world as contours of F_E, i, and d. Figures 7.8 through 7.10 illustrate such maps for a portion of the Western Hemisphere. If world maps are examined carefully, the contour patterns appear somewhat similar to those that would be produced by a dipole. In fact, the Earth's field can be quite well approximated by placing a small dipole of large moment at the Earth's center and tilting the dipole at an angle of 11.5° to the Earth's axis of rotation (Figure 7.11).

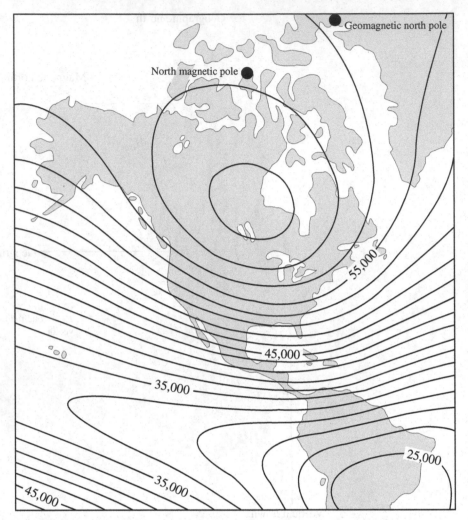

FIGURE 7.8 Values of total intensity for the geomagnetic field (F_E) for a portion of the Western Hemisphere. Geomagnetic epoch 1985.0. Contour interval = 2000 nT. (Based on information in U.S. Geological Survey Total Intensity Chart GP-987-F.)

The points where an extension of the axis of this imaginary dipole intersects the Earth's surface are referred to as the *geomagnetic north and south poles*. Note that these do not coincide with the dip poles, which are referred to as the *north and south magnetic poles*.

Although treating the Earth's field as dipolar in nature is useful in deriving a number of relationships that we will investigate, the fact that the geomagnetic and dip poles do not coincide plus the irregularities evident on maps of inclination, declination, and total-field intensity demonstrate that this field cannot be accurately portrayed by one dipole. Actually, about 10 percent of the field cannot

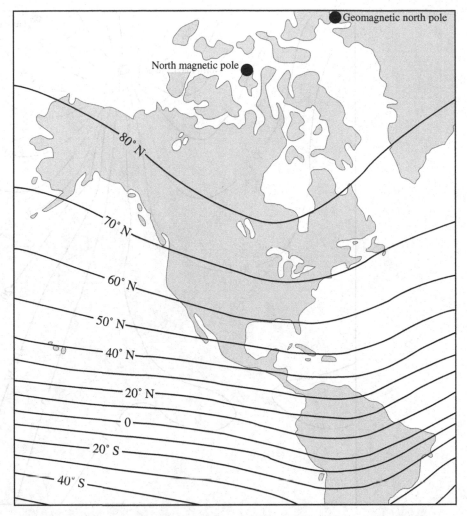

FIGURE 7.9 Values of inclination (*i*) of the geomagnetic field for a portion of the Western Hemisphere. Geomagnetic epoch 1985.0. (Based on information in U.S. Geological Survey Inclination Chart GP-987-I.)

be accounted for by a single dipole, and it is necessary to employ numerous other dipoles of smaller magnetic moments at various other positions near the outer boundary of the Earth's outer core to accurately model the existing field.

7.2.3 Variations of the Earth's Field

As we acquire magnetic data at various points on the Earth's surface, intensity and orientation of the total field vary. In this section we examine the reasons for such variations with position and also consider variations with time.

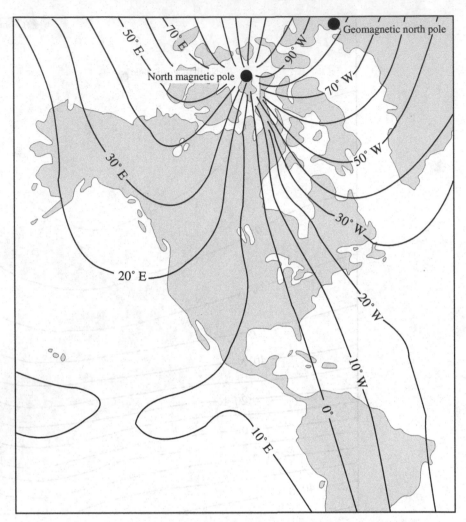

FIGURE 7.10 Values of declination (*d*) of the geomagnetic field for a portion of the Western Hemisphere. Geomagnetic epoch 1985.0. (Based on information in U.S. Geological Survey Declination Chart GP-987-D.)

By far the major portion of the geomagnetic field arises from currents within the Earth's fluid core. As discussed, this *main magnetic field* consists of a primary dipole portion and a secondary nondipole component that produce variations in the total-field vector over the Earth's surface. In addition, this main field is not constant but changes slowly in both intensity and direction. This slow or *secular variation* occurs over long periods and can be ignored when we conduct exploration over intervals of days or weeks. However, we must adjust data for secular variation when compiling data from different years or decades. To facilitate such adjustments, geomagnetic reference maps are produced by various government agencies

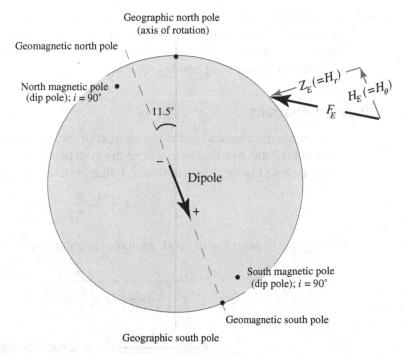

FIGURE 7.11 Some important features of the Earth's magnetic field.

at five-year intervals. These maps are constructed from a network of magnetic observatories from which repeat readings are obtained. Such readings used in conjunction with very sophisticated models for the dipolar and nondipolar components of the geomagnetic field result in the International Geomagnetic Reference Field (IGRF), which is presented in the form of maps and data tables. The direction and intensity of the main field at any position on the surface can be obtained from this information. Figures 7.8 to 7.10 are based on information taken from a 1985 calculation of the IGRF. IGRF information also is critical for adjusting field values from magnetic surveys that encompass more than a very limited area. Such adjustments are discussed in the sections about field procedures.

Flow of charged particles in the Earth's ionosphere produces an *external magnetic field*. The external field is small in contrast to the main field (normally tens of nanotesla compared with tens of thousands of nanotesla). This flow of particles toward the magnetic poles is affected by the gravitational attraction of the Sun and Moon and varies with time. Such *diurnal variations* must be determined and corrected for during surveys because modern magnetometers are capable of precisions of 1 nT, and we naturally want to maintain this level of precision. Occasionally, and unpredictably, solar activity increases substantially, which causes an abrupt increase in ionized particles arriving in the ionosphere. These *magnetic storms* can produce variations of hundreds of nanotesla. Due to this large and erratic variation in magnetic intensities, fieldwork usually halts during such activity.

Of course, variations in magnetic susceptibilities of rocks leads to local variations in induced magnetization, which affect total-field values. This *anomalous field* is the one we are interested in isolating—just as we strive to isolate gravity anomalies by reducing gravity observations.

7.2.4 Dipole Equations

The magnetic potential is a convenient approach to describe the magnetic field at a point P due to a dipole. Consider the system illustrated in Figure 7.12(a). We assume that r is much larger than l. Using Equation 7.8, the potential at P is

$$V = \frac{m}{r_1} - \frac{m}{r_2}. \tag{7.10}$$

Using our assumption of $r >> l$, we have the relations

$$V = \frac{m}{r - \left(\frac{l}{2}\cos\theta\right)} - \frac{m}{r + \left(\frac{l}{2}\cos\theta\right)},$$

$$V = \frac{ml\cos\theta}{r^2 - \left(\frac{l}{2}\right)^2 \cos^2\theta}, \text{ and, approximating,}$$

$$V = \frac{ml\cos\theta}{r^2} = \frac{M\cos\theta}{r^2}. \tag{7.11}$$

For our purposes at this time, we wish to derive the radial and tangential components of the field at P (see Figure 7.12(b)). Recalling that we can determine the magnetic field in a given direction by taking the negative of the derivative of the potential in that direction and noting that θ is in radians, we see that

$$H_r = -\frac{dV}{dr} = \frac{2M\cos\theta}{r^3} \tag{7.12}$$

and

$$H_\theta = -\frac{1}{r}\frac{dV}{d\theta} = \frac{M\sin\theta}{r^3}. \tag{7.13}$$

Because slightly more than 90 percent of the Earth's field can be represented by a dipole at the Earth's center, these equations provide a good approximation of some of the properties of this field. By consulting Figure 7.11 you should see that the radial field is equivalent to the vertical field ($H_r = Z_E$) and that the tangential field is equivalent to the horizontal field ($H_\theta = H_E$). Given that we are using cgs units, the magnetic moment (M) of the imaginary dipole responsible for the major

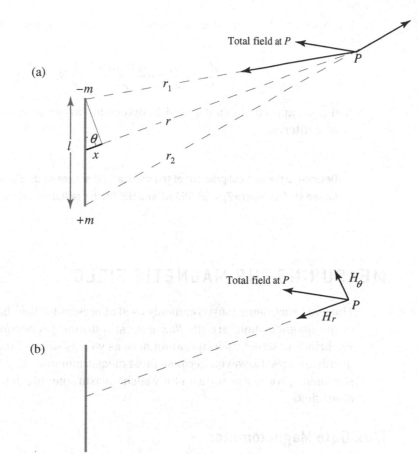

FIGURE 7.12 Parameters used in deriving the dipole equations.

portion of the Earth's field is 8×10^{25} emu. Noting that r must be expressed in centimeters and that θ is equivalent to colatitude, we can use Equations 7.12 and 7.13 for a rough approximation of the vertical and/or horizontal components of the Earth's field (and also, therefore, the total field) at any point on the surface. Resulting field strength is expressed in oersteds, which must be multiplied by 10^5 to express the strength in nT.

> Determine if Equations 7.12 and 7.13 provide a reasonable approximation for the horizontal and vertical components of the Earth's field at the equator and poles.

Equations 7.12 and 7.13 also are useful in determining gradients of magnetic field components at various locations on the Earth's surface. Assume we are interested in the vertical gradient of the vertical component of the field. We use Equation 7.12 to obtain

$$\frac{dZ_E}{dr} = \frac{dH_r}{dr} = -\frac{6M\cos\theta}{r^4} = -\frac{3}{r}H_r = -\frac{3}{r}Z_E.$$ (7.14)

A similar approach can be used to determine approximations for other gradients of interest.

Determine the vertical gradient of the vertical component of the Earth's field at a point on the surface where $Z_E = 52{,}950$ nT and the Earth's radius equals 6.3×10^6 m.

7.3 MEASURING THE MAGNETIC FIELD

The two instruments most commonly used at present for field-based measurements of the magnetic field are the *flux-gate* and *proton-precession* magnetometers. We briefly describe their operation here as well as some of their advantages and disadvantages. However, because most measurements today are total-field measurements, we concentrate on how such measurements relate to the local anomalous field.

7.3.1 Flux-Gate Magnetometer

In the common design for flux-gate magnetometers, two bars of metal with high permeability are placed close together and oriented parallel to one another. Each bar is wound with a primary coil but the sense of the windings is reversed, so that the magnetic fields induced when alternating current is applied to the coils will be opposite in orientation. A secondary coil surrounds the primary coils. In the absence of an external field, the voltage induced in the secondary coil will be zero when current is passed through the primary coils due to the opposite magnetizations they induce. If the metal cores are aligned parallel to a component of a weak external field (such as the Earth's field), one core will produce a field in the same sense as the Earth's field and reinforce it, whereas the other will be in opposition to the field. This difference is sufficient to induce a voltage in the secondary core, which is proportional to the strength of the external field.

Note that this instrument measures only relative changes in the Earth's field, but it can measure the component of the Earth's field parallel to which the metal bars are aligned. We therefore can obtain readings for F_E, Z_E, and H_E. With modern versions of this instrument it is routine to rapidly obtain readings for F_E that are accurate to ± 1 nT. However, without elaborate orientation efforts values for Z_E and H_E typically are not accurate to more than a few nT.

7.3.2   Proton-Precession Magnetometer

The "sensor" component of a proton-precession magnetometer is a cylindrical container filled with a liquid rich in hydrogen atoms and surrounded by a coil. The sensor is connected by a cable to a small unit in which is housed the power supply and other necessary electronics such as an amplifier and a frequency-measuring device. When power is applied to the coil, a magnetic field is created that is aligned parallel to the coil axis. The hydrogen nuclei (protons), which behave as minute spinning dipoles, become aligned along the direction of this field. Power then is removed, and the hydrogen nuclei precess around the Earth's total field. This precession induces a small alternating current to flow in the coil at the precession frequency. Because the frequency of precession is proportional to the strength of the total field and the constant of proportionality is the well-known gyromagnetic ratio of the proton, we can accurately determine the total-field intensity.

One important advantage of the proton-precession magnetometer is its ease of use and reliability. Sensor orientation need only be at a high angle to the Earth's field, which is easily achieved. No other leveling or orientation requirements exist. The lack of moving parts ensures generally trouble-free operation. Readings can be obtained very rapidly. Equally as important is that a precision of ±1 nT is routinely achieved. Limitations of this magnetometer are minor but must be kept in mind during fieldwork. A large magnetic field gradient on the order of 600 nT/m will differentially affect regions of the sensor and seriously degrade readings. Also, because the signal generated by precession is small, the instrument cannot be operated close to AC power sources.

Although the flux-gate magnetometer can measure the horizontal and vertical components of the total field as well as F_E itself, the proton-precession magnetometer measures only the absolute strength of the Earth's total field. Note that this gives us only the absolute intensity of the total field and not its direction. This is somewhat a disadvantage because total-field measurements tend to be more difficult to interpret than are vertical component measurements. However, its ease of use (both on ground and in aircraft), the speed with which a survey can be completed, and its precision all contribute to the predominance of the proton-precession magnetometer in modern surveys. Thus we must concentrate on understanding what these total-field measurements encompass and how to interpret them.

7.3.3   Total-Field Anomalies

As a first step in this process, let's decide on our terminology. We already chose H_E, Z_E, and F_E as references to the Earth's main field. We use H_A, Z_A, and F_A in a similar sense for a magnetic field induced in a geologic body by the Earth's field. We normally derive the vertical and horizontal anomalous-field components (H_A and Z_A) to compute the induced field due to various subsurface geometries. F_A is the resultant of these components and is computed from them just as F_E is determined from H_E and Z_E (Equation 7.9).

At any point on the surface at which a total-field measurement is taken, the measurement includes the main field plus the anomalous field (plus the very small contribution from the external field, which is discussed in the following section). For instance, in the situation illustrated in Figure 7.13(a) the total-field reading from a magnetometer would equal 55,005 nT, which includes the undisturbed main-field reading F_{E_U} (55,000 nT) plus the contribution of the anomalous field (5 nT). We refer to this total as F_{E_T}. However, if we do not know the orientation of F_{E_T}, how do we proceed with interpretations? First, we assume that $F_{E_U} >>> F_A$, which is almost always the case, especially in most shallow subsurface surveys, unless the survey is close to large steel objects or encounters rocks with very high susceptibilities. If this assumption holds, then the total-field vector F_{E_T} that includes the contribution from the anomalous field is for all practical purposes

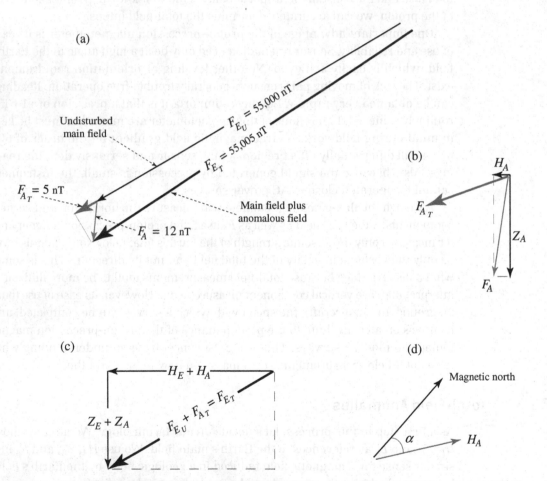

FIGURE 7.13 Relationships surrounding the meaning of the total-field anomaly. (a) Vectors of the main field and anomalous field. (b) Components of the anomalous field. (c) Components of the undisturbed main field and the anomalous field. (d) Correction for the horizontal component of the anomalous field when it does not lie along a magnetic meridian.

parallel to the direction of the main field in the absence of the anomaly (the undisturbed main field F_{E_U}). *Thus the effect of the anomaly on the magnetometer reading essentially is the component of the anomaly F_A in the direction of the undisturbed main field.* We refer to this anomaly component as F_{A_T} (see Figure 7.13(b)). This relationship allows us to develop equations to determine total-field anomalies F_{A_T} from H_A and Z_A and to reduce total-field magnetometer readings for interpretation.

In general, for simplicity, we will continue to refer to the undisturbed main field as F_E and use the more specific terminology introduced in the previous paragraph only when we need to make such specific distinctions. Study Figure 7.13(b) once again and note our goal. If we derive values for Z_A and H_A, it is straightforward to determine F_A, but this is of no use in interpreting or computing total-field values. We want to determine F_{A_T}. Now consider Figure 7.13(c). If the anomalous field is oriented so that H_A is directed toward magnetic north (that is, the H_A–Z_A plane is parallel to a magnetic meridian or, equivalently, α in Figure 7.13(d) = 0) and given the assumptions in the preceding paragraph, we have the relationships depicted in Figure 7.13(c). Then

$$(F_E + F_{A_T})^2 = (Z_E + Z_A)^2 + (H_E + H_A)^2$$

and

$$F_E^{\,2} + 2F_{A_T} F_E + F_{A_T}^{\,2} = Z_E^{\,2} + 2Z_E Z_A + Z_A^{\,2} + H_E^{\,2} + 2H_E H_A + H_A^{\,2}. \qquad (7.15)$$

Because $F_E >>> F_A$, we ignore $F_{A_T}^{\,2}$, $H_A^{\,2}$, and $Z_A^{\,2}$, which gives

$$F_E^{\,2} + 2F_{A_T} F_E = Z_E^{\,2} + 2Z_E Z_A + H_E^{\,2} + 2H_E H_A. \qquad (7.16)$$

But $F_E^{\,2} = Z_E^{\,2} + H_E^{\,2}$, so Equation 7.16 reduces to $F_{A_T} F_E = Z_E Z_A + H_E H_A$, and

$$F_{A_T} = Z_A \left(\frac{Z_E}{F_E} \right) + H_A \left(\frac{H_E}{F_E} \right). \qquad (7.17)$$

Finally, we apply the relationships among the geomagnetic elements (Equation 7.9) to arrive at

$$F_{A_T} = Z_A \sin i + H_A \cos i. \qquad (7.18)$$

If H_A does not lie along a magnetic meridian, we use the component of H_A parallel to the meridian because this is the only effect of H_A on the total anomaly. In such a case

$$F_{A_T} = Z_A \sin i + H_A \cos \alpha \cos i. \qquad (7.19)$$

We return to these relationships in our discussion of magnetic anomalies and interpretation of magnetic data. First, however, we take a slight diversion to summarize field techniques relevant to the collection of such field data.

7.4 BASIC FIELD PROCEDURES

In general, the requirements for ground-based magnetic surveying are not as demanding as those related to gravity exploration. Nevertheless, you must follow a number of procedures to compile useful data with an accuracy comparable to the instrument used. At present most magnetic exploration surveys are conducted from aircraft. Such surveys impose a number of considerations in addition to those detailed here; but because this text is concerned chiefly with the shallow subsurface, we do not include aeromagnetic survey procedures. However, aeromagnetic maps constitute an extremely valuable resource for certain types of shallow subsurface investigations, and you always should check the availability of such maps for any area in which you work. If you are interested in this topic, an excellent introductory summary is presented in Robinson and Coruh (1988).

7.4.1 Magnetic Cleanliness

When using the sensitive portable magnetometers so common today, we cannot be too obsessed with magnetic cleanliness. Seemingly innocent objects (belt buckles, metal eyeglasses, clipboards, watches, pens, and the like) can disrupt readings if they are too close to the instrument in use. The best rule is for the person holding the instrument or sensor to divest herself or himself of all possible contaminants.

Nearby objects can be just as problematic or even more so. Cars, fences, metal poles, AC power lines, metal pipes near the surface, buildings with metal beams or doors, and many other items must be avoided. Although it might be somewhat inconvenient to move 20 meters or more from such objects, the final integrity of the data is worth the effort.

When we use any magnetometer, the sensor never should be closer to the ground than 1 meter and in most cases should be at least 3 meters above ground level. This is to ensure that local accumulations of magnetite such as heavy-mineral pockets in sands do not affect readings.

7.4.2 Diurnal Corrections

Because of the short-term variations in the external field that often cause variations of several nanotesla in main-field intensities, it is necessary to correct magnetic field data in much the same way that we correct gravity data for drift and tidal effects.

A common approach for surveys of limited extent or resources is to reoccupy base stations at regular intervals and follow the same procedure used for gravity drift corrections. Returns to base ideally should occur about every hour so we can be reasonably certain that fluctuations in the external field are not missed. Figure 7.14 illustrates a hypothetical, but not unrealistic, record for diurnal variations. Note that much of the record contains linear changes that would be appro-

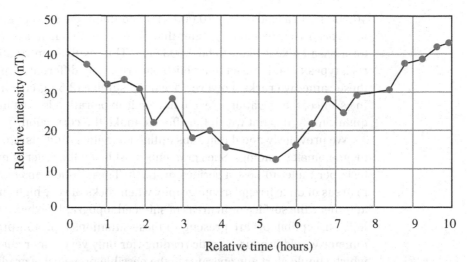

FIGURE 7.14 Representative variation in magnetic field intensity compiled by returning to a base station at short intervals during a day's work.

priately managed by hourly returns to base. However, occasional fluctuations of several nanotesla would be missed by such hourly visits.

A better approach is to place a continuously recording magnetometer at a base station, begin the day's activities at this site, and then return to base at the end of the day. Corrections to magnetometer readings then consist simply of adjusting field readings by the fluctuations observed on the magnetometer at the base station. Even this procedure is not sufficient during magnetic storms because variations in magnitude as great as several hundred nanotesla occur, and differences on the order of 5 to 10 nT are likely between a base and a field position just a few kilometers away. Fieldwork obviously should be suspended during such periods.

7.4.3 Elevation Corrections

Magnetic field data normally are not corrected for elevation differences among recording sites. If you worked out the answer to the second boxed question in Section 7.2.4, you discovered that the vertical gradient of the vertical component of the Earth's magnetic field at 42° N latitude is 0.025 nT/m. Although this gradient clearly varies with position, it gives us an idea of how the magnetic field varies with elevation. Recall that we made a free-air correction for gravity data that amounted to a factor of roughly 0.3 mGal/m in a total field of some 983,000 mGal, and yet we do not make a correction for a variation of roughly 0.03 nT/m in a total field of some 70,000 nT. It certainly is legitimate to ask why such a correction is not made.

The answer lies in the variation of the main field and the relative susceptibilities of rocks. The total variation of gravity amounts to only some 5,200 mGal in a total field of 983,000 mGal, whereas the magnetic field varies by some

40,000 nT in a field of only 70,000 nT. In addition, the susceptibility of a sedimentary rock rarely would be greater than 0.0003 emu, but an average susceptibility for a basalt dike is likely to be 0.003 emu. The relative contributions of these two rock types to an induced magnetic field are vastly different. Typical densities for these same two rocks, however, are quite similar (2.65 g/cm^3 versus 2.8 g/cm^3). In essence, the elevation effect in magnetic exploration is "lost in the noise," and, quite simply, it is not worth the effort to make the correction.

We previously noted that substantial magnetic gradients can negatively affect magnetometer readings. Such gradients must be on the order of hundreds of nanotesla per meter to pose a serious problem. These conditions normally arise only in areas of quite irregular topography when rocks of very high susceptibilities are at or near the surface or in areas of subdued topography when rocks of extremely high susceptibilities are present. In these situations a proton-precession magnetometer would produce erratic readings for only very minor changes in position, which should alert any operator to the possible presence of gradients. If substantial gradients exist due to topography, corrections can be made, but the process is more complex than terrain corrections in gravity work. An approach to magnetic terrain corrections is described in Gupta and Fitzpatrick (1971).

7.4.4   Correcting for Horizontal Position

How can we determine the rate at which the magnetic field varies with position? One approach is to use the same technique as we did for elevation changes. Let's determine the change of the vertical field as horizontal position changes along a magnetic meridian. To do this we once again resort to the dipole equation and the derivative of the vertical field with respect to horizontal position (θ). Thus,

$$\frac{1}{r}\frac{dZ_E}{d\theta} = \frac{1}{r}\frac{-2M\sin\theta}{r^3} = \frac{-2M\sin\theta}{r^4} = -2\frac{H_E}{r}. \tag{7.20}$$

Using appropriate values for 42° N latitude ($r = 6.3 \times 10^6$ m and $H_E = 18,200$ nT), we obtain 0.00578 nT/m or roughly 6 nT/km. This gradient appears sufficiently great to warrant a correction for any but very local surveys; however, note that positioning requirements will not be as severe as in gravity work.

The best way to correct for horizontal position is to consult IGRF data for the area in which a field survey is planned. Such data are available from several sources (such as the U.S. Geological Survey, Branch of Global Seismology and Geomagnetism, Box 25046, MS 968, Denver Federal Center, Denver CO 80225). For example, the total intensity values used in Figure 7.15 were obtained over a computer link to the USGS source for the four specified latitude and longitude coordinates. These coordinates were chosen to extend just beyond the planned survey area (Figure 7.15).

Let's assume we are collecting total-field data. Each field site is noted on a topographic base map, and we obtain accurate latitude and longitude values by

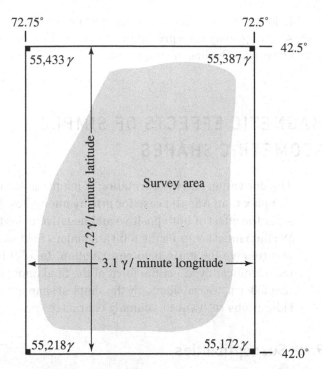

FIGURE 7.15 Variations in total-field intensity as a function of longitude and latitude for a given survey area.

following the overlay method described in the gravity correction discussion. We might designate the F_{E_U} control value for the survey to be 55,172 nT because this value is located in the southeastern corner of the rectangle in Figure 7.15. The main field increases to the north and to the west from this position at the rates noted in Figure 7.15. As an example of such a position correction, assume we recorded a total-field value (F_{E_T}) of 55,193 nT at a site located 0.45 minute north and 0.18 minute west of the control point. The correction should be

$$F_{E_U} = 55{,}172 \text{ nT} + 0.45 \text{ minute}(7.2 \text{ nT/minute}) + 0.18 \text{ minute}(3.1 \text{ nT/minute}) = 55{,}176 \text{ nT}.$$

If we now subtract this value from the total-field reading, we have a total-field anomaly of 17 nT. Of course, if we are recording vertical-field intensities, we follow the same procedure using IGRF vertical-field values.

At this point we've corrected field data for short-term variations of the external magnetic field and for variations of the undisturbed field with geographic latitude and longitude. In general we can ignore the effect of elevation differences in sampling localities; and because most modern instruments record absolute values of the magnetic field, we do not have to tie our readings into base stations as we did in gravity surveying. What now remains is the anomalous magnetic field. To interpret these results, we follow a similar path as we did in our discussion of gravity interpretation and begin by considering the magnetic effects of subsurface bodies of simple shape.

7.5   MAGNETIC EFFECTS OF SIMPLE GEOMETRIC SHAPES

The derivation and interpretation of magnetic anomalies are considerably more complex than was the case for gravity anomalies. This is because we must consider the effect of both positive and negative poles to derive the anomalous field. We then must compute the total anomalous field so that during data reduction we can remove the main field contribution. In addition, remanent magnetization can significantly affect the form of the final anomaly. Therefore, in the material that follows, we consider only the simplest shapes to emphasize the important considerations relevant to anomaly character.

7.5.1   Rock Susceptibilities

Before embarking on an analysis of magnetic anomalies, we pause for a moment to reconsider the topic of magnetic susceptibility and to review several other considerations relevant to computation of magnetic intensities. Recall that we elected to use the cgs system of units in this chapter. Although susceptibility is dimensionless, values normally are expressed as cgs or emu (electromagnetic units) to call attention to the fact that cgs units are in use rather than SI units. The susceptibility values given here can be converted to SI units by multiplying by 4π.

Although several familiar minerals have high susceptibilities (magnetite, ilmenite, and pyrrhotite), magnetite is by far the most common. Rock susceptibility almost always is directly related to the percentage of magnetite present. The true susceptibility of magnetite varies from 0.1 to 1.0 emu depending on grain size, shape, and impurities. Average susceptibility as quoted in several sources ranges from 0.2 to 0.5 emu, so we will adopt a value of $k = 0.35$ emu for our computation. Average susceptibilities for ilmenite and pyrrhotite are 0.15 and 0.125 emu, respectively.

As you might guess, rock susceptibilities vary over quite a range even within a given rock type. The only reliable approach is to measure k in rocks in an exploration area if detailed information is necessary. Representative values for a selection of common rocks and minerals are presented in Telford, Geldart, and Sheriff (1990, 74). For our purposes we believe it is preferable to concentrate on average susceptibilities for common rock groups. This is sufficient for the following discussions of anomaly interpretation and will make representative values easier to remember, but keep in mind that these values vary widely. Approximate averages for common rock groups are as follows:

Sedimentary rocks	0.00005 emu
Metamorphic rocks	0.0003 emu
Granites and rhyolites	0.0005 emu
Gabbros and basalts	0.006 emu
Ultrabasic rocks	0.012 emu

Application of magnetic exploration to the shallow subsurface also may encompass archaeological targets or searches for metallic objects. Susceptibilities of many buried objects (such as stone walls) will, of course, possess the susceptibility of the rock itself. It is not uncommon for soils to have susceptibilities on the order of 0.0001 emu due to the selective sorting of heavy minerals (including magnetite) and the possible presence of maghemite, which has a relatively high k-value. Many iron and steel objects possess a susceptibility ranging from 1 to 10 emu.

For our purposes the k-values here are sufficient. You should know, however, that the apparent susceptibility k_a may be quite different from the true susceptibility k_t when the latter value is greater than 1. In such cases a shape factor η may be present, which reduces the value of induced magnetization that would be predicted on the basis of k_t alone. This is known as the *demagnetization* effect and can be computed if η is known using the relationship

$$k_a = \frac{k_t}{(1+\eta k_t)}. \tag{7.21}$$

Because η varies between 0 and 4π, let's compute the demagnetization effect for a thin horizontal sheet (with true susceptibility = 0.0001 emu) magnetized perpendicular to its surface (in which case $\eta = 4\pi$):

$$k_a = \frac{k_t}{(1+\eta k_t)} = \frac{0.0001}{(1+0.0001\eta)} = 0.00009987 \text{ emu}$$

In this case it is obvious that the demagnetization effect is so small that it can be ignored.

Is there a serious demagnetization effect for a sphere (true susceptibility = 0.001 emu and $\eta = 4\pi/3$)?

Finally, we wish to emphasize once again that remanent magnetization I_r is not considered in the material that follows. You already are aware that the effect of I_r may be substantial, especially if it is close in magnitude to the induced magnetization I_i and oriented in a different direction. In such cases the resulting anomaly is the vector sum of the two magnetizations, which clearly can be very different than I_i alone. In basalts I_r / I_i often is 10 or greater, so you should be especially alert to the role of remanent magnetization when working near basaltic rocks. Other approximate values for the ratio I_r / I_i are ~1.0 in granitic rocks, < 1.0 in metamorphic rocks, and < 0.1 in sedimentary rocks.

7.5.2   Magnetic Effect of an Isolated Pole (Monopole)

Our first effort focuses on the magnetic field above a single pole. Although such an isolated pole cannot exist, the exercise is convenient because of its simplicity. In addition, a very long, thin body oriented vertically and magnetized along its length essentially functions as a monopole because the top surface has a pole strength of $-m$ and the bottom surface $(+m)$ is sufficiently far removed for its effect to be negligible. In our derivation we use Equations 7.5, 7.7, and 7.8. Refer to Figure 7.16 for relevant relationships and notation. These considerations produce

$$V = \frac{m}{r}, \quad m = IA = kF_E A, \quad \text{and} \quad r = (c^2 + z^2)^{\frac{1}{2}} = (x^2 + y^2 + z^2)^{\frac{1}{2}}.$$

Remembering that we determine the magnetic field in a given direction by taking the negative of the derivative of the potential in that direction, we can state

$$Z_A = -\frac{dV}{dz} = -\frac{2z(-\frac{1}{2})(kF_E A)}{(x^2 + y^2 + z^2)^{\frac{3}{2}}} = \frac{z(kF_E A)}{(x^2 + y^2 + z^2)^{\frac{3}{2}}} \tag{7.22}$$

where the inducing field is F_E and A is cross-sectional area. When we determine the horizontal field due to the monopole, it is convenient to orient our coordinate system so that the $+x$ axis is oriented toward magnetic north (Figure 7.16). This orients the horizontal components of the anomalous field H_{A_x} and H_{A_y} parallel to X and Y of the Earth's field. Vertical down is the $+z$ axis. Thus magnetic field components oriented in the sense of the black arrows in Figure 7.16 are considered positive. Now we can determine H_{A_x} and H_{A_y} using the same approach:

$$H_{A_x} = -\frac{dV}{dx} = \frac{x(kF_E A)}{(x^2 + y^2 + z^2)^{\frac{3}{2}}} \tag{7.23}$$

and

$$H_{A_y} = -\frac{dV}{dy} = \frac{y(kF_E A)}{(x^2 + y^2 + z^2)^{\frac{3}{2}}}. \tag{7.24}$$

The total anomalous field is calculated using the form of Equation 7.18. Because H_{A_x} represents the component of the horizontal anomalous field in the direction of magnetic north, we have

$$F_{A_T} = Z_A \sin i + H_{A_x} \cos i. \tag{7.25}$$

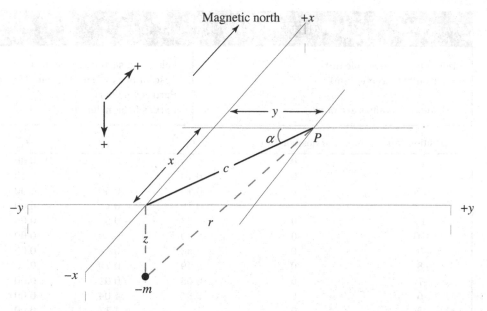

FIGURE 7.16 Relationships and notation used to derive the magnetic effect of a single pole.

Following our practice in previous chapters, let's put these equations into a dynamic table (Table 7.1) and analyze the results. The table as illustrated provides results for a traverse along the x axis but lets us obtain results for any position in x–y space.

Figure 7.17 is a graph of the vertical, horizontal, and total anomaly fields for the traverse values in Table 7.1. The vertical-field curve is quite straightforward: You would expect intensity to be greatest directly over the monopole and to decline equally in both directions away from this position. Usually we are interested only in the vertical and total fields because these are the fields measured during magnetic exploration. However, we present the horizontal-field component here so you can visualize (with the aid of Equation 7.18) the contribution it makes to the total field. The horizontal-field components are easy to understand if you refer to Figure 7.17(b). Small inserts illustrate orientations of the vertical- and horizontal-field vectors for $-x$ and $+x$ values. Based on the convention we adopted in Figure 7.16, the vertical-field component always is positive, but the horizontal-field component is positive for $-x$ values and negative for $+x$ values. You can best understand the total-field values and curve form by noting the curves for Z_A and H_A in addition to the value of i and referring once again to Equation 7.18.

7.5.3 Magnetic Effect of a Dipole

The next step up in complexity is to derive the magnetic field due to a dipole without the constraints of distance and dipole length imposed in our previous

TABLE 7.1	MAGNETIC EFFECT OF A MONOPOLE				

Depth to negative pole (m)	**5**		Pole cross-sectional area (m²)	**0.5**	
Horizontal increment (m)	**1**		Susceptibility of pole (cgs emu)	**0.003**	
			Earth's field (nT)	**55000**	
(All anomaly values are in nT.)			Earth's field inclination (°)	**70**	

x position (m)	y position (m)	Z_A	H_{A_x}	H_{A_y}	F_{A_T}
−15	0	0.10	0.31	0.00	0.21
−14	0	0.13	0.35	0.00	0.24
−13	0	0.15	0.40	0.00	0.28
−12	0	0.19	0.45	0.00	0.33
−11	0	0.23	0.51	0.00	0.40
−10	0	0.30	0.59	0.00	0.48
−9	0	0.38	0.68	0.00	0.59
−8	0	0.49	0.79	0.00	0.73
−7	0	0.65	0.91	0.00	0.92
−6	0	0.87	1.04	0.00	1.17
−5	0	1.17	1.17	0.00	1.50
−4	0	1.57	1.26	0.00	1.91
−3	0	2.08	1.25	0.00	2.38
−2	0	2.64	1.06	0.00	2.84
−1	0	3.11	0.62	0.00	3.14
0	0	3.30	0.00	0.00	3.10
1	0	3.11	−0.62	0.00	2.71
2	0	2.64	−1.06	0.00	2.12
3	0	2.08	−1.25	0.00	1.53
4	0	1.57	−1.26	0.00	1.05
5	0	1.17	−1.17	0.00	0.70
6	0	0.87	−1.04	0.00	0.46
7	0	0.65	−0.91	0.00	0.30
8	0	0.49	−0.79	0.00	0.19
9	0	0.38	−0.68	0.00	0.12
10	0	0.30	−0.59	0.00	0.08
11	0	0.23	−0.51	0.00	0.04
12	0	0.19	−0.45	0.00	0.02
13	0	0.15	−0.40	0.00	0.01
14	0	0.13	−0.35	0.00	0.00
15	0	0.10	−0.31	0.00	−0.01

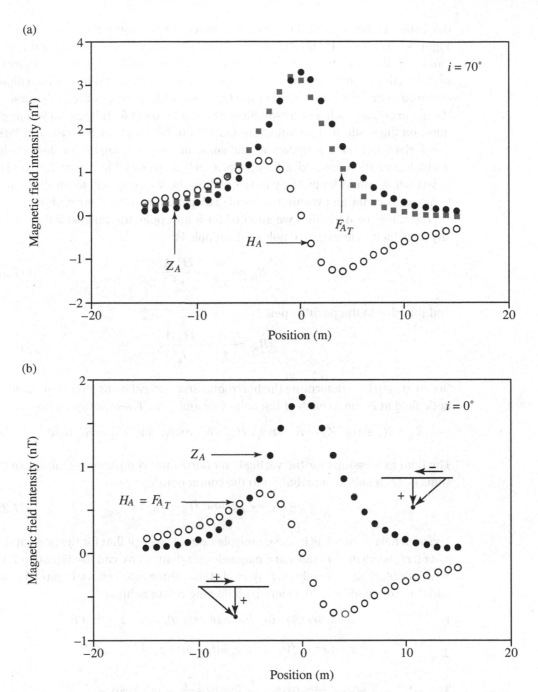

(a)

(b)

FIGURE 7.17 Intensities of vertical (Z_A), horizontal (H_{A_x}), and total (F_{A_T}) magnetic field anomalies for a monopole. Traverse directly over the monopole in a direction parallel to a magnetic meridian. (a) Inclination of Earth's field = 70°. (b) Inclination of Earth's field = 0°.

derivation of dipole equations. For the present derivation we use Equation 7.2. Figures 7.18(a) and (b) illustrate the general geometry and notation for the derivation. In this case we assume the dipole is magnetized along its axis (parallel to its length). If the magnetization is induced, such a situation requires both dipole orientation and the orientation of the Earth's field F_E to be parallel. Of course, in the general case such a correspondence would be most fortuitous, so you might question the wisdom of pursuing this derivation. We do so for two reasons. First, the derivation is straightforward, and some of the resulting anomalies provide useful information. Second, although the dipole as pictured in Figure 7.18 and its polarization would be unlikely in the real world, we soon will learn that dipole behavior is contained within bodies of geologic interest. So bear with us.

Recalling the definition we adopted for a unit pole, the magnetic field intensity at P due to the negative pole of the dipole is

$$R_{A_n} = +\frac{m}{r_n^2} = +\frac{kF_E A}{r_n^2} \tag{7.26}$$

and that due to the positive pole is

$$R_{A_p} = -\frac{m}{r_p^2} = -\frac{kF_E A}{r_p^2} \ . \tag{7.27}$$

Our next step is to determine the horizontal and vertical components of the magnetic field at P due to each of the poles ($-m$ and $+m$). These components are

$$Z_{A_n} = R_{A_n} \sin\phi_1, \ Z_{A_p} = R_{A_p} \sin\phi_2, \ H_{A_n} = R_{A_n} \cos\phi_1, \ \text{and} \ H_{A_p} = R_{A_p} \cos\phi_2. \tag{7.28}$$

The final expressions for the vertical and horizontal components simply are the sums of each pole's contribution to the component or

$$Z_A = Z_{A_n} + Z_{A_p} \quad \text{and} \quad H_A = H_{A_n} + H_{A_p}. \tag{7.29}$$

The total field is found as in the monopole example except that for the present derivation H_A is oriented parallel to a magnetic meridian, so we can use Equation 7.18. Now let's place these results in a dynamic table. Once again check Figures 7.18(a) and (b), from which we develop the following relationships:

1. $\qquad a = L\cos(180 - \theta), \quad b = L\sin(180 - \theta), \quad \text{and} \quad z_p = z_n + b$

2. $\qquad r_n = (x^2 + z_n^2)^{\frac{1}{2}}, \quad \text{and} \quad r_p = ((x-a)^2 + z_p^2)^{\frac{1}{2}}$

3. $\qquad \sin\phi_1 = \frac{z_n}{r_n}, \quad \cos\phi_1 = \frac{x}{r_n}, \quad \sin\phi_2 = \frac{z_p}{r_p}, \quad \text{and} \quad \cos\phi_2 = \frac{(x-a)}{r_p}$

Equations 7.26 through 7.29 and the relationships noted here all are used in Table 7.2. To save space several cells are not shown in Table 7.2 but are included

(a)

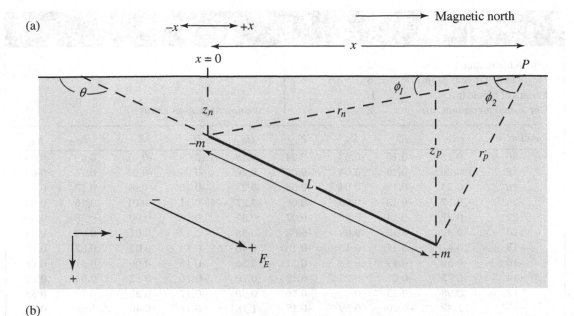

(b)

FIGURE 7.18 Relationships and notation used to derive the magnetic field of a dipole.

in the spreadsheet template on the resources website. These include the inducing field F_E (55,000 nT), the susceptibility k (.003 emu), and the cross-sectional area A (1.0 m²). These all may be changed if you wish.

Following our standard procedure, let's graph the values in Table 7.2 and then investigate some other relationships facilitated by this table. Figure 7.19 illustrates the vertical, horizontal, and total magnetic field anomalies as represented in Table 7.2. The angle of inclination of the dipole (120°) represents an inclination of the Earth's field of 60° (see Figure 7.18). Immediately evident is the fact that the vertical-field anomaly has negative as well as positive values (as does the

TABLE 7.2   MAGNETIC EFFECT OF A DIPOLE

Dipole inclination (°)	120
Depth to negative pole (m)	5
Length of dipole (m)	15
Horizontal increment (m)	2

(All anomaly values are in nT.)

Position (m)	R_{A_n}	R_{A_p}	Z_{A_n}	Z_{A_p}	H_{A_n}	H_{A_p}	Z_A	H_A	F_{A_T}
−30	0.18	−0.10	0.03	−0.04	0.18	−0.09	−0.01	0.09	0.03
−28	0.20	−0.10	0.04	−0.05	0.20	−0.09	−0.01	0.11	0.04
−26	0.24	−0.11	0.04	−0.05	0.23	−0.10	−0.01	0.13	0.06
−24	0.27	−0.13	0.06	−0.06	0.27	−0.11	−0.01	0.16	0.07
−22	0.32	−0.14	0.07	−0.07	0.32	−0.12	0.00	0.20	0.10
−20	0.39	−0.15	0.09	−0.08	0.38	−0.13	0.01	0.25	0.13
−18	0.47	−0.17	0.13	−0.10	0.46	−0.14	0.03	0.32	0.18
−16	0.59	−0.19	0.18	−0.11	0.56	−0.15	0.06	0.41	0.26
−14	0.75	−0.21	0.25	−0.13	0.70	−0.16	0.12	0.54	0.37
−12	0.98	−0.23	0.38	−0.16	0.90	−0.17	0.22	0.73	0.55
−10	1.32	−0.26	0.59	−0.19	1.18	−0.18	0.40	1.00	0.85
−8	1.85	−0.29	0.98	−0.22	1.57	−0.19	0.76	1.38	1.35
−6	2.70	−0.33	1.73	−0.26	2.08	−0.20	1.47	1.88	2.21
−4	4.02	−0.36	3.14	−0.30	2.51	−0.19	2.84	2.32	3.62
−2	5.69	−0.40	5.28	−0.35	2.11	−0.19	4.93	1.93	5.23
0	6.60	−0.43	6.60	−0.40	0.00	−0.17	6.20	−0.17	5.29
2	5.69	−0.47	5.28	−0.45	−2.11	−0.14	4.84	−2.25	3.06
4	4.02	−0.49	3.14	−0.48	−2.51	−0.09	2.66	−2.61	1.00
6	2.70	−0.51	1.73	−0.50	−2.08	−0.04	1.23	−2.12	0.00
8	1.85	−0.51	0.98	−0.51	−1.57	0.01	0.47	−1.56	−0.37
10	1.32	−0.50	0.59	−0.50	−1.18	0.07	0.09	−1.11	−0.47
12	0.98	−0.48	0.38	−0.47	−0.90	0.12	−0.09	−0.78	−0.47
14	0.75	−0.45	0.25	−0.42	−0.70	0.15	−0.17	−0.55	−0.42
16	0.59	−0.42	0.18	−0.38	−0.56	0.18	−0.20	−0.38	−0.37
18	0.47	−0.38	0.13	−0.33	−0.46	0.19	−0.20	−0.26	−0.31
20	0.39	−0.34	0.09	−0.28	−0.38	0.20	−0.19	−0.18	−0.25
22	0.32	−0.31	0.07	−0.24	−0.32	0.19	−0.17	−0.12	−0.21
24	0.27	−0.28	0.06	−0.20	−0.27	0.19	−0.15	−0.08	−0.17
26	0.24	−0.25	0.04	−0.17	−0.23	0.18	−0.13	−0.05	−0.14
28	0.20	−0.22	0.04	−0.15	−0.20	0.17	−0.11	−0.03	−0.11
30	0.18	−0.20	0.03	−0.12	−0.18	0.16	−0.09	−0.02	−0.09

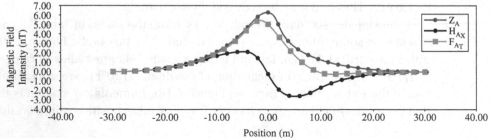

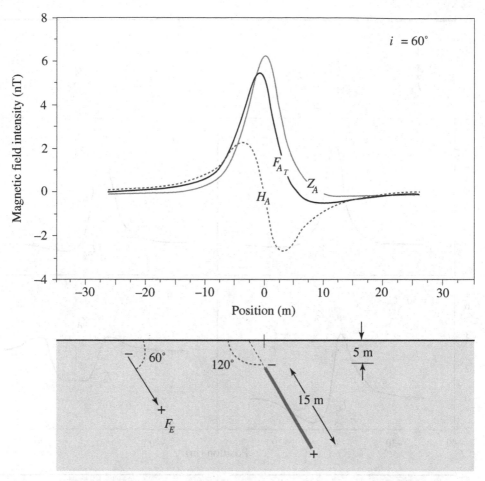

FIGURE 7.19 Intensities of vertical, horizontal, and total magnetic field anomalies over a dipping dipole polarized along its axis. The dipole is oriented parallel to a magnetic meridian.

total-field anomaly). You now should begin to realize that magnetic anomalies are going to be more difficult to interpret because in most cases both positive and negative poles will affect final curve form. Note also that the total-field curve differs to a fair degree from the vertical-field curve. The reason for this should be evident from the form of the horizontal-field curve and Equation 7.18. Maxima and minima are not in the same position, nor do they have the same values. We already mentioned that vertical-field curves are simpler to work with, so this disparity likely will increase the difficulty of quantitatively interpreting magnetic field curves.

In general, total-field curves and vertical-field curves for anomalies are similar at high magnetic latitudes (approximately 70° and greater), so for normal purposes we can analyze total-field curves as if they were vertical-field curves if our data come from such latitudes. Figure 7.20 illustrates vertical and total anomaly curves for several inclinations of the Earth's field. Of course, the total-field

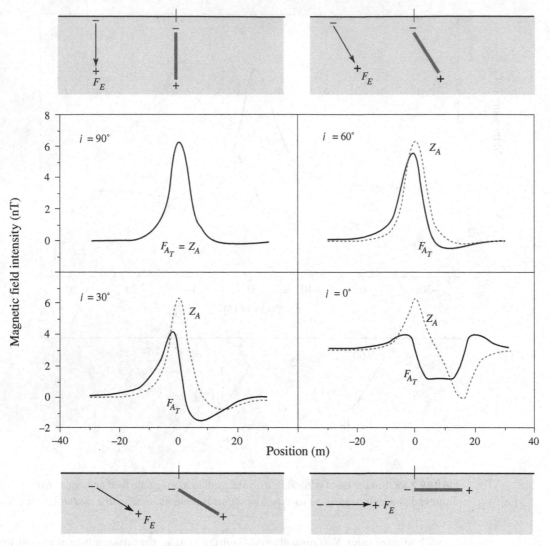

FIGURE 7.20 Total (solid) and vertical (dashed) magnetic anomalies over a dipole polarized along its axis for various inclinations of the Earth's field. In each case the dipole is parallel to a magnetic meridian.

and vertical-field curves are identical for an inclination of 90°. For $i = 60°$ the two curves are somewhat close in form and in position and magnitude of maxima and minima (as just noted). These correspondences decrease progressively as i is decreased to 30° and 0°.

Due to complexity, we generally do not derive equations or provide diagrams of magnetic anomalies as they appear on maps. However, in the dipole case, equations for x–y–z space are convenient to generate. Figure 7.21 provides the general relationships. Most quantities are the same as in Figure 7.18. The dipole

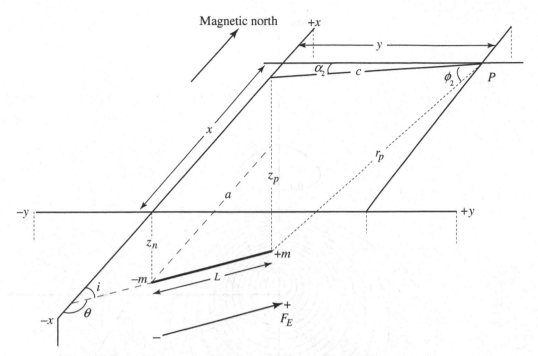

FIGURE 7.21 Relationships and notation used to derive the magnetic effect of a dipole oriented parallel to a magnetic meridian at any point on an x–y surface.

still is contained within the x–z plane. The only adjustment we need to make is to realize that at each point P we compute H_x and H_y components. Similarly to the previous derivation, we generate H_x by determining H_{x_n} and H_{x_p} and adding these latter two quantities. We determine H_y in the same way. Final values can be calculated in a spreadsheet similar to Table 7.2, but we do not include it here due to file size and the amount of space available on the enclosed disks. The results of such a derivation are illustrated in Figure 7.22.

7.5.4 Magnetic Effect of a Sphere

The derivation of equations to determine the magnetic effect of a sphere is more complex than our derivation for a dipole. Our purpose here is to present the general steps in this derivation to give you a sense of the procedure. With this in mind, we consider the simplest case: that of a vertically polarized sphere. We follow this with the case of a sphere with inclined polarization. Figure 7.23 serves as a reference for both derivations. In the case of the vertically polarized sphere, $i = 90°$. Thus this derivation is strictly applicable only to cases in which a spherical mass is located where the Earth's field has an inclination of 90°.

The crucial relationship for this derivation is referred to as *Poisson's relation* (Dobrin and Savit 1988, 640–41), which states that the magnetic potential V is

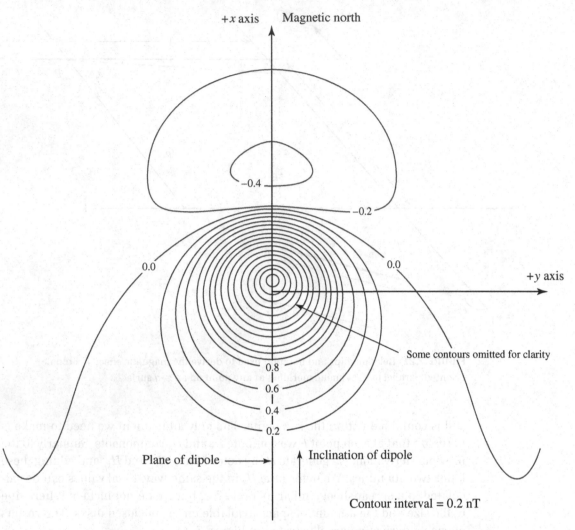

FIGURE 7.22 Contours of total-field anomaly intensities above a dipole polarized along its axis. Input parameters are similar to those in Table 7.2.

proportional to the derivative of the gravity potential U in the direction of magnetization or

$$V = -\frac{I}{\rho G}\frac{dU}{dw} \tag{7.30}$$

where w is the direction of magnetic polarization, assuming the body has uniform susceptibility and density. Remembering that the direction of magnetization for our derivation is vertical or z, we define vertical- and horizontal-field anomalies, Z_A and H_A, as

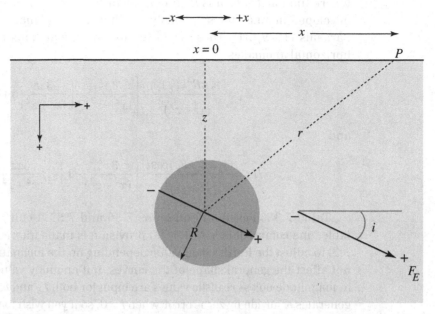

FIGURE 7.23 Notation used for the derivation of magnetic field anomalies over a uniformly magnetized sphere.

$$Z_A = -\frac{dV}{dz} = \frac{I}{\rho G}\frac{d^2U}{dz^2} \tag{7.31}$$

and

$$H_A = -\frac{dV}{dx} = \frac{I}{\rho G}\frac{d}{dx}\left(\frac{dU}{dz}\right). \tag{7.32}$$

Because the gravitational potential of a sphere is

$$U = \frac{GM}{r} = \frac{G\tfrac{4}{3}\pi R^3\rho}{r} = \frac{G\tfrac{4}{3}\pi R^3\rho}{(x^2+z^2)^{\tfrac{1}{2}}}, \tag{7.33}$$

then

$$\frac{d^2U}{dz^2} = \frac{(G\tfrac{4}{3}\pi R^3\rho)(2z^2-x^2)}{(x^2+z^2)^{\tfrac{5}{2}}}. \tag{7.34}$$

This gives us the vertical anomaly:

$$Z_A = \left(\frac{I}{G\rho}\right)\frac{(G\tfrac{4}{3}\pi R^3\rho)(2z^2-x^2)}{(x^2+z^2)^{\tfrac{5}{2}}} = \frac{(\tfrac{4}{3}\pi R^3 I)(2z^2-x^2)}{(x^2+z^2)^{\tfrac{5}{2}}}. \tag{7.35}$$

The horizontal anomaly follows similarly.

If we approach the more general case, that of a sphere uniformly magnetized where the Earth's field is inclined, the derivation is more complex but can be developed in much the same fashion. Thorough explanations are presented in Parasnis (1979, 31). These yield the following equations for the vertical and horizontal anomalies:

$$Z_A = \frac{(\tfrac{4}{3}\pi R^3 kF_E)\sin i}{(x^2+z^2)^{\frac{3}{2}}}\left[\frac{3z^2}{(x^2+z^2)}-\left(\frac{3xz}{(x^2+z^2)}\cot i\right)-1\right] \qquad (7.36)$$

and

$$H_A = \frac{(\tfrac{4}{3}\pi R^3 kF_E)\cos i}{(x^2+z^2)^{\frac{3}{2}}}\left[\left(\frac{3x^2}{(x^2+z^2)}-1\right)-\left(\frac{3xz}{(x^2+z^2)}\tan i\right)\right] \qquad (7.37)$$

Table 7.3 is based on Equations 7.36 and 7.37. Before we work with this table, one comment is in order. No provision is made (nor is one made in Table 7.2) to adjust the Earth's field value depending on the inclination value. This does not affect the general shape of the curves, but anomaly values will be incorrect in magnitude unless realistic values are input for both F_E and i. Also, Equation 7.36 generates a "divide by zero" error when $i = 0$, so if you wish to use such a low inclination value, substitute $i = 0.001$.

Curves based on the values in Table 7.3 (Earth field inclination of 60°) are illustrated in Figure 7.24. Compare Figure 7.24 with Figure 7.19. They are amazingly similar! If we continue this comparison by graphing anomaly curves for the sphere with inducing field inclinations of 0°, 30°, and 90°, we finish with the results in Figure 7.25. Again this figure is remarkably similar to Figure 7.20. If we were even more careful and ensured that the magnetic moment of the sphere was the same as that of the dipole, these curves would have similar magnitudes as well as form.

It now seems warranted to conclude that, if a sphere is uniformly magnetized by an inducing field, it produces anomalies similar to those resulting from a *dipole magnetized along its axis* (see Figure 7.23). As in gravity exploration, a sphere can represent a somewhat equidimensional mass, which is much more likely in the subsurface than a dipole that happens to be in the same orientation as that of the inducing field. Thus our use for the dipole is clearly valuable even though it achieves its status through the intermediary of the sphere. Studying Figure 7.25 once again supports our earlier contention that vertical and total anomalies are similar in form at high magnetic latitudes. In addition, note how different the total field is from the vertical field at low magnetic latitudes. If we substitute a dipole for the sphere, it's quite easy to understand the vertical anomaly for an inclination of low value, but it takes a bit more effort to work out the reason for the form of the total-field anomaly.

Although we presented the case of the horizontal cylinder in the previous chapter, we do not do so in this chapter. As long as the strike length is long rel-

TABLE 7.3   MAGNETIC EFFECT OF A SPHERE

Position (m)	Z_A	H_A	F_{A_T}
−15	0.00	0.01	0.00
−14	0.00	0.01	0.00
−13	0.00	0.01	0.00
−12	0.00	0.01	0.00
−11	0.00	0.01	0.01
−10	0.00	0.02	0.01
−9	−0.01	0.03	0.01
−8	−0.01	0.04	0.02
−7	−0.01	0.06	0.03
−6	0.00	0.09	0.05
−5	0.01	0.15	0.10
−4	0.05	0.26	0.20
−3	0.19	0.52	0.48
−2	0.76	1.02	1.24
−1	2.65	1.21	2.81
0	3.45	−1.45	1.71
1	0.20	−1.71	−0.94
2	−0.47	−0.44	−0.64
3	−0.29	−0.06	−0.26
4	−0.16	0.02	−0.11
5	−0.09	0.03	−0.05
6	−0.05	0.03	−0.03
7	−0.03	0.02	−0.01
8	−0.02	0.02	−0.01
9	−0.02	0.01	0.00
10	−0.01	0.01	0.00
11	−0.01	0.01	0.00
12	−0.01	0.01	0.00
13	−0.01	0.01	0.00
14	0.00	0.00	0.00
15	0.00	0.00	0.00

Depth to sphere center (m)	**1.75**
Sphere radius (m)	**0.4**
Susceptibility (cgs emu)	**0.001**
Horizontal increment (m)	**1**

Earth's field *(F_E)* (nT)	**45000**
Earth's field inclination (°)	**50**

(All anomaly values are in nT.)

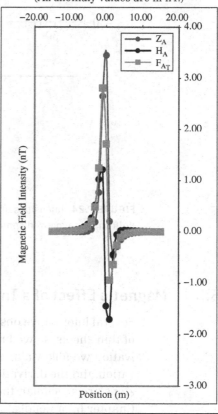

ative to cross-sectional radius, we can treat the cylinder as a two-dimensional case. The curves are qualitatively similar to those of the sphere. At this point we change gears and revert to a previous derivation to understand magnetic anomalies due to thin sheets.

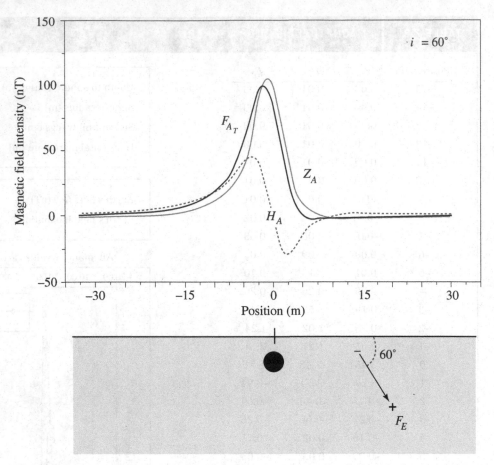

FIGURE 7.24 Intensities of vertical, horizontal, and total magnetic field anomalies over a uniformly magnetized sphere.

7.5.5   Magnetic Effect of a Thin, Horizontal Sheet

Several informative observations stem from a consideration of the magnetic effect of thin sheets, so we briefly present how such effects are calculated. In this derivation we follow almost precisely the same path we used for the Bouguer derivation and the derivation of the gravity effect of a horizontal sheet. Because the steps are so similar, the presentation here is abbreviated, so be sure to refer to Chapter 6 for details.

In all considerations of horizontal sheets presented here, we assume the inducing field is vertical. If we consider a uniform strip of negative poles (Figure 7.26(a)), then, using Equation 7.2, the magnetic field intensity at point P due to a small area of the strip is

$$R_A = \frac{(m/_{area})\,dx\,dy}{r^2}. \tag{7.38}$$

Following the Bouguer effect derivation, we can say that the vertical field at P due to the strip of poles is

$$Z_A = \int_{x=-\infty}^{x=+\infty} \frac{(m/_{area})\,dx\,dy}{r^2}\cos\theta, \tag{7.39}$$

$$Z_A = \int_{-\pi/2}^{\pi/2} \frac{(m/_{area})\,dy}{d}\cos\theta\, d\theta, \tag{7.40}$$

and

$$Z_A = \frac{2(m/_{area})\,dy}{d}. \tag{7.41}$$

We next sweep out a sheet of poles using the strip. Recall that in the Bouguer derivation, we rotated our view with respect to the coordinate axes for the second

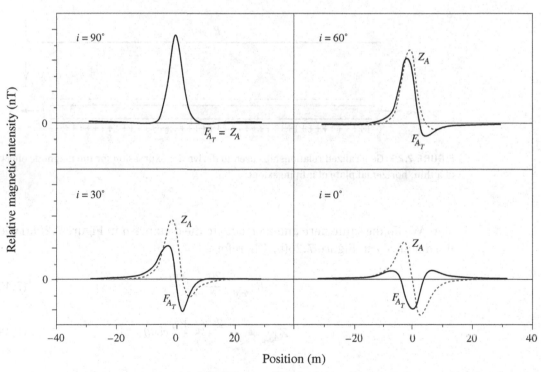

FIGURE 7.25 Total (solid) and vertical (dashed) magnetic anomalies over a uniformly magnetized sphere for various inclinations of the Earth's field.

(a)

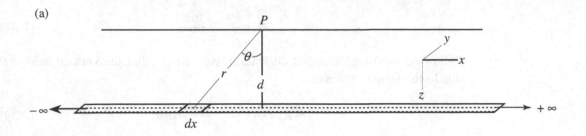

(b)

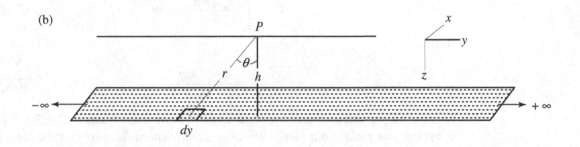

(c)

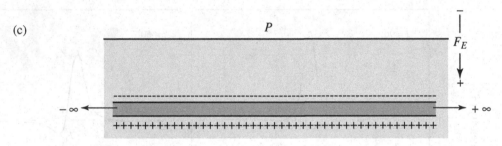

FIGURE 7.26 Generalized relationships used to derive the expression for the magnetic effect of a thin, horizontal plate of infinite extent.

step. We do the same here and then equate the distance d in Figure 7.26(a) with the distance r in Figure 7.26(b). Therefore,

$$R_{A_{sheet}} = \frac{2(m/_{area})dy}{r}, \tag{7.42}$$

$$Z_{A_{sheet}} = \int_{y=-\infty}^{y=+\infty} \frac{2(m/_{area})dy}{r} \cos\theta, \tag{7.43}$$

$$Z_{A_{sheet}} = \int_{-\pi/2}^{\pi/2} 2(m/_{area})d\theta, \tag{7.44}$$

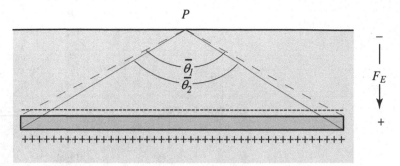

FIGURE 7.27 Notation used to derive the expression for a thin plate of limited extent.

and

$$Z_{A_{sheet}} = 2(m/_{area})\pi = 2\pi I . \qquad (7.45)$$

We can apply Equation 7.45 to estimate the magnetic anomaly due to such a thin, horizontal plate of infinite extent. For such a case and with a vertical inducing field, negative poles are induced on the upper surface of the plate and positive poles on the bottom as illustrated in Figure 7.26(c). The magnetic anomaly due to such a feature is

$$Z_A = Z_{A_{top}} - Z_{A_{bottom}} = 2\pi I - 2\pi I = 0.$$

Our next step is to truncate the slab as illustrated in Figure 7.27. Similarly to our gravity example, we can demonstrate that the total included angle can be substituted for π in Equation 7.45, so that the vertical-field anomaly for the truncated thin plate is

$$Z_A = Z_{A_{top}} - Z_{A_{bottom}} = 2I\bar{\theta}_1 - 2I\bar{\theta}_2 = 2I(\bar{\theta}_1 - \bar{\theta}_2). \qquad (7.46)$$

One of the most interesting applications of these relationships is to determine magnetic anomalies due to basement that is overlain by nonmagnetic material. One example of such a situation is a thick sequence of Paleozoic sedimentary rocks overlying Precambrian metamorphic and igneous rocks in the midcontinent region of the United States. Another example consists of a thin cover of Pleistocene sediments overlying Mesozoic sedimentary rocks cut by basaltic volcanics, which is similar to the geology of the Connecticut Valley of Massachusetts and Connecticut.

Let's look at a representation of this latter geology and assume a vertical inducing field. Consider Figure 7.28(a). If the basement units are very thick, the effective result is a sheet of negative poles induced on the basement surface (the positive poles effectively are removed from consideration). The susceptibility of the Pleistocene sediments is so low that we can ignore their magnetic effect. The *effective susceptibility k'* of the basalt is 0.0045. Using Equation 7.46 and

(a)

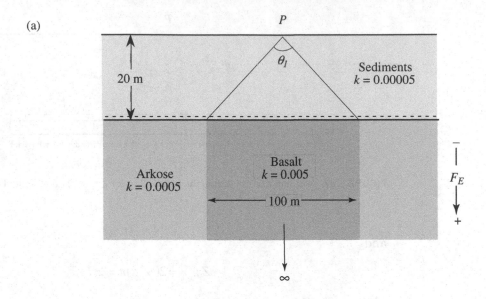

(b)

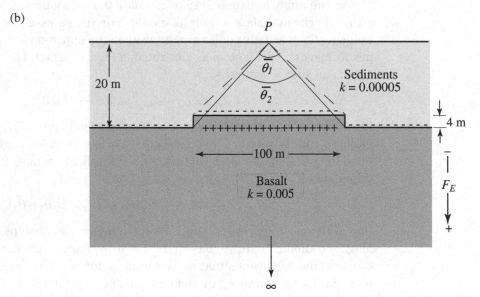

FIGURE 7.28 Comparison of the magnetic effects of a lithologic contrast and structural relief.

remembering that the included angle is in radians, we see that the vertical-field anomaly at P is

$$Z_A = Z_{A_{top}} - Z_{A_{bottom}} = [2(0.0045 \text{ emu})(60,000 \text{ nT})(2.38)] - 0 = 1286 \text{ nT}.$$

Figure 7.28(b) diagrams an instance of structural relief on basement which, although only 4 m, is significant considering the thickness of the overlying sediments. Due to this relief, we also have to contend with a strip of positive poles at the level of the deeper basement surface. Therefore, the anomaly at P due to this structural relief is

$$Z_A = Z_{A_{top}} - Z_{A_{bottom}} = [2(0.005 \text{ emu})(60{,}000 \text{ nT})(2.52)] - [2(0.005 \text{ emu})(60{,}000 \text{ nT})(2.38)]$$

$$= 1513 \text{ nT} - 1429 \text{ nT} = 84 \text{ nT}.$$

Such results demonstrate that for situations similar to that diagrammed in Figure 7.28, magnetic exploration is much more effective in mapping lithologic contrasts (assuming sufficient k-values are present) than in determining structural relief. Of course, relief on the order of that in Figure 7.28 in rocks with such a high k-value should be quite evident.

As a final exercise in deriving equations for magnetic effects, we once again apply Equation 7.46 to the problem of a thin, semi-infinite plate or sheet. The geometry is presented in Figure 7.29 and is almost identical to a similar derivation we presented for the gravity effect of such a model. The only difference in the magnetic case is that we are dealing with two included angles due to negative and positive poles. Once again, refer to the derivation in Chapter 6 for details. In this case

$$\bar{\theta}_1 = \frac{\pi}{2} + \tan^{-1}\left(\frac{x}{z}\right), \quad \bar{\theta}_2 = \frac{\pi}{2} + \tan^{-1}\left(\frac{x}{(z+t)}\right),$$

and so

$$Z_A = 2I(\bar{\theta}_1 - \bar{\theta}_2) = 2kF_E\left[\frac{\pi}{2} + \tan^{-1}\left(\frac{x}{z}\right) - \frac{\pi}{2} - \tan^{-1}\left(\frac{x}{z+t}\right)\right],$$

or

$$Z_A = 2kF_E\left[\tan^{-1}\left(\frac{x}{z}\right) - \tan^{-1}\left(\frac{x}{z+t}\right)\right]. \tag{7.47}$$

Finally, we place this equation in a dynamic table, Table 7.4. This table is organized the same way as Table 6.9, so we not only can produce curves for

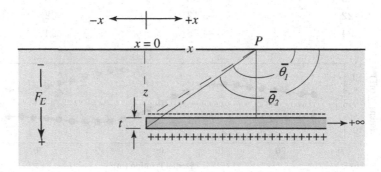

FIGURE 7.29 Relationships and notation for derivation of semi-infinite sheet equation.

TABLE 7.4   MAGNETIC EFFECT OF THIN, SEMI-INFINITE SHEETS

	Sheet A		Sheet B
Depth to sheet (m)	5	Depth to sheet (m)	10
Sheet thickness (m)	1	Sheet thickness (m)	1
Susceptibility (emu)	0.001	Susceptibility (emu)	0.001
Horizontal increment (m)	2	Horizontal increment (m)	2

Horizontal Position (m)	Z_A (nT)	Horizontal Position (m)	Z_A (nT)
24	4.75	−24	−4.20
22	5.13	−22	−4.44
20	5.58	−20	−4.70
18	6.10	−18	−4.97
16	6.71	−16	−5.24
14	7.42	−14	−5.49
12	8.26	−12	−5.67
10	9.21	−10	−5.71
8	10.19	−8	−5.51
6	10.88	−6	−4.93
4	10.41	−4	−3.81
2	7.05	−2	−2.11
0	0.00	0	0.00
−2	−7.05	2	2.11
−4	−10.41	4	3.81
−6	−10.88	6	4.93
−8	−10.19	8	5.51
−10	−9.21	10	5.71
−12	−8.26	12	5.67
−14	−7.42	14	5.49
−16	−6.71	16	5.24
−18	−6.10	18	4.97
−20	−5.58	20	4.70
−22	−5.13	22	4.44
−24	−4.75	24	4.20

(a)

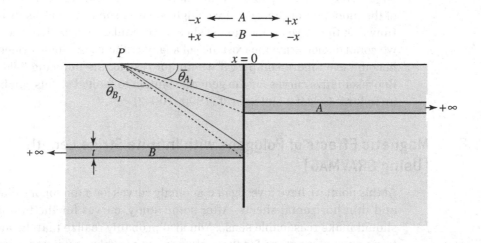

(b)

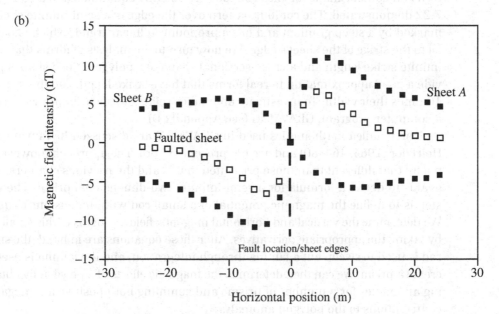

FIGURE 7.30 Using semi-infinite sheets to simulate a faulted horizontal bed. (a) Notation used in Table 7.4 to derive magnetic anomaly values. (b) Graph illustrating effects of sheet A, sheet B, and sum of A and B (the faulted relationship shown in (a)).

a semi-infinite sheet as in Figure 7.29 but also can simulate faulted sheets (Figure 7.30) as we did in the gravity case. For the representation of Table 7.4 here, the magnitude of the vertical inducing field is set at 60,000 nT.

The results from Table 7.4 are illustrated in Figure 7.30. Let's focus on the curve for sheet A for a moment. Note the pronounced asymmetry (remember that this is for the vertical-field anomaly Z_A) and the steep slope directly over the edge of the sheet. The form of the curve for Z_A is sufficiently different from that of the monopole or dipole that we might suspect some sheetlike subsurface form. However, the curve for the faulted sheet is so similar in form that it seems unlikely we could demonstrate the existence of a fault without additional evidence. Matters become even more complex if we do not restrict the inducing field to vertical. Representative curves for the general solution to faulted sheets can be consulted in Telford, Geldart, and Sheriff (1990, 100–2).

7.5.6 Magnetic Effects of Polygons with Infinite Strike Length (Using GRAVMAG)

At this point we have investigated anomaly curves for a monopole, dipole, sphere, and thin horizontal sheets. After some study, curves for the first three forms should make reasonable sense. You also probably realize that the anomaly patterns on contour maps for these shapes are roughly equidimensional as Figure 7.22 demonstrated. The contour pattern over the edge of a semi-infinite sheet is marked by a steep gradient and has a pronounced linear trend, which is parallel to the strike of the sheet's edge. We now turn to an analysis of forms that have infinite strike length and a cross-sectional shape of a polygon. Such shapes provide a good approximation to real forms that have strike lengths on the order of 10 times their width. To investigate anomalies related to such shapes we turn to a computer program, GRAVMAG (see Appendix D).

The detailed mathematics used in GRAVMAG are discussed in Talwani and Heirtzler (1964, 464–80) and are not presented here. The approach, however, is not all that different than those presented in this and the previous chapters. The goal is to derive a formula for a semi-infinite, two-dimensional prism. The first step is to define the magnetic potential of a small rod with dimensions $dxdydz$. We determine the vertical and horizontal magnetic field strengths of this small rod by taking the appropriate derivatives. After these equations are in hand, the small rod is used to sweep out a lamina through integration, and the lamina is used to create a prism. We can then determine the magnetic effects of a polygon by obtaining anomalies for a number of prisms and summing both positive and negative contributions to the polygon anomaly.

In our preceding discussion of magnetic anomalies we investigated the difference in curve form for vertical-, horizontal-, and total-field anomalies and learned how these curves vary with inclination of the Earth's field. We also made a number of simplifying arrangements such as orienting our dipole parallel to a magnetic meridian. Of course, in the real world we won't be so fortunate, so our main emphasis in this section is to study the effect of model orientation on anomaly magnitude and form. We begin with a vertical prism and primarily are concerned with the relationship between prism strike and the direction of magnetic north. To keep everything relatively simple, we present curves only for Z_A. After you understand the following material, create total-field curves for the same input parameters and see if you can explain resulting curve forms. For the material that follows we use several files included on the resources website: *zero magnetics data—2 km, shallow vertical prism*, and *inclined shallow prism*.

First let's orient the vertical prism so its strike is parallel to a magnetic meridian. We choose a total-field value of 52,000 nT and an inclination of 65°. The resulting curve for the vertical-field anomaly is presented in Figure 7.31(a). The curve is quite symmetric about the midpoint of the prism top and resembles that for a vertical dipole. If we now rotate the prism through 90° so that its strike is west–east, curve shape changes and is clearly more asymmetric. Note that the only parameter changed is prism strike. Why does this produce a curve shape change and noticeable differences in field intensities? The most direct answer involves a consideration of the horizontal and vertical components of the inducing field (Figure 7.32(a)). Z_E acts normal to the top and bottom of the prism, and

(a)

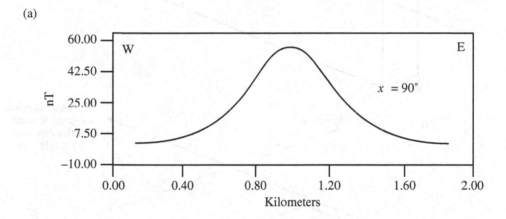

(b)

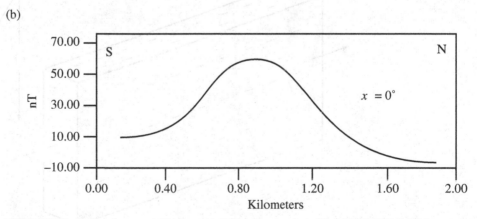

FIGURE 7.31 Anomaly curves (Z_A) for a vertical prism produced by GRAVMAG. (a) Prism strike is parallel to magnetic north (x angle = 0°; traverse direction is west–east); (b) prism strike is perpendicular to magnetic north (x angle = 90°; traverse direction is south–north).

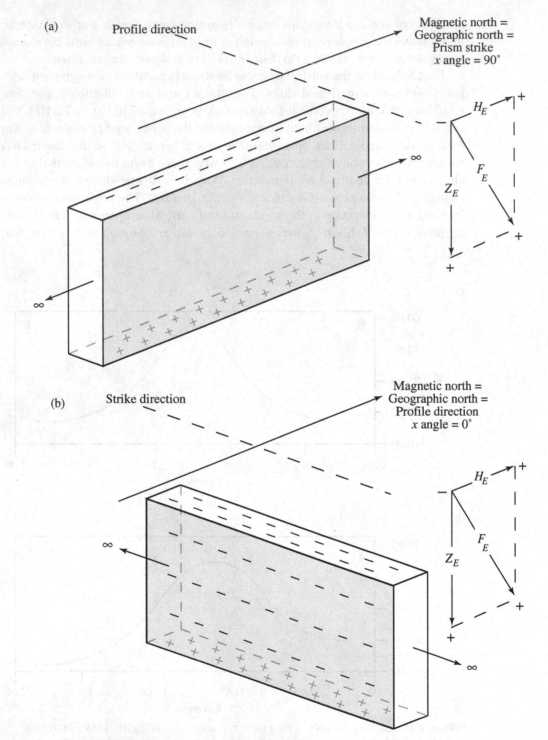

FIGURE 7.32 Prism orientations responsible for curves in Figure 7.31. (a) corresponds to Figure 7.31(a) and (b) corresponds to Figure 7.31(b).

it therefore induces a magnetization that can be presented as an accumulation of negative poles on the upper surface of the prism and positive poles on the bottom surface. H_A, however, acts normal to the ends of the prism that are infinitely far removed from the position of measurement. These pole accumulations therefore do not affect local field intensities, which are due to the effects produced by the top and bottom of the prism; hence the symmetric form of the curve.

In the case of a west–east striking prism, the horizontal component of the Earth's field does have an effect. Pole accumulations are induced on the top and bottom surfaces of the prism as before. Now, however, the broad, vertical side of the prism is perpendicular to H_E, and negative and positive pole accumulations

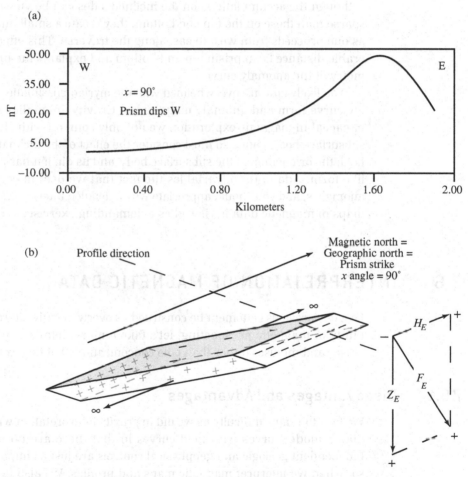

FIGURE 7.33 Effects of an inclined prism. (a) Anomaly curve (Z_A) for an inclined prism produced by GRAVMAG. (b) Prism strike is parallel to magnetic north (x angle = 0°; traverse direction is west–east).

will be induced on these sides. Negative poles are indicated on the side closest to an observer (the shaded surface in Figure 7.32(b)), but the positive poles are not shown on the far side so as not to clutter the diagram. These pole accumulations are not as dense as those produced by Z_A because H_A is significantly less than Z_A when the inclination is 65°. For a north–south traverse over this prism, the pole accumulations on the vertical sides add to the effect of those on the top and bottom and produce the observed asymmetry.

Next let's take the prism with north–south strike (and symmetric curve) and incline it to the west (Figure 7.33(b)). The curve form now becomes markedly asymmetric with a long "tail" in the direction of dip. What is different? We can analyze this in the same qualitative way that we just used for prism strike. H_A still has no effect for this prism orientation. However, Z_A now induces pole accumulations on the inclined sides of the prism as well as on its top and bottom. Although the accumulations on the inclined sides can be viewed as much more sparse than those on the top and bottom, they create a small but detectable effect as one proceeds from west to east along the traverse. This effect is felt a considerable distance from prism top and bottom and explains the pronounced asymmetry of the anomaly curve.

You likely now are overwhelmed with the myriad possibilities that affect anomaly curve form and anomaly magnitudes. Gravity anomalies may appear as a panacea! In magnetic exploration we not only contend with shape and depth of subsurface bodies but also must consider the effect of the inclination of the Earth's field, the orientation of the subsurface body, and its dip if it has a tabular or sheetlike form. Add to these variables the fact that we deal mostly with total-field anomalies, and you should appreciate why a detailed interpretation of profiles and maps of magnetic data is viewed as a demanding exercise.

7.6   INTERPRETATION OF MAGNETIC DATA

Lest our previous statement be construed as overly negative in regard to the likely utility of magnetic prospecting, let's take a more detailed look at some of the pluses and minuses (other than pole concentrations!) of this exploration method.

7.6.1   Disadvantages and Advantages

We face the same difficulty as we did in gravity interpretation when attempting to match model curves with field curves in that there are no unique solutions. Independent geologic and geophysical controls are just as important, if not more so, when we interpret magnetic maps and profiles. We also face a difficulty not present in gravity surveys: that of remanent magnetization. If sufficient remanence is present, the anomaly we view is a combination of this remanence and the induced magnetization and may be very different in form than if no remanence existed. Because data for remanent magnetization characteristics most often are absent, a degree of uncertainty commonly is added to many interpretations.

Not only is the range of magnetic susceptibility quite variable even for one rock type, but there is no guarantee that the minerals responsible for the susceptibility are uniformly distributed throughout the rock mass. In a basalt flow with which our classes worked, susceptibility was quite high due to abundant magnetite. However, the magnetite was concentrated in pockets distributed throughout the flow and would give somewhat discontinuous anomalies if mapped in great detail on a large scale. On the other hand, these pockets were distributed throughout the extent of the flow, so the overall magnetic signature of this unit was clear.

The common practice of working with total-field measurements does nothing to reduce the difficulties of interpreting magnetic data. Because the direction of the magnetic vector is not known when these data are collected, a number of data processing techniques used in gravity work either cannot be applied or are much more difficult to apply. In concluding this list of disadvantages, we emphasize once again the dependence of anomaly characteristics on direction of magnetization.

Possibly the greatest advantage of the magnetic method is cost relative to area surveyed. This is especially true when aeromagnetic surveying is considered but is also true for ground surveys. This low relative cost coupled with its high precision makes it a very attractive method for several purposes even if quantitative results are difficult to achieve.

Although the local orientation of the Earth's field affects anomaly form, this orientation is, of course, essentially constant for all surveys in a given area even when the area is quite large. A geophysicist therefore can become thoroughly familiar with characteristic curve shapes for the range of geologic model forms typically employed in interpretation methods.

Because magnetic susceptibilities are low in a wide variety of rock types, it usually is easier to focus on the likely source material when large anomalies are encountered since high susceptibilities belong to such a restricted range of rocks. Of course, some external geologic information is necessary; but a reduced set of such information often is sufficient as compared to the task of correlating gravity anomalies with specific rock types. In addition, the simple result of finding a large anomaly (such as in an area of suspected mineralization) often is sufficient, and more elaborate interpretation procedures are not necessary. Good examples of using anomalies to focus other exploration techniques without analysis of the anomalies themselves are presented during the discussion of case studies in a following section. Marked susceptibility contrasts also lend themselves to geologic mapping, which relies heavily on tracing of high magnetic gradients but not necessarily on detailed analysis of anomaly form.

You already are aware of the utility of Poisson's relation from our discussion of a sphere's magnetic effect. This ability to relate the magnetic and gravitational potentials makes it possible to transform magnetic data into a *pseudogravitational* field. Not only should this field be less complex to analyze than the magnetic field from which it was derived, but it can be compared with gravity data directly if these are available. A good correlation between anomalies in the pseudogravitational and gravitational fields suggests a similar source for both and more narrowly defines the interpretation process.

Many data processing techniques used in gravity work can be adapted for magnetic work, often with little additional effort. The separation of regional trends from local anomalies continues to be an important step in the interpretation process. Second derivative methods emphasize local anomalies and remove more regional trends. Upward continuation removes anomalies from shallow sources and is particularly useful when survey objectives include basement depth determinations. Downward continuation enhances shallow anomalies and can be used to separate overlapping anomalies from nearby shallow sources. Now might be a good time to review the descriptions of these approaches in Chapter 6.

7.6.2 Quantitative Interpretation Techniques

7.6.2.1 HALF-MAXIMUM TECHNIQUES
As in the case of gravity, half-maximum techniques can provide a reasonably good value with which to begin other modeling endeavors if we have some information about the body responsible for the observed anomaly. The procedure for developing such half-maximum relationships is identical to those followed for gravity anomalies; but owing to the complexity of anomaly shape from even simple subsurface shapes, the process provides a rougher approximation than in the gravity case.

As a straightforward example, consider the slender vertical rod with bottom far removed from the observer. Recall that this shape can be represented by a monopole. If we assume our traverse is directly over the rod, Equation 7.22 simplifies to

$$Z_A = \frac{z(kF_E A)}{(x^2 + z^2)^{\frac{3}{2}}}. \tag{7.48}$$

Because the maximum is directly over the center of the rod, one-half this maximum value is

$$Z_{A_{\frac{1}{2}\max}} = \frac{(kF_E A)}{2z^2}. \tag{7.49}$$

At a position where the maximum is one-half its value, we have

$$Z_{A_{\frac{1}{2}\max}} = \frac{(kF_E A)(z^2 + x^2_{\frac{1}{2}\max})^{\frac{1}{2}}\cos\theta}{(z^2 + x^2_{\frac{1}{2}\max})^{\frac{3}{2}}} \tag{7.50}$$

because $\cos\theta = \dfrac{z}{r}$ and $r = (z^2 + x^2_{\frac{1}{2}max})^{\frac{1}{2}}$ (see Figure 7.16).

As before, we equate Equations 7.49 and 7.50 such that

$$\frac{(z^2 + x^2_{\frac{1}{2}max})^{\frac{1}{2}} \cos\theta}{(z^2 + x^2_{\frac{1}{2}max})^{\frac{3}{2}}} = \frac{1}{2z^2}. \tag{7.51}$$

This reduces to

$$2z^3 = (z^2 + x^2_{\frac{1}{2}max})^{\frac{3}{2}} \tag{7.52}$$

and, finally,

$$0.766z = x_{\frac{1}{2}max} \quad \text{or} \quad z = x_{\frac{1}{2}max} / 0.766. \tag{7.53}$$

Relationships for several other shapes include the following (Telford, Geldart, and Sheriff 1990, 85–102):

- *Sphere:* Total width of anomaly curve at $Z_{A_{max}}/2$ roughly equals depth to sphere center.c
- *Cylinder:* Total width of anomaly curve at $Z_{A_{max}}/2$ roughly equals depth to cylinder center.
- *Semi-infinite sheet:* Depth roughly equals one-half the distance from $Z_{A_{max}}$ to $Z_{A_{min}}$.

7.6.2.2 SLOPE METHODS Figure 7.34(a) pictures Z_A curves for three vertical prisms with tops at different depths. As you no doubt would expect, the prism closest to the surface produces the curve with the greatest amplitude and greatest slopes. We've already observed similar effects in our study of gravity anomalies, so perhaps you are wondering if some precise relation exists between curve shape characteristics and source depth. Many geophysicists have explored this possibility and have developed a number of useful techniques for analysis of anomaly curves. As you also might expect, due to the complexities inherent in magnetic anomaly analysis, most of these methods provide only approximations, work only in somewhat restricted circumstances, and are not universally applicable. Nevertheless, slope methods are widely applied in efforts to determine source depths and do give useful first approximations. These approximations may be sufficient or may be used as a starting point for more sophisticated analysis. Two of the most frequently cited approaches are those of Peters (1949) and Vacquier et al. (1951). To provide some sense of how these methods work, we present a summary of Peters's solution to the depth problem.

Peters's slope method uses Z_A curves related to vertical prisms by first identifying the maximum slope on the anomaly curve (Figure 7.35). The next step is

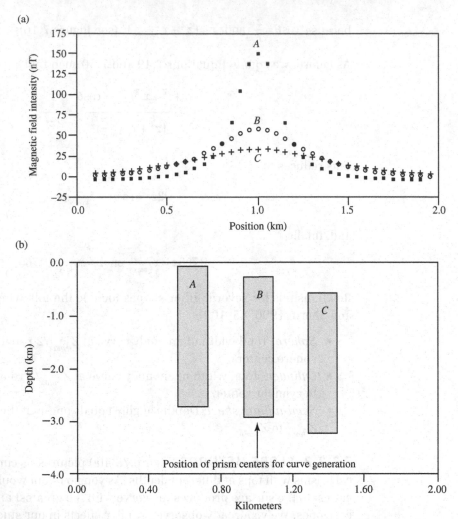

FIGURE 7.34 (a) Magnetic anomaly curves (Z_A) for vertical prisms with tops at various depths. (b) Depths of prisms used to generate curves in (a).

to create a line that has a slope equal to one-half the maximum slope value. This "one-half maximum slope" line is moved about on the anomaly curve until its two points of tangency with the curve are identified (Figure 7.35). The distance on the horizontal axis between these two points of tangency is denoted as d. The normal relationship between depth to the top of the prism z and d is

$$z = \frac{d}{1.6}. \tag{7.54}$$

In Figure 7.35 $d = 0.43$ km, which results in a value for z of 0.3 km. This corresponds to the depth of the top of prism B in Figure 7.34(b), which was used to generate the anomaly curve.

This approach assumes that prism length is great compared to width, that the depth to prism top is the same order of magnitude as prism width, and that the prism has infinite strike length and is oriented parallel to a magnetic meridian. Naturally, all of these constraints rarely are realized, so that in most cases Equation 7.54 provides at best an approximation. Peters notes various adjustments that can be made to this equation, such as changing the value 1.6 to 1.2 if prism width is very narrow.

The Peters slope method has been widely applied in spite of its constraints and limitations. For instance, total-field anomalies can be analyzed at high magnetic latitudes because they are so similar to vertical-field anomalies in these regions. In this as in other similar methods, a great deal of experience is necessary to apply slope approaches with an expectation of useful estimates to anomaly depths.

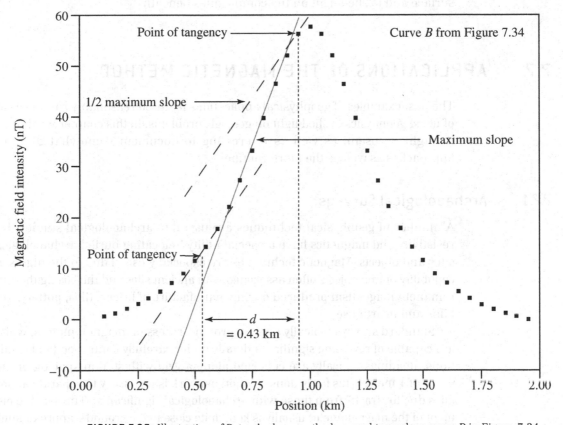

FIGURE 7.35 Illustration of Peters's slope method as used to analyze curve B in Figure 7.34.

7.6.2.3 COMPUTER MODELING As in the interpretation of gravity anomalies, modeling programs such as GRAVMAG or ones that accept three-dimensional models often are used for magnetic interpretation. We compare anomaly values calculated by the model with measured values to evaluate the model that presumably is based as much as possible on geologic and/or geophysical controls.

When a satisfactory fit is achieved, the model represents one possible interpretation of the subsurface configuration responsible for the observed magnetic measurements. We emphasize again in conclusion that many different models can produce similar magnetic anomalies. Unique solutions do not exist. As geologic and geophysical controls diminish, the number of solutions increases rapidly. Thus interpretative techniques such as those outlined here are extremely useful in supplying appropriate boundary conditions. However, the most useful and reliable analyses will utilize every bit of information available from well logs and surface mapping as well as from other geophysical studies. This is why it is so important to be a good geologist as well as a good geophysicist and why no geophysical approaches should be neglected in the quest to penetrate beyond the surface and to shed light on the complexities beneath.

7.7 APPLICATIONS OF THE MAGNETIC METHOD

The past examples of geophysical applications featured reasonably routine cases of using geophysics to shed light on geologic problems. In this chapter we thought it might be useful as well as interesting to document somewhat different approaches as well as the more routine.

7.7.1 Archaeological Surveys

A number of geophysical techniques are useful in archaeological studies, but resistivity and magnetics have a special utility in locating buried archaeological sites and objects. Magnetic techniques are especially useful due to the high susceptibility of iron objects often associated with ancient sites and the strong thermoremanent magnetism produced during manufacture of bricks, tiles, pottery, and kilns and hearth use.

Standard surveys typically employ proton-precession magnetometers, which are capable of revealing significant details under carefully controlled field conditions. Readings typically are collected along a grid with station spacing on the order of 1 meter. This fairly dense station network is necessary to separate anomalies due to "trash" from those with archaeological significance. The sensing element of the magnetometer usually is kept quite close to the ground—approximately 0.5 m. Because of the station density and number of readings collected (due to the ease of fieldwork with the proton-precession magnetometer) computer data stor-

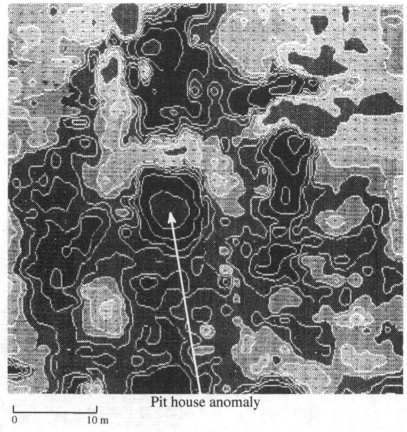

Pit house anomaly

0   10 m

FIGURE 7.36 Total-field magnetic map from the Dolores Archaeological Program, Colorado. Contour interval is 1 nT. The main anomaly is from a burned and filled pit house. (From John W. Weymouth. 1986. Archaeological site surveying program at the University of Nebraska. *Geophysics*, v. 51, 538–52.)

age normally is mandatory. Figure 7.36 illustrates the result of such a survey. The major anomaly, that of a burned and filled pit house, has an average anomaly of 27 nT (Weymouth 1986, 546).

An example of an especially detailed survey is highlighted in Figure 7.37. This survey employed an adaptation of the flux-gate magnetometer we described previously. Two detectors were rigidly mounted and aligned to measure the gradient of the vertical-field component. Readings were taken continuously along a specified traverse direction and plotted on an *x–y* recorder. If numerous traverses are aligned side-by-side as pictured in Figure 7.37, an extremely detailed site map is produced. This amount of detail makes it much easier to differentiate anomalies of true archaeological significance from peaks due to recent rubbish, small accumulations of magnetic minerals in soils, and other disturbing influences. In Figure 7.37 the two large circles are ditches of Bronze Age round

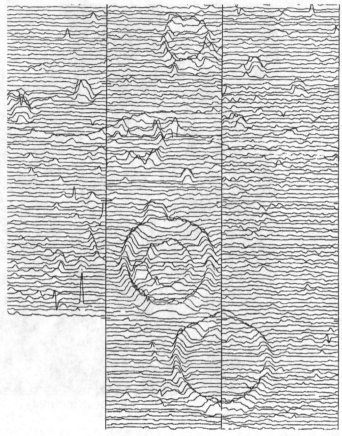

FIGURE 7.37 Results of a flux-gate magnetometer gradiometer survey at Radly, Oxfordshire, England. The major circular features are ditches of Bronze Age burial mounds. (From Anthony J. Clark. 1986. Archaeological geophysics in Britain. *Geophysics*, v. 51, 1404–13.)

barrows. The upper circle contains an inner ditch and a central grave pit (Clark 1986, 1407). Such wonderful detail illustrates what magnificent dividends can result from a little ingenuity and careful attention to the basic fundamentals of geophysical exploration techniques.

7.7.2 Detection of Voids and Well Casings

At first it might not seem logical to search for caves or voids using the magnetic method. However, if rocks have sufficient magnetic susceptibility, then a void within these rocks produces a negative susceptibility contrast. An excellent example of a study located in an area with such a contrast is presented in Arzate et al. (1990). This study describes a magnetic survey at Teotihuacan, Mexico, conducted to map tunnels in the Oztoyohualco area. Standard magnetic survey procedures included the use of a proton-precession magnetometer, measurements

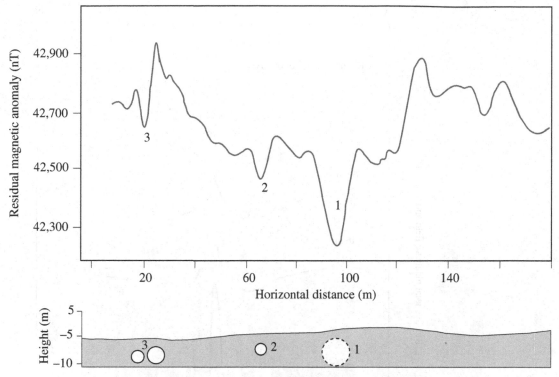

FIGURE 7.38 Residual magnetic anomaly profile in nT demonstrating correlation of magnetic lows with known tunnels (solid circles at positions 3 and 2) and revealing an undiscovered tunnel (dashed circle at position 1). (Modified from J.A. Arzate, L. Flores, R.E. Chávez, L. Barba, and L. Manzanilla. 1990. Magnetic prospecting for tunnels and caves in Teotihuacan, México. In *Geotechnical and environmental geophysics, volume 3: Geotechnical,* ed. Stanley H. Ward, 155–62. Society of Exploration Geophysicists Investigations in Geophysics, no. 5.)

taken 2 m above ground level, and removal of a regional field. Residual magnetic lows such as those illustrated in Figure 7.38 (positions 2 and 3) correlate well with known tunnel locations. Another pronounced low (position 1) is believed to be due to a previously unknown tunnel location. The susceptibility of the pyroclastic rocks in which the tunnels are located is approximately 0.006 emu. Using a susceptibility contrast of −0.006 emu and an average magnitude of 300 nT for the anomaly at position 1, the authors of the study arrived at a tunnel size of 4.5 m radius and depth to center of 7.7 m (Arzate et al. 1990, 159).

Such interpretations appear straightforward, but note that they are possible only because previous independent knowledge of the subsurface was available to enable correlation of anomalies with known sources. Even so, the identification of the anomaly at position 1 in Figure 7.38 would be considerably strengthened by conducting an additional survey with a geophysical method known to be capable of detecting voids (such as the studies described involving electrical resistivity and gravity).

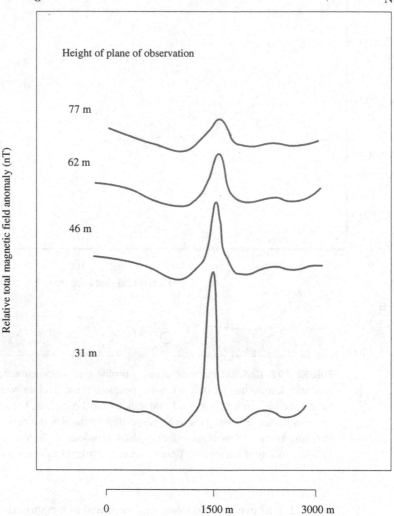

S   N

Height of plane of observation

77 m

62 m

46 m

31 m

Relative total magnetic field anomaly (nT)

0   1500 m   3000 m

FIGURE 7.39 Total magnetic field anomaly profiles in nT measured at various elevations above a cased well at Piney Creek, Colorado. (Modified from F.C. Frischknecht and P.V. Raab. 1984. *Location of abandoned wells with geophysical methods*. Environmental Protection Agency 600/4-84-085.

Magnetic methods are, as you might expect, very efficient at detecting iron and steel objects buried at depths of only a few meters. Often buried barrels and old well casings are of great concern because these frequently pose a significant contamination hazard. Frischknecht and Raab (1984) demonstrated that abandoned well positions can be identified by the magnetic method. Figure 7.39 is taken from this study and confirms that well casings provide an identifiable magnetic signature even at distances of several tens of meters above the well.

7.7.3   Defining Landfill Geometry

In the preceding chapter we described a detailed investigation of the Thomas Farm landfill designed to assess the effectiveness of the gravity method in studying landfills. As part of this same study, magnetic data were collected using a proton-precession magnetometer (Roberts, Hinze, and Leap 1990). Stations were located on a 2 m by 2 m grid; observations were at 1 m above ground level; and data were corrected for diurnal variations. The magnetic pattern is characterized by intense magnetic variations. Figure 7.40 provides a sample of the amplitudes of these variations, which are due to buried iron and steel objects. Some variations also are due to surface objects beyond landfill boundaries. A 6 m upward continuation of the magnetic data also is plotted in Figure 7.40. Because this upward continuation subdues the effect of shallow sources, the 6 m curve provides a more generalized view of landfill character. This should be apparent when you compare the form of the 6 m curve and the location of landfill boundaries (see Figure 7.40).

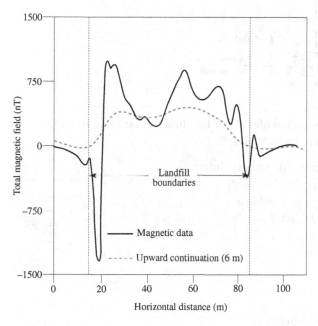

FIGURE 7.40 Total magnetic field profile in nT. The profile is oriented along an east–west line and crosses the southern third of the Thomas Farm landfill. (Modified from R.L. Roberts, W.J. Hinze, and D.I. Leap. 1990. Data enhancement procedures on magnetic data from landfill investigations. In *Geotechnical and environmental geophysics, volume 2: Environmental and groundwater,* ed. Stanley H. Ward, 261–66. Society of Exploration Geophysicists Investigations in Geophysics, no. 5.)

Before closing this chapter, we urge you to reread Chapter 1 and to pay particular attention to the discussions about fundamental considerations, defining objectives, limitations of geophysical methods, and the advantage of using multiple methods. These words perhaps will be more significant now than when you first read them. Consider the admonitions carefully, and above all, remember the importance of linking geophysical interpretations to geologic information.

PROBLEMS

7.1 What is the horizontal gradient in nT/m of the Earth's vertical field (Z_E) in an area where the horizontal field (H_E) equals 20,000 nT and the Earth's radius is 6.3×10^8 cm?

7.2 Use Equation 7.2 as a basis for deriving an equation for the magnetic effect of a thin, vertical, infinitely long rod magnetized along its axis (monopole).

7.3 A buried stone wall constructed from volcanic rocks has a susceptibility contrast of 0.001 emu with its enclosing sediments. The wall is very long compared to its other dimensions. Approximate the wall with a cylinder and assume the anomaly from a sphere is sufficiently close to that of a cylinder to allow a reasonable estimate of field magnitudes. Determine the wall's detectability with a typical proton-precession magnetometer when $i = 50°$.

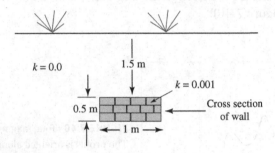

7.4 How large would an iron artifact ($k = 1.0$) have to be to be detected with a proton-precession magnetometer?

7.5 Could a total-field magnetic survey detect the burial chamber (spherical void) illustrated here in a region where $F_E = 55,000$ nT and $i = 70°$?

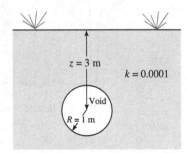

7.6 Determine the vertical-field anomaly (Z_A) for the intrusive diagrammed here. The intrusive has a very long strike length. F_E is vertical and equal to 55,000 nT.

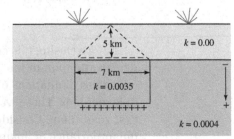

7.7 The magnetic data graphed below represent vertical-field measurements (Z_A) in an area where shallow crystalline basement is overlain by non-magnetic sediments. The basement gneisses are intruded by numerous thin kimberlite pipes. Both gneisses and kimberlite are eroded to a common level surface. Determine the likely depth to basement. $F_E = 58,000$ nT and $i = 80°$.

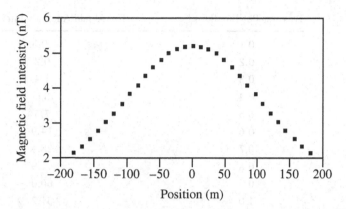

7.8 The curve illustrated below is from a magnetic traverse in an area of numerous basalt flows. A 25-meter deep drill hole at position –15 m encountered no basalt. Comment as fully as possible on the near-surface geology.

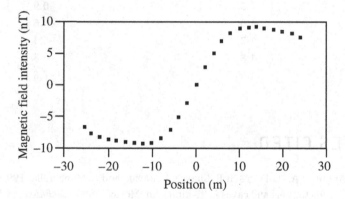

7.9 Discuss the wisdom of using the Peters slope method to determine depth to the top of a vertical prism in the following situation. The vertical prism is 2 km wide and has a large strike length and large vertical length. The prism strikes N70W and has a susceptibility contrast of 0.0013 emu. Data for the Earth's field at the prism location are $F_E = 50,000$ nT, $d – 0°$, and $i = 70°$. Only a proton-precession magnetometer is available for survey use.

7.10 The data given here represent total-field anomaly intensities in a region where tilted basalt flows are known to exist. Important magnetic field information includes $F_E = 55{,}000$ nT, $i = 65°$, $d = 20°$, and $k = .001$ emu. The basalt flows generally strike N70E, so the traverse direction (x angle) for the total-field measurements is 340°. Comment on the direction of dip and likely dip amount of the flow over which the data were gathered.

Traverse Position (km)	Total-Field Anomaly Intensity (nT)
0.1	−0.5
0.2	0.2
0.3	1.4
0.4	3.6
0.5	7.5
0.6	14.9
0.7	28.4
0.8	47.9
0.9	60.0
1.0	51.7
1.1	32.9
1.2	17.9
1.3	8.7
1.4	2.9
1.5	−0.9
1.6	−3.6
1.7	−5.4
1.8	−6.8
1.9	−7.8

REFERENCES CITED

Arzate, J.A., L. Flores, R.E. Chávez, L. Barba, and L. Manzanilla. 1990. Magnetic prospecting for tunnels and caves in Teotihuacan, México. In *Geotechnical and environmental geophysics, volume 3: Geotechnical,* ed. Stanley H. Ward, 1–30. Society of Exploration Geophysicists Investigations in Geophysics, no. 5.

Clark, Anthony J. 1986. Archaeological geophysics in Britain. *Geophysics,* v. 51, 1404–13.

Dobrin, Milton B., and Carl H. Savit. 1988. *Introduction to geophysical prospecting,* 4th ed. New York: McGraw-Hill.

Frischknecht, F.C., and P.V. Raab. 1984. *Location of abandoned wells with geophysical methods.* Environmental Protection Agency 600/4-84-085.

Gupta, V.K., and M.M. Fitzpatrick. 1971. Evaluation of terrain effects in ground magnetic surveys. *Geophysics,* v. 36, 582–89.

Parasnis, D.S. 1979. *Principles of applied geophysics*, 3rd ed. London: Chapman and Hall.

Peters, Leo J. 1949. A direct approach to magnetic interpretation and its practical application: *Geophysics*, v. 14, 290–320.

Roberts, R.L., W.J. Hinze, and D.I. Leap. 1990. Data enhancement procedures on magnetic data from landfill investigations. In *Geotechnical and environmental geophysics, volume 2: Environmental and groundwater,* ed. Stanley H. Ward, 261–66. Society of Exploration Geophysicists Investigations in Geophysics, no. 5.

Robinson, Edwin S., and Cahit Coruh. 1988. *Basic exploration geophysics*. New York: John Wiley & Sons.

Talwani, M., and J.R. Heirtzler. 1964. Computation of magnetic anomalies caused by two-dimensional structures of arbitrary shape. In *Computers in the mineral industries, part 1,* 464–80. Stanford University Publication.

Telford, W.M., L.P. Geldart, R.E. Sheriff, and D.A. Keys. 1976. *Applied geophysics*. Cambridge, England: Cambridge University Press.

Telford, W.M., L.P. Geldart, and R.E. Sheriff. 1990. *Applied geophysics*, 2nd ed. Cambridge, England: Cambridge University Press.

Vacquier, Victor, Nelson C. Steenland, Roland G. Henderson, and Isadore Zietz. 1951. *Interpretation of aeromagnetic maps*. Geological Society of America Memoir 47.

Weymouth, John W. 1986. Archaeological site surveying program at the University of Nebraska. *Geophysics*, v. 51, 538–52.

SUGGESTED READING

Breiner, S. 1973. *Applications manual for portable magnetometers*. Palo Alto, CA: Geometrics.

Burazer, M., M. Grbovic, and V. Zitko. 2001. Magnetic data processing for hydrocarbon exploration in the Pannonian Basin, Yugoslavia. *Geophysics*, v. 66, no. 6, 1669–79.

Mathe, V., and F. Leveque. 2003. High-resolution magnetic survey for soil monitoring: Detection of drainage and soil tillage effects. *Earth and Planetary Science Letters*, v. 212, no. 1–2, 241–51.

Nagata, Takesi. 1961. *Rock magnetism*. Tokyo: Maruzen Publishing Company Ltd.

Nettleton, L.L. 1971. *Elementary gravity and magnetics for geologists and seismologists*. Society of Exploration Geophysicists Monograph Series, no. 1.

Sparlin, M.A., and R.D. Lewis. 1994. Interpretation of the magnetic anomaly over the Omaha oil field, Gallatin County, Illinois. *Geophysics*, v. 59, no. 7, 1092–99.

Stern, D.P. 2002. A millennium of geomagnetism. *Reviews of Geophysics*, v. 40, no. 3, article no. 1007.

CHAPTER 8

Electromagnetic Surveying

Electromagnetic methods are finding increasing use in environmental, engineering, and archaeological applications because of their high resolution and speed and ease of use. However, the basic physics behind electromagnetic methods is more complicated than for most other geophysical methods. In this chapter we try to give you an understanding and appreciation of the basic physics behind EM induction, rather than presenting these tools as a black box method that is often used but seldom understood. In addition to geologic applications, the use of electromagnetic induction is common in everyday life. Electromagnetic induction is used by metal detectors at airports, by magnetic tape recorders, by digital computer memory, and by transformers that transmit electric power over large distances.

Electromagnetic methods are sensitive to variations in electrical properties of subsurface materials and can map out regions with enhanced conductivity because of the presence of fluids, metals, or other variations. Electromagnetic induction instruments induce currents in conductors within the Earth without having to make direct contact with the ground. Such instruments can collect data rapidly and have the additional advantage that no cables are needed. Ground-penetrating radar methods are much like seismic reflection methods but are sensitive to variations in electrical properties of subsurface layers. They have high resolution but limited depth penetration. With GPR detailed shallow subsurface radar reflection images can be made rapidly.

In this chapter we begin with an introduction to electromagnetic wave propagation and then discuss the basic physics of electromagnetic induction. We describe the physical properties that electromagnetic waves are sensitive to, including conductivity and dielectric permittivity. Factors common to electromagnetic wave propagation, such as absorption, attenuation, and skin depth, are explained. Specific inductive electromagnetic techniques are then covered, followed by ground-penetrating radar methods and case studies using EM induction and GPR methods.

8.1 ELECTROMAGNETIC WAVES

Electromagnetic waves include light, radar, radio, microwaves, gamma rays, X-rays, and others, all of which involve the propagation of electric and magnetic fields through space with velocity $c = 3 \times 10^8$ m/sec. Electromagnetic waves are produced by the acceleration of electrons or other charged particles. The differences between the various types of electromagnetic waves are related to their frequency and wavelength. Figure 8.1 shows the electromagnetic spectrum and the names associated with frequency and wavelength ranges, which often are not well defined and sometimes overlap. For example, EM waves with wavelength about 10^{-10} m are called gamma rays if their origin is nuclear and X-rays if their origin is atomic. Humans experience EM waves differently depending on their wavelengths: Our eyes detect wavelengths between 4 and 7×10^{-7}m (visible light), whereas our skin detects longer wavelengths (infrared). The sun emits not only visible light but IR (infrared) and UV (ultraviolet) as well. The portion of the EM spectrum used in geophysical exploration covers a wide range. On the high end of the spectrum are EM waves used for ground-penetrating radar (GPR), ranging from approximately 100 MHz to 1 GHz. Simple metal detectors use the 1–10 kHz range. The Geonics EM31 ground conductivity meter operates at 9.8 kHz, and the Geonics EM34 operates at frequencies of 0.4–6.4 kHz. At the lower-frequency end of the EM spectrum are induced polarization techniques, which use frequencies of approximately 10 Hz.

A complete description of EM waves is based on the laws of electricity and magnetism as expressed in an important set of partial differential equations called *Maxwell's equations*. Although solving differential equations is beyond the scope of this book, derivatives and fields changing with time are concepts that are at a level appropriate for undergraduate geophysics courses, and these are useful to contemplate to better understand how EM induction works. Maxwell's equations include the following:

1. Gauss's Law, which relates an electric field (the E field) to its sources

$$div \, D = q \qquad (8.1)$$

where D is the electric field displacement and q is the electric charge density.

2. A similar law for the magnetic field (the H field)

$$div \, B = 0 \qquad (8.2)$$

where B is the magnetic induction or flux density.

3. Faraday's Law, which states that an electric field is produced by a changing magnetic field:

$$curl \, E = -\frac{\partial B}{\partial t}. \qquad (8.3)$$

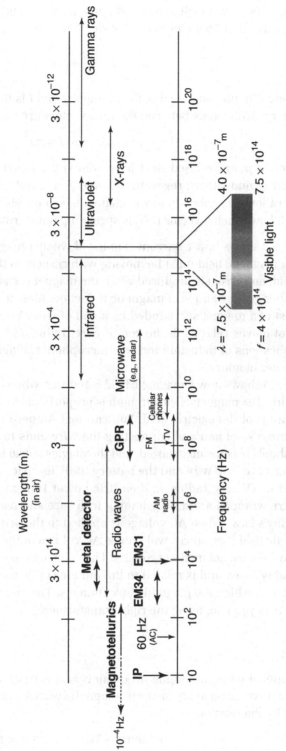

FIGURE 8.1 Electromagnetic spectrum. Wavelengths of geophysical interest are marked in boldface.

4. Ampere's Law, which states that a magnetic field is produced by an electric current or by a changing electric field

$$curl\ H = \frac{\partial D}{\partial t} + I \tag{8.4}$$

where H is the magnetizing field intensity and I is the current density. Other relationships between the field vectors are

$$D = \varepsilon E;\ B = \mu H;\ \text{and}\ I = \sigma E \tag{8.5}$$

where ε, μ, and σ are dielectric permittivity, magnetic permeability, and electric conductivity, respectively. Divergence and curl, which are vector operations described in most calculus-based physics courses, have to do with how much a vector field is spreading out or rotating in space.

Figure 8.2 shows how a current is induced when a magnet is moved toward a coil. The magnetic field must be moving with respect to the coil to produce an electric field; no current is produced when the magnet is stationary. A seismometer is another good example of magnetic induction. Most mechanical seismometers consist of a magnet surrounded by a coil of wire. When the Earth shakes, the magnet moves relative to the coil of wire, inducing a current in the wire. Other applications of induction include microphones, magnetic tape recording, and computer memory.

Figure 8.3 shows how a magnetic field is induced when current flows through a coil of wire. The magnetic field strength is proportional to the current and to the rate of change of the electric field. You can test Ampere's Law by wrapping a wire around a steel nail and connecting the wire ends to a small battery. The steel nail should become magnetized, and the magnetic field strength will vary with the number of coils of wire and the battery EMF. Electric transformers, such as those used in TV sets, radios, and electric power transmission, consist of two coils of wire wrapped around a common iron core and use both Ampere's Law and Faraday's Law. When AC voltage is applied to the primary coil, the changing magnetic field it produces will induce AC voltage in the secondary coil, even though the coils are not in direct contact. The voltage in the secondary coil comes from Faraday's Law and is related to the number of turns in the secondary coil and the rate at which the magnetic flux changes. Thus you can step AC voltage up or down as you like, using the right transformer.

8.1.1   Wavelengths

The concepts of wave amplitude, frequency, and period, outlined in Chapter 2 for seismic waves, also apply to electromagnetic waves. Wavelength is related to frequency by the relation

$$\text{Velocity} = \text{Frequency} \times \text{Wavelength}\ (v = f \times \lambda). \tag{8.6}$$

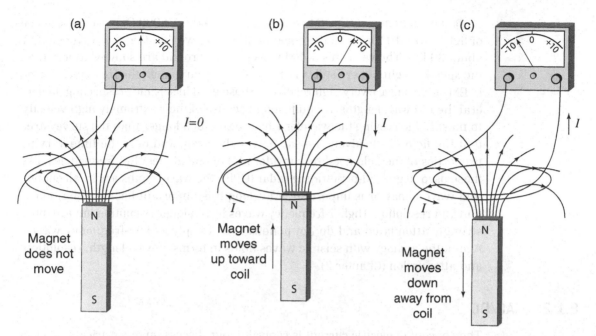

FIGURE 8.2 (a) No current is induced if the magnet does not move relative to the coil. (b) A current is induced when a magnet is moved toward a coil. (c) The induced current is opposite when the magnet is moved away from the coil.

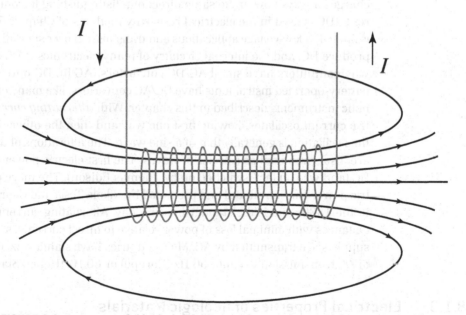

FIGURE 8.3 A magnetic field (black solid lines with arrows) created due to current (I) through a coil.

For a typical frequency for EM equipment of 10 kHz (10,000 Hz), using the speed of light of 3×10^8 m/s and the equation $v = f \times \lambda$, we find that the wavelength is about 30 km. The velocities of EM waves in the ground are somewhat less than the speed of light but are still very high. Thus, even though the frequencies used in EM surveying are very high relative to those used in seismic prospecting, in general the EM wavelengths are quite long because of their extremely high velocity. In most EM surveying the wavelengths are so much longer than the survey area that the field is effectively the same over the area, and only the time-varying properties of the field need be considered. For the highest EM frequencies, like those used in ground-penetrating radar (GPR), the wavelengths are small enough that wave behavior is important. Wavelength is important because of attenuation and resolution. Higher-frequency waves have higher resolution but are more strongly attenuated and do not penetrate as deeply as low-frequency waves. Again, the analogy with seismic waves holds in terms of wavelength, resolution, and attenuation (Chapter 2).

8.1.2   AC/DC

The concept of electric current is central to our discussion of electrical and electromagnetic techniques. There are two main types of electrical current: direct current (DC) and alternating current (AC). *Direct current (DC or continuous current)* is the continuous flow of electric current in one direction through a conductor, such as a wire or the Earth, from high to low potential. In direct current, the electric charges always flow in the same direction, distinguishing it from alternating current. (DC is used in the electrical resistivity methods of Chapter 5.) Direct current is limited to low-voltage applications and transmission over short distances. Batteries produce DC, and the internal circuitry of many electronics is DC as well, which is why computers have small AC/DC converters (AC in, DC out). Likewise, many battery-operated instruments have DC/AC converters, like many of the electromagnetic instruments described in this chapter. With *alternating current (AC)* the electric current oscillates, flowing first one way and then the other. The waveform of the oscillation is generally that of a sine wave. The alterations of the magnetic field are also spread out as a series of waves. The first electric power distribution was DC (and was, in fact, switched on by Thomas Edison). The more efficient AC electric power was introduced and promoted by Nikola Tesla and George Westinghouse in the 1880s and is the standard used today. Alternating currents can flow large distances with minimal loss of power relative to direct currents, so domestic power supplies are transmitted by AC. Most countries have standardized their frequency of AC transmission to either 50 Hz (Europe) or 60 Hz (United States).

8.1.3   Electrical Properties of Geologic Materials

Electromagnetic wave propagation is influenced by the electrical properties of the material through which the current flows. These material properties affect the efficiency of propagation, the loss of energy within the medium, and the speed of wave propagation.

8.1.3.1 ELECTRICAL RESISTIVITY AND CONDUCTIVITY *Electrical resistivity* ρ and its reciprocal *electrical conductivity* σ are physical properties of material that describe the mobility of an electric charge in the presence of an electric field. In a conductive material electric charge moves freely; in a resistive material electric charge is impeded and does not propagate efficiently. An example of a good conductor is copper wire, which is often used in electrical wiring. An example of a good resistor is rubber, which is used as an insulator on electrical wiring. According to Ohm's Law, resistance R is given by V/I, where V is the potential difference and I is applied current. The physical property of resistivity can be determined from resistance via $\rho = RA/l$, where A is cross-sectional area and l is the length of the resistor, giving the dimensions of the resistor. Resistivities are measured in ohm·meters (Ω·m). Conductivities are given in Siemens/m or mho/m, where 1 Siemen = 1 mho = 1/(ohm).

Electromagnetic surveys are more sensitive to conductive bodies than to resistors, and results of EM surveys are usually described in terms of conductivity. For most geologic materials conductivities range over about five orders of magnitude, from seawater and metallic ores (most conductive) to dry quartz sands and granite (most resistive). (See Table 8.1.) Highly conductive geologic materials include shales, clays, contaminated water, sulfide and metallic ores, and saltwater. Metals and metallic sulfides conduct electricity efficiently by flow of electrons. Geologic materials with low conductivity include limestone, dolomite, evaporites, and quartz-rich rocks. Most rock-forming minerals are poor conductors, and ground currents are largely carried by ions in pore waters. The variability in conductivity between mineral grains is small relative to the variability in conductivities between minerals and fluid (electrolyte) solutions. The electrical conductivity of pore fluids depends on the presence of dissolved salts. Clay minerals are ionically active and conduct well even if only slightly moist. Conductivity generally increases with porosity and with the conductivity of pore fluid.

A relation between electrical resistivity and rock formation porosity is given by Archie's Law, which for clean granular rocks (no clay) is written as

$$F = \rho/\rho_w = a\phi^{-m} \tag{8.7}$$

where ρ is electrical resistivity of the formation rocks, ρ_w is the resistivity of the formation water, ϕ is porosity, F is formation factor, and a and m are constants related to saturation and cementation for a particular rock type. Tables are available for a and m constants for different lithologies. This empirical (based on observation rather than physics) relation between electrical resistivity and formation porosity is widely used by hydrologists and environmental geophysicists.

8.1.3.2 DIELECTRIC PROPERTIES The electromagnetic properties of geologic materials vary with the frequency of an applied alternating current. Low-frequency

Material	Conductivity (mS/m)	Relative Permittivity	Radar Velocity (m/ns)
Air	0	1	0.3
Freshwater	0.5	80	0.033
Salt water	3000	81–88	0.01
Dry sand	0.01	3–10	0.15
Wet sand	0.1–1	20–30	0.06
Limestone	0.5–2	4–8	0.12
Shale	1–100	5–15	0.09
Clay	2–1000	5–40	.06–.17
Granite	0.01–1	4–6	0.13
Ice	0.01	3–4	0.16
Concrete	0.01–10	6	0.09

Values are approximate and come from various sources including Schultz (2002); Milsom (2003); Davis and Annan (1989); and Conyers (2004).

inputs provide constraints on conductivity structure. The redistribution of charge that occurs at high frequencies (10–1000 MHz) yields information about dielectric properties. The high-frequency material properties are most important for GPR studies. For a given electric field strength, the *relative dielectric permittivity ε_r* (also called the *dielectric constant*) measures the ability of a material to store a charge when an electric field is applied. Relative dielectric permittivity (unitless) is given by ε_r ($= \varepsilon/\varepsilon_0$) and is the ratio of the dielectric permittivity of the medium to the dielectric permittivity of free space (8.85×10^{-12} F/m). A *farad* (F) is the SI unit of capacitance and is the capacitance at which one volt of potential causes the storage of one coulomb. The more electrically conductive a material, the less dielectric it is, and energy will attenuate at a much shallower depth.

The relative dielectric permittivity varies between 1 (air) and 80 (seawater) (Table 8.1). Most dry sediments have a dielectric constant of 4 to 10, and the range of most geologic materials is 3 to 40. In general, dielectric permittivity increases with water content, and water content has a significant influence on the propagation of radar waves.

Magnetic permeability is a measure of the ability of a medium to become magnetized when an electromagnetic field is imposed on it. Relative magnetic permeability μ_r (= μ/μ_0) is the ratio between magnetic permeability of the medium and magnetic permeability of free space ($\mu_0 = 4\pi \times 10^{-7}$ H/m). A *henry* (H) is the SI unit of electrical inductance, the inductance where one volt is induced by a current change of one ampere per second. Most soils and sediments are only slightly magnetic and have a low magnetic permeability. The higher the magnetic permeability, the more electromagnetic energy will be attenuated during its transmission. Media that contain magnetite, iron oxide cement, or iron-rich soils can have a high magnetic permeability and transmit radar energy poorly. Conveniently, μ_r is close to unity for most rock materials (except a few strongly magnetic rocks) and therefore is often not shown explicitly in radar equations.

Radar wave velocities are influenced by dielectric permittivity and magnetic permeability of the material that they travel through, and are given by

$$V = c/\left[(\mu_r \, \varepsilon_r)^{\frac{1}{2}}\right]. \tag{8.8}$$

8.1.4 Absorption and Attenuation

Like seismic waves, electromagnetic waves lose energy as they propagate through geometrical spreading, scattering, and absorption. Geometrical (or spherical) *spreading* is the decrease in wave strength (energy per unit area of wave front) with distance as a result of the spreading out of wave energy in all directions from its source. The energy spreads over the surface area of a sphere ($A = 4\pi r^2$, r = sphere radius) that increases as the wave front propagates. Thus the energy per unit area decreases as a function of $1/r^2$. Because wave amplitude is proportional to the square root of wave energy, wave amplitude decreases as a function of $1/r$. *Scattering* is irregular dispersion of energy caused by inhomogeneities in the medium through which a wave is traveling.

Absorption is the conversion of wave energy into heat. The use of microwaves (high-frequency radio waves) to cook food is a good example. Absorption is a physical property of the material and, for EM waves, is related to electrical conductivity and dielectric permittivity. EM waves can travel indefinitely in a vacuum but are absorbed in conducting media. For example, EM radio waves are absorbed by rock, which is why radios and cell phones do not work in tunnels without special repeaters.

Attenuation is the decrease in intensity of a wave as a result of absorption of energy and of scattering out of the path to the detector, but does not include the reduction due to geometric spreading. Attenuation is a loss of energy per cycle. Higher frequencies have a larger number of cycles in a given amount of time than low frequencies, so high frequencies are attenuated more rapidly than low frequencies transmitted through the same material. Attenuation is not an important issue for many EM methods because the wavelength of the waves is much larger than the distance between the transmitter and receiver; but attenuation is important at the high frequencies of GPR prospecting.

The EM attenuation factor, α, is given by

$$\alpha = \omega \left\{ \frac{\mu\varepsilon}{2} \left[\left(1 + \frac{\sigma^2}{\varepsilon^2\omega^2} \right)^{\frac{1}{2}} - 1 \right] \right\}^{\frac{1}{2}} \tag{8.9}$$

(Sharma, 1997) where $\omega \, (= 2\pi f)$ is the angular frequency of the field, ε is the dielectric permittivity (F/m), μ is the magnetic permeability (H/m), and σ is the electric conductivity (S/m) of the medium. There are two important extreme cases of propagation of EM waves: the inductive regime and the radar regime. The inductive regime is dominant at low frequencies ($f < 10^5$ Hz), as used in ground conductivity surveys. In this case ($\sigma^2/\varepsilon^2\omega^2 \gg 1$), and in the inductive regime Equation 8.9 simplifies to

$$\alpha = (\omega\mu\sigma/2)^{\frac{1}{2}}. \tag{8.10}$$

At the other extreme is the radar regime at high frequencies ($f > 10^7$ Hz). In this case ($\sigma/\varepsilon\omega \ll 1$), and attenuation for radar waves is given by

$$\alpha = (\sigma/2)\left[(\mu/\varepsilon)^{\frac{1}{2}} \right]. \tag{8.11}$$

For nonmagnetic media $\mu \cong \mu_0 (= 4\pi \times 10^{-7}$ H/m) and Equation 8.11 simplifies to

$$\alpha = 1685\sigma/(\varepsilon^{\frac{1}{2}}) \tag{8.12}$$

in dB/m. A decibel (dB) is a measure of input relative to output. One dB = $10 \log_{10}$(output/input). A 10 dB gain corresponds to a tenfold increase in signal power. Negative values correspond to losses.

Attenuation increases with material conductivity. Materials with high attenuation for radar waves include clays, silts, shales, and seawater. Materials with low attenuation include distilled water, air, dry sand, and ice (Table 8.1). The low attenuation of ice makes it nearly transparent to radar waves and thus very dangerous to fly over when we are using radar to avoid collision with the terrain.

Given a desired target depth and an estimate of the electrical properties of the subsurface medium, we can choose the frequency in an EM survey to most accurately penetrate and sample that depth. The depth at which wave amplitude drops to $1/e$ (about one-third) of its original value is called the *skin depth, δ*. Skin depth is sometimes referred to as the *depth of penetration* or the *maximum depth sensitivity* of an EM wave. The skin depth decreases with increased conductivity and frequency, and in the conductive regime (as for EM prospecting for conductive bodies) is given by the inverse of the attenuation factor

$$\delta = 1/\alpha = \left(\frac{2}{\mu_0\sigma\omega} \right)^{\frac{1}{2}} \tag{8.13}$$

$$= 504(1/\sigma f)^{\frac{1}{2}}, \text{ in meters}$$

(Sharma, 1997) where σ is the material conductivity in S/m, $\omega \, (= 2\pi f)$ is the angular frequency of the EM field, and μ is the magnetic permeability (H/m) where

TABLE 8.2	EM AND GPR SKIN DEPTH

Conductive Regime ($f < 10^5$ Hz)		Radar Regime ($f > 10^7$ Hz)	
Frequency (Hz)	2×10^4	Frequency (Hz)	2×10^8
Material conductivity (S/m)	0.01	Material conductivity (S/m)	0.01
		Relative dielectric permittivity	15
Skin depth (m) = 36		Skin depth (m) = 0.23	

$\mu \cong \mu_0 (= 4\pi \times 10^{-7})$ for nonmagnetic media. Skin depths are shallower for higher frequencies and higher conductivities. Thus a highly conductive surface layer can severely impair the depth penetration and utility of EM surveys.

At high frequencies (radar regime) the skin depth parameter is expressed as

$$\frac{1}{\alpha} = \delta = \left(\frac{2}{\sigma}\right)\left(\frac{\varepsilon}{\mu_0}\right)^{\frac{1}{2}} \tag{8.14}$$

$$= (\sqrt{\varepsilon_r})/(1685\sigma), \text{ in meters}$$

(Sharma, 1997). Dynamic Table 8.2 can be used to calculate skin depth given frequency, material conductivity, and dielectric permittivity. Equation 8.13 is used for frequencies on the conductive regime, and equation 8.14 is used for frequencies on the radar regime.

8.2   EM SOUNDING

EM sounding methods measure electric or magnetic field components induced in the Earth by a current source, which can be natural or man-made. Electromagnetic soundings consist of a transmitter, which transmits alternating current into the Earth, and a receiver, which receives the EM fields resulting from the interaction of the transmitted current with the Earth. EM instruments operate on the principle of electromagnetic induction, which is based on three fundamental laws of physics.

The first is Ampere's Law (Figure 8.3), which states that a magnetic field is produced by an electric current or by a changing electric field. An electric current I has an associated magnetic field B. The magnetic field strength is proportional to the strength of the electric current and has the same phase. This means that whatever their relative amplitudes, the two fields increase and decrease in step with each other. Figure 8.4 illustrates the concept of phase. *In phase* means that two phases have the same ups and downs at the same times. *Out of phase* means

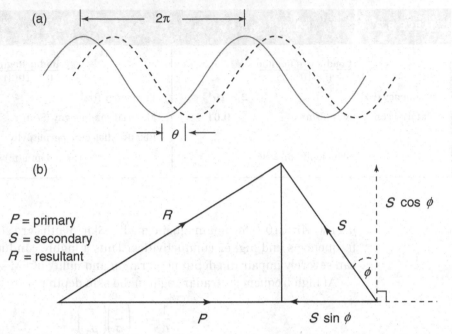

FIGURE 8.4 (a) The phase difference θ between two waveforms. (b) Vector diagram illustrating phase and amplitude relationships between primary, secondary, and resultant EM fields. (Keary, Brooks, and Hill, 2002.)

that one field is delayed relative to the other. The phase difference can be described by a time or more commonly by an angle. A quarter-wavelength phase delay is a $\pi/2$ or 90° phase delay.

The next is Faraday's Law (Figure 8.2), which states that an electric field is produced by a changing magnetic field. The strength of the induced electric field is proportional to the rate of change (dB/dt) of the magnetic field. For any circuit penetrated by the varying magnetic field a voltage V' will be generated. A conductor in the subsurface is a *circuit*. If the magnetic field does not change with time, no voltage is induced. If the magnetic field is sinusoidally alternating, the induced voltage will alternate at the same frequency but will lag the magnetic field by a quarter of a cycle (90°).

Ohm's Law relates voltage to current flow and conductivity (Chapter 5). If the circuit just described is a conductor, the voltage V' generated in the circuit by the alternating magnetic field will cause current to flow. The current, which we will denote as I' to distinguish it from our original current I, will obey Ohm's Law

$$I' = V'/R' \tag{8.15}$$

where I' is current, V' is voltage, and R' is resistance. Because R' is a constant, I' will be in phase with V', thus lagging the inducing field B and its generating current I by 90°. The induced current I' will generate its own magnetic field B', which will be in phase with I' and out of phase with B. The circuit is now altering the fields that excited it in the first place, a process known as *self-induction*. Self-induction is a property of circuits with high conductivity. The response field B' lags the transmitting field B by up to 180° for very high-conductivity circuits (such as metal pipes and ore bodies).

Electromagnetic prospecting is based on the laws of physics just described. The basic components of electromagnetic induction prospecting are shown schematically in Figure 8.5 and described here. A transmitter (usually a coil) is connected to a source of AC (some sort of generator). The current through the transmitter produces an alternating magnetic field about the transmitter according to Ampere's Law. The alternating magnetic field produced by the transmitter spreads out both above and below the ground (solid black lines in Figure 8.5). This field is called the *primary field*. The receiver (also a coil) is set up at some distance from the transmitter. The receiver has a voltage V_p (p for primary) induced in it according to Faraday's Law, and this is the primary signal as seen at the receiver.

If an electrical conductor (target) exists near or between the transmitter and receiver, then by Faraday's Law an alternating voltage is induced about this conductor by the primary field that intersects it. Because this conductor has a resistance R, the induced voltage will create an alternating current (*eddy current*) in the conductor, governed by Ohm's Law. The eddy currents are shown as gray lines in Figure 8.5. The eddy current will produce a magnetic field by Ampere's Law

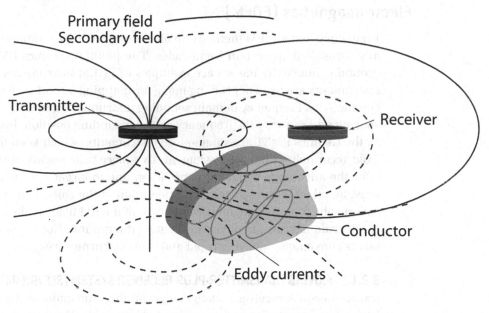

FIGURE 8.5 Generalized picture of electromagnetic induction prospecting. The primary magnetic field produced by the transmitter is shown in solid black lines; the secondary magnetic field generated by eddy currents induced in a subsurface conductor is shown as dashed lines. (Modified from Grant and West, 1965.)

and a voltage V_s (s for secondary) by Faraday's Law. The magnetic field (dashed lines in Figure 8.5) generated by the eddy currents differs from the primary field in both amplitude and phase. The receiver records the voltage associated with the secondary field (V_s) as well as the primary V_p. The voltage V_s has an amplitude that depends on the frequency of the induced fields, the area of the target, the conductivity of the target, and the coils' orientation to each other and the ground.

The primary field is much larger than the secondary field, and the technological challenge in EM sounding is separating the primary field (the transmitter or source signal) from the secondary field (the signal that we are interested in from a conductor in the Earth). Bodies with high conductivity, such as ore bodies or saltwater intrusions, produce strong secondary fields. Objects with low conductivity produce a more subtle secondary field. The difference between the transmitted and received EM fields provides information about the geometry, size, and electrical properties of the subsurface conductor. The two main classes of techniques for separating the secondary field from the primary field are *continuous wave* or *frequency domain electromagnetic* methods (*FDEM*) and *time domain electromagnetic* methods (*TDEM*). FDEM techniques separate the components of the response field that are in phase and out of phase (quadrature) with respect to the primary field. TDEM techniques observe the transient EM field present after the transmitter is turned off.

8.2.1 Near-Field Continuous-Wave Methods of Frequency Domain Electromagnetics (FDEM)

Frequency domain EM methods involve broadcasting a continuous signal, usually sinusoidal, at certain frequencies. The popular Geonics EM31 and EM34 ground conductivity meters are examples of FDEM instruments. Most of these instruments are used in profiling mode for mapping lateral variations in ground conductivity. Frequency domain sounding instruments generally use a fixed coil spacing and can vary the frequency at each sounding position. Instruments such as the Geonics EM31 are single-piece instruments, which keep the coil spacing fixed accurately within the instrument. Two-piece instruments, such as the EM34, offer the advantage of deeper penetration due to larger coil spacings, but care must be taken in the spacing and orientation of the coils. The induction of current results from the magnetic component of the EM field, so the transmitter and receiver do not have to physically contact the ground. Not needing ground contact is convenient, allowing rapid and even airborne surveys.

8.2.1.1 MOVING TRANSMITTER-PLUS-RECEIVER SYSTEM (SLINGRAM) In the moving transmitter-plus-receiver system (Slingram) the transmitter coil and receiver coil have a fixed separation and are connected by a cable (Figure 8.6). The transmitter and receiver are of the same design and are moved simultaneously along a traverse. Current is delivered to the transmitter coil, usually by a battery. The angle of the loop to the magnetic field affects the magnetic flux, and the orientation of the

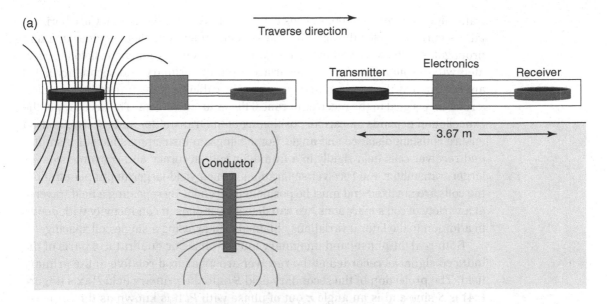

FIGURE 8.6 (a) Generalized picture of moving transmitter-plus-receiver Slingram system. The primary field produced by the transmitter is shown in black lines centered on the transmitter; eddy currents induced in a subsurface conductor are shown as gray lines centered on the conductor. For EM31 the ground conductivity meter transmitter and receiver are at a fixed distance of 3.67 m from each other and connected together in a rigid casing (After Mussett and Khan, 2000.) (b) The EM31 ground conductivity meter in use in the field.

transmitter and receiver loops needs to be the same. In most cases they are horizontal. As current passes through the transmitter, a magnetic field is generated that propagates both above and below the ground. A secondary magnetic field is generated within conductors in the subsurface. Both the primary and secondary fields are recorded by the receiver. In many cases the cable from the transmitter supplies a reference signal to the receiver to cancel the induced current. The accuracy of the cancellation depends on careful positioning of the transmitter and receiver coils to ensure constant distance and angle. Some Slingram instruments have transmitter and receiver coils held rigidly in a frame for both accuracy and convenience. For larger transmitter and receiver separations, which yield larger depth penetration, the coils are not fixed and must be positioned carefully. By repeating a field traverse at a variety of coil separations, we can detect variations in conductivity with depth, in addition to the lateral variations obtained rapidly using a single coil spacing.

Both real (in-phase) and imaginary (out-of-phase or quadrature) parts of the induced signal as recorded at the receiver are measured relative to the primary field. The projection of the secondary field S onto the primary field P axis (Figure 8.4) is $S \sin\phi$ and is an angle π out of phase with P. It is known as the in-phase or real component of S. The vertical projection is $S \cos\phi$, $\pi/2$ out of phase with P, and is known as the out-of-phase, imaginary, or quadrature component. For a good conductor, such as a pipe or a metallic object, ϕ will approach $\pi/2$ (Figure 8.4). For a poor conductor (most geologic materials), ϕ will be almost zero. Thus the in-phase mode is sensitive to high-conductivity features, such as pipes and metallic objects, and the out-of-phase (quadrature) mode is most sensitive to objects with low conductivity, such as soil variations. Modern instruments can split the secondary electromagnetic field into its real and imaginary components.

Most traverses are made perpendicular to the strike of the target, and the readings (H_s/H_p, where H_s is the secondary magnetic field at the receiver coil and H_p is the primary magnetic field at the receiver coil) are plotted for the midpoint between the transmitter and receiver (Figure 8.7). The maximum signal occurs when the midpoint between the transmitter and receiver is over the target. For a vertical target there is no signal when either the transmitter or receiver is directly over the target. This occurs because the horizontal coils of the receiver record only the vertical field lines. When the transmitter is directly over a vertical target, no secondary magnetic field is produced because there is no magnetic flux through the target. When the receiver is directly over a vertical target, the secondary field is horizontal and does not pass through the receiver coils. A dipping target produces an asymmetric anomaly profile. Quantitative interpretation of Slingram data is possible through type curves generated for sample geometries.

8.2.1.2 NONCONTACTING GROUND CONDUCTIVITY MEASUREMENTS Noncontacting ground conductivity meters are portable Slingram systems that directly measure conductivity σ averaged over shallow depths. No ground contact is required, so measurements can be made rapidly; these systems offer considerable advantages

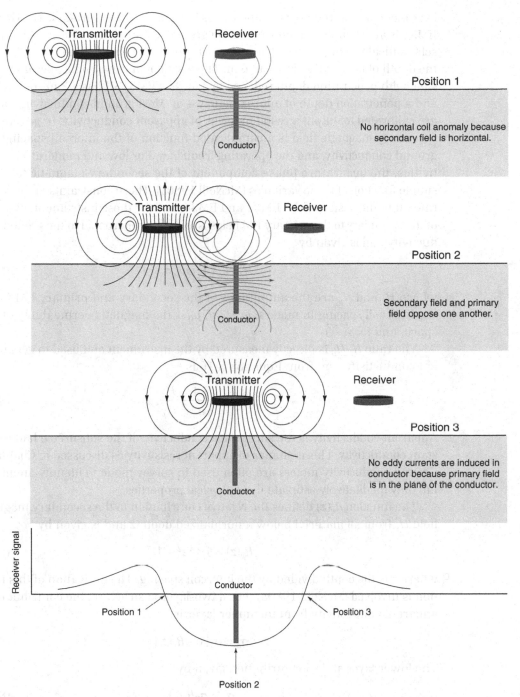

FIGURE 8.7 Slingram survey over vertical conductor. The primary magnetic field produced by the transmitter is shown in solid black lines centered on the transmitter; the secondary field from eddy currents induced in a subsurface conductor is shown in broken gray lines centered on the conductor. The bottom panel shows relative receiver signal (H_s/H_p) plotted at the midpoint between transmitter and receiver antenna. (After Milsom, 2003.)

over methods that require coils and cables in contact with the ground. The Geonics EM31 is an example of a noncontacting ground conductivity meter and is shown schematically in Figure 8.6. The coil separation in the EM31 is fixed at 3.7 m and the depth of penetration is about 6 m. The instrument can also be operated on its side, with penetration depth reduced to 3 m. The EM31SH has a 2 m coil spacing and a penetration depth of approximately 4 m. Most ground conductivity meters are calibrated to output a result in terms of apparent conductivity. In general the secondary magnetic field is a complicated function of the intercoil spacing, the ground conductivity, and the operating frequency. For low and moderate conductivities, the quadrature phase component of the secondary magnetic field is a simple function of these variables (McNeill 1980a). These constraints are incorporated into the design of the EM31 and EM34, and for these instruments the ratio of the secondary to the primary EM fields is linearly proportional to the ground conductivity and is given by

$$H_s/H_p \cong (i \, \omega \mu_0 \, \sigma \, s^2)/4 \tag{8.16}$$

where H_s and H_p are the amplitudes of the secondary and primary EM fields, s is intercoil spacing in meters, $\omega = 2\pi f$, μ_0 is the magnetic permeability of free space, and $i = \sqrt{(-1)}$.

The ratio H_s/H_p is directly measured by the instrument and used to get apparent conductivity, and from Equation 8.16 is

$$\sigma_a = \frac{4}{\omega \mu_0 s^2} \frac{H_s}{H_p}. \tag{8.17}$$

Apparent conductivity is the subsurface conductivity if the subsurface had a constant conductivity. This is similar to apparent resistivity as discussed in Chapter 5. Ground conductivity meters are often used in survey mode to identify anomalies but only qualitatively estimate their physical properties.

The function $R_v(z)$ defines the relative contribution to the secondary magnetic field H_s from all material below a normalized depth z and is given by

$$R_v(z) = 1/(4z^2 + 1)^{\frac{1}{2}} \tag{8.18}$$

where z is the depth divided by the intercoil spacing. The derivation of this function is involved (McNeill 1980a). For a two-layered structure, the contribution to apparent conductivity from the upper layer is

$$\sigma_a = \sigma_1[1 - R_v(z)]. \tag{8.19}$$

The lower layer adds a contribution given by

$$\sigma_a = \sigma_2 R_v(z). \tag{8.20}$$

The actual instrument reading will be a sum of these two quantities:

$$\sigma_a = \sigma_1[1 - R_v(z)] + \sigma_2 R_v(z). \tag{8.21}$$

Thus given a model with subsurface layer conductivities σ_1 and σ_2, we can calculate the apparent conductivity σ_a that would be measured in an EM field survey.

As an example, let's calculate the EM31 apparent conductivities that would be measured across the buried river valley shown in Figure 8.8. The buried river valley consists of a surface layer of till (glacial sediments) over a layer of shale. The shale was incised by the river channel, and the till subsequently filled in the old river channel. The apparent conductivity measurements will vary with different thicknesses of till, revealing, we hope, the location of the buried channel. At point A the till thickness is 8 m. We will use Equations 8.18 and 8.21 to solve for apparent conductivity. For the EM31 the coil separation s is 3.67 m, and we will use $R(z)$ corrected for instrument operation at waist (1 m) height. Although it is most convenient and rapid to operate the EM31 at waist height, measurements made at ground level are more reproducible. At height 1 m above surface point A,

$$z = (\text{Depth} + \text{Instrument height})/(\text{Intercoil spacing}) = (8+1)/3.67 = 2.45, \quad (8.22)$$

$$R_v(z) = 1/(4z^2 + 1)^{\frac{1}{2}} = 1/(4(2.45)^2 + 1)^{\frac{1}{2}} = 0.2, \text{ and} \quad (8.23)$$

$$\sigma_a = \sigma_1[1 - R_v(z)] + \sigma_2 R_v(z) = 10 \text{ mS/m} \times [1 - 0.2] + 50 \text{ mS/m} \times (0.2) = 18 \text{ mS/m.} \quad (8.24)$$

At height 1 m above surface point B,

$$z = (\text{Depth} + \text{Instrument height})/(\text{Intercoil spacing}) = (5+1)/3.67 = 1.63, \quad (8.25)$$

$$R_v(z) = 1/(4z^2 + 1)^{\frac{1}{2}} = 1/(4(1.63)^2 + 1)^{\frac{1}{2}} = 0.3, \text{ and} \quad (8.26)$$

$$\sigma_a = \sigma_1[1 - R_v(z)] + \sigma_2 R_v(z) = 10 \text{ mS/m} \times [1 - 0.3] + 50 \text{ mS/m} \times (0.3) = 22 \text{ mS/m.} \quad (8.27)$$

And at height 1 m above surface point C,

$$z = (\text{Depth} + \text{Instrument height})/(\text{Intercoil spacing}) = (1+1)/3.67 = 0.54, \quad (8.28)$$

$$R_v(z) = 1/(4z^2 + 1)^{\frac{1}{2}} = 1/(4(0.54)^2 + 1)^{\frac{1}{2}} = 0.67, \text{ and} \quad (8.29)$$

$$\sigma_a = \sigma_1[1 - R_v(z)] + \sigma_2 R_v(z) = 10 \text{ mS/m} \times [1 - 0.67] + 50 \text{ mS/m} \times (0.67) = 37 \text{ mS/m.} \quad (8.30)$$

For a three-layered Earth (McNeill, 1980a) we use the same procedure to determine the apparent conductivity, with the relation

$$\sigma_a = \sigma_1[1 - R(z_1)] + \sigma_2[R(z_1) - R(z_2)] + \sigma_3 R(z_2). \quad (8.31)$$

As an example, let's calculate the apparent conductivity measured for the three-layered structure shown in Figure 8.9, with the top layer fixed at 0.5 m thickness and the second layer thickness varying from 0.5 to 1.0 m. The conductivity of layer 1 (the top layer) is 10 mS/m, of layer 2 (the middle layer) is 1 mS/m, and of layer 3 (the bottom layer) is 10 mS/m.

For intercoil spacing of 3.67 m, top layer thickness of 0.5 m, and instrument height of 1 m,

$$z_1 = (0.5 \text{ m} + 1 \text{ m})/3.67 \text{ m} = 0.4 \text{ and} \quad (8.32)$$

$$R_v(z_1) = 1/(4z_1^2 + 1)^{\frac{1}{2}} = 1/(4(0.4)^2 + 1)^{\frac{1}{2}} = 0.8. \quad (8.33)$$

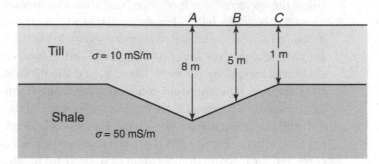

Buried river valley

FIGURE 8.8 Subsurface conductivity structure associated with a buried river valley. In this case high-conductivity shale, incised by an ancient river, is overlain by lower-conductivity glacial till.

For layer 2 at 0.5 m thickness, the depth of interest is then 0.5(layer 1 thickness) + 0.5(layer 2 thickness):

$$z_2 = (\text{Depth} + \text{Instrument height})/3.67 \text{ m} = (1 + 1)/3.67 = 0.54 \text{ and} \qquad (8.34)$$

$$R_v(z_2) = 1/(4z_2^2 + 1)^{\frac{1}{2}} = 1/(4(.54)^2 + 1)^{\frac{1}{2}} = 0.68. \qquad (8.35)$$

And for layer 2 at 1 m thickness, the depth of interest is 0.5(layer 1 thickness) + 1(layer 2 thickness):

$$z_2 = (\text{Depth} + \text{Instrument height})/3.67 \text{ m} = (1.5 + 1)/3.67 = 0.68 \text{ and} \qquad (8.36)$$

$$R_v(z_2) = 1/(4z_2^2 + 1)^{\frac{1}{2}} = 1/(4(.68)^2 + 1)^{\frac{1}{2}} = 0.59. \qquad (8.37)$$

In case 1 (layer 2 thickness = 0.5 m),

$$\sigma_a = \sigma_1[1 - R(z_1)] + \sigma_2[R(z_1) - R(z_2)] + \sigma_3 R(z_2) \qquad (8.38)$$

$$= 10 \text{ mS/m} \times [1 - 0.8] + 1 \text{ mS/m} \times [0.8 - 0.68] + 10 \text{ mS/m} \times [0.68]$$

$$= 8.92 \text{ mS/m}.$$

In case 2 (layer 2 thickness = 1 m),

$$\sigma_a = \sigma_1[1 - R(z_1)] + \sigma_2[R(z_1) - R(z_2)] + \sigma_3 R(z_2) \qquad (8.39)$$

$$= 10 \text{ mS/m} \times [1 - 0.8] + 1 \text{ mS/m} \times [0.8 - 0.59] + 10 \text{ mS/m} \times [0.59]$$

$$= 8.11 \text{ mS/m}.$$

We can perform such calculations to prepare for surveys and to interpret the data collected. With the EM31 the most robust results are the relative conductivity differences rather than absolute values of conductivity.

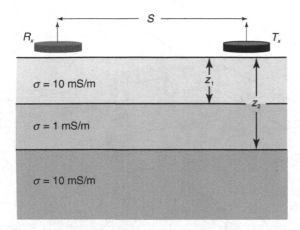

FIGURE 8.9 Subsurface conductivities for a three-layer structure. Note that depths z_1 and z_2 are depths over which conductivity structure is integrated, and can include multiple layers.

8.2.2 Other FDEM Systems

Many FDEM systems exist beyond the few described here. The methods differ in size, spacing, and orientation of the transmitter and receiver. Larger transmitters allow increased signal and depth penetration, and the lack of a cable connecting transmitter and receiver can allow greater flexibility in the terrain sampled (Figure 8.10). Several techniques called *tilt-angle* or *dip-angle* methods measure spatial variations in the angle of the electromagnetic field. Very-low-frequency (VLF) methods use low-frequency (15–25 kHz) EM waves generated by radio transmitters used in communications and navigation. Metal detectors used in airports are FDEM systems. Electromagnetic induction logs are commonly used in geophysical well logging and can give conductivity of rock formations within a drill hole with depth. Induction logs are typically used with resistive muds, such as oil-based or freshwater muds. For conductive muds, such as salty muds, DC resistivity (laterologs) is used to determine formation resistivity.

8.2.3 Time Domain Electromagnetics (TDEM)

A common problem with FDEM methods is accurate removal of the large primary field so we can measure the small secondary field induced by interaction of the primary field with the target. One way to avoid the large primary field signal is to use a primary field that is pulsed rather than continuous, and measure the secondary field when the primary field is turned off (Figure 8.11). Time domain electromagnetic (TDEM) methods use transient source signals rather than continuous waves, and they measure the decay of the secondary EM field after the primary field has been turned off. With highly conductive bodies, the eddy currents decay more slowly than for weak conductors. Measurement of the decay rate of the eddy currents can be used to locate subsurface conductors and estimate their conductivity. TDEM methods use a transmitter coil and a receiver coil, with a variety of transmitter and receiver coil sizes and geometries used.

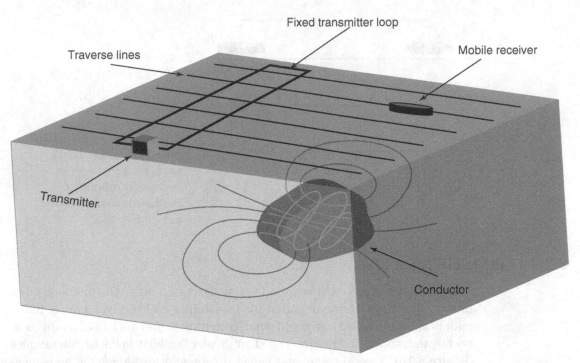

FIGURE 8.10 Schematic of FDEM system with large fixed transmitter loop and small mobile receiver. The larger transmitter permits larger signal and depth of penetration at the expense of decreased resolution. The variation in geometry as the receiver moves independently of the transmitter results in complicated equations to relate observed readings to subsurface conductivity structure. (After McNeill, 1980b.)

The loop–loop configuration in TDEM has transmitter and receiver coils moved together along a transect, similar to the FDEM Slingram method. The distance between transmitter and receiver can be varied depending on the target, with larger intercoil spacing having a greater depth of penetration but lower resolution, and smaller intercoil spacing providing higher-resolution sounding but only at shallower depths. The position of the conductor can be estimated from the variation in amplitude of the secondary field as measured at selected time intervals, usually on the order of microseconds to tens of milliseconds.

Another geometry used in TDEM measurements has a large central transmitting loop and small receiver coil (Figure 8.12). The Geonics EM42 uses a transmitting loop of 500×500 m, and the Geonics EM47 uses a transmitting loop of 30×30 m. In this case the depth of investigation is a function of delay time, with deeper targets having a larger delay time in the secondary field. This method is convenient for sounding in a given region because the transmitter and receiver do not have to be moved. At each sounding location repeated measurements of the decaying field can be made at fixed times (channels) and stacked to improve signal-to-noise ratio.

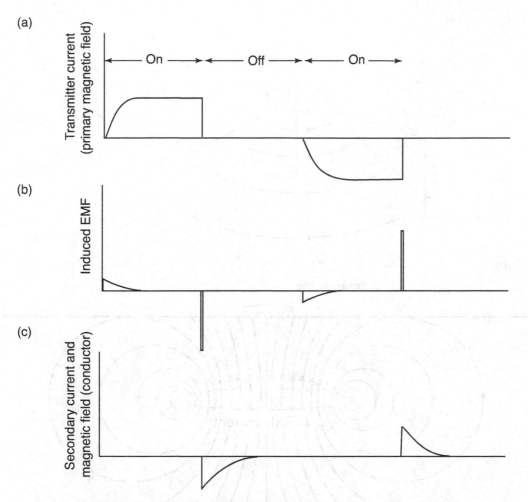

FIGURE 8.11 TDEM transmitter output and receiver signal as a function of time. (a) Transmitter current as a function of time. (b) Induced EMF. (c) Secondary current. (After McNeill, 1980b.)

8.3   EM FIELD TECHNIQUES

8.3.1   Profiling versus Sounding

Most electromagnetic field surveys are performed in profile mode. Here we describe electromagnetic sounding and profiling. These general survey modes are also used for electrical resistivity surveys.

8.3.1.1 SOUNDING The goal of electrical or electromagnetic sounding is to estimate conductivity structure as a function of depth in one place. This is achieved by increasing the spacing between the transmitting and receiving antennas (or,

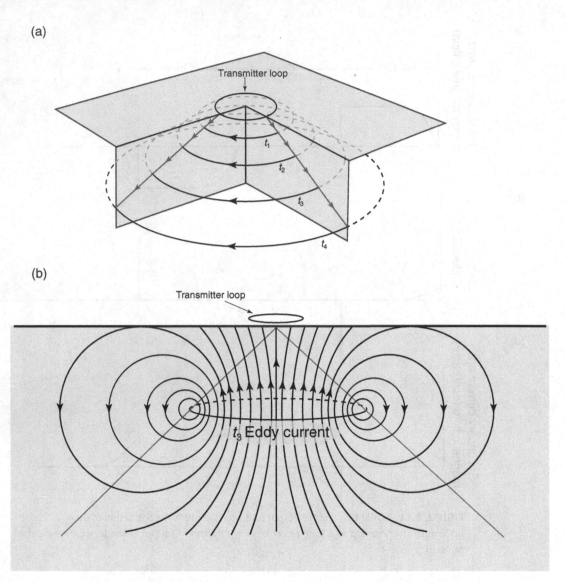

FIGURE 8.12 TDEM central transmitter. (a) Depth and width of penetration increase with time, as shown by location of maximum circulating current at increasing delay times t_1, t_2, t_3, and t_4. (After Reynolds, 1997.) (b) Equivalent current loop at time t_3 after termination of current in transmitter loop. Arrows are on lines of magnetic field. (After Milsom, 2003.)

in the case of electrical resistivity sounding, the current and potential electrodes), keeping the center point between transmitting and receiving antennas fixed. Depth of penetration increases with distance between the antennas. The data are plotted as apparent conductivity versus antenna spacing, which forms a curve with one or more inflection points. The inflection points denote layer boundaries. The curves can be modeled using a set of master curves, or inverted for conductivity structure using a computer program.

8.3.1.2 PROFILING Electromagnetic profiling consists of making measurements at many different places with the distance between source and receiver fixed. This is the most common type of electromagnetic surveying and is sometimes referred to as "mowing the lawn." The operator walks across the area of interest along a single line or a set of grid lines. If a single line is traversed, the data are plotted as apparent conductivity versus distance along the line. If a grid of lines is covered, the data can be plotted as a contour map of apparent conductivity. These maps are useful for a variety of applications, including finding buried pipes or utility lines, finding walls of buried structures in archaeological studies, and mapping the lateral extent of geologic features of such as igneous intrusions.

8.3.2　Interpretation

The mathematics and physics behind electromagnetic methods are more difficult than for the other geophysical methods described in this book, and consequently the interpretation of electromagnetic data varies between the two extremes of qualitative black box analysis and sophisticated but complex quantitative modeling. Many electromagnetic instruments, such as metal detectors, are used successfully without an in-depth understanding of the physics involved. Developing new methods, or rigorous interpretation of the data, requires a solid grounding in physics and mathematics. The main modes of interpretation follow.

1. *Contouring data:* Apparent conductivity data collected on grid lines can be contoured by hand or using computer programs. The contoured data often reveal buried structures of interest. The display of relative variations in conductivity structure as revealed by a contour map is often sufficient for most exploration purposes.
2. *Forward modeling:* Apparent conductivity can be calculated for simple layered structures and geometries, as shown in Equations 8.18 to 8.21 and with the EM option in the program RESIST (Chapter 5). The calculated apparent conductivity can be compared to the observed data and the subsurface conductivity model altered to better fit the observed data.
3. *Inversion:* Inversion is the process of finding a model that best fits the observed data by minimizing the difference (similar to fitting a line to data using least squares) between the observations and synthetic data produced from a model. To develop an inversion you must know the equations that relate the data to the model parameters that you wish to solve for. Menke (1989) and Aster et al. (2004) provide a good summary of the basics of geophysical inversion methods, and Rodi and Mackie (2001) and Fitterman and Yin (2004) provide examples of the application of inverse theory to electromagnetic data. Both forward modeling and inverse modeling for EM data are nonunique, which means many solutions will fit the data equally well. Creating a model that fits the data perfectly is not always meaningful: Take care to

ensure that your interpretations make sense based on what you already know about the subsurface structure and geology. The use of known constraints can be of great help in interpretation and modeling of EM data.

8.4 GROUND-PENETRATING RADAR

Radar uses radio waves to detect the presence and location of objects. The word *radar* is derived from *radio detection and ranging*, which convey the two purposes of detection and location. World War II code letters are used for the different frequency bands of radar (L, S, C, and so on). Radar has many applications on scales that range from a few centimeters, such as measuring the thickness of walls, to long-range systems probing planets across the solar system. Ground-penetrating radar was initially developed to probe glaciers and was applied to soils and the ground subsurface in the 1960s and 1970s. Ground-penetrating radar has been popular in engineering and archaeological studies since the mid-1980s. Applications include searching for land mines, pipes, reinforcing bars, tunnels and mineshafts, and archaeological and geologic structures and strata.

Radio waves are part of the electromagnetic spectrum (Figure 8.1) and in a vacuum travel at the speed of light. A radar system consists of a signal generator, a transmitting antenna that generates radio waves, a receiving antenna, and a receiver consisting of a recording or printing output device. The transmitter emits a radio signal, and the radio receiver receives the small amount of energy that is scattered back.

Because radar waves travel at the speed of light, travel times associated with radar waves are extremely short. For example, the time for a radar wave to reflect from an object half a meter away is

$$\text{Time} = \text{(Two-way distance)/Velocity} = (1 \text{ m})/(3 \times 10^8 \text{ m/s}) = 3 \times 10^{-9} \text{ s} = 3 \text{ ns.} \qquad (8.40)$$

In comparison, for a seismic air wave,

$$\text{Time} = \text{Distance/Velocity} = (1 \text{ m})/(300 \text{ m/s}) = .0033 \text{ s} = 3.3 \text{ ms.} \qquad (8.41)$$

The wavelengths of radar waves used in applied geophysics are usually quite short compared to seismic wavelengths, resulting in higher resolution than seismic studies. For example, for a radar wave in granite, using a radar frequency of 200 MHz,

$$\lambda = v/f = (1.3 \times 10^8 \text{ m/s})/200 \text{ MHz} = (1.3 \times 10^8 \text{ m/s})/(200 \times 10^6 \text{ Hz}) = 0.65 \text{ m.} \qquad (8.42)$$

For a compressional seismic wave in granite, assuming a seismic frequency of 100 Hz,

$$\lambda = v/f = (5000 \text{ m/s})/100 \text{ Hz} = 50 \text{ m.} \qquad (8.43)$$

The display of ground-penetrating radar data (Figure 8.13) is analogous to a reflection seismic section (Figure 4.48). Processing techniques from seismic reflection, such as common midpoint stacking and migration, can be applied to GPR data. Signal stacking is frequently done within the GPR instrument as the data are collected. Migration processing is not as commonly used for GPR as it is for seismic reflection, and with GPR it is more common to recognize the characteristic signals from diffracted waves rather than to correct for them using migration.

Ground-penetrating radar can be a good choice for high-resolution subsurface imaging in the right conditions. However, ground-penetrating radar does not work well in all environments. It does not work well in the presence of high surface conductivity, such as regions with wet clays near the surface or conductive fluids like seawater. In such cases we should consider other geophysical techniques. Ground-penetrating radar typically produces good results in regions of ice, snow, or dry sandy soil, and over concrete.

8.4.1 Radar Velocity

In a vacuum, radar waves travel at the speed of light (c). In other materials, radar waves travel at speeds less than c, and the speed of a radar wave is governed by the relative dielectric permittivity (ε_r) and the relative magnetic permeability (μ_r) of the material through which it passes. In general, dielectric permittivity increases with water content, and water content can have a significant influence on the propagation of radar waves. Radar signal velocities are given by

$$V = c / \left[(\mu_r \, \varepsilon_r)^{\frac{1}{2}} \right].$$

(8.44)

Relative magnetic permeability is assumed to be 1 for most geologic materials, except those that are highly magnetic, and is often not stated explicitly in the radar velocity equation.

For example, let's calculate radar wave velocity through dry sand and compare this to seismic P-wave velocity for dry sand. From Table 8.1 we see that the relative permittivity ε_r of dry sand is 3–10. Using

$$V = c / \left[(\mu_r \, \varepsilon_r)^{\frac{1}{2}} \right] - (3 \times 10^8 \text{ m/s}) / \left[(1 \times 4)^{\frac{1}{2}} \right],$$

(8.45)

radar velocities for dry sand are

$$1.5 \times 10^8 \text{ m/s} \times \text{s}/10^9 \text{ns} = .15 \text{ m/ns}.$$

(8.46)

From Table 2.2, seismic P-wave velocity through dry sand is 200–1000 m/s. So the radar wave velocity is many thousands of times faster than the seismic P-wave velocity.

The size of the returned radar signal depends on the contrast in electrical properties across an interface or into a scatterer. Examples of scatterers are lampposts, fences, and trees, which bounce radar energy back to unshielded antennas. Because radar energy goes out in all directions from an unshielded

FIGURE 8.13 (a) Ground-penetrating radar profile showing inclined gravel. Time converted to depth using velocity of 0.14 m/ns. (Smith and Jol, 1992.) (b) GPR field equipment in use: Ramac system with 200 MHz antenna.

antenna, even the radar system operator can be a scatterer. The radar reflection coefficient is analogous to the seismic case (Chapter 4) and is given by the ratio of the amplitude of reflected waves (A_{rfl}) to incident waves (A_i). At vertical incidence the reflection coefficient is given by

$$R = A_{rf}/A_i = ((\varepsilon_2)^{\frac{1}{2}} - (\varepsilon_1)^{\frac{1}{2}})/((\varepsilon_2)^{\frac{1}{2}} + (\varepsilon_1)^{\frac{1}{2}}) = (V_2 - V_1)/(V_2 + V_1) \qquad (8.47)$$

where ε_1 and ε_2 are the relative dielectric permittivities, and V_1 and V_2 are the radar wave velocities in the respective media. If the contrast in dielectric permittivity across a layer boundary is small, then V_1 and V_2 are nearly the same and the numerator approaches zero, giving a very small reflection coefficient R. The strongest reflections come from layer boundaries with a strong contrast, such as a water–rock interface.

Similar to seismic reflection, EM waves of different frequencies have different resolution and typical depth of penetration. Frequency is related to wavelength by $v = f \times \lambda$, and the smallest layer that can be resolved in reflection is typically assumed to be of thickness $\lambda/4$. Although higher frequencies have higher resolution, they attenuate more rapidly than longer-period waves and thus have a smaller depth of penetration. Low-frequency waves give a coarser view of the subsurface but can penetrate to greater depth. Typical frequencies used for ground-penetrating radar range from 25 MHz to 1 GHz.

For example, let's calculate the smallest layer that can be resolved using a 1 GHz GPR antenna versus a 100 MHz GPR antenna, assuming the radar velocity in the subsurface media is $v = 0.8 \times 10^8$ m/s. The smallest layer that can be resolved is of thickness $\lambda/4$. For the 1 GHz antenna,

$$\lambda = v/f = (0.8 \times 10^8 \text{ m/s})/1 \text{ GHz} = (0.8 \times 10^8 \text{ m/s})/(1 \times 10^9 \text{ Hz})$$

$$= 0.08 \text{ m} = 8 \text{ cm, and } \lambda/4 = 2 \text{ cm.} \qquad (8.48)$$

For the 100 MHz antenna,

$$\lambda = v/f = (0.8 \times 10^8 \text{ m/s})/100 \text{ MHz} = (0.8 \times 10^8 \text{ m/s})/(100 \times 10^6 \text{ Hz})$$

$$= 0.8 \text{ m} = 80 \text{ cm, and } 1/4 \, \lambda = 20 \text{ cm.} \qquad (8.49)$$

Thus the 1 GHz antenna can resolve a layer as thin as 2 cm for this particular subsurface velocity, and the 100 MHz antenna can resolve a layer of 20 cm thickness. For material with a higher radar velocity, such as ice, the resolution is even higher (thinner layers can be resolved).

8.4.2 Data Acquisition

The acquisition of ground-penetrating radar data is similar to that of reflection seismology. Rather than a hammer, explosion, or other seismic source, the GPR source is a (usually small) antenna that broadcasts radio waves within a narrow range of frequencies. Rather than geophones, the GPR receiver is an antenna that looks similar or identical to the transmitting antenna. GPR surveys are

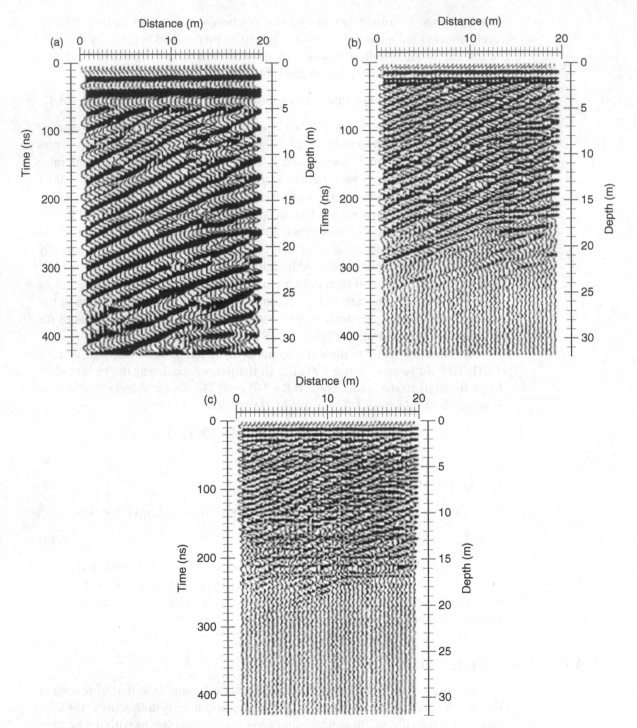

FIGURE 8.14 Ground-penetrating radar data collected using three different antenna frequencies. (a) 50 MHz. (b) 100 MHz. (c) 200 MHz. (Smith and Jol, 1992.)

performed at much higher frequencies, and usually much smaller spatial scales, than reflection seismology. The frequencies used for GPR range from approximately 25 MHz to 1 GHz, with the higher frequencies offering higher resolution but less depth of penetration. We collect GPR data by moving a transmitter and receiver antenna over the area of interest, continuously broadcasting and receiving the radar waves. The transmitter and receiver antenna are usually separated by a fixed distance and moved together in a constant-offset profile. Because GPR surveys are noninvasive, the ground is not disturbed by impact or insertion of electrodes. The noninvasive aspect of GPR surveys makes them popular for archaeological and structural engineering applications.

The most common mode of GPR data acquisition is *reflection profiling* (also called *zero-offset profiling*) (Figure 8.15 (a)). In GPR reflection profiling, many zero or near-zero offset traces are plotted sequentially. GPR reflection profiles can be gathered rapidly—at the speed of a slow walk—and can provide an immediate view of subsurface layering. This mode works well with horizontal or near-horizontal interfaces. In the presence of steeply dipping reflectors or point reflectors, artifacts such as diffraction hyperbolas will be present. For most shallow environmental and engineering applications, the diffraction hyperbolas are recognized on reflection profiles and used as a guide for finding sharp reflectors, but they are not often corrected for.

Wide angle reflection and refraction (WARR) is radar reflection profiling performed by keeping one antenna, usually the transmitter, in place and moving the receiving antenna progressively farther away from the transmitter (Figure 8.16). We often use the WARR method to determine subsurface velocity from the resulting move-out curves (Figure 8.16). As with the seismic reflection case, the air wave and direct wave (ground wave) velocities can be determined from the slope of the time versus source–receiver distance (antenna separation) graph, and reflection wave velocity can be determined from the $x^2 - t^2$ method. Figure 8.16 shows an example of a WARR measurement recorded on loamy sand with a 225 MHz antenna. The air wave and ground wave are marked.

Common midpoint (CMP) or *common depth point* (CDP) data acquisition for GPR is the same as CMP stacking for reflection seismology (Chapter 4). For GPR CMP is used more for velocity analysis than as a stacking tool. *CMP stacking* refers to combining reflection waveforms that have the same midpoint between source and receiver (Figure 8.15(b), Figure 8.17). For flat layered media, data that have the same midpoint between source and receiver will also sample the same point in depth beneath the midpoint (CDP). Before CMP stacking, we need to make a normal move-out (NMO) correction to account for the difference in time between data with different distances between the source and receiver. For example, a midpoint reflection from a source and receiver that are 3 m apart will arrive later than the midpoint reflection for a source–receiver separation of 0.5 m. With the NMO, all data are time shifted to have timing equivalent to a zero-offset trace (no distance between source and receiver). The subsurface velocity can be estimated

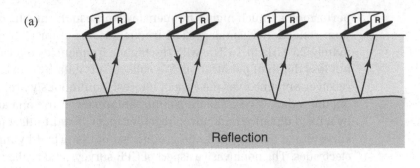

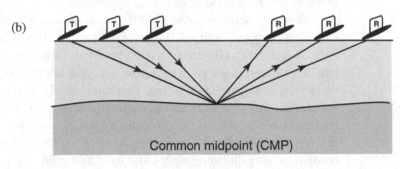

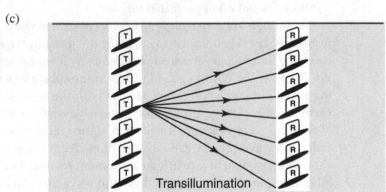

FIGURE 8.15 Some common modes of ground-penetrating radar data acquisition. (a) Reflection profiling. (b) Common midpoint profiling. (c) Transillumination. (After Annan, 1992.)

from the NMO curves. CMP stacking is not used as commonly in GPR as it is in seismic reflection because adequate signals can usually be resolved with the zero-offset reflection profiling in GPR, and postprocessing adds time and expense.

Transillumination or radar tomography refers to the case where radar transmitter and receiver are located on separate sides of a target, such as a wall or a mine pillar, and moved sequentially to have many wave paths crossing through the target (Figure 8.15(c)). Radar tomography can be performed in boreholes or mines, or for nondestructive testing of concrete. With tomography variations in physical properties can be found in two or three dimensions, rather than a simple

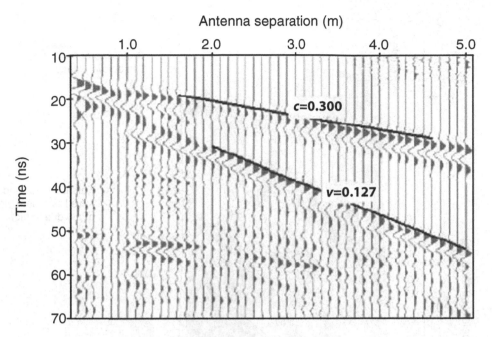

FIGURE 8.16 Wide angle reflection and refraction (WARR) GPR data collected using 225 MHz antenna. The velocity of the ground wave is v and that of the air wave is c, both given in meters per nanosecond. (Huisman et al., 2003.)

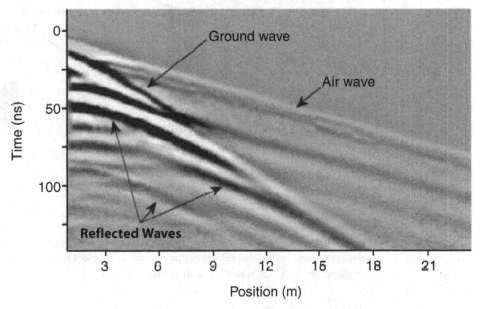

FIGURE 8.17 Common midpoint (CMP) GPR data collected using 100 MHz antenna. Air wave, ground wave, and reflections are marked. Position (m) is antenna separation in meters. (Huisman et al., 2003.)

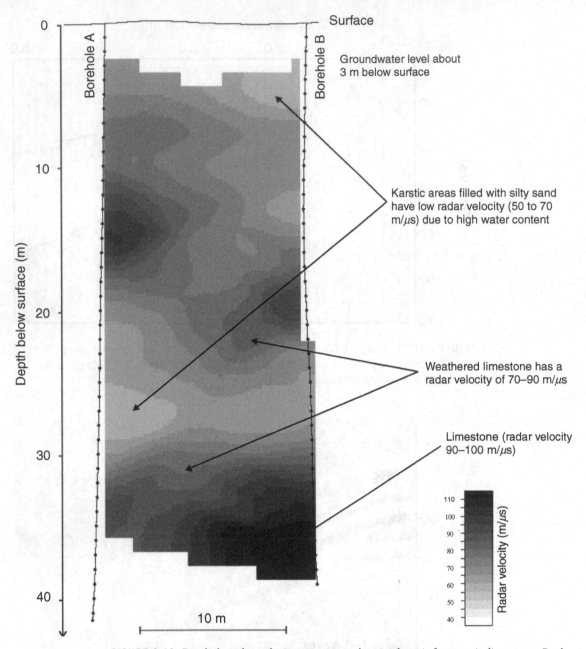

FIGURE 8.18 Borehole radar velocity tomogram showing karstic features in limestone. Dark colors represent high radar velocities (limestone); light colors represent low radar velocities (karstic areas with high water content). (Corin et al., 1997.)

one-dimensional layered structure as is usually determined with reflection processing. With cross-borehole radar tomography, the radar transmitter is placed down one borehole and the radar receiver antenna down another borehole (Figure 8.15(c)). The transmitter and receiver are moved up and down the borehole to maximize the number of transmitter–receiver paths crossing between the boreholes. The data are processed on a computer to determine a pattern of lateral subsurface radar velocity and/or attenuation variations that best fits the variations in radar travel time and amplitude observed for the collection of transmitter–receiver geometries. Variations in travel time determine velocity, and variations in radar amplitude determine lateral variations in attenuation. An example of a borehole radar tomography image (tomogram) is shown in Figure 8.18.

8.4.3   GPR Velocity Analysis

We must know radar wave velocity to convert radar reflections, which are recorded as a function of time, to depth. Three common methods of determining subsurface radar velocity are described here.

8.4.3.1 BURIAL OF KNOWN OBJECT

Because ground-penetrating radar studies are usually aimed at shallow targets, and metal provides such a strong radar reflection, a piece of metal such as a pipe or reinforcing rod (rebar) is often buried at a known depth and imaged using GPR. We determine velocity by dividing the known depth of the object by half of the observed two-way time to the object. Be cautious with this method because velocity can be altered—usually increased because of more porosity (more air) and less water (soil drying out from exposure)—by disturbing the soil to bury the known object.

Figure 8.19 shows an example of such a burial test. An aluminum pipe has been buried at a known depth of 57 cm, and GPR data have been collected over the pipe. The reflection arrives at 7 ns (indicated by the crosshairs in Figure 8.19). The distance to the bar and back is $0.57 \text{ m} \times 2 = 1.14 \text{ m}$, and the two-way time is 7 ns. Velocity = Distance/Time = 1.14 m/7 ns = 0.16 m/ns.

8.4.3.2 WALKAWAY TEST

In the walkaway test the source antenna is kept in one place, and we walk away from the source antenna while pulling the receiver antenna and recording (Figure 8.20). This is the same geometry as the WARR GPR method described in section 8.4.2. We can estimate the subsurface radar velocity from the variation in reflection time as a function of source–receiver distance. The walkaway test can be used with simple layered velocity structures.

The travel-time equations for radar waves are the same as those for seismic reflection (Chapter 4). Figure 8.20(a) shows the geometry for a radar air wave, ground wave, and reflected wave. The travel-time equation for the air wave is

$$\text{Time} = \text{Distance/Velocity} = EB/v = x/c \tag{8.50}$$

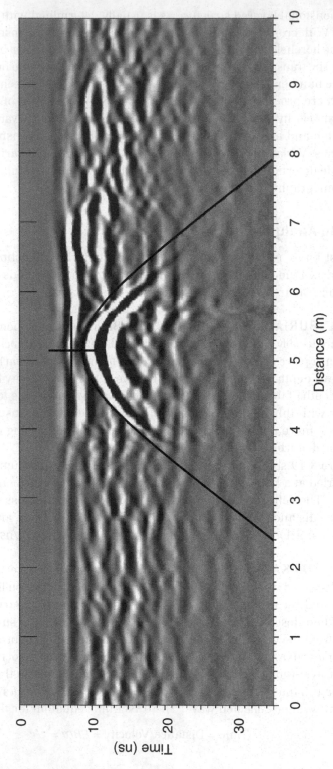

FIGURE 8.19 GPR data collected with 500 MHz antenna over pipe. The target is a hollow aluminum pipe 5 cm in diameter and 88 cm long. Burial was 57 cm to its top. The pipe was buried horizontally with the long axis perpendicular to the tow direction. Source and receiver antennas are at a constant offset of 0.8 m. (Courtesy of J. Lucius, USGS.)

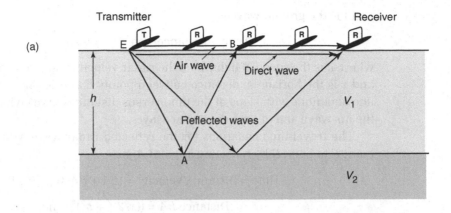

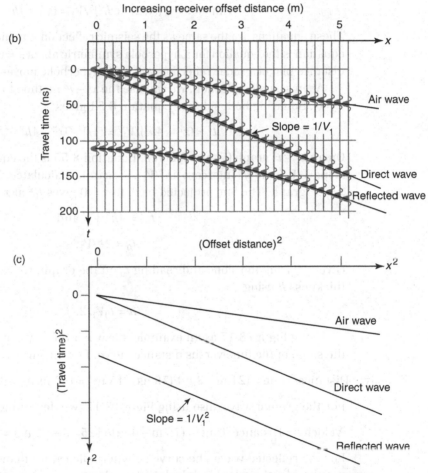

FIGURE 8.20 (a) Generalized diagram illustrating GPR ray paths for air wave, direct wave (ground wave), and reflected wave for material with two layers of contrasting dielectric permittivity. (b) GPR record section showing time–distance relationships for the air wave, direct wave, and reflected waves. (c) $x^2 - t^2$ graph from GPR record section in (b).

and for the ground wave is

$$\text{Time} = \text{Distance/Velocity} = EB/V_1 = x/V_1 \tag{8.51}$$

where c is the speed of light, V_1 is the radar velocity of the first subsurface layer and x is the horizontal distance between points E and B. These are both simple line equations with slope of the time versus distance curve giving the velocity of the air wave and of the near-surface layer.

The travel-time equations for the reflected radar wave, which travels from point E to point B by reflecting at point A, are

$$\text{Time} = \text{Distance/Velocity} = (EA + BA)/V_1 = 2EA/V_1, \tag{8.52}$$

$$\text{Distance } EA = ((x/2)^2 + h^2)^{\frac{1}{2}}, \text{ and} \tag{8.53}$$

$$\text{Time} = t = 2\,((x/2)^2 + h^2)^{\frac{1}{2}}/V_1 = (x^2 + 4h^2)^{\frac{1}{2}}/V_1. \tag{8.54}$$

These equations are the same as the seismic reflection Equation 4.1. The reflection equation is the equation for a hyperbola symmetric about $x = 0$, and the time versus distance plot of a radar reflection will have hyperbolic move-out (Figure 8.20(b)).

As in seismic reflection, we can use the $x^2 - t^2$ method to determine the velocity V_1. Squaring both sides of Equation 8.54 gives

$$t^2 = (x^2 + 4h^2)/V_1{}^2 = (1/V_1{}^2)x^2 + 4h^2/V_1{}^2. \tag{8.55}$$

On an $x^2 - t^2$ graph (Figure 8.20(c)), Equation 8.55 is the equation for a line with slope $1/V_1{}^2$ and y^2 intercept $4h^2/V_1{}^2$. V_1 can be calculated from the slope of the $x^2 - t^2$ graph. The line projected back to $x = 0$ gives $t_0{}^2$ at $x = 0$:

$$t_0{}^2 = 4h^2/V_1{}^2 \text{ and} \tag{8.56}$$

$$t_0 = 2h/V_1. \tag{8.57}$$

Given V_1 from the slope and reading t_0 off the graph, we can calculate the layer thickness h using

$$h = t_0 V_1/2. \tag{8.58}$$

Using Figure 8.17 as an example, air wave velocity can be determined from the slope of the time versus distance curve. For the air wave,

$$\text{Velocity} = \text{Distance/Time} = (21 \text{ m} - 3 \text{ m})/(75 \text{ ns} - 15 \text{ ns}) = 0.3 \text{ m/ns} = 3 \times 10^8 \text{ m/s}. \tag{8.59}$$

For the ground wave, also using Figure 8.17, we determine

$$\text{Velocity} = \text{Distance/Time} = (16 \text{ m} - 4 \text{ m})/(125 \text{ ns} - 25 \text{ ns}) = 0.12 \text{ m/ns}. \tag{8.60}$$

For the reflected wave, the curve is hyperbolic rather than linear, and we must use the $x^2 - t^2$ method to calculate the velocity:

$$V_1{}^2 = (x_2{}^2 - x_1{}^2)/(t_2{}^2 - t_1{}^2) = (12 \text{ m}^2 - 3 \text{ m}^2)/(100 \text{ ns}^2 - 50 \text{ ns}^2) = 0.018 \text{ m}^2/\text{ns}^2 \text{ and}$$

$$V_1 = (0.018 \text{ m}^2/\text{ns}^2)^{1/2} = 0.134 \text{ m/ns}. \tag{8.61}$$

In this case the velocities of the ground wave and the reflected wave are not identical. For a homogeneous and uniform two-layered structure, they should be the same. With actual field data the velocities of the ground wave and the first reflected wave often differ due to variation in velocity with depth. The ground wave samples near the surface, whereas the reflected wave samples the whole layer.

8.4.3.3 DIFFRACTION HYPERBOLA

Radar waves go out in all directions, not just straight down into the Earth and back up. Thus strong reflections can arrive from features that are not directly below the radar antennas. A good example is a metal fencepost, which produces a strong radar reflection. The reflection will be seen from a wide range of distances. A reflection arrival will be recorded as we approach the fencepost, will appear progressively earlier as we come closer to the fencepost, will be at its earliest arrival time when we are closest to the fencepost, and will continue to be recorded at progressively later times as we move away from the fencepost. A plot of the arrival times versus distance from the fencepost will form a hyperbola, and we can use the difference in time versus distance to determine velocity. In the fencepost case, the velocity should turn out to be the speed of an EM wave in air (that is, the speed of light). Likewise, a diffraction hyperbola from an object buried in the subsurface will produce a diffraction hyperbola, with move-out velocity corresponding to the radar wave velocity of the subsurface medium. Figure 8.21 shows an example of radar reflection data collected over a laterally varying anomalous subsurface zone. At point E, reflections are received from the anomalous zone and from the subsurface bedrock. As we near the anomalous zone, the time for the reflection to arrive from the anomalous zone decreases, and it is at a minimum when we are directly over the anomalous zone (position B). As we pass over the anomalous zone, the time of reflection again increases (position C).

The equations for a diffraction hyperbola are similar to those from seismic reflection (Chapter 4) except that in this case the source–receiver distance is kept fixed and we are considering reflections from a point reflector rather than from a flat layer. Figure 8.21 shows the geometry for a radar survey over a point reflector.

The travel-time equations for a radar wave traveling from point E to point A and reflecting back to E (Figure 8.21) are

$$\text{Time} = \text{Distance/Velocity} = 2EA/V_1 \tag{8.62}$$

$$\text{Distance } EA = (x^2 + h^2)^{\frac{1}{2}} \tag{8.63}$$

$$\text{Time} = t = 2(x^2 + h^2)^{\frac{1}{2}}/V_1 \tag{8.64}$$

where x is the horizontal distance between surface points E and B, and h is the depth to the anomalous zone.

These equations are similar to the seismic reflection Equation 4.1, except that in this case the horizontal distance x is between the source and the nearest surface point to the subsurface reflector A, and in the seismic case the distance x is

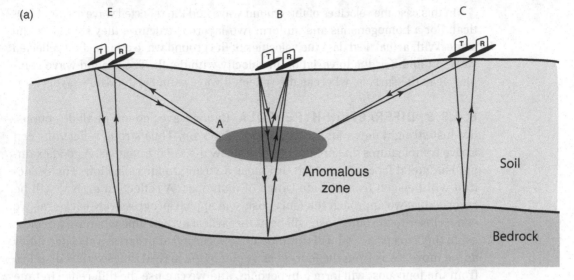

Soil

Anomalous
zone

Bedrock

Horizontal position

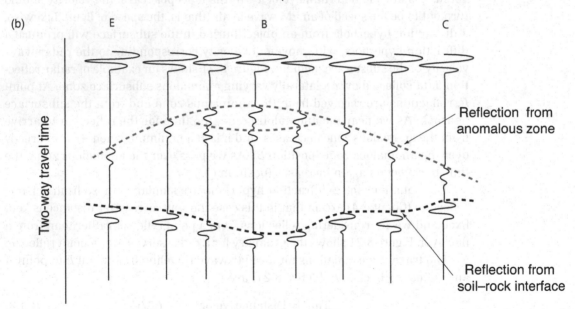

Reflection from
anomalous zone

Reflection from
soil–rock interface

FIGURE 8.21 (a) Generalized diagram illustrating GPR ray paths for data collected using fixed antenna separation. The transect is over a structure with an anomalous zone of high radar velocity at the center of the transect and a near-horizontal bedrock layer. (b) GPR record section showing reflected arrivals from the anomalous zone and from the soil–rock interface. (After Davis and Annan, 1989.)

between the source and receiver, which are on opposite sides of the reflector point A (x in Equation 8.64 is $x/2$ in Equation 4.1).

As in seismic reflection, we can use the $x^2 - t^2$ method to determine the velocity V_1. Squaring both sides of Equation 8.64 gives

$$t^2 = 4(x^2 + h^2)/V_1^2 = (4/V_1^2)x^2 + 4h^2/V_1^2. \qquad (8.65)$$

On an $x^2 - t^2$ graph, Equation 8.65 is the equation for a line with slope $4/V_1^2$ and y^2 intercept $4h^2/V_1^2$. The line projected back to $x = 0$ gives t_0^2 at $x = 0$:

$$t_0^2 = 4h^2/V_1^2 \qquad (8.66)$$

$$t_0 = 2h/V_1 \qquad (8.67)$$

We can calculate V_1 from the slope of the $x^2 - t^2$ graph. Given V_1 and reading t_0 off the graph, we calculate the layer thickness h using

$$h = t_0 V_1/2. \qquad (8.68)$$

Although a diffraction hyperbola might not excite many people other than authors of geophysics textbooks, we know of at least one case in which a diffraction hyperbola was a lifesaver. University of Colorado professor Konrad Steffen uses GPR in his studies of annual snow accumulation in Greenland, and he pulls the GPR unit on a sled behind a snowmobile. While driving the snowmobile in a whiteout with near-zero visibility, Steffen noticed a radar reflection on his data display. As he drove forward, the timing of the arrival decreased. Steffen prudently stopped his snowmobile because he recognized that the radar reflection was a diffraction hyperbola and that a possible source of this hyperbola was a large glacial crevasse. In fact the hyperbola was produced by a crevasse, big enough to swallow both a snowmobile and a determined geophysics professor.

8.5 APPLICATIONS OF ELECTROMAGNETIC SURVEYING

Electromagnetic surveys are useful for a range of applications, including detailed archaeological surveys, environmental applications, geologic exploration, and mapping of snow and ice for climate studies.

8.5.1 Archaeological Surveys

8.5.1.1 AZTEC RUINS, NEW MEXICO EM surveys are particularly attractive for investigation of archaeological sites because of their noninvasive nature, which minimizes disturbance of sensitive locations. Results from geophysical exploration permit more efficient and focused excavation of targets. Figure 8.22 shows results of an EM31 ground conductivity survey at Aztec Ruins National Monument in New Mexico. Pueblo sites of the eleventh to thirteenth centuries

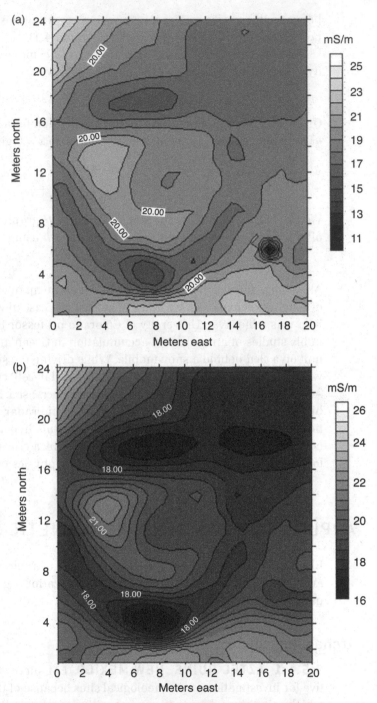

FIGURE 8.22 Contour map of apparent conductivity from EM31 survey of Aztec Ruins site, New Mexico. Contours are in mS/m with darker colors indicating low conductivity. A circular ring of low conductivity is shown, likely associated with resistive walls of an ancient Pueblo kiva house. (a) Raw gridded data. (b) Filtered to remove contouring artifact at 18E, 6N.

have been excavated in the region. The goal of the study was to determine whether a depressed area of ground was the site of a great kiva (ceremonial structure). Geophysical work on similar depressions had either confirmed or disproved suggested interpretations. EM31 ground conductivity data were collected over a 24 m × 20 m grid at 1 m intervals in quad phase mode. The quad phase mode is most sensitive to objects with low conductivity. The in-phase mode is sensitive to high-conductivity features like pipes and metallic objects. A circular pattern of low conductivity was revealed (Figure 8.22), likely associated with walls of a kiva house. The magnitude of conductivity variations was fairly small (range in conductivity was 10–25 mS/m), but the pattern of conductivity variations associated with the narrow circular outline of the walls of the structure was clear. GPR and magnetics surveys were also performed in the same area, with less success. Often we apply multiple geophysical methods to initial studies of an archaeological site, with more detailed follow-up using the most successful methods. Although it is possible to estimate which geophysical techniques will be most successful based on expected variations in physical or electrical properties, in practice it is risky to rely on a single method. In the Aztec Ruins case, it was thought that GPR would work well based on the dry sandy soil. However, GPR did not produce strong reflections. This may have been due to uneven terrain and the presence of vegetation or to salts in the desert soil, which increase the soil conductivity and attenuate the radar energy. The magnetic survey did not find clear anomalies, suggesting a lack of magnetic susceptibility contrast between the kiva walls and the soil. Interference from a nearby power line also introduced noise into the magnetics survey.

8.5.1.2 CEREN, EL SALVADOR At the Joya de Ceren site in El Salvador, geophysical methods were used to explore a prehistoric village buried by volcanic ash (Sheets 1979). A magnetometer survey of a nearby site was unsuccessful because of the strongly magnetic nature of the underlying lava flows. Therefore, electrical DC resistivity in a Wenner electrode configuration was used, resulting in the discovery of a number of cultural features that were later excavated. By far the most successful geophysical method at this particular site was GPR. A strong contrast in electrical properties exists between the high-conductivity adobe walls of the structures and the low-conductivity volcanic tephra. Figure 8.23 shows an example GPR profile from the Ceren site. Reflections from the agricultural furrows are shown by white diamonds. Structures imaged with GPR were confirmed by excavations. An EM34 survey was performed, but the depth of main sensitivity (7 to 15 m) was below the cultural horizons of interest. Preliminary surveys using EM31 did not reveal significant anomalies. A summary of the initial geophysical investigations at Ceren is given in Conyers and Spetzler (2002), and more details of the GPR processing are described in Conyers (2004). Results of these surveys indicate that some geophysical methods are more effective in the rainy season when the ground is saturated and that others work better in the dry season. Electrical

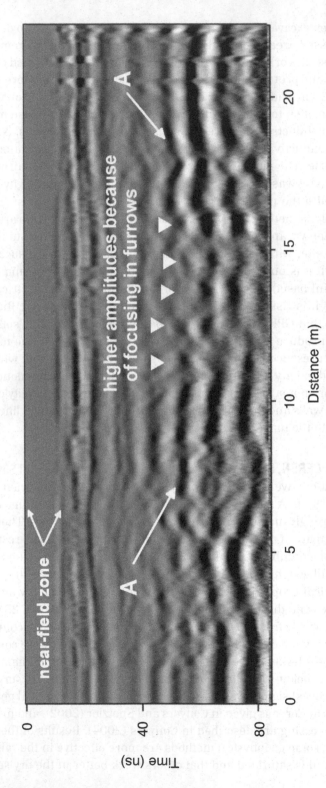

FIGURE 8.23 GPR radargram from Ceren site, El Salvador. Reflection surface A is Ceren village floor surface. Ancient agricultural ridges and furrows from Ceren village are revealed by varying high and low radar amplitudes, indicated by white arrows. (Figure courtesy of L. Conyers.)

resistivity works well when the ground is wet and the most conductive. GPR works best when the ground is dry and less signal is attenuated in the near surface.

8.5.2   Geologic Applications

Geologic applications of electromagnetic methods include determining subsurface stratigraphy, finding the extent of buried faults, exploring for minerals, and studying aquifers.

Smith and Jol (1992) demonstrate how GPR can be used to determine stratigraphic parameters including bed thickness, dip, and internal structure, and that high-quality reflections from depths of tens of meters are possible in appropriate geologic conditions. Their study at the Brigham, Utah, gravel pit detected gravel facies at 32 m depth using a 50 MHz GPR antenna. Figure 8.24 shows a schematic diagram of stratigraphic layering and a corresponding radargram from 100 MHz antennas. The deep reflections found in this GPR survey suggest that the quartz-rich sands mined in the quarry operation extend below the floor of the gravel pit and remain relatively free of silt and clay (which would attenuate the signal). Comparison of three radar profiles using antenna frequencies of 50 MHz, 100 MHz, and 200 MHz (Figure 8.14) illustrates the trade-off between signal penetration and resolution.

Electromagnetic studies have long been used in mineral exploration. Power, Belcourt, and Rockel (2004) describe the application of a suite of geophysical measurement tools to explore for diamond-bearing igneous kimberlites in northern Canada. Near the surface, kimberlites are found in clusters of pipelike bodies. Most kimberlites intrude into granitic and metamorphic rocks, and contrasts in physical properties such as electrical resistivity and dielectric permittivity can be used to distinguish the kimberlites from the host rocks. Most kimberlites are more conductive than their host rocks, mainly due to serpentization and clay alteration in the kimberlite. Airborne EM is often used in reconnaissance surveys, with ground follow-up at higher resolution. Kimberlites have higher dielectric permittivity than host rocks, mainly due to the higher liquid concentrations in the clays of the kimberlites compared to the relatively dry host rocks. Figure 8.25 illustrates the GPR response of a kimberlite pipe beneath a lake. The survey was run in the winter by towing the radar across the frozen lake on a snowmobile. The kimberlite reflection in the middle of the figure shows a clear reflector with strong attenuation. The surrounding granite is fractured, and this difference in texture is evident in the many diffraction hyperbolas and tails in the radargram.

Schultz (2002) describes a combined electromagnetic study of a coastal aquifer in Georgia. Coastal areas are characterized by regions of high conductivity (high salinity) adjacent to lower-conductivity regions (freshwater). Electromagnetic methods including both ground conductivity measurements and GPR, along with DC resistivity, can be used to determine the geometry of the freshwater–saltwater interface. The intrusion of saline waters into freshwater aquifers is an important

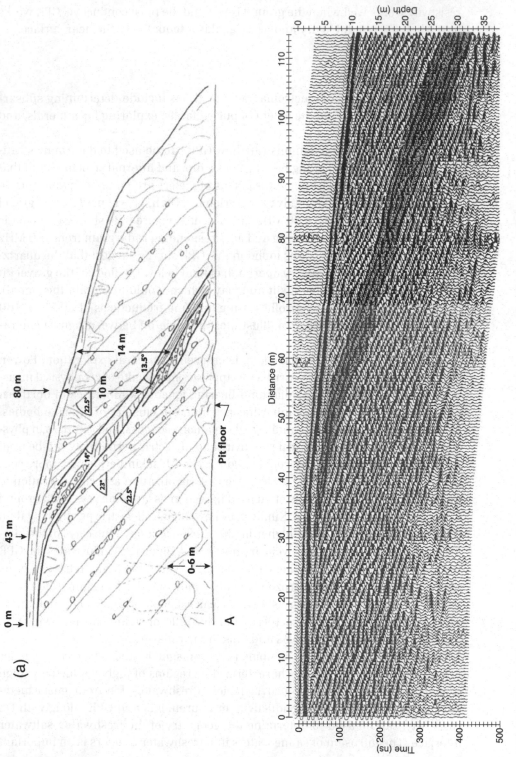

FIGURE 8.24 (a) Schematic diagram of stratigraphic layering at Brigham, Utah, gravel pit. (b) GPR data collected using a 100 MHz antenna at this gravel pit. Time has been converted to depth using a velocity of 0.14 m/s, determined from a CMP survey. Radar profile shows inclined reflections due to dipping sandy unit. (Smith and Jol, 1992.)

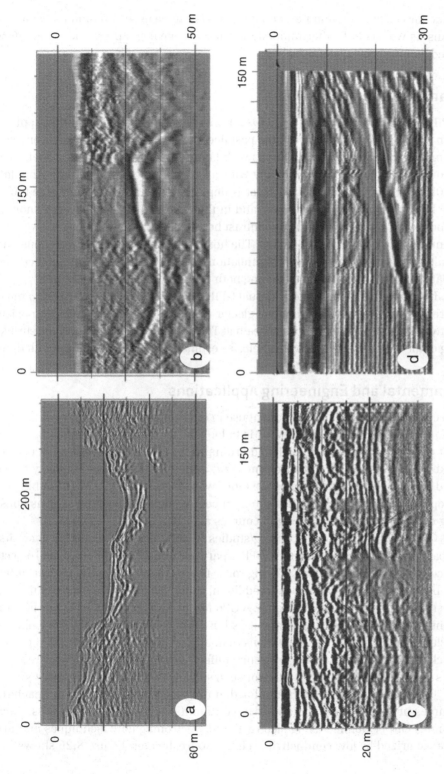

FIGURE 8.25 GPR radargrams from kimberlite exploration in Canada. (a) Response of a kimberlite pipe beneath a lake. Antenna frequency = 25 MHz. Top of pipe is smooth planar reflector with evident attenuation. Surrounding granite is fractured and displays many diffraction hyperbolas, generating a contrast in texture with the smooth kimberlite. Lake bottom sediments are also visible. (b) Response of a kimberlite pipe beneath 40 m of boulder till. Antenna frequency = 12.5 MHz. The kimberlite produces a strong reflection, its lateral position is 30–180 m on the plot, and its depth is 26-38 m. (c) GPR response of 25 m wide vertical dike intruding into bedded limestone. Antenna frequency = 25 MHz. Diffractions from margins of the dike assist in defining the walls of the intrusion. (d) Response of shallow dipping kimberlite dike. Antenna frequency = 25 MHz. (Power, et al., 2004.)

issue for coastal communities, and geophysics can be used to monitor such intrusions as well as to further understand the hydrogeologic properties of aquifers used as municipal water supplies.

8.5.3 Snow and Ice Mapping

GPR was first developed in the 1930s for snow and ice imaging, and that application has found increasing use in the past decade. In Greenland GPR has been used to map snow and ice accumulations with lateral range on the scales of meters to kilometers and depth range mainly within several meters (Steffen and Abdalati 2005). Each winter a new snow layer is deposited. During the summer the top of the layer of snow metamorphoses, and in the next winter a new layer of snow is deposited. There is a permittivity contrast between the ice and soft snow, revealing annual snow layers in a GPR image. The horizons are verified using snow pits and shallow ice cores. The annual accumulation of snow and subsequent melt can be related to the Arctic climate, including both precipitation and temperature.

In another GPR application, annual GPR studies have been performed to measure the thickness of the Arapahoe glacier in Colorado. A strong reflection exists between the ice and the bedrock beneath it. These annual studies quantify the shrinking of the Arapahoe glacier, which helps us estimate variations in climate with time.

8.5.4 Environmental and Engineering Applications

Electromagnetic work finds frequent use in environmental and engineering studies. A common use for EM surveys is to look for land mines or bombs that have not yet exploded (called unexploded ordinance or UXO). UXO studies are undertaken in active and former war zones and also at the sites of old military bases and bombing ranges. Such sites are more widespread than you might think, and geophysical surveys for UXO are common before putting in facilities such as housing developments, malls, and golf courses as urban boundaries expand.

GPR can be used in engineering studies to find buried pipes, utility conduits, rebar, and other metallic materials. GPR is particularly useful for finding metal objects in concrete, such as rebar reinforcing rods. There is a strong contrast in permittivity between concrete and metal. In addition, concrete has very low attenuation, so the radar waves propagate efficiently. A GPR test pad was built at the Denver Federal Center with a number of known objects buried beneath the concrete. The objects included rebar at various orientations (flat, angled, vertical) and metal and plastic buckets. This facility is useful for testing different GPR units and antenna frequencies to determine which is optimal for such shallow engineering purposes.

EM studies can also be useful for denoting the extent of landfills. Leachate plumes originating from landfills can contaminate groundwater supplies. Such plumes can be defined using electrical and electromagnetic techniques and are characterized by low conductivity relative to freshwater. Figure 8.26 shows the

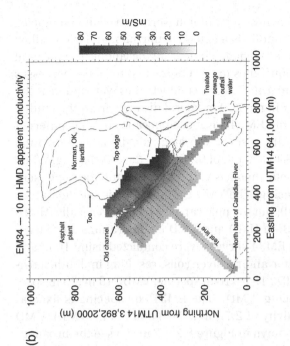

(b)

EM34 — 10 m HMD apparent conductivity

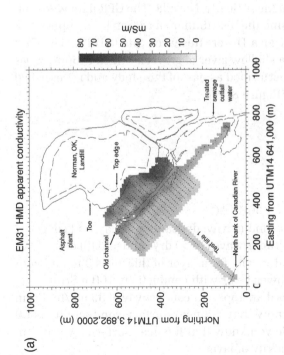

(a)

EM31 HMD apparent conductivity

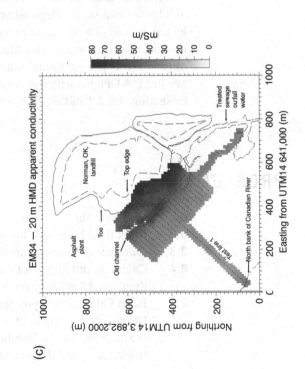

(c)

EM34 — 20 m HMD apparent conductivity

FIGURE 8.26 Results of EM induction surveys near the Norman, Oklahoma, landfill (Lucius and Bisdorf 1995). Data shown were collected with the coils vertical (horizontal magnetic dipole or HMD). (a) EM31 response, HMD, depth sensitivity to approximately 3 m. Black indicates high conductivity and is interpreted as leachate; white is low conductivity. High-conductivity zone is observed at the SW edge of the landfill. (b) EM34 response with 10 m coil spacing, HMD, depth sensitivity to 7.5 m. Black is high-conductivity plume. (c) EM34 response with 20 m coil spacing, HMD. Depth sensitivity to 15 m. High-conductivity zone, interpreted as leachate plume, has broadened with depth.

results of an EM31 and EM34 study near a landfill in Norman, Oklahoma (Lucius and Bisdorf, 1995). The Norman landfill is located on alluvium, and the shallow alluvial aquifer is contaminated by a leachate plume from the landfill. Geophysical mapping of the leachate plume contributes to our understanding of the processes of contaminant distribution and migration. High conductivities were found at the landfill and its boundaries, extending to 200 m lateral distance from the landfill boundaries. Both the EM31 and the EM34 were used because of their different depth sensitivity. When the coils are vertical, the magnetic dipole is horizontal (HMD), and the depth sensitivity is to a depth of 0.75 times the coil spacing. When the coils are horizontal, the magnetic dipole is vertical (VMD), and the depth sensitivity is to 1.5 times the coil spacing. The VMD configuration is the default for the EM31 (holding pole horizontal), but the pole can also be held vertically (HMD) for shallower depth resolution. Both HMD and VMD configurations were used for the EM31 and the EM34. The EM34 surveys were conducted using 10 m and 20 m spacings between transmitter and receiver coils, resulting in depth sensitivities of 7.5 m (10 m spacing, HMD), 15 m (10 m spacing, VMD), 15 m (20 m spacing, HMD), and 30 m (20 m spacing, VMD). The EM31 coil spacing is fixed at 3.67 m, resulting in depth sensitivity of 2.75 m (HMD) and 5.5 m (VMD). HMD results for all configurations are shown in Figure 8.26. You can see the broadening of the landfill leachate plume with depth and extent to the SW direction by comparing Figures 8.26(a) and 8.26(c).

Another example of environmental application of geophysics is a GPR study of a hazardous waste disposal area near Beatty, Nevada. The GPR data were collected as part of a study to determine the distribution of tritium in the upper two meters of sediments at the Amargosa Desert Research Site (ADRS). The GPR objective was to identify and map a shallow layer of gravel in which tritium values are high. A GPR velocity test was performed as part of this study and is described in section 8.4.3.1 of this chapter (Figure 8.19).

PROBLEMS

8.1 Calculate the wavelength of a 60Hz AC.

8.2 Rank from highest to lowest conductivity: dry sand, wet clay, and concrete.

8.3 Calculate the skin depth for radar waves in (a) dry sand and (b) damp clay.

8.4 Calculate the apparent conductivity for a layer of thickness 3 m and conductivity 15 mS/m over a lower layer with conductivity 100 mS/m.

8.5 In an EM31 survey you record an apparent conductivity of 35 mS/m. From DC resistivity work, you know that the near-surface conductivity is 60 mS/m, and from a borehole you know that the near-surface layer is 2 m thick. Calculate the conductivity of layer 2.

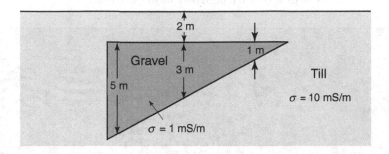

Gravel deposit

8.6 Calculate the apparent conductivity readings that will be measured at points A, B, and C over the gravel deposit shown above.

8.7 Estimate variations in depth to bedrock given apparent conductivity readings of 6.9, 15.9, and 20.1 mS/m, surface layer (soil) conductivity of 30 mS/m, and bedrock (granite) conductivity of 0.1 mS/m.

8.8 Calculate the travel time for a radar wave to reflect in air from an object 10 m away. Compare this to seismic travel times.

8.9 Calculate the wavelength for a 200 MHz radar wave in (a) air and (b) granite. Use the radar velocities from Table 8.1. Calculate the wavelength for a compressional seismic wave at 200 Hz through the same materials, and discuss.

8.10 Calculate the radar wave velocity through dry sand.

8.11 Of the following, which would produce the best radar target, and why? Damp sand over dry sand, dry clay over wet sand, or wet clay over dry sand.

8.12 Calculate the smallest layer that can be resolved using a 1 GHz GPR antenna versus a 100 MHz GPR antenna. Assume $v = 0.8 \times 10^8$ m/s.

8.13 Calculate the air wave and ground wave velocities for the WARR geometry shown in Figure 8.16. Show your work.

8.14 Calculate the velocities of the air wave, ground wave, and reflected waves of the CMP measurement shown in Figure 8.17. Show your work.

8.15 Calculate the subsurface velocity from the move-out on the reflection hyperbola in Figure 8.19. Compare to the velocity determined using the burial test, and discuss.

8.16 Why isn't the burial of known object method used to calibrate *seismic* velocity structure?

8.17 Make a graph of distance versus arrival time from a reflection from a metal light pole, which ranges from 2 m to 10 m distance to your profile line, plotting the reflection arrival time every 0.2 m along your line.

8.18   Interpret the linear arrivals marked *A* in the radar figure shown below, and discuss. (Figure courtesy of L. Conyers.)

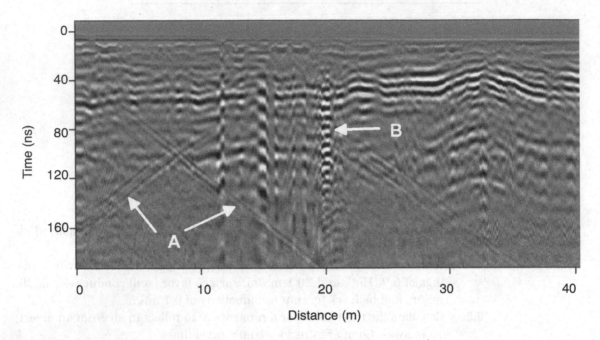

8.19   Sample records of GPR data collected over a test pad are shown on pages 551 and 552. The test pad was constructed by placing materials of environmental or engineering interest (like buckets of different construction and contents and rebar) in loose sand at different orientations and then covering everything uniformly with a layer of concrete.

Your goal is to use the data to find and map some of the items buried beneath the test pad. The data were collected on a grid with 10 north–south lines and 20 east–west lines, first using a 500 MHz antenna, and then using a 900 MHz antenna. Line 0.5E is the westernmost north–south line, and line 0.5N is the southernmost east–west line. Lines 4.5N, 5.5N, 3.5E, and 4.5E are shown. Time was converted to depth using a velocity of 11.33 cm/ns.

Mark the GPR data to show likely features, transfer this information to a location map, and discuss your results. Make separate maps for the 500 and 900 MHz data. Discuss the differences between the data collected with 500 and 900 MHz antennas. (Figures courtesy of J. Lucius.)

8.20   Describe some of the advantages and disadvantages of the EM31 electromagnetic instrument.

8.21   Give an example of a target for which (a) DC resistivity is superior to EM31 or (b) EM31 is superior to DC resistivity.

U.S. Bureau of Reclamation DFC GPR Test Pad   Line   4.5 N

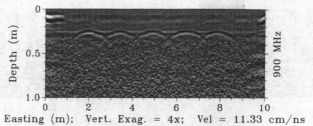

Easting (m);   Vert. Exag. = 4x;   Vel = 11.33 cm/ns

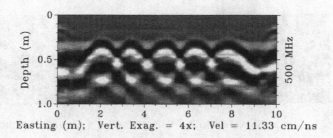

Easting (m);   Vert. Exag. = 4x;   Vel = 11.33 cm/ns

U.S. Bureau of Reclamation DFC GPR Test Pad   Line   5.5 N

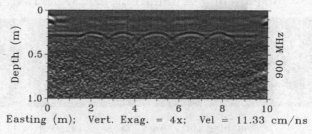

Easting (m);   Vert. Exag. = 4x;   Vel = 11.33 cm/ns

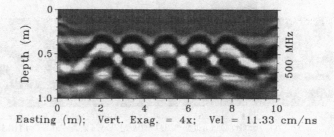

Easting (m);   Vert. Exag. = 4x;   Vel = 11.33 cm/ns

PROBLEM 8.19 *(continued on page 552)*

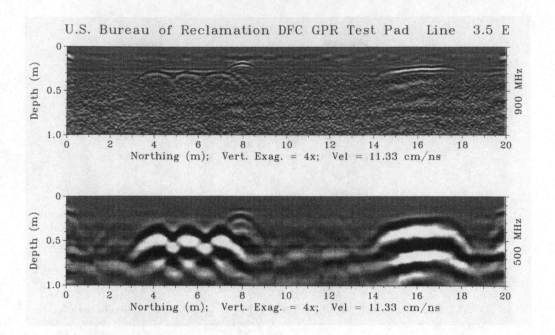

U.S. Bureau of Reclamation DFC GPR Test Pad   Line   3.5 E

Northing (m);   Vert. Exag. = 4x;   Vel = 11.33 cm/ns

Northing (m);   Vert. Exag. = 4x;   Vel = 11.33 cm/ns

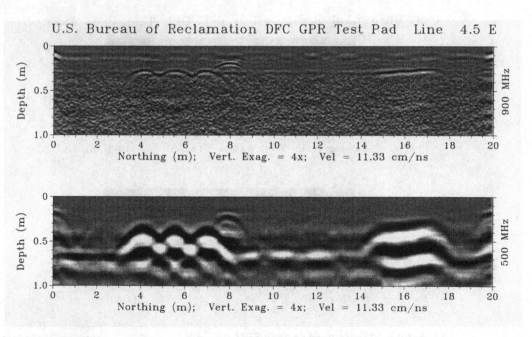

U.S. Bureau of Reclamation DFC GPR Test Pad   Line   4.5 E

Northing (m);   Vert. Exag. = 4x;   Vel = 11.33 cm/ns

Northing (m);   Vert. Exag. = 4x;   Vel = 11.33 cm/ns

PROBLEM 8.19 *(continued)*

REFERENCES CITED

Annan, A.P. 1992. *Ground penetrating radar workshop notes*. Mississauga, Ontario, Canada: Sensors and Software, Inc.

Aster, R., B. Borchers, and C. Thurber. 2004. *Parameter estimation and inverse problems*. Elsevier Academic Press.

Conyers, L.B. 2004. *Ground-penetrating radar for archaeology*. Walnut Creek, CA: Altamira Press.

Conyers, L.B., and H. Spetzler. 2002. Geophysical exploration at Ceren. In *Before the volcano erupted: The ancient Ceren village in Central America,* ed. P. Sheets, 24–32. Austin: University of Texas Press.

Corin, L., I. Couchard, B. Dethy, L. Halleaux, A. Monjoie, T. Richter, and J. P. Wauters. 1997. *Radar tomography applied to foundation design in a karstic environment*. London: Geological Society, Engineering Geology Special Publication no. 12, 167–73.

Davis, J.L., and J.P. Annan. 1989. Ground penetrating radar for high-resolution mapping of soil and rock stratigraphy. *Geophysical Prospecting*, v. 37, 531–51.

Fitterman, D.V., and C.C. Yin. 2004. Effect of bird maneuvers on frequency-domain helicopter EM response. *Geophysics*, v. 69, no. 5, 1203–15.

Grant, F.S., and G.F. West. 1965. *Interpretation theory in applied geophysics*. New York: McGraw-Hill.

Huisman, J.A., S.S. Hubbard, J.D. Redman, and A.P. Annan. 2003. Measuring soil water content with ground penetrating radar: a review. *Vadoze Zone Journal*, vol. 2, 476–91.

Keary, P., M. Brooks, and I. Hill, 2002. *An introduction to geophysical exploration*. Oxford, Blackwell Science, Ltd.

Lucius, J.E., and R.J. Bisdorf. 1995. Results of geophysical investigation near the Norman, Oklahoma, municipal landfill, 1995. *U.S. Geological Survey Open-file Report 95-825*.

McNeill, J.D. 1980a. Electromagnetic terrain conductivity measurement at low induction numbers. *Geonics Limited Technical Note TN-6*.

McNeill, J.D., 1980b. *Applications of transient electromagnetic techniques*. Geonics Limited Technical Note TN-7.

Menke, W. 1989. *Geophysical data analysis: Discrete inverse theory*. San Diego: Academic Press.

Milsom, J. 2003. *Field geophysics*, 3rd ed. Chichester: John Wiley and Sons.

Mussett, A.G., and M.A. Khan, 2000. *Looking into the Earth: An introduction to geological geophysics*. Cambridge: Cambridge University Press.

Power, M., G. Belcourt, and E. Rockel. 2004. Geophysical methods for kimberlite exploration in northern Canada. *The Leading Edge*, v. 23, 1124–29.

Reynolds, John M. 1997. *An introduction to applied and environmental geophysics*. West Sussex, England: John Wiley and Sons.

Rodi, W., and R.L. Mackie. 2001. Nonlinear conjugate gradients algorithm for 2-D magnetotelluric inversion. *Geophysics*, v. 66, no. 1, 174–87.

Schultz, G.M. 2002. *Hydrologic and geophysical characterization of spatial and temporal variations in coastal aquifer systems*. Georgia Institute of Technology, Ph.D. thesis.

Sharma, P.V., 1997. *Environmental and engineering geophysics*. Cambridge: Cambridge University Press.

Sheets, P.D. 1979. Maya recovery from volcanic disasters—Ilopango and Ceren. *Archaeology*, v. 32, no. 3, 32–42.

Smith, D.G., and H.M. Jol. 1992. Ground penetrating radar investigation of a Lake Bonneville delta, Provo level, Brigham City, Utah. *Geology*, v. 20, 1083–86.

Steffen, K., and W. Abdalati. 2005. Accumulation variability due to surface undulation on the Greenland ice sheet. *J. Geophys. Res.,* submitted.

SUGGESTED READING

Annan, A.P. 1973. Radio interferometry depth sounding: Part 1—Theoretical discussion. *Geophysics*, v. 38, no. 3, 557–80.

Cagnoli, B., and J.K. Russell. 2000. Imaging the subsurface stratigraphy in the Ubahebe hydrovolcanic field (Death Valley, California) using ground-penetrating radar. *Journal of Volcanology and Geothermal Research*, v. 96, 45–56.

Cunningham, K.J. 2004. Application of ground-penetrating radar, digital optical borehole images, and cores for characterization of porosity hydraulic conductivity and paleokarst in the Biscayne aquifer, southeastern Florida USA. *Journal of Applied Geophysics*, v. 55, 61–76.

Knight, R., 2001. *Ground penetrating radar for environmental applications*. Annu. Rev. Earth Planet Sci., 29:229–255.

Liu, L., and Y. Li. 2001. Identification of liquefaction and deformation features using ground-penetrating radar in the New Madrid seismic zone, USA. *Journal of Applied Geophysics*, v. 47, 199–215.

Schultz, G., and C. Ruppel. 2005. Inversion of inductive electromagnetic data in highly conductive terrains. *Geophysics*, v. 70, no. 1, G-16–G-18.

Sheriff, R.E., 2002. *Encyclopedic dictionary of exploration geophysics,* 4th ed. Tulsa, OK: Society of Exploration Geophysicists.

Telford, W.M., L.P. Geldart, and R.E. Sheriff. 1990. *Applied geophysics*, 2nd ed. Cambridge, England: Cambridge University Press.

Ward, S.H. 1990. Geotechnical and environmental geophysics. Tulsa, OK: Society of Exploration Geophysicists, *Investigations in Geophysics,* no. 5.

Instructions for Using REFRACT

REFRACT produces time–distance data for a subsurface model you define. REFRACT also lets you fit lines to time–distance data you have collected and then calculates subsurface structure based on your line positions.

PRODUCING TIME–DISTANCE DATA FOR A SUBSURFACE MODEL

This is the default condition for REFRACT. When you first start the program, the *Model Table* window appears (Figure A.1). The upper half of the *Model Table* window contains all the elements you need to design a subsurface model. You can add or delete layers by clicking the buttons with the appropriate labels.

You can edit layer thickness, velocity, and dip angle using normal editing techniques. Use the *Tab* key to cycle quickly through each entry. Similarly, you can edit the field design by changing the number of geophones, shot offset, and geophone spacing.

After you have completed your setup, click the *Update Times and Plots* button to update the table in the bottom half of the *Model Table* window.

Normally you should open the *Plot* and *Cross section* windows (Figure A.2) before designing models so you can see the results of your efforts more graphically. These can be accessed from the *Window* menu (Figure A.3).

The remaining options for producing time–distance data from a model can be viewed in Figure A.3. These consist of using different units and saving your models and are straightforward.

Although the *Model Table* window makes creating a model quite easy, it is possible to create a model as a text file in a word processing program and then to open that model. The best way to select the correct format for the text file is to save a model in REFRACT and then examine the resulting model file.

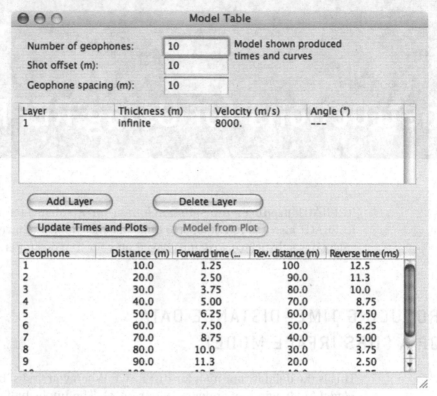

FIGURE A.1 *Model Table* window.

DETERMINING SUBSURFACE STRUCTURE FROM TIME–DISTANCE DATA

The most direct way to enter time–distance data for analysis is to first set up the correct field parameters (the number of geophones, shot offset, and geophone spacing) in the *Model Table* window. Next, adjust the number of layers that you expect to be present based on your inspection of field data. Finally, open the *Data* window in REFRACT (Figure A.4) and enter your data.

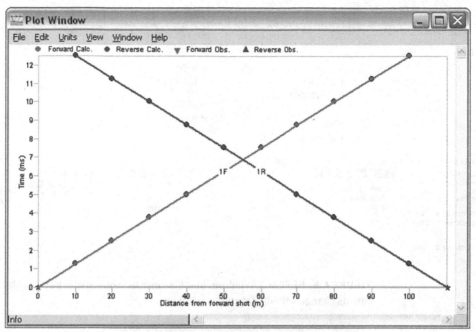

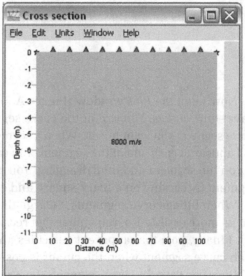

FIGURE A.2 *Plot* and *Cross section* windows.

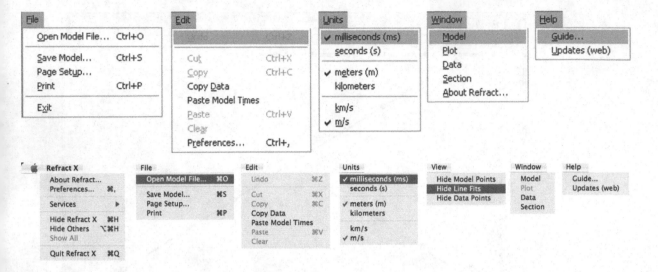

FIGURE A.3 REFRACT Windows and Mac menus. Some menu items in the *File* and *Edit* menus change with the frontmost window.

Now open the *Plot* window (Figure A.5) to see your data points as well as travel-time curves. You can fit the curve segments to your data by clicking on a curve segment you want to edit. When a segment is active, two small black squares will appear at each end of the segment. You can move the entire segment by clicking on the segment and then dragging. You can change the slope and extent of a segment by clicking on a black square and dragging.

When fitting curve segments, enlarge the *Plot* window to make your adjustments more easily. Also, remember that reciprocal times must be equal (or nearly so). Finally, note that apparent velocities and intercept times are displayed for each curve segment when the cursor is positioned on that segment.

Once you have the curve segments the way you want them, click the *Model from Plot* button in the *Model Table* window. The *Layer* information in the *Model Table* window will change to show the results of your curve fits. You will need to click the *Update Times and Plots* button in the *Model Table* window to update the arrival-time information in the *Model Table* window and the subsurface structure in the *Cross section* window.

If, after trying to fit curve segments to your data, you decide a travel-time curve for an additional layer is needed, simply return to the *Model Table* window and click the *Add Layer* button.

You can save your travel-time data by selecting *Save Data* in the *File* menu when the *Data* window is active. You can save your model data when the *Model Table* window is active. As was the case for model data, you can read your travel-time data from a text file, if you wish. Once again, save a sample data file generated within REFRACT and then open it to view the proper format for generating a text file of travel-time data.

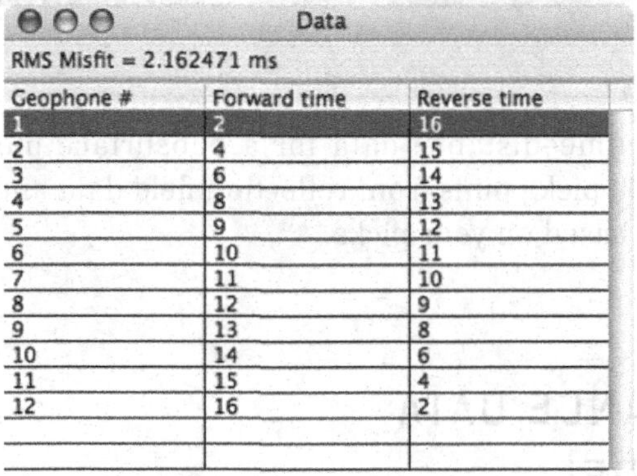

Data

RMS Misfit = 2.162471 ms

Geophone #	Forward time	Reverse time
1	2	16
2	4	15
3	6	14
4	8	13
5	9	12
6	10	11
7	11	10
8	12	9
9	13	8
10	14	6
11	15	4
12	16	2

FIGURE A.4
Data window with time–distance data entered.

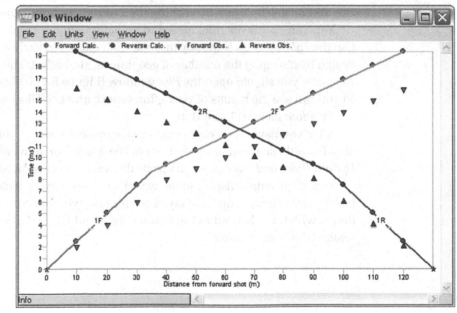

FIGURE A.5
Plot window with data points.

Instructions for Using REFLECT

REFLECT produces time–distance data for a subsurface model you define. REFLECT also lets you pick points from reflection field data and then calculates subsurface structure based on your picks.

PRODUCING TIME–DISTANCE DATA
FOR A SUBSURFACE MODEL

This is the default condition for REFLECT. When you first start the program, the *Model Table* window appears (Figure B.1). The upper half of the *Model Table* window contains all the elements you need to design a subsurface model. You can add or delete layers by clicking the buttons with the appropriate labels.

You can edit layer thickness and velocity using normal editing techniques. Use the *Tab* key to cycle quickly through each entry. Similarly, you can edit the field design by changing the number of geophones, shot offset, and geophone spacing. Normally you should open the *Plot* window (Figure B.2) before designing models so you can see the results of your efforts more quickly. This can be accessed from the *Window* menu (Figure B.3).

After you have completed your setup, choose the waves you would like to display from those listed at the bottom of the *Model Table* window. Then click the *Update Times and Plots* button to update the waveforms displayed in the *Plot* window.

To see the entire display of the waveforms you have selected, you might have to adjust the time range displayed in the *Plot* window. Simply edit the display times, which are located at the bottom right and left of this window, or click the *Show All Arrivals* button.

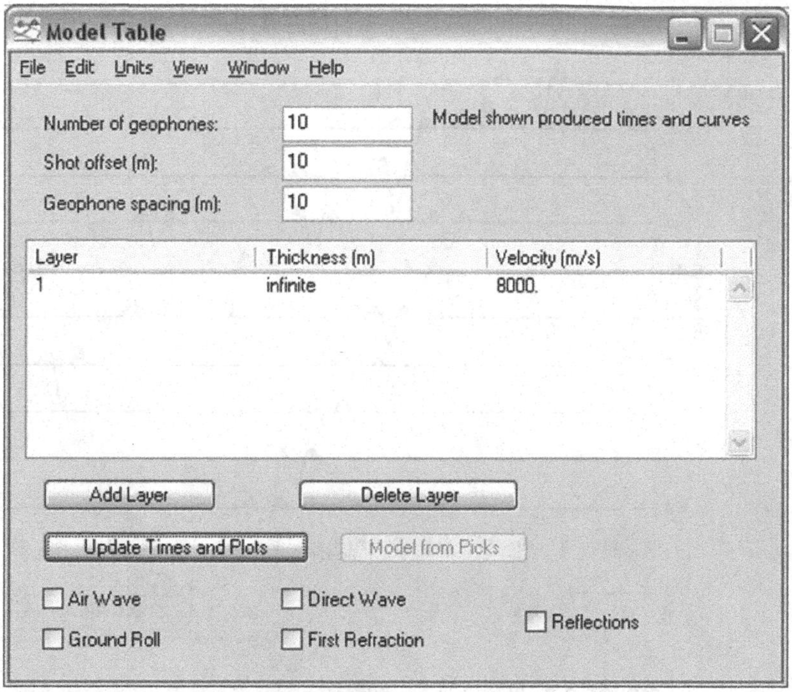

FIGURE B.1 *Model Table* window.

The remaining options for producing time–distance data from a model can be viewed in Figure B.3. These consist of using different units and saving your models and are straightforward.

Although the *Model Table* window makes creating a model quite easy, it is possible to create a model as a text file in a word processing program and then to open that model. The best way to select the correct format for the text file is to save a model in REFLECT and then examine the resulting model file.

The *Data* window (Figure B.4) illustrates time–distance data for a model and waveforms designated in the *Model Table* window.

DETERMINING SUBSURFACE STRUCTURE FROM TIME–DISTANCE DATA

The most direct way to enter time–distance data for analysis is to first set up the correct field parameters (the number of geophones, shot offset, and geophone spacing) in the *Model Table* window. Next, adjust the number of layers that you expect to be present based on your inspection of the field data. Finally, open the *Data* window in REFLECT (Figure B.4) and enter your picks from a field record.

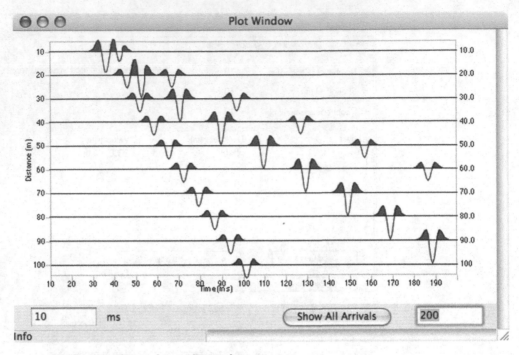

FIGURE B.2 *Plot* window with waveforms.

Now observe the *Plot* window (Figure B.5), and you should see your data points displayed as small triangles superimposed on the model you previously created. When everything is ready, open the *Fit* window, and your picks will appear on an x^2 - t^2 graph (Figure B.6(a)). Click the *Invert Picks* button to have lines fit to your field data (Figure B.6(b)). You can adjust the line(s) shown by clicking on them and dragging the handles (black boxes) at either end of the line. Because the Dix equation is most accurate for small source-receiver offsets, you might choose to have the line pass closer to points with small offsets. This also results in the calculation of two-way travel times, RMS velocities, interval velocities, and unit thicknesses. You can view the first two quantities by clicking the *Fit Statistics* button, which also will present the RMS errors for the line fits; the latter two quantities appear in the *Model Table* window once the *Model from Picks* button is pressed (Figure B.7(a)). Access the *Cross section* window for a diagrammatic view of the calculated subsurface structure (Figure B.7(b)).

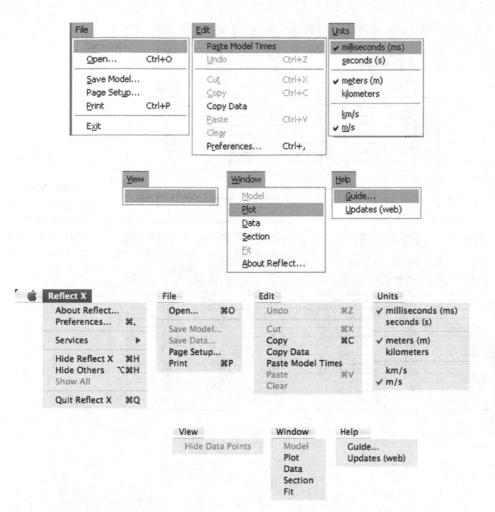

FIGURE B.3 REFLECT Windows and Mac menus. Some items differ depending on the frontmost window.

Next, click the *Update Times and Plots* button in the *Model Table* window to see the waveform pattern in the *Plot* window for the model calculated based on your picks and the line fits in the *Fit* window.

You can save your travel-time data by selecting *Save Data* in the *File* menu when the *Data* window is active. You can save your model data when the *Model Table* window is active. As was the case for model data, you can read your travel-time data from a text file, if you wish. Once again, save a sample data file generated within REFLECT and then open it to view the proper format for generating a text file of travel-time data.

(a)

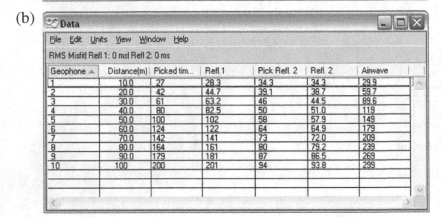

FIGURE B.4

(a) *Data* window for a selected model.
(b) *Data* window with time–distance picks.

(b)

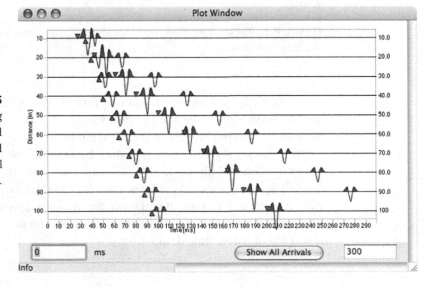

FIGURE B.5
Plot window illustrating picks from a field record superimposed on specified model parameters.

(a)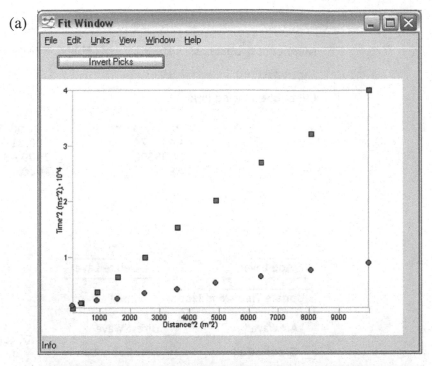

(b)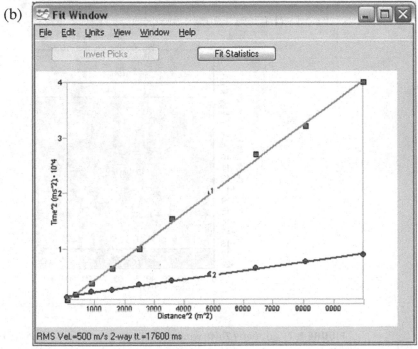

FIGURE B.6 (a) *Fit* window with x^2 - t^2 data from field record picks. (b) *Fit* window with lines fit to x^2 - t^2 data.

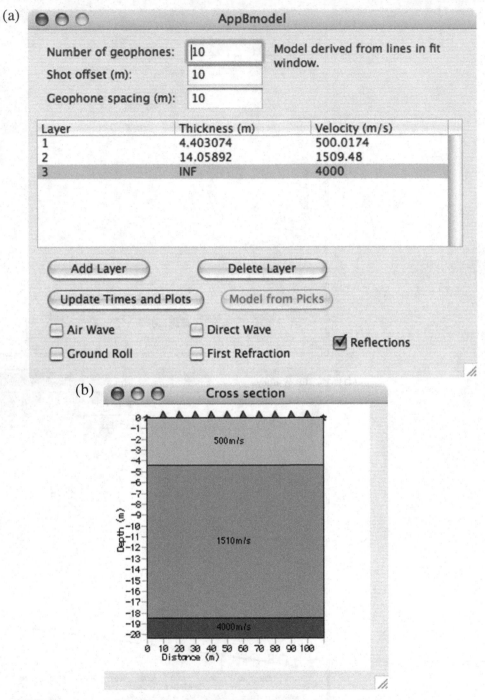

FIGURE B.7 (a) *Model Table* window with solution from *Fit* window. (b) *Cross section* window illustrating subsurface velocity structure based on picks.

Instructions for Using RESIST

RESIST calculates apparent resistivities at preselected electrode spacings for an input model consisting of layer thicknesses and resistivities. RESIST will accept field observations acquired with either the Wenner or Schlumberger electrode configuration and will attempt to derive a subsurface model that fits these data.

PRODUCING APPARENT RESISTIVITIES FROM A SUBSURFACE MODEL

When RESIST is first opened, the *Model Table* window appears (Figure C.1). Field geometry parameters default to those shown at the top of the table. All can be adjusted except the spacings/decade value, which is fixed at 6.

You enter a subsurface model simply by adding or deleting layers using the available buttons and by entering thicknesses and resistivities. When you are satisfied with a model, click the *Calculate from model* button. Apparent resistivities will appear in the right column at the bottom of the *Model Table* window, and the apparent conductivity measured with an EM31-type device will be displayed just above the sounding measurements. A curve based on the calculated values will appear in the *Plot* window (Figure C.2) if the window is open.

DETERMINING SUBSURFACE STRUCTURE FROM APPARENT RESISTIVITY DATA

Figure C.1 illustrates the *Model Table* window with observed field data already entered. Use the *Tab* key to move from row to row when entering data. Cells can be left blank if data for that spacing are suspect or missing.

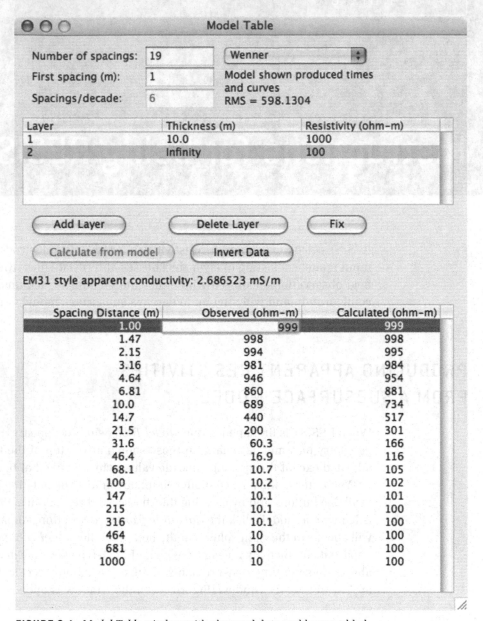

FIGURE C.1 *Model Table* window with observed data and layers added.

When attempting to match apparent resistivity data with a subsurface model, it usually is best to use your knowledge to provide a likely model. Simply devise a model and click the *Calculate from model* button. Your observed data and the curve for your model will appear in the *Plot* window (Figure C.2(a)), and an RMS value for the fit will appear in the *Model Table* window. As used in RESIST, the RMS misfit is the RMS of one minus the ratio of calculated resistivity to observed resistivity.

(a)

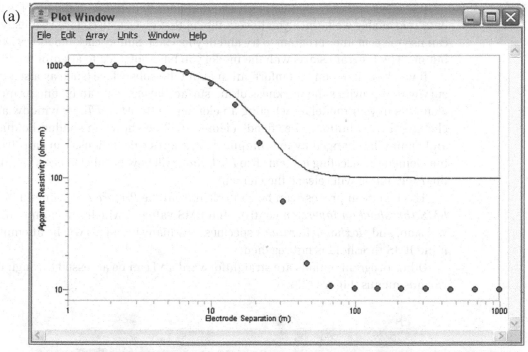

(b)

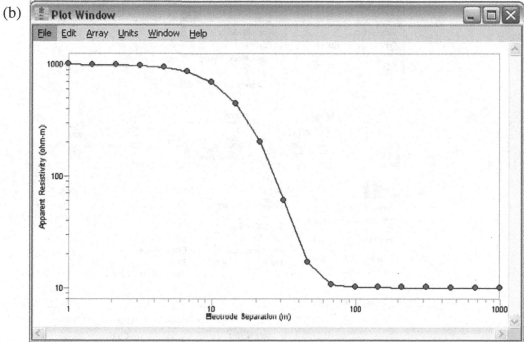

FIGURE C.2 (a) *Plot* window illustrating data calculated from a model (solid curve) and observed data (circles). (b) *Plot* window illustrating curve fit to observed data after the *Invert Data* button has been clicked.

It is quick and easy to attempt several iterations to derive a model that fits the data reasonably well. Once you have a model that gives satisfactory results, you can use the computer program to try and fit your data. Simply click the *Invert Data* button. The program starts with the model you have supplied to attempt the fit.

If you have independent information about the subsurface (such as a seismic survey that provides the thickness of the surface layer), you can fix one or more elements in your model by selecting an element in the *Model Table* window and clicking the *Fix* button. The "fixed" element will be shown in shaded boldface and cannot be changed by the computer program's calculations. You can "free" the element by selecting it again. The *Fix* button will now be labeled *Free*. Clicking the *Free* button will release the element.

The inversion process can be controlled from the *Preferences...* menu item. *RMS threshold on inversion* controls the RMS value at which the computation will stop, and *Maximum iterations* specifies how many iterations will be attempted if the RMS threshold is not reached.

Other program options are straightforward and can be accessed through the various menus (Figure C.3).

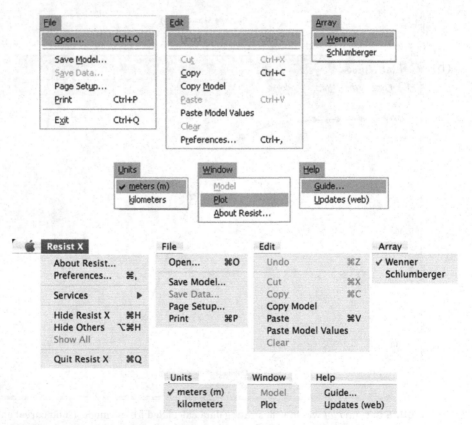

FIGURE C.3 RESIST Windows and Mac menus. Some items differ depending on the frontmost window.

Instructions for Using GRAVMAG

GRAVMAG consists of gravity and magnetics modeling routines. The gravity routine calculates gravity values for solids of infinite strike length and polygonal cross sections. The magnetics routine calculates induced vertical-, horizontal-, and total-field anomaly values for similar solids and also can incorporate remanent magnetics. Both routines use the same display windows, data windows, and input techniques. Gravity effects or magnetic effects can be explored individually, or both can function simultaneously.

These routines can be used simply to investigate gravity and magnetic anomalies produced by buried bodies of various geometries and physical properties. Or you can incorporate observed field values to explore the possible subsurface configurations responsible for observed anomalies.

GRAVITY MODELING

When GRAVMAG is first opened, the *Model Table* window is presented and is configured for gravity modeling only. Figure D.1 illustrates the *Model Table* window after the *Add Body* button has been clicked once; the *Cross section* window also was opened by selecting it from the *Window* menu (Figure D.2).

Note that when you click in the upper half of the *Model Table* window, you can edit the number, density, and color of a selected body, and the *Add Body* and *Delete Body* buttons are available. When you click in the lower half of the *Model Table* window, you can edit the parameters governing the shape of a body selected in the upper half of the window. You also can add or delete points by clicking either the *Add Point* or *Delete Point* button. Added points are placed halfway between the selected point and the following point.

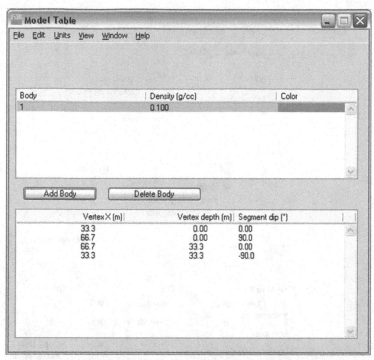

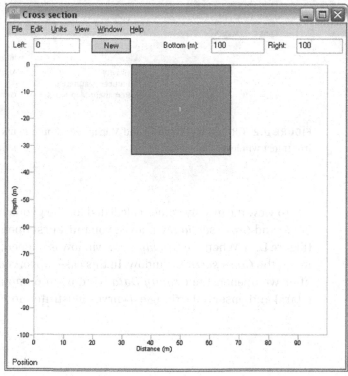

FIGURE D.1 *Model Table* window and *Cross section* window configured for gravity modeling.

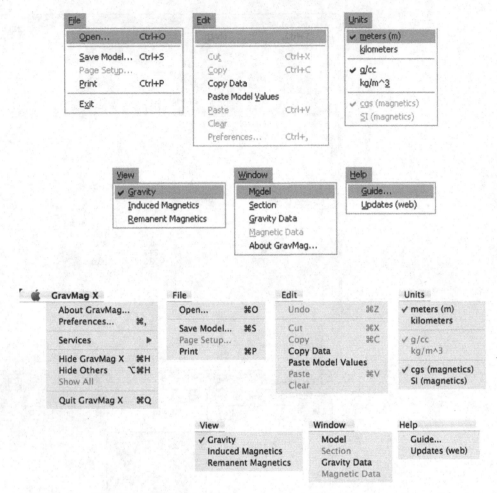

FIGURE D.2 GRAVMAG Windows and Mac menus. Some items differ depending on the frontmost window.

To view the gravity values calculated for the geometries displayed in the *Model Table* and *Cross section* windows, you must first open the *Gravity Data* window (Figure D.3). When the *Gravity Data* window is opened, the gravity values are plotted in the *Cross section* window. In this case, observed field values were entered after we opened the *Gravity Data* window to demonstrate both the calculated (stars) and observed (triangles) curves illustrated in Figure D.4.

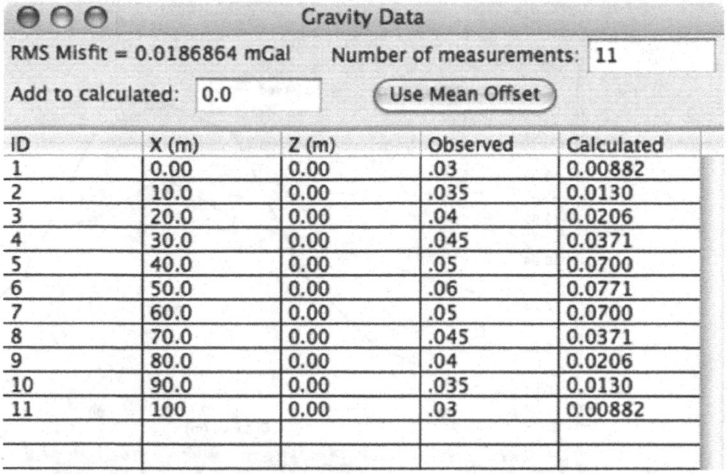

Gravity Data

RMS Misfit = 0.0186864 mGal   Number of measurements: 11

Add to calculated: 0.0   (Use Mean Offset)

ID	X (m)	Z (m)	Observed	Calculated
1	0.00	0.00	.03	0.00882
2	10.0	0.00	.035	0.0130
3	20.0	0.00	.04	0.0206
4	30.0	0.00	.045	0.0371
5	40.0	0.00	.05	0.0700
6	50.0	0.00	.06	0.0771
7	60.0	0.00	.05	0.0700
8	70.0	0.00	.045	0.0371
9	80.0	0.00	.04	0.0206
10	90.0	0.00	.035	0.0130
11	100	0.00	.03	0.00882

FIGURE D.3 *Gravity Data* window.

You can adjust the position and shape of any bodies in the *Cross section* window by clicking and dragging or by selecting a corner and dragging. You also can add another body simply by clicking the *New* button in the *Cross section* window. Draw the body with the mouse, and double-click to complete the task. As new bodies are added and as parameters are changed, the calculated curves and values automatically adjust.

MAGNETIC MODELING

Magnetic modeling functions for the most part exactly like gravity modeling. When GRAVMAG is started and the *Model Table* window appears, select *Induced Magnetics* from the *View* menu. Because gravity modeling is the default, you should deselect it in the *View* menu if you don't want a *Density* column in the *Model Table* window.

The only major difference is the addition of a drop-down menu to the *Magnetic Data* window that contains options for the magnetic field component (total, vertical, or horizontal) that will appear in the *Calculated* and *Observed* columns in the *Magnetic Data* window and be used in the magnetics pane of the *Cross section* window. You can enter observations for all three components of the magnetic field and cycle through both observations and calculations by repeatedly selecting an item from the drop-down list. You should enter the profile's azimuth (relative to magnetic north) and latitude in the *Model Table* window. Alternatively, the intensity and orientation of the inducing field can be set directly by changing the appropriate checkbox in the *Preferences* window.

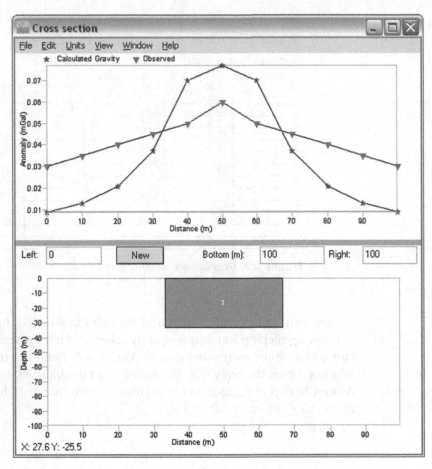

FIGURE D.4 *Cross section* window with observed (triangles) and calculated (stars) gravity curves.

To include the effects of remanent magnetization in all calculations, simply choose *Remanent Magnetics* from the *View* menu when the *Model Table* window is active and then enter the appropriate values in the *Model Table* cells.

MISCELLANEOUS ITEMS

If you wish, you can simultaneously view magnetic and gravity anomalies for a given model. To do this, open both the *Gravity Data* and *Magnetic Data* windows. Both gravity and magnetics curves will then plot in the *Cross section* window.

When the *Cross section* window is active, you can use arrow keys to move a selected polygon or polygon point pixel by pixel. This is handy for making fine adjustments.

When calculating magnetic values, the program uses a simple dipole. If you want to specify the inducing field to be used, select *Preferences...* in the *GravMag* menu, and check *Enter Magnetic Field Directly*. Inducing field parameters can then be specified in the *Model Table* window.

When you open the *Gravity Data* or *Magnetic Data* window, the default number of measurements is 11. Of course, if you are working with field measurements, you will use the number of measurements you have collected and enter the correct coordinates and values. However, if you simply want to view the magnetic or gravity values (or both) for a given model, you might prefer many measurement points to produce a smoother curve. Simply typing in (for example) 100 will give the desired number of calculation points, but they will not be evenly distributed along the section profile. A work-around is as follows:

1. Change the profile length in the *Model Table* window.
2. Type the number 1 for *Number of measurements* in a data window and click somewhere in the data entry area of the same window.
3. Finally, type the number you really want for *Number of measurements* and click again in the data entry area. This will result in the desired number of points, evenly distributed, in the *Cross section* window (Figure D.5).

It is possible to adjust the width of columns in the *Model Table* window and data windows by clicking on the divider at the top of the table and dragging to the desired width (as you would do in Microsoft Excel to change column widths). Similarly, you can change the position of the divider between the graph and model areas of the *Cross section* window.

Error bars are available if you check this option in the *Preferences...* window.

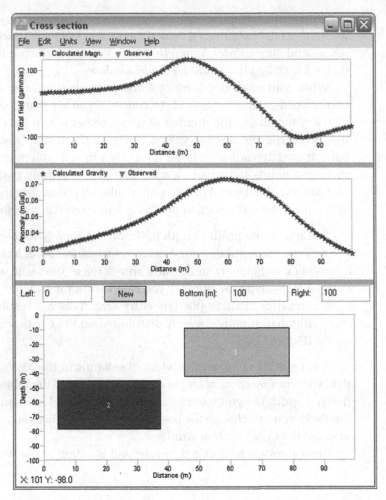

FIGURE D.5 *Cross section* window with 100 measurement points for gravity and magnetic values.

Instructions for Using DIFFRACT

DIFFRACT produces time–distance data for a subsurface model you define. DIFFRACT also lets you pick points from GPR field data and then display those picks against your models.

PRODUCING TIME–DISTANCE DATA
FOR A SUBSURFACE MODEL

This is the default condition for DIFFRACT. When you first start the program, the *Model Table* window appears (Figure E.1). The upper half of the *Model Table* window contains all the elements you need to design a subsurface model. You can add or delete layers by clicking the buttons with the appropriate labels.

You can edit layer thickness and velocity using normal editing techniques. Use the *Tab* key to cycle quickly through each entry. Similarly, you can edit the field design by changing the number of soundings, transmitter–receiver offset, and sounding spacing. To add diffractors, check the *Reflections/Diffractions* box and then select the list box at the bottom of the *Model Table* window. The *Add Layer* button becomes *Add Diffractor*. Unlike layers, you can change the label on diffractors. Note that often GPR profiles have many soundings; for clarity, you probably will want to use a subset of soundings here.

Normally you should open the *Plot* (Figure E.2) and *Cross section* windows before designing models so you can see the results of your efforts more quickly. These can be accessed from the *Window* menu (Figure E.3).

After you have completed your setup, choose the waves you would like to display from those listed at the bottom of the window. Then click the *Update Times and Plots* button to update the waveforms displayed in the *Plot* window.

To see the entire display of the waveforms you have selected, you might have to adjust the time range displayed in the *Plot* window. Simply edit the display

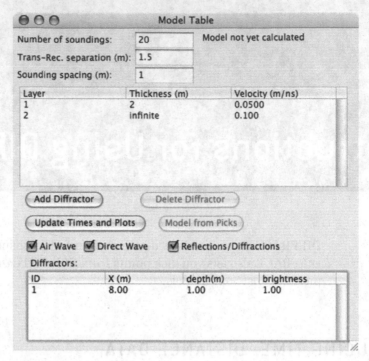

FIGURE E.1 *Model Table* window.

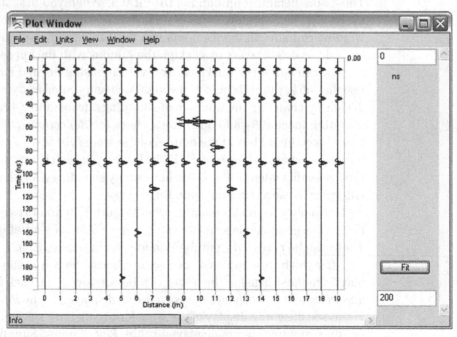

FIGURE E.2 *Plot* window with waveforms.

times, which are located at the top and bottom right of the window, or click the *Fit* button near the lower right corner of the window.

The remaining options for producing time–distance data from a model can be viewed in Figure E.3. These consist of using different units and saving your model and are straightforward.

Although the *Model Table* window makes creating a model quite easy, it is possible to create a model as a text file in a word processing program and then to open that model. The best way to select the correct format for the text file is to save a model in DIFFRACT and then examine the resulting model file.

The *Data* window (Figure E.4) illustrates time–distance data for a model and waveforms designated in the *Model Table* window.

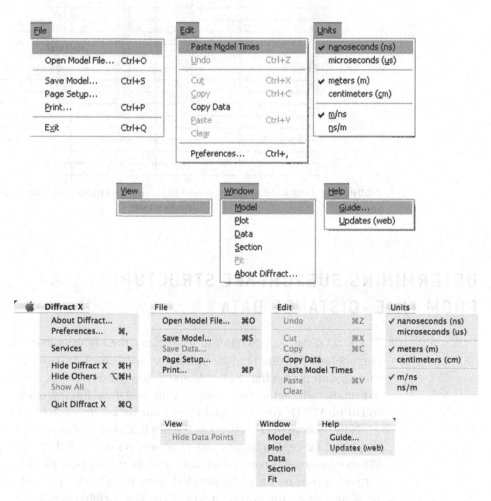

FIGURE E.3 DIFFRACT Windows and Mac menus. Some items differ depending on the frontmost window.

RMS Misfit| Diff 1: 2.24 ns
Airwave time: 5.00 ns
Direct wave time: 30.0 ns
Reflection wave 1 time 85.4 ns

Sounding	Distance(m)	Picked time (ns)	Diff.1
1	0.00	321	322
2	1.00	282	283
3	2.00	242	243
4	3.00	203	204
5	4.00	164	165
6	5.00	125	126
7	6.00	88	89.4
8	7.00	56	56.6
9	8.00	41	50.0
10	9.00	57	56.6
11	10.0	88	89.4
12	11.0	127	126
13	12.0	166	165
14	13.0	205	204
15	14.0	244	243
16	15.0	282	283
17	16.0	321	322
18	17.0	363	362
19	18.0	403	402
20	19.0	443	442

FIGURE E.4 *Data* window for the model in Figure E.1 with time–distance picks from a field record.

DETERMINING SUBSURFACE STRUCTURE FROM TIME–DISTANCE DATA

The most direct way to enter time–distance data for analysis is to first set up the correct field parameters (the number of soundings, transmitter–receiver offset, and sounding spacing) in the *Model Table* window. Next, enter your diffractor expectations based on inspection of your field data. Finally, open the *Data* window in DIFFRACT (Figure E.4) and enter your picks from a field record.

Now observe the *Plot* window (Figure E.5), and you should see your data points displayed as small triangles superimposed on the model you previously created. When everything is ready, open the *Fit* window, and your picks for the first diffractor will appear on an $x^2 - t^2$ graph (Figure E.6(a)). Picks for other diffractors can be shown using the pop-up menu in the lower right corner. You must adjust the x position of the diffractor in the edit box at the bottom of the plot; an initial guess

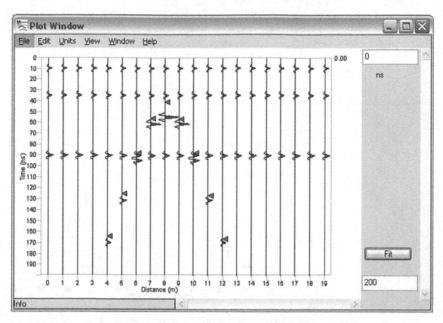

FIGURE E.5 *Plot* window illustrating picks from a field record superimposed on the model parameters shown in Figure E.1.

based on the earliest arrival is provided. The best value will result in your points lying along a line. Click on the *Invert Picks* button to have a line fit to your field data (Figure E.6(b)). This also results in the calculation of two-way travel times, RMS velocities, velocities at the diffractors, and diffractor depths. You can view the first two quantities by moving the cursor over the line and noting the values in the lower left corner of the window, or by clicking the *Fit Statistics* button, which also will present the RMS errors for the original line fits. The latter two quantities appear in the *Model Table* window after you click the *Model from Picks* button (Figure E.7(a)).

Next, click the *Update Times and Plots* button in the *Model Table* window to see the waveform pattern in the *Plot* window for the model calculated based on your picks and the line fits in the *Fit* window. Access the *Cross section* window for a diagrammatic view of the calculated subsurface structure (Figure E.7(b)).

You can save your travel-time data by selecting *Save Data* in the *File* menu when the *Data* window is active. You can save your model data when the *Model Table* window is active. As was the case for model data, you can read your travel-time data from a text file, if you wish. Once again, save a sample data file generated within DIFFRACT and then open it to view the proper format for generating a text file of travel-time data.

(a)

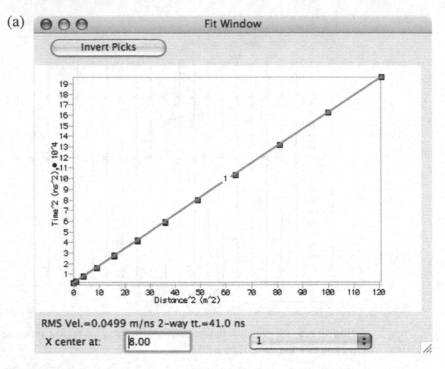

(b)

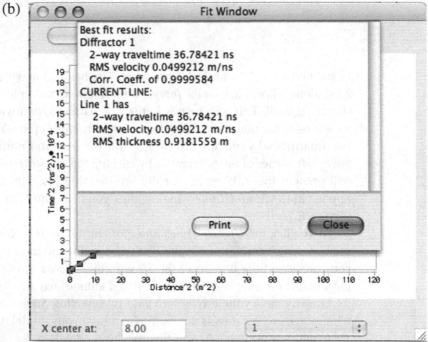

FIGURE E.6 (a) *Fit* window with line fit to the field data picks. (b) *Fit* window with fit statistics displayed. Statistics for both the original best fit line and for the line as presently displayed (i.e., after any user modifications) are shown.

(a)

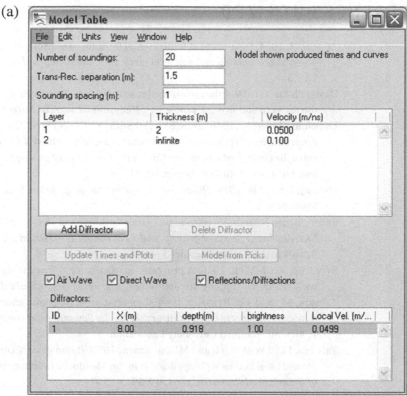

(b)

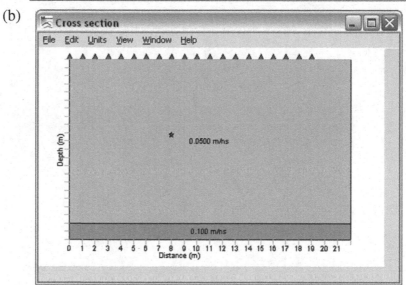

FIGURE E.7 (a) *Model Table* window with solution from *Fit* window. Note the addition of a *Local Velocity* column indicating the velocity of the layer the diffractor is within, as determined from the inversion. (b) *Cross section* window illustrating subsurface structure including the diffractor at the star.

REFERENCES UTILIZED

Adachi, R. 1954. On a proof of fundamental formula concerning refraction method of geophysical prospecting and some remarks. *Kumamoto Journal of Science, Series A,* v. 2, 18–23.

Davis, Philip A. 1979. Interpretation of resistivity data: Computer programs for solutions to the forward and inverse problems. *Minnesota Geological Survey Information Circular 17.*

Lankston, Robert M. 1986. SEIS: SYNTHESEIS, X2T2, INTERVEL—Three microcomputer programs for exercises with the seismic reflection method. In *Personal computer software for geological education: National association of geology teachers special publication no. 1,* ed. H. Robert Burger, 64–81.

Mooney, Harold M. 1977. *Handbook of engineering geophysics.* Minneapolis: Bison Instruments.

Salem, Bruce B., John W. Harbaugh, Claude G. Abry, and Henry Crichlow. 1973. *Interactive CRT display of Bouguer gravity models in computer-assisted instruction in geology.* Report to the National Science Foundation.

Sverdrup, Keith Allen. 1986. A program for calculating refraction times for layered models with arbitrarily dipping discontinuities. *Journal of Geological Education,* v. 34, 262–64.

Talwani, M., and J.R. Heirtzler. 1964. Computation of magnetic anomalies caused by two-dimensional structures of arbitrary shape. In *Computers in the mineral industries, part 1,* 464–80. Stanford University Publication.

Talwani, M., J.W. Worzel, and M. Landsman. 1959. Rapid gravity computations for two-dimensional bodies with application to the Mendocino submarine fracture zone. *Journal of Geophysical Research,* v. 64, 49–59.

Index

Absolute gravity stations, establishing, in exploration area, *374*
Absorption, 40–42, 499, 507
  energy losses due to, *42*
Acoustical impedance, 156
Adachi, R., 100
Adachi equations, 100–102, *101*
Aeromagnetic maps, value of, 450
Air guns, 48, 49
Air wave, 20, 36, *36*
Alternating current (AC), 504
Amargosa Desert Research Site (ADRS), GPR study at, 548
Ammeter, 315, 316
Amperes, 268
  Ampere's Law, 502, 509, 511
Amplitude, 9, 11, 502
Amplitude adjustments, 244
Analog to digital converter (A/D converter), 57
Angle of incidence, 23, *30*
Anomalous field, 444, *448*
Antiferromagnetic materials, magnetic domains in, 434, *435*
Apparent conductivity, 516–17
Apparent resistivity:
  for constant-spread traverse, oriented perpendicular to vertical contact, *308*
  curves for, drawn by RESIST, *300*
  electrical resistivity and, 287, 302
  electrode spacing and, *289, 290*
  plotting of, against electrode spacing, *299*
  quantitative interpretation of apparent resistivity curves:
    automated curve matching, 329–31
    curve matching, 326–29
    electrical resistivities of geologic materials, 324–25
    empirical methods, 325–26
  variation of, with electrode spacing for single horizontal interface, 298*t*
Apparent resistivity curve:
  comparison of, for constant-spread traverse across dipping interfaces, *315*

for constant-spread traverse over faults in Illinois, *337*
for constant-spread traverse over karst topography, Hardin County, Illinois, *337*
Apparent resistivity curve types, for two horizontal interfaces, *303*
Apparent velocity, 91
Applied currents, 265
Aquifers
  electrical-resistivity surveying and, 332–35
  reflective seismology and, 248–49
Archaeological surveys
  electromagnetic surveying and, 539–43
  using the magnetic method, 488–90
Archie's Law, 505
Arrival times
  direct and refracted waves, 70–71, 72*t*
  reflection seismology, 156–57, *157*
Arzate, J. A., 490
a-spacing, 298
Aster, R., 523
Attenuation, 499, 504, 507, 508
Automated curve matching, 329–31
Automatic gain control (AGC), 55
Aztec Ruins National Monument, New Mexico
  contour map of apparent conductivity from EM31 survey of, *540*
  electromagnetic survey of, 539–41

Bar magnet, magnetic lines of force produced by, 430, *430*
Barnes Layer Method, 325, 326
Base stations, establishing, 373
Bedrock, Bouguer gravity and depth to, along street in Reading, Massachusetts, *416*
Bedrock depths, determining using gravity method, 415–16
Bedrock reflection, optimum window and, *202*
Bedrock refractions, *133*

Belcourt, G., 543
Benchmarks, 374
Body waves, 17
Borehole radar velocity tomogram, karstic
    features in limestone shown in, *532*
Bouguer, P., 360
Bouguer anomaly, 363, 378, 380, 383, 390,
    393, 399–409, 413–15
Bouguer anomaly curves, for horizontal
    cylinders with various radii, *385*
Bouguer anomaly maps, for Easthampton
    and Mt. Holyoke Quadrangles (west-
    central Massachusetts), *405, 406*
Bouguer correction, 360–64
  relationships and notation used in
    derivation of, *361*
  terrain correction and, *365*
Bouguer gravity anomalies, over sphere and
    horizontal cylinder, *382*
Branham, K. L., 252
Brigham, Utah gravel pit, schematic diagram
    of stratigraphic layering of, *544*
Brooks, M., 217
Bulk modulus, 12, *13*, 18
Burial of known object, GPR velocity
    analysis and, 533
Buried stream channel, mapping, with use
    of resistivity, *334*
Butler, D. K., 416

Cavendish, Lord, 350
Cavity detection, using reflective seismology,
    249–52
CDP gathers, stacking, 238–39, 241, *241*
Chandler, W. E., Jr., 412
Chargeability, 339
Circuit, 510
Coal seams, faults cutting, 252, *252*
Common-depth-point (CDP) data acquisition,
    for GPR, 529
Common-depth-point (CDP) method, 219,
    225–29
Common-depth-point (CDP) profiling,
    227–28
Common midpoint (CMP) data acquisition,
    for GPR, 529
Common midpoint (CMP) GPR data
    collection, with 100 MHz antenna,
    531
Common midpoint (CMP) profiling, *530*
Common midpoint (CMP) stacking, 529, 530
Common offset data, 232
Common-offset method, 219, 221–25, *223,
    224*
Common offset profiling, *221*
Commutators, 316
Compressional waves, 15, 19, 36, 43
Computer modeling, interpretation of
    magnetic data, 488
Computer processing of data, reflective
    seismology:
  correcting for normal move-out, 232–38

  migration, 241–44
  stacking CDP gathers, 238–41
  static correction, 229–32
  time and depth sections, 245–48
  waveform adjustments, 244–45
Conductivity, electrical, 505, 506*t*
Confined aquifer, determining thickness of,
    with seismic refraction and electrical
    resistivity, 5, *6*
Conklin, H. R., 266
Constant-spread traverse, 306–9
  across two vertical contacts, *313*
Contamination:
  electrical-resistivity surveying and, 335
  refraction seismology and, 136–37, *137*
Continuous current, 504
Continuous wave method, 512
Contouring data, 523
Contour map, of apparent conductivity from
    EM31 survey of Aztec Ruins site, *540*
Contours, of equipotential surfaces, 274, *274*
Conyers, L. B., 541
Cook, K. L., 266, 305, 309
Coruh, C., 170, 217, 355, 450
Corwin, R. F., 340
Cotten, S. A., 252
Coulombs, 268
Coulomb's Law, 431
Couple, magnitude of, 432
Critical angle, 29
Critical distance, 74–76, 85, 87
  calculating, *74*
  for two horizontal interfaces, 86*t*
  values for various thicknesses, 75*t*
Critically refracted ray, symbols used in
    derivation of time of travel for, *69*
Critical refraction, 28–29
Crosby, I. B., 266
Cross-borehole radar tomography, 533
Crossover distance, 73–74
Current, 268
Current density, 268, *269*, 284–87
Current distribution, 280–84
  for single horizontal interface, *281*
Current electrodes, two, potential at a point
    with, 273*t*
Current flow (homogeneous, isotropic earth):
  point current source, 270–71
  two current electrodes, 271–76
  two potential electrodes, 276–80
Current flow lines, 284–87
  qualitative distribution of, *286*
  refraction of, *285*
Curve matching, 339
  basic procedures for, *327–28*
  techniques, 326–29
    automated, 329–31
Cylinder, half-maximum relationship for,
    485

de Fermat, P., 22
Delay time, defined, 120

Delay-time depths, refractor surface
   constructed from, *126*
Delay-time method, 120, 122–25
Delay-time relationships, symbols used in
   derivation of, *121*
Delay times, refractor depths computed
   with, 125*t*
Demagnetization effect, 455
Density contrast, 379
Depth/current electrode separation, *276*
Depth of penetration, 508
Depth profiling, 306
Depth sections, 245–48, *246, 247*
Diamagnetic minerals, 434
Dielectric constant, 506
Dielectric properties, 505–7
DIFFRACT, A24–A30
Diffraction, 29, *34, 34–35, 35,* 209–17, *212,*
   214, 215
Diffraction hyperbola, 537, 539
Diffraction travel times, 216*t*
Digital recording, 56–59
Dip-angle method, 519
Dip move-out (DMO), 197, *198,* 199
Dipole-dipole array, 317, *318,* 319, 339
Dipole equations, 444–46
   parameters used in derivation of, *445*
Dipole magnetized along its axis, 468
Dipole(s), 430
   contours of total-field anomaly intensities
      above, *466*
   magnetic effect of, 457–65
   magnetic field of, 462*t*
   magnet represented as assemblage of, *431*
   relationships and notation used to derive
      magnetic effect of, *465*
   relationships and notation used to derive
      magnetic field of, *461*
   total and vertical magnetic anomalies
      over, *464*
Dipping dipole, intensities of vertical,
   horizontal, and total magnetic field
   anomalies over, polarized along its
   axis, *463*
Dipping interface, reflection time-distance
   values for, using split-spread
   arrangement, 185*t*
Dipping interface, single (electrical resistivity
   method), 314
Dipping interface, single (reflection
   seismology):
   derivation of travel-time equation, 183–86
   determining dip, thickness, and velocity,
      186–92, 197–99
   normal move-out, 192–96
Dipping interface, single (refraction
   seismology):
   analyzing the problem, 91–96
   derivation of travel-time equation, 97–99
   determining thickness, 99–100
Dipping interface(s), multiple (refraction
   seismology):

analyzing field seismograms, 102, 105
   travel-time equation, 100–102
Direct current (DC), 504
Direct-current resistivity method, 265
Direct rays:
   arrival times for, 70–71, 72*t*
   time-distance relationships for, *70*
Direct wave, 36, *36*
   reflection waveform superimposed on, *39*
Dispersion, 18
Diurnal corrections, magnetic surveying
   and, 450–51
Diurnal variations, in magnet field, 443
Dix, C. H., 170
Dix equation, 170, 173
Dix method, 170, *172,* 175–81, *182*
Dobrin, M. B., 18, 136, 217
Dolores Archaeological Program, Colorado,
   total-field magnetic map of, *489*
Down-dip, 97, 98, 99
Downward continuation, 403–7, 484
Drift, 355, 370
Drift curves, *375*
   for gravity observations, *370*
Drift effects, 370–73
Drill rate log, 140, *141*
Dynamic corrections, 232
Dynamic table, 19, 20

Earth, fundamental considerations in study
   of, 2–3
Earthquakes, 7, 8
Earth's magnetic field:
   dipolar nature of, 439–41
   dipole equations, 444–46
   elements of, *439*
   field elements, 438
   some important features of, *443*
   variations of, 441–44
Eaton, G. P., 324, 325
Eddy current, *11,* 511
Edison, T., 504
Elastic coefficients, 11–14
   for selected common rocks, 14*t*
Electrical circuit, *268*
Electrical conductivity, 505, 506*t*
Electrical drilling, 306
Electrical mapping, 306
Electrical resistivity, 505
   determining thickness of confined aquifer
      with, 5, *6*
Electrical resistivity method:
   applications of:
      in mapping karst and geologic
         structures, 335
      related to aquifers, 332–35
      related to contamination, 335
      related to geothermal fields, 338
   applied currents, 265
   basic electricity, 267–70
   current flow in a homogeneous, isotropic
      earth:

Electrical resistivity method *(continued)*:
   point current source, 270–71
   two current electrodes, 271–72, 274–76
   two potential electrodes, 276–80
  dipping interfaces, 314
  field procedures, 314, 323
   electrode configurations, 317–20
   equipment, 315–17
   surveying strategies, 321–22
  hemispherical structures, 313–14
  history of, 266–67
  induced potential (IP), 338–39
  multiple horizontal interfaces, 302–5
  natural currents, 266
  quantitative interpretation of apparent
    resistivity curves:
   automated curve matching, 329–31
   curve matching, 326–29
   electrical resistivities of geologic
    materials, 324–25
   empirical methods, 325–26
  single horizontal interface, 280
   apparent resistivity, 287
   current distribution, 280–84
   current flow lines and current density,
    284–87
   qualitative development of the resistivity
    pattern over a horizontal interface,
    288
   quantitative development of the
    resistivity pattern over a horizontal
    interface, 290–97
   using RESIST, 297–301, A13–A16
  spontaneous potential (SP), 339–40
  telluric and magnetotelluric methods, 340
  two vertical contacts, 312
  vertical contacts, 305–6
   constant-spread traverse, 306–9
   expanding-spread traverse, 309–10
Electrical resistivity surveys, 5
Electrical surveying system, main elements
   of, *316*
Electrodes, potential, determining potential
   difference at, *278*
Electrode spacing, 298
  apparent resistivity and, *289, 290*
  apparent resistivity plotted against, *299*
  hemispherical sink with diameter greater
   than, *314*
  variation of apparent resistivity with, for
   single horizontal interface, 298*t*
Electromagnetic (EM) induction, 499
  prospecting, *511*
Electromagnetic (EM) surveying, 265
Electromagnetic skin depth, 509*t*
Electromagnetic spectrum, *501*
Electromagnetic surveying, 499
  applications
   archaeological surveys, 539–43
   environmental and engineering studies,
    546–48
   geologic, 543–46

   snow and ice mapping, 546
electromagnetic waves, 500–502
  absorption and attenuation, 507–9
  AC/DC, 504
  dielectric properties, 505–7
  electrical properties of geologic
   materials, 504–5
  electrical resistivity and conductivity, 505
  wavelengths, 502, 504
EM sounding, 509–12
  near-field continuous-wave methods of
   frequency domain electromagnetics,
   512–19
  tilt-angle or dip-angle methods, 519
  time domain electromagnetics, 519–20
  field procedures:
   interpretation, 523–24
   profiling *versus* sounding, 521–23
  ground-penetrating radar, 524–25
   data acquisition, 527–33
   radar velocity, 525–27
   velocity analysis, 533–39
Electromagnetic waves, 500–502
  absorption and attenuation, 507–9
  AC/DC, 504
  electrical properties of geologic materials,
   504–7
  dielectric properties, 505–7
  electrical resistivity and conductivity, 505
  types of, 500
  wavelengths, 502–4
Electromotive force (emf), 267
Elevation corrections, magnetic surveying,
   451–52
Elevations, determining, 373–75
Ellipsoid of revolution, 356
Elongation, 12
Energy partitioning, 42–46
  from incident P-wave, *45*
Energy sources:
  considerations related to, 49–50
  types of, 47–49
Equation constant, 367
Equipotential surfaces, 270–71, *271*
  current lines of flow and, *277*
  patterns of, 274, *274*
Equivalence, 329, *330*
Expanding-spread, Wenner configuration
   field curves, *333*
Expanding-spread survey, across vertical
   contact, 310*t*
Expanding-spread traverse, 309–10
  apparent resistivity for, oriented
   perpendicular to vertical contact, *311*
Explosive energy sources, 47–49
Extending-spread traversing, 306
External magnetic field, 443

Falling-body method, 355
Fan shooting, *132*
Faraday's Law, 500, 502, 510, 511, 512
Farad (F), 506

Faulted horizontal sheet
  gravity effects of, *396*
  gravity over, 395*t*
Fermat's principle, 22–23, *24*, 26, *27*
Ferrimagnetic materials
  hysteresis curve for, in presence of
    magnetizing field, *436*
  magnetic domains in, 435, *435*
Ferromagnetic minerals, magnetic domains
    in, 434, *435*
Field procedures:
  electrical resistivity, 314–15
    electrode configurations, 317–20
    equipment, 315–17
    surveying strategies, 321–22
  electronic surveying:
    interpretation, 523–24
    profiling, 523
    sounding, 521–22
  gravity exploration:
    determining elevations, 373–75
    determining horizontal position, 375
    drift and tidal effects, 370–73
    establishing base stations, 373
    selecting a reduction density, 376
    survey procedure, 376, 378
  magnetic surveying:
    corrections for horizontal position,
      452–54
    diurnal corrections, 450–51
    elevation corrections, 451–52
    magnetic cleanliness, 450
  reflective seismology:
    equipment, 217–19
    geophone spreads, 219–29
  refractive seismology:
    corrections to data, 134–36
    equipment considerations, 130–31
    geophone spread geometries and
      placements, 131–34
    site selection and planning
      considerations, 129–30
Field seismograms
  from Connecticut Valley, Massachusetts,
    76, 88, 103, 104, 166, 205
  constructing travel-time curve from,
    76–77, *78*
  with multiple reflections, from Connecticut
    Valley, Massachusetts, *181*
  from Whately, Massachusetts, *138*
Field survey objectives, defining, 3–4, 6
Fitterman, D. V., 523
Fitzpatrick, M. M., 452
Flux-gate magnetometer, 446
Flux-gate magnetometer gradiometer
  survey, results of, at Radly,
    Oxfordshire, England, *490*
Formation resistivity, determining, 519
Forward modeling, 4, 523
Forward travel-time curve, effect of vertical
    velocity discontinuity on, *115*
Forward traverse, 93, *93*, 94, *95*, *103*

ray paths and resulting travel-time curves
    for, over vertical step, *117*
Fox, R. W., 266
Free-air anomaly, 359
Free-air correction, 358–59
Frequency, 10, 11, 502
  in vertical and horizontal resolution,
    217–19, *218*
Frequency adjustments, 244, 245
Frequency domain electromagnetics
    (FDEM), 512
  near-field continuous-wave methods of:
    moving transmitter-plus-receiver
      system, 512, *513*, 514
    noncontacting ground conductivity
      measurements, 514–18
  system with large fixed transmitter loop
    and small mobile receiver, *520*
Frequency spectrum, sampled by 100 Hz
    geophone and 200 Hz filter, *56*
Fresnel zone, 218
Frischknecht, F. C., 297, 492

Gammas, 432
Gamma waves, 500
Gauss's Law, 500
Geldart, L. P., 43, 55, 149, 170, 177, 305,
    393, 454
Generalized reciprocal method (GRM), 129,
    138, *139*
Geodetic reference system formula of 1967
    (GRS67), 357
Geoid, 357
Geologic applications, with electromagnetic
    methods, 543–46
Geologic materials
  common, densities of, 379*t*
  electrical conductivity, relative permittivity,
    and radar velocity of, 506*t*
  electrical properties of:, 504–5
    dielectric properties, 505–7
    electrical resistivity and conductivity, 505
Geomagnetic north and south poles, 440
Geometrical (or spherical) spreading, 507
Geometric shapes:
  gravity effects of:
    of a horizontal cylinder, 383–86
    of a horizontal sheet, 390–97
    of inclined rod, 388–90
    rock densities, 378
    of semi-infinite sheets, 390–97
    of a sphere, 379–83
    using GRAVMAG, 397–99, A17–A23
    of a vertical cylinder, 386–88
  magnetic effects of:
    of a dipole, 457, 460–65
    of isolated pole (monopole), 456–57
    of polygons with infinite strike length,
      478–82
    rock susceptibilities, 454–56
    of a sphere, 465–69
    of a thin horizontal sheet, 470–78

Geonics ground conductivity meters:
  EM31, 500, 512, 516, 518
  EM34, 512, 516
  EM42, 520
  EM47, 520
Geophone interval, normal move-out and, *168*
Geophones, 36, 50–53, *51, 52,* 60, 65
  analog output of, *58*
  common depth point, 225–29
  common offset, 221–25
  comparison of model parameters with
    Green and Dix solutions for three
    spreads, 179*t, 180*
  fan shooting, *132*
  in-line spread, 131
  split spreads, 219–21
  spread arrangements, *132*
  spread geometries and placements,
    131–34
Geophysical methods:
  limitations of, in study of subsurface, 4–5
  multiple, 5–6
Geophysics:
  defined, 2
  value of, 1
Geothermal fields, electricity-resistivity
    surveying and, 338
Gish, O. H., 266
Global Positioning System (GPS), 375
Gochioco, L. M., 252
Gravimeter, relative measurements taken
    with, 352–55
Gravitational acceleration, 350–51
Gravity, measuring:
  absolute measurements, 355
  International Gravity Standardization Net
    1971 (IGSN71), 355–56
  using a gravimeter, 352–55
  using a pendulum, 351–52
Gravity anomaly, 358
Gravity exploration method, 349–50
  adjusted observed gravity, 356
    Bouguer correction, 360–64
    correcting for the latitude effect, 357–58
    free-air correction, 358–59
    isostatic anomaly, 369
    terrain correction, 364–69
    variation in *g* as a function of latitude,
      356–57
  analyzing anomalies, 399
    filtering, 409
    regionals and residuals, 399–402
    second derivatives, 407–9
    trend surfaces, 402–3
    upward and downward continuation,
      403–7
  application of:
    bedrock depths, 415–16
    landfill geometry, 417–20
    subsurface voids, 416–17
  field procedures:
    determining elevations, 373–75

determining horizontal position, 375–76
    drift and tidal effects, 370–73
    establishing base stations, 373
    selecting a reduction density, 376
    survey procedure, 376, 378
  fundamental relationships, gravitational
    acceleration, 350–51
  gravity effects of simple geometric shapes:
    of a horizontal cylinder, 383–86
    of a horizontal sheet, 390–97
    of inclined rod, 388–90
    rock densities, 378
    of semi-infinite sheets, 390–97
    of a sphere, 379–83
    of a vertical cylinder, 386–88
  gravity interpretation, 409–10
    Bouguer anomaly values, 413–15
    half-maximum technique, 410–11
    second-derivative technique, 411–13
GRAVMAG, 397–99, 488, A17–23
measuring gravity:
  absolute measurements, 355
  International Gravity Standardization
    Net 1971 (IGSN71), 355–56
  using a gravimeter, 352–55
  using a pendulum, 351–52
Gravity reduction, example of, 364*t*
Gravity traverse, data for, on Connecticut
    River floodplain, 377*t*
Gravity values, variation in, due to tidal
    effects, *371*
GRAVMAG, 397–99, 409, 413, 488, A17–23
  anomaly curves for vertical prism
    produced by, *479*
  calculating gravity anomalies due to
    irregular shapes with, *398*
  magnetic effects of polygons with infinite
    strike length, and use of, 478–79,
    481–82
Green, C. H., 163
Green method, 169
Ground-penetrating radar (GPR), 499, 504,
    524–25
  data acquisition, 527–33
  data collection, with three different
    antenna frequencies, *528*
  data collection with 500 MHz antenna
    over pipe, *534*
  profile showing inclined gravel, *526*
  ray paths for air wave, direct wave and
    reflected wave, *535*
  ray paths for data collected using fixed
    antenna separation, *538*
  skin depth, 509*t*
  velocity analysis:
    burial of known object, 533
    diffraction hyperbola, 537, 539
    walkaway test, 533, 536–37
  velocity of, 525–27
Ground roll, 18, 36, *36*
Gupta, V. K., 452
Gutenberg, B., 65

Half-maximum techniques:
  interpretation of gravity data, 410–11
  interpretation of magnetic data, 484–85
Hammer, S., 365, 366
Head wave, 29, *33, 36*
  arrival times for irregular refraction, 124*t*
Hemispherical sink, with diameter greater
    than electrode spacing, *314*
Hemispherical structures, 313–14
Henry (H), 507
Hertz (Hz), 10
Hinze, W. J., 417, 419
Homogeneous, isotropic earth, percentage of
    current's penetration of, 275*t*
Hookean behavior, 11
Horizontal cylinder:
  Bouguer anomaly curves for, with various
    radii, *385*
  Bouguer gravity anomalies over, *382*
  gravity effect of, 383–86
  gravity over, 384*t*
Horizontal interface, single (electrical
    resistivity method), 280
  apparent resistivity, 287
  current distribution, 280–84
  current flow lines and current density,
    284–87
  percentage of current penetrating below,
    283*t*
  qualitative development of the resistivity
    pattern over a horizontal interface,
    288–90
  quantitative development of the resistivity
    pattern over a horizontal interface,
    290–97
  using RESIST, 297–301, A13–A16
Horizontal interface, single (reflective
    seismology):
  analysis of arrival times, 156–59
  applying $x^2 - t^2$ method to field
    seismogram, 165–67
  derivation of travel-time equation, 151–56
  determining velocity and thickness, 163–64
  normal move-out, 160–62
  using REFLECT, 150–51, A6–A12
Horizontal interface, single (refraction
    seismology):
  analysis of arrival times, 70–71, 72*t*
  constructing travel-time curve from field
    seismogram, 76–77, *78*
  critical distance, 74–76
  crossover distance, 73–74
  derivation of travel-time equation, 67–70
  determining thickness, 72–73
  Mohorovicic discontinuity, 79–81
  using REFRACT, 78–79, A1–A5
Horizontal interface(s), four apparent
    resistivity curve types for, *303*
Horizontal interface(s), multiple (electrical
    resistivity method), 302–5
Horizontal interface(s), multiple (reflective
    seismology), 167–70

analyzing a field seismogram containing
    multiple reflections, 181–82
  determining thicknesses, 174–75
  determining velocities, 170–74
  Dix equation, 170, 175–81
Horizontal magnetic dipole (HMD)
    configurations, for Norman,
    Oklahoma landfill, *547*
Horizontal plate of infinite extent,
    generalized relationships used to
    derive expression for magnetic effect
    of, *472*
Horizontal position, determining, 375
Horizontal sheet
  gravity effect of, 390–97
  magnetic effect of, 470–78
Hubbert, M. K., 335, 397, 398
Hunter, J. A., 203, 252
Huygens, C., 21
Huygen's principle, 21–22, 23, *23,* 25, 26,
    26, 34, 35
Hydrophones, 52
Hyperbolic curve window, velocity analysis
    method of comparing waveforms
    within, *240*
Hysteresis curve, 437
  for ferrimagnetic material, *436*

Ice mapping, 546
Image point, 183, *184,* 186, *188, 189*
Imaginary component, 514
Incident ray, *34*
Inclined prism, magnetic effects of, *481*
Inclined reflector, templates for calculating
    thickness, dip, and velocity for, *190*
Inclined rod
  gravity effect of, 388–90
  gravity over, 389*t*
  notation used to derive gravity effect of,
    388
Induced polarization (IP), 265
Induced potential (IP)
  in the frequency domain, 339
  in the time domain, 338–39
Induction logs, electromagnetic, 519
In-line spread, geophones, 131, *132*
In phase, 509, 514
Instrumental drift, 370
Intensity of magnetization, 433–34
Intercept time, 72
Interface discontinuities, 116–20
International Association of Geodesy, 357
International Geomagnetic Reference Field
    (IGRF), 443
International Gravity Standardization Net
    1971 (IGSN71), 355–56, 373
Interval velocity, 173, 174
Inverse filtering, 245
Inverse modeling, 4
Inversion, 523–24
Irregular refraction, head wave arrival times
    for, 124*t*

Isolated pole, magnetic effect of, 456–57
Isopach map, of Thomas Farm landfill,
    Indiana, *419*
Isostasy, 369
Isostatic anomaly, 369

Jol, H. M., 543
Joya de Ceren site, El Salvador:
    electromagnetic survey of, 541, 543
    GPR radargram from, *542*

Karst, mapping using electricity-resistivity
    surveying, 335
Karst topography (Hardin County, Illinois),
    apparent resistivity curve for
    constant-spread traverse over, *337*
Kearey, P., 217
Keller, G. V., 297
Kick, J. F., 415
Kimberlite exploration, Canada, GPR
    radargrams from, *545*
Knapp, R. W., 249
Koehler, F., 194

LaCoste-Romberg gravimeter, 417
Landfill geometry:
    defining using gravity methods, 417–20
    defining using magnetic methods, 493
Landfills, electromagnetic studies of,
    546–48
Landsman, M., 397
Lankston, R. W., 138
Laterally varying velocity, 114–16
Laterologs, 519
Latitude, normal gravity, and position
    requirements, 358*t*
Latitude effect, correcting for, 357
Law of Cosines, 183
Leachate plumes, generalized diagram of, in
    association with landfills, *336*
Leap, D. I., 417, 419
Least time principle, 22
Lee modification, 320
Leonardon, E. G., 266
Levin, F. K., 181
Linear instrument drift, variation of gravity
    values and, *372*
Lithology log, *140,* 141
Longitudinal wave, 14, 15
Long-path multiples, 206, *207*
Looping, 370
Love, A. E. H., 17
Love waves, 17, *17,* 18, 20
Low-pass filter, 41
Low-velocity layer, 107–10

Mabey, D. R., 324, 325
Mackie, R. L., 523
Magnet, as assemblage of small dipoles, *431*
Magnetic cleanliness, 450
Magnetic dip poles, 438
Magnetic domains, 434, *435*

Magnetic effects, of a lithologic contrast and
    structural relief, *474*
Magnetic equator, 438
Magnetic field:
    creation of, due to current through coil, *503*
    external, 443
    main, 442
    measuring:
        flux-gate magnetometer, 446
        proton-precession magnetometer, 447
        total-field anomalies, 447–49
    strength of, 431–32
Magnetic force, 431
    lines of, 430, *430*
Magnetic method, 429
    applications of:
        archaeological surveys, 488–90
        defining landfill geometry, 493
        detection of voids and well casings,
            490–92
    earth's magnetic field:
        dipolar nature of, 439–41
        dipole equations, 444–46
        field elements of, 438
        variations of, 441–44
    field procedures:
        correcting for horizontal position,
            452–54
        diurnal corrections, 450–51
        elevation corrections, 451–52
        magnetic cleanliness, 450
    fundamental relationship, 430–31
        intensity of magnetization, 433–34
        magnetic field strength, 431–32
        magnetic force, 431
        magnetic moment, 432–33
        magnetic potential, 437–38
        magnetic susceptibility, 434–37
    interpretation of magnetic data:
        computer modeling, 488
        disadvantages and advantages, 482–84
        half-maximum techniques, 484–85
        slope methods, 485–87
    magnetic effects of simple geometric
            shapes:
        of a dipole, 457, 460–65
        of isolated pole (monopole), 456–57
        of polygons with infinite strike length
            (using GRAVMAG), 478–82
        rock susceptibilities, 454–56
        of a sphere, 465–69
        of a thin horizontal sheet, 470–78
    measuring the magnetic field:
        flux-gate magnetometer, 446
        proton-precession magnetometer, 447
        total-field anomalies, 447–49
Magnetic moment, 432–33, *433*
Magnetic permeability, 431, *432,* 506, 507,
    508
Magnetic poles, 430
Magnetic potential, 437–38
Magnetic storms, 443

Magnetic susceptibility, 434–37
Magnetotelluric method, 266, 340
Main magnetic field, 442, *448*
Mapping:
  electrical, 306
  karst and geologic structures, 335
  of snow and ice, 546
Master curves, 328–29
Maximum depth, travel-time curve used in
    calculation of, to second interface,
    113
Maximum depth sensitivity, 508
Maxwell's equations, 500
McElwee, C. D., 249
Medford Caves, Florida, gravity study at,
    416–17, *418*
Meers fault, Oklahoma, 249, *250*
Menke, W., 523
Mesozoic sedimentary rocks, 413, 414
Meter drift, 355
Migration, 241–44, *243*
Miller, R. D., 252
Modeling
  forward, 4
  inverse, 4
Mohorovicic, A., 65, 79
Mohorovicic discontinuity, 79–81, *80*
Monopole:
  intensities of vertical, horizontal, and total
      magnetic field anomalies for, *459*
  magnetic effect of, 456–57, *458t*
  relationship and notation used to derive
      magnetic effect of, *457*
Mooney, H. M., 120, 288
Moore Cumulative Resistivity Method, 325,
    326
Moving transmitter-plus-receiver system
    (Slingram), 512, *513, 514*
Multiple reflections, 206–9, *207, 210, 211*
  relative energy comparisons among, 209t

Nanotesla, 432
Natural currents, 266
Near-field continuous-wave methods of
    frequency domain electronics:
  moving transmitter-plus-receiver system
      (Slingram), 512, *513, 514*
  noncontacting ground conductivity
      measurements, 514–18
Negative current electrode (sink), 271, *272*
Negative pole of dipole, magnetic field
    intensity due to, 460
Nettleton, L. L., 376
Newton's laws:
  second law of motion, 350
  universal gravitation, 350
Noise, 5
Noncontacting ground conductivity
    measurements, 514–18
Normal move-out (NMO), 154t, 160, *162*,
    192–95, 195t, *196,* 209, *211*, 213,
    229, 529

corrections:, 232–33
  velocity analysis A, 233, 235
  velocity analysis B, 235, 237
  velocity analysis C, 238
  for different velocities and/or depths,
      161t
Norman, Oklahoma landfill:
  EM studies of, 546–48
  results of EM induction surveys near, *547*
North and south magnetic poles, 440

Objectives, defining, 3–4
Oersted, 432
Offset spread, geophones, *132*
Ohm, G. S., 268
Ohm's Law, 268, 505, 510
One-way vertical travel time, 170, 173
Optical analog method, 291, *291, 294,* 295
Optimum offset method, 221
Optimum window, 200–206, *201, 202*
  position of far side of, *204*
Oscilloscope display, on seismic unit, 59, *59*
Out of phase, 509, 514

Paleozoic rocks, 413, 414, 415
Paramagnetic minerals, 434
Particle displacements, hammer blow and, *9*
Pendulum, relative measurements taken
    with, 351–52
Period, 9, 502
Peters, L. J., 485
Peters's slope method, 485
Plouff, D., 399
Point current source, 270–71
Poisson's ratio, 12, *12,* 18
Poisson's relation, 465, 484
Polygons, magnetic effects of, 478–82
Polynomials, fit to Bouguer anomaly data,
    404
Porous pot, 316
Positive current electrode (source), 271,
    272
Positive pole of dipole, magnetic field
    intensity due to, 460
Potential difference
  determining, at two potential electrodes,
      278
  measuring, *279*
Power, M., 543
Primacord®, 47
Primary field, 511, 512, 514
Primary reflections, 206, *210, 211*
  relative energy comparisons among, 209t
Principle of least time, 22
Prism
  inclined, magnetic effects of, *181*
  orientations, anomaly curves and, *480*
  vertical, anomaly curves produced by
      GRAVMAG, *479*
Profiling, 220
  sounding *versus,* 521–23
Proton-precession magnetometer, 447

Pseudogravitational field, 484
Pullan, S. E., 203, 252
P-waves (primary waves), 15, *15*, 17, 18, 19, 36, 41, 43, 60
  angles of reflection and refraction for, 32*t*
  Mohorovicic discontinuity and, 80
  reflection coefficients for, at normal incidence, *44*
  velocities, 20*t*

Quadrature phase component, 514

Raab, P. V., 492
Radar, 524
Radar tomography, 530
Radly, Oxfordshire, England, results of flux-gate magnetometer gradiometer survey at, *490*
Rayleigh, Lord J. W. S., 17
Rayleigh waves, 17, *17*, 18
Ray paths, actual, in multilayered sequence *vs.* straignt-line ray paths, *169*
Ray paths in layered materials, 21–39
  critical refraction, 28–29
  diffraction, 29, 34–35
  Fermat's principle, 22–23, *24*, 26, *27*
  Huygen's principle, 21–22, *22*, *23*
  reflection, 23–25
  refraction, 25–27
  Snell's law, 27–28
  wave arrivals at surface, 35–36, 38
Rays, 11
Ray tracing, 129
Reciprocal time, 120
Reciprocity, 94, 122
Redpath, B. B., 123
Reduction density, selecting, 376
Reference ellipsoid, 357
REFLECT, 150–51, 152, *155*, 164, 166, 167, 174, 181, 203, A6–A12
Reflected ray, *34*
  symbols used in derivation of travel-time equation for, *184*
Reflected waves, travel times for, 154*t*
Reflection, 23–25, *36*, 60
  angle of, *30*
  angles of, for P- and S-waves, 32*t*
  symbols used in derivation of time of travel for, *152*
  waveforms superimposed on direct wave, *39*
Reflection arrivals, variations in presentation of, *155*
Reflection coefficient, 291
  for P-wave at normal incidence, *44*
Reflection paths, for two layer thicknesses, *159*
Reflection profiling, 529, *530*
Reflection ray paths, for source-receiver offsets, *175*
Reflection seismology, 149–50
  application of, 248–53

    cavity detection, 249–52
    detailing shallow structures, 249
    related to aquifers, 248–49
    related to hazardous waste facility, 252–53
  computer processing of data:
    correcting for normal move-out, 232–33, 235, 237–38
    migration, 241–44
    stacking CDP gathers, 238–41
    static correction, 229–32
    time and depth sections, 245–48
    waveform adjustments, 244–45
  dipping interface, 182–83
    derivation of travel-time equation, 183–86
    determining dip, thickness, and velocity, 186–92, 197–99
    normal move-out, 192–96
  field procedures:
    equipment, 217–19
    geophone spreads, 219–29
  multiple horizontal interfaces, 167, 169–70
    analyzing field seismogram containing multiple reflections, 181–82
    determining thicknesses, 174–75
    determining velocities, 170–71, 173–74
    Dix equation, 170, 175–77, 179, 181
  optimum window, 200, 202–3, 206
  shallow interfaces, 200
    diffractions, 209, 211–13, 217
    multiple reflections, 206, *207*, 208–9
  single horizontal interface:
    analysis of arrival times, 156–57
    applying $x^2 - t^2$ method to field seismogram, 165–67
    derivation of a travel-time equation, 151–54, 156
    determining velocity and thickness, 163–64
    normal move-out, 160, *161*, *162*
    using REFLECT, 150–51, A6–A12
Reflection time-distance data, plot of, *153*
Reflection time-distance values, for refracted ray paths in multilayered sequence, 176*t*
Reflection travel-time data, time-distance plots showing effect of depth and velocity on curvature of, *158*
REFRACT, 102, 109, 111, 112, 114, 130, 131, 134, 150, 151
  travel-time curve, *87*
  using, 78–79, A1–A5
Refracted ray, *34*
Refracted waves
  arrival times for, 70–71, 72*t*
  time-distance relationships for, *70*
Refraction, 25–27, 60
  angle of, *30*
  angles of, for P- and S-waves, 32*t*

critical, 28–29, 29
of current flow lines, 285
Refraction seismology:
  applications:
    related to contamination, 136–37, 137
    related to Superconducting Super
      Collider, 139–41
    related to waste disposal site, 138
    related to water-table elevations,
      137–38
  delay-time method, 120, 121, 122–25
  determining thickness, 99–100
  dipping interfaces:
    analyzing the problem, 91–96
    derivation of travel-time equation,
      97–99
  field procedures:
    corrections to data, 134–36
    equipment considerations, 130–31
    geophone spread geometries and
      placements, 131–34
    site selection and planning
      considerations, 129–30
  generalized reciprocal method, 129
  homogeneous subsurface, 65–67
  multiple dipping interfaces:
    analyzing field seismograms, 102, 105
    travel-time equation, 100–102
  multiple horizontal interfaces:
    analyzing field seismogram, 87, 89–90
    critical distance, 85, 86t, 87
    derivation of travel-time equation,
      81–83
    determining thickness, 83–85
  multiple interfaces, 90–91
  nonideal subsurface:
    hidden zones: low-velocity layer,
      107–10
    hidden zones: thin layer, 111–14
    interface discontinuities, 116–20
    laterally varying velocity, 114–16
  ray tracing method, 129
  single horizontal interface:
    analysis of arrival times, 70–72
    constructing travel-time curve from field
      seismogram, 76–78
    critical distance, 74–76
    crossover distance, 73–74
    derivation of travel-time equation,
      67–70
    determining thickness, 72–73
    Mohorovicic discontinuity, 79–81
    using REFRACT, 78–79, A1–A5
  wave front method, 127, 128, 129
Regional trends, 399–402
Relative dielectric permittivity, 506
Relative energy, 49
Remanent magnetization, 436, 437, 482,
  483
Residual anomalies, 399–402
Residual gravity anomaly map, of Thomas
  Fram landfill, Indiana, 420

RESIST, 297–301, 302, 317, 320, 324, 325,
  326, 331, 523, A13–A16
  apparent resistivity curves drawn by, 300
Resistance-length resistivity units, 270
Resistivity, 270
  electrical, 505
Resistivity meter, 317
Resistivity profiling, 306
Resistivity sounding, 306
Resistors, of different lengths, 269, 269–70
Resolution limits, 5
Reverse travel-time curve, effect of vertical
    velocity discontinuity on, 115
Reverse traverse, 93, 93, 94, 95, 104
  ray paths and resulting travel-time curves
    for, over vertical step, 117
Rigidity modulus, 12, 18
River valley (buried), subsurface
    conductivity structure associated with,
    518
Roberts, R. L., 417, 419
Robinson, E. S., 170, 217, 355, 450
Rock densities, 378
Rockel, E., 543
Rock susceptibilities, 454–56
Rodi, W., 523
Rooney, W. J., 266
Root-mean-square velocity, 170

Sagoci, H. F., 44
Salt domes, 149
Savit, C. H., 18, 136, 217
Schlumberger, C., 266
Schlumberger array, 317, 318, 319, 320,
  321, 339
Schlumberger expanding-spread traverse,
    apparent resistivity values illustrating
    effect of changing MN spacing during,
    320
Schultz, G. M., 543
Secondary field, 512, 514
Secondary waves, 15
Second derivative curve, for gravity traverse
    across fault block near Amherst,
    Massachusetts, 413
Second derivatives, examples of, 408
Second-derivative techniques, 407–9
  interpretation of gravity data, 411–13
Second intercept time, 83
Secular variation, in magnetic field, 442
Seismic equipment:
  signal conditioning, 53, 55–56
  signal detection, 50, 51–52
  signal recording, 56–59
Seismic reflection surveys, 5
Seismic refraction, determining thickness of
    confined aquifer with, 5, 6
Seismic section record:
  near Shawville, Quebec, 225
  near Val Gagné, Ontario, 226
Seismic velocities, for selected common
    rocks, 14t

Seismic waves, 7–9, 14–18
  attenuation and amplitude:
    absorption, 40–42
    energy partitioning, 42–46
    spherical spreading, 40
  elastic coefficients, 11–14
  energy sources:
    considerations related to, 49–50
    types of, 47–49
  equipment:
    signal conditioning, 53, 55–56
    signal detection, 50, 51–52
  hammer blow and, 8, *8*
  ray paths in layered materials:
    critical refraction, 28–29
    diffraction, 29, 34–35
    Fermat's principle, 22–23, *24,* 26, *27*
    Huygen's principle, 21–22, *22, 23, 26*
    reflection, 23–25
    refraction, 25–27
    Snell's law, 27–28
    wave arrivals at the surface, 35–39
  relative amplitude of, *54*
  signal recording, 56–59
  terminology related to, 9–11
  time-distance values for, *37*
  velocities of, 18–21
Seismograms, 59, 60
  second, refraction method analysis, 87,
    89–90
Seismographs, 55, *55,* 59
Seismometers, 50
Self-induction, 511
Self-potential (SP) method, 266
Semi-infinite sheet equation, relationships
    and notation for derivation of, *475*
Semi-infinite sheet relationships,
    relationships and notation for
    derivation of, *391*
Semi-infinite sheets:
  gravity effect of, 390–96
  gravity over, *392t*
  half-maximum relationship for, 485
  magnetic effect of, 470–75, *476t,* 477–78
  using, to simulate a faulted horizontal
    bed, *394, 477*
Shah, P. M., 181
Shallow interfaces, 200
  diffractions, 209–17
  multiple reflections, 206–9
  optimum window, 200–206
Shallow refraction surveys, 130
Shear strain, 12
Shear stress, 12
Shear waves, 15
Sheriff, R. E., 43, 55, 149, 170, 177, 305,
    393, 454
Short-path multiples, 206, *207*
Shotgun energy sources, 47, *48,* 56
  filtering on relative energies, for frequency
    spectrum produced by, *57*
Siemens, 505

Signal conditioning, 53, 55–56
Signal detection, 50, 51–52
Signal enhancement seismographs, 253, *253*
Signal recording, 56–59
Sinkhole formation, Reno County, Kansas,
    250, *251*
Skin depth, 499, 508–9
  EM and GPR, *509t*
Slaine, D. P., 252
Slingram (moving transmitter-plus-receiver
    system), 512, *513,* 514
Slingram survey, over vertical conductor, *515*
Slope methods:
  illustration of, *487*
  interpretation of magnetic data with,
    485–87
Slotnick, M. M., 27, 183, 199
Smith, D. G., 543
Snell's law, 27–28, 81, 107, 230
Snow and ice mapping, Greenland, 546
Sounding, profiling *versus,* 521–23
Sounding methods, electromagnetic, 509–12
  near-field continuous-wave methods of
    frequency domain electromagnetics:
    moving transmitter-plus-receiver
    system, 512–14
    noncontacting ground conductivity
    measurements, 514–18
  tilt-angle or dip-angle methods, 519
  time domain electromagnetics, 519–20
Source-receiver distance, importance of, in
    reflection surveying, *201*
Sparkers, 48–49
Spencer, T. W., 44
Spetzler, H., 541
Sphere:
  Bouguer gravity anomalies over, *382*
  buried, notation used in deriving gravity
    effect of, *380*
  diagram of gravity anomaly over, *411*
  gravity effect of, 379–83
  gravity over, *381t*
  half-maximum relationship for, 485
  magnetic effect of, 465–69, *469t*
  uniformly magnetized:
    intensities of vertical, horizontal, and
    total magnetic field anomalies over, *470*
    notation used for derivation of magnetic
    field anomalies over, *467*
    total and vertical magnetic anomalies
    over, for various inclinations of
    Earth's field, *471*
Spherical spreading, 40
  energy losses due to, *41,* 41t
Split spreads, 219–21
Spontaneous potential (SP) method, 266,
    339–40
Spot elevations, 374
Stacking velocities, 233
Stanley, J. M., 412
Stanley's second derivative technique,
    notation used for, *412*

Static correction, 230, *231*
Steeples, D. W., 249, 252
Steffen, K., 539
Straight-line ray paths, actual reflection ray paths in multilayered sequence *vs.*, *169*
Strain, 11
Stress, 11
Strip-chart recorder, 57
Subsurface:
    approaching, 1–6
    fundamental considerations in study of, 2–3
    limitations with geophysical methods in study of, 4–5
    multiple geophysical methods in study of, 5–6
    options in approaching, 1–2
Subsurface conductivities:
    of buried river valley, *518*
    for three-layer structure, *519*
Subsurface radar velocity, determining, 533–39
Subsurface voids, 416–17
Superconducting Super Collider, 139–41
    seismic investigation of proposed site for, *140*
Suppression, 329, *330*
Surface waves, 17, 18
S-waves (secondary or shear waves), 15, 16, *16*, 18, 19, 28, 36, 41, 43
    angles of reflection and refraction for, *32t*
    Mohorovicic discontinuity and, 80, 81

Talwani, M., 397, 398
Taner, M. T., 194
Telford, W. M., 43, 55, 305, 393, 454
Telluric currents, 266
Telluric methods, 340
Teotihuacan, Mexico:
    magnetic prospecting for tunnels and caves in, *491*
    magnetic survey at, 490
Terrain correction, 364–69, *368t*, 387
    representing masses for, *366*
Terrain correction template, placement of, on topographic map, *367*
Tesla, N., 504
Thermoremanent magnetization, 437
Thickness:
    computation of, for low-velocity second layer, *110t*
    reflection seismology, 163–64
    refraction seismology, 72–73, 83–85, 97, 98, 99–100
    two-layer, reflection paths for, *159*
Thin layer, 111–14
    intermediate, not detected by refraction, *112*
    refraction times for, *111t*
Thin plate of limited extent, notation used to derive expression for, *473*
Thomas Farm landfill (Indiana):
    isopach map of, *419*

residual gravity anomaly map of, *420*
    total magnetic field profile for, *493*
Thornburgh, H. R., 127
Three-layer case, 81
    refraction times for, with low-velocity intermediate layer, *109t*
    symbols used:
        in derivation of critical distance for ray critically refracted in, *86*
        in derivation of time of travel for ray critically refracted in, *82*
Three-layer curves, comparison of two-layer curves and, *304*
Tidal effects, 370–73
    variation of gravity values due to, *371*
    variation of gravity values due to combination of tidal effects and linear instrument drift, *372*
Tilt-angle method, 519
Time-distance data plot, *108*
Time-distance graph, *68*
Time-distance values:
    for seismic waves, *37*
    for seismogram, *77*
Time domain electromagnetics (TDEM), 512, 519–20
    central transmitter, *522*
    transmitter output and receiver signal as function of time, *521*
Time sections, 245–48, *246*, *247*
Tooley, R. D., 44
Topography, apparent resistivity measurements and, 323
Total-field anomalies, 447–49
Total-field magnetic intensity, as function of longitude and latitude for given survey area, *453*
Total-field magnetic map, of Dolores Archaeological Program, Colorado, *489*
Totally reflected, 29
Total magnetic field anomaly profiles, measurements above cased well at Piney Creek, Colorado, *492*
Total magnetic field profile, for Thomas Farm landfill, *493*
Transillumination, 530, *530*
Transmission adjustments, 244, 245
Transmission coefficient, 292
Transverse waves, 14
Travel-time curve, 38, 65, *66*, *68*
    constructing from field seismogram, 76–77, *78*
    correlation of, with geophone positions above single dipping interface, *95*
    correlation of, with wave paths to geophones at equal distances from energy source, *93*
    field seismogram data, Connecticut Valley, Massachusetts, *105*
    important features of, for single dipping interface, *96*

Travel-time curve *(continued)*, 38, 65, *66, 68*
  for reflections from horizontal and
    inclined interfaces, *188*
Travel-time equation:
  derivation of, 67–70
  reflection seismology, 151–56, 183–86
  refraction seismology, 67–70, 81–83,
    97–99, 100–102
  symbols used in derivation of, for single
    dipping interface, *97*
Trend surfaces, 402–3
Tsuboi, C., 355, 357
Two-layer curves, three-layer curves
  compared with, *304*

Unexploded ordinance (UXO) studies,
  electromagnetic surveys and, 546
Up-dip, 97, 98, 99
Upward continuation, 403–7, 484
  simulation of, *406*
Upward continuation map, for Easthampton
  and Mt. Holyoke Quadrangles (west-
  central Massachusetts), *407*

Vacquier, V., 485
Van Nostrand, R. G., 266, 305, 309
Velocity, 20, 170–71, 173–74, 233–38
  P-waves, 20*t*
  of seismic waves, 18–21
Velocity analysis, using V_{st} *and* t_o for
  automating of, 234*t*
Vertical contact(s), 305–6
  apparent resistivity for constant-spread
    traverse, oriented perpendicular to,
    308
  apparent resistivity for expanding-spread
    traverse, oriented perpendicular to,
    311
  constant-spread survey across, 307*t*
  constant-spread traverse and, 306–9
  expanding-spread survey across, 310*t*
  expanding-spread traverse and, 309–10
  two, 312
    constant-spread traverse across, *313*
Vertical cylinder:
  gravity effect of, 386–88
  notation used to derive gravity effect of, *387*
Vertical magnetic dipole (VMD)
    configurations, for Norman,
    Oklahoma landfill, 548
Vertical prisms:
  anomaly curves for, produced by
    GRAVMAG, *479*
  magnetic anomaly curves for, at various
    depths, *486*
Vertical rods, gravity over, 389*t*
Very-low-frequency (VLF) methods, 519
Vibratory energy sources, 47, 48
Vibroseis, 48
Voids:
  detection using the gravity method,
    416–17

detection using the magnetic method,
  490–92
Voltmeter, 315, 316
Volt (V), 267

Walkaway test, 533, 536–37
Ward, S. H., 339
Waste-disposal site, refraction seismology
  and, 138
Waters, K. H., 244
Water-table elevations, mapping, using
  refraction seismology, 137–38
Wavefront, 11
Wavefront method, 127, *128, 129*
Wavelength, 9, 11
Wavelength filtering, 409
Wavelets, 21, 22
Wave propagation, 7–9
Waves. *See also* Electromagnetic waves;
  Seismic waves
Weathering correction, 133, 135, 229–30, *231*
Weight-drop energy sources, 47, 48, *48*
Well logs, 136, *137*, 141
Wells, detection of using the magnetic
  method, 490–92
Wenner, F., 266, 319
Wenner array, 317, *318*, 319, 320, 321
  apparent resistivity values obtained with,
    321
Wenner/Lee electrode geometry, *318*
Western Hemisphere:
  values of declination of geomagnetic field
    for portion of, *442*
  values of inclination of geomagnetic field
    for portion of, *441*
  values of total intensity for geomagnetic
    field for portion of, *440*
Westinghouse, G., 504
Wide angle reflection and refraction
    (WARR), 529
  data collection with 225 MHz antenna, *531*
Woollard, G. P., 356
Worden gravimeter:
  interior of, *353*
  photographs of, *354*
Worzel, J. W., 397

$x^2 - t^2$ method, 163, *164, 165*, 165–67, *167*,
  177, *178*, 179, *182*

Yin, C. C., 523
Young's modulus, 12

Zero-length spring, 353
Zero-offset profiling, 529, 530
Zoeppritz, K., 43
Zoeppritz's equations, *43*, 43–44
Zohdy, A. A. R., 324, 325, 331
Zone, 365

Printed in the United States
by Baker & Taylor Publisher Services